XXII International Conference on Integrable Systems and Quantum Symmetries 2014

(ISQS-22)

Journal of Physics: Conference Series Volume 563

Prague, Czech Republic
23-29 June 2014

Editors:

Cestmir Burdik
Ondrej Navratil
Severin Posta

ISBN: 978-1-63439-671-4
ISSN:1742-6588

Printed from e-media with permission by:

Curran Associates, Inc.
57 Morehouse Lane
Red Hook, NY 12571

Some format issues inherent in the e-media version may also appear in this print version.

Printed by Curran Associates, Inc. (2015)

For permission requests, please contact the Institute of Physics
at the address below.

Institute of Physics
Dirac House, Temple Back
Bristol BS1 6BE UK

Phone: 44 1 17 929 7481
Fax: 44 1 17 920 0979

techtracking@iop.org

Additional copies of this publication are available from:

Curran Associates, Inc.
57 Morehouse Lane
Red Hook, NY 12571 USA
Phone: 845-758-0400
Fax: 845-758-2634
Email: curran@proceedings.com
Web: www.proceedings.com

Table of contents

Volume 563

2014

XXII International Conference on Integrable Systems and Quantum Symmetries (ISQS-22)
23–29 June 2014, Prague, Czech Republic

Accepted papers received: 10 November 2014
Published online: 26 November 2014

Preface

Papers

Journal of Physics: Conference Series **563** (2014) 011001 doi:10.1088/1742-6596/563/1/011001

Preface

The XXII International Conference on Integrable Systems and Quantum symmetries (ISQS-22), organized by the Department of Mathematics, Faculty of Nuclear Sciences and Physical Engineering, Czech Technical University Prague and the Bogoliubov Laboratory of Theoretical Physics of the Joint Institute for Nuclear Research, belongs to the successful series of conferences held at the Czech Technical University which began in 1992 and is devoted to problems of mathematical physics related to the theory of integrable systems, quantum groups and quantum symmetries. During the last 20 years, each of the conferences gathered around 60 scientists from all over the world. 33 papers of plenary lectures and contributions presented at ISQS-22 are published in the present issue of Journal of Physics: Conference Series.

Čestmír Burdík, Ondřej Navrátil, Severin Pošta
Editors

Journal of Physics: Conference Series **563** (2014) 011002 doi:10.1088/1742-6596/563/1/011002

Peer review statement

All papers published in this volume of *Journal of Physics: Conference Series* have been peer reviewed through processes administered by the proceedings Editors. Reviews were conducted by expert referees to the professional and scientific standards expected of a proceedings journal published by IOP Publishing.

XXII International Conference on Integrable Systems and Quantum Symmetries (ISQS-22) IOP Publishing
Journal of Physics: Conference Series **563** (2014) 012001 doi:10.1088/1742-6596/563/1/012001

Virasoro representations with central charges $\frac{1}{2}$ and 1 on the real neutral fermion Fock space $F^{\otimes\frac{1}{2}}$

Iana I. Anguelova
Department of Mathematics, College of Charleston, Charleston SC 29424, USA
E-mail: anguelovai@cofc.edu

Abstract. We study a family of fermionic oscillator representations of the Virasoro algebra via 2-point-local Virasoro fields on the Fock space $F^{\otimes\frac{1}{2}}$ of a neutral (real) fermion. We obtain the decomposition of $F^{\otimes\frac{1}{2}}$ as a direct sum of irreducible highest weight Virasoro modules with central charge $c = 1$. As a corollary we obtain the decomposition of the irreducible highest weight Virasoro modules with central charge $c = \frac{1}{2}$ into irreducible highest weight Virasoro modules with central charge $c = 1$. As an application we show how positive sum (fermionic) character formulas for the Virasoro modules of charge $c = \frac{1}{2}$ follow from these decompositions.

1. Introduction

This paper is a continuation of [Ang14], and is a part of a series studying various particle correspondences in 2 dimensional conformal field theory from the point of view of chiral algebras (vertex algebras) and representation theory. The study of fermionic oscillator representations of the Virasoro algebra (Vir) is long-standing, dating back to [Fre81], [FF83], [GOS85], [GNO85], [GKO86], [KR87], [FFR91], and many others. What was new in [Ang14] is that we used 2-point-local (local at both $z = w$ and $z = -w$ as opposed to just local at $z = w$) Virasoro fields generating the fermionic oscillator representations. In particular, in [Ang14] we constructed a 2-parameter family (depending on parameters $\lambda, b \in \mathbb{C}$) of 2-point-local Virasoro fields with central charge $-2 + 12\lambda - 12\lambda^2$ on the fermionic Fock space $F^{\otimes\frac{1}{2}}$ of the real fermion. In this paper we study the nature of these representations depending on the parameters, and their decomposition into irreducible modules. We show that for particular choices of the parameters (λ, b) these two-point local Virasoro field representations can produce each one of the well known discrete series of Virasoro representations. Another important and interesting particular case of the parameters is that of $\lambda = \frac{1}{2}$, i.e., central charge 1, and $b \in \frac{1}{\sqrt{2}}\mathbb{Z}$. For this choice of (λ, b) we obtain the decomposition of $F^{\otimes\frac{1}{2}}$ into irreducible highest weight modules for Vir of central charge 1. It is well known (going back to D. Friedan and I. Frenkel) that $F^{\otimes\frac{1}{2}}$ is a natural highest weight module for Vir of central charge $\frac{1}{2}$ via a one-point local Virasoro field, and as such it decomposes into two irreducible highest weight modules for Vir of central charge $\frac{1}{2}$ (these are well known as two of the Ising minimal models). We show how each of these irreducible Vir highest weight modules of central charge $\frac{1}{2}$ decomposes into irreducible Vir highest weight modules of central charge 1. As an application, this allows us to directly write a positive sum (fermionic) character formula for these irreducible highest weight modules for Vir of central charge $\frac{1}{2}$. There has been extensive research on positive sum (fermionic) Virasoro character formulas, see for instance [KKMM93], [FQ97], [BF00], [Wel05], [FFW08]. Fermionic-type character formulas are known for all the Virasoro minimal models (the discrete Virasoro series), although the vertex algebra (field theory) foundation for such character formulas is still lacking in the general case. It is new here that we obtain these specific fermionic formulas as a direct result of the decomposition of the charge $\frac{1}{2}$ "1-point-local" modules into charge 1 "2-point-local" modules (i.e., we explicitly use multi-local fields and twisted vertex algebra techniques). We hope that such approach is possible also more generally for the discrete Virasoro series by using quasi-particle Fock spaces in the place of a single real fermionic Fock space of central charge $\frac{1}{2}$. After bosonising such quasi-particle Fock spaces, the idea is then to obtain a decomposition of the irreducible modules representing the one-point local Virasoro field of discrete-series-type central charge into irreducible modules representing multi-local Virasoro fields of charge 1.

2. Notation and background

We use the term "field" to mean a series of the form

$$a(z) = \sum_{n\in\mathbf{Z}} a_{(n)}z^{-n-1}, \quad a_{(n)} \in \mathrm{End}(V), \quad \text{such that} \quad a_{(n_v)}v = 0 \quad \text{for any } v \in V,\ n_v \gg 0. \tag{2.1}$$

Let

$$a(z)_- := \sum_{n\geq 0} a_n z^{-n-1}, \quad a(z)_+ := \sum_{n<0} a_n z^{-n-1}. \tag{2.2}$$

Definition 2.1 (Normal ordered product) *Let $a(z), b(z)$ be fields on a vector space V. Define*

$$:a(z)b(w):= a(z)_+b(w) + (-1)^{p(a)p(b)}b(w)a_-(z). \tag{2.3}$$

One calls this the normal ordered product of $a(z)$ and $b(w)$.

Remark 2.2 Let $a(z), b(z)$ be fields on a vector space V. Then $:a(z)b(\lambda z):$ and $:a(\lambda z)b(z):$ are well defined fields on V for any $\lambda \in \mathbb{C}^*$.

Definition 2.3 *([ACJ14])* **(2-point-local fields)** *We say that a field $a(z)$ on a vector space V is* **even** *and 2-point self-local at $(1;-1)$, if there exist $n_0, n_1 \in \mathbb{N}$ such that*

$$(z-w)^{n_0}(z+w)^{n_1}[a(z), a(w)] = 0. \tag{2.4}$$

In this case we set the **parity** *$p(a(z))$ of $a(z)$ to be 0.*
We set $\{a, b\} = ab + ba$. We say that a field $a(z)$ on V is 2-point self-local at $(1;-1)$ and **odd** *if there exist $n_0, n_1 \in \mathbb{N}$ such that*

$$(z-w)^{n_0}(z+w)^{n_1}\{a(z), a(w)\} = 0. \tag{2.5}$$

In this case we set the **parity** *$p(a(z))$ to be 1. For brevity we will just write $p(a)$ instead of $p(a(z))$.*
Finally, if $a(z), b(z)$ are fields on V, we say that $a(z)$ and $b(z)$ are 2-point mutually local at $(1;-1)$ if there exist $n_0, n_1 \in \mathbb{N}$ such that

$$(z-w)^{n_0}(z+w)^{n_1}\Big(a(z)b(w) - (-1)^{p(a)p(b)}b(w)a(z)\Big) = 0. \tag{2.6}$$

For a rational function $f(z, w)$, with poles only at $z = 0$, $z = \pm w$, we denote by $i_{z,w}f(z, w)$ the expansion of $f(z, w)$ in the region $|z| \gg |w|$ (the region in the complex z plane outside the points $z = \pm w$), and correspondingly for $i_{w,z}f(z, w)$. The mathematical background of the well-known and often used (both in physics and in mathematics) notion of Operator Product Expansion (OPE) of product of two fields for case of usual locality ($N = 1$) has been established for example in [Kac98], [LL04]. The following lemma extended the mathematical background to the case of 2-point locality (in fact to N-point locality, for $N \in \mathbb{N}$):

Lemma 2.4 *([ACJ14])* **(Operator Product Expansion (OPE))**
Let $a(z)$, $b(w)$ be 2-point mutually local. Then exists fields $c_{jk}(w)$, $j = 0, 1; k = 0, \ldots, n_j - 1$, such that we have

$$a(z)b(w) = i_{z,w} \sum_{k=0}^{n_0-1} \frac{c_{0k}(w)}{(z-w)^{k+1}} + i_{z,w} \sum_{k=0}^{n_1-1} \frac{c_{1k}(w)}{(z+w)^{k+1}} + :a(z)b(w): . \tag{2.7}$$

We call the fields $c_{jk}(w)$, $j = 0, 1; k = 0, \ldots, n_j - 1$ OPE coefficients. We will write the above OPE as

$$a(z)b(w) \sim \sum_{k=0}^{n_0-1} \frac{c_{0k}(w)}{(z-w)^{k+1}} + \sum_{k=0}^{n_1-1} \frac{c_{1k}(w)}{(z+w)^{k+1}}. \tag{2.8}$$

The $\sim$ signifies that we have only written the singular part, and also we have omitted writing explicitly the expansion $i_{z,w}$, which we do acknowledge tacitly. Often also the following notation is used for short:

$$\lfloor ab \rfloor = a(z)b(w) - :a(z)b(w): = [a(z)_-, b(w)], \tag{2.9}$$

i.e., the *contraction* of any two fields $a(z)$ and $b(w)$ is in fact also the $i_{z,w}$ expansion of the singular part of the OPE of the two fields $a(z)$ and $b(w)$.

The OPE expansion in the multi-local case allowed us to extend the Wick's Theorem (see e.g., [BS83], [Hua98]) to the case of multi-locality (see [ACJ14]), and we will use it in what follows, together with the Taylor Expansion Lemma (see [ACJ14]).

3. The Fock space $F^{\otimes \frac{1}{2}}$ and 2-point-local field representations of Vir

We recall the definitions and notations for the Fock space $F^{\otimes \frac{1}{2}}$ as in [Fre81], [DJKM81a], [Kac90], [Wan99b]; in particular we follow the notation of [Wan99b], [Wan99a].

Consider a single odd self-local field $\phi^D(z)$, which we index in the form $\phi^D(z) = \sum_{n\in \mathbb{Z}+\frac{1}{2}} \phi^D_n z^{-n-\frac{1}{2}}$. The OPE of $\phi^D(z)$ is given by

$$\phi^D(z)\phi^D(w) \sim \frac{1}{z-w}. \tag{3.1}$$

This OPE completely determines the commutation relations between the modes ϕ^D_n, $n \in \mathbb{Z} + \frac{1}{2}$:

$$\{\phi^D_m, \phi^D_n\} := \phi^D_m\phi^D_n + \phi^D_n\phi^D_m = \delta_{m,-n}1. \tag{3.2}$$

and so the modes generate a Clifford algebra Cl_D. The field $\phi^D(z)$ is usually called a "real neutral fermion field". The Fock space, denoted by $F^{\otimes \frac{1}{2}}$, of the field $\phi^D(z)$ is the highest weight module of Cl_D with vacuum vector $|0\rangle$, so that $\phi^D_n|0\rangle = 0$ for $n > 0$. This well known Fock space is often called the Fock space of the free real neutral fermion (see e.g. [DJKM81b], [FFR91], [KW94], [KWY98], [Wan99a], [Wan99b], [RT12]). $F^{\otimes \frac{1}{2}}$ has basis

$$\{\phi^D_{-n_k-\frac{1}{2}} \cdots \phi^D_{-n_2-\frac{1}{2}}\phi^D_{-n_1-\frac{1}{2}}|0\rangle, |0\rangle \mid \text{where } n_k > \cdots > n_2 > n_1 \geq 0, n_i \in \mathbb{Z}_{\geq 0}, i = 1, 2, \ldots, k\} \tag{3.3}$$

We recall the various gradings on $F^{\otimes \frac{1}{2}}$. The space $F^{\otimes \frac{1}{2}}$ has a well known $\mathbb{Z}_2$ grading given by k *mod* 2,

$$F^{\otimes \frac{1}{2}} = F^{\otimes \frac{1}{2}}_{\bar{0}} \oplus F^{\otimes \frac{1}{2}}_{\bar{1}},$$

where $F^{\otimes \frac{1}{2}}_{\bar{0}}$ (resp. $F^{\otimes \frac{1}{2}}_{\bar{1}}$) denote the even (resp. odd) components of $F^{\otimes \frac{1}{2}}$. This $\mathbb{Z}_2$ grading can be extended to a $\mathbb{Z}_{\geq 0}$ grading $\tilde{L}$, called "length", by setting

$$\tilde{L}(\phi^D_{-n_k-\frac{1}{2}} \cdots \phi^D_{-n_2-\frac{1}{2}}\phi^D_{-n_1-\frac{1}{2}}|0\rangle) = k. \tag{3.4}$$

Using the $\mathbb{Z}_2$ grading the space $F^{\otimes \frac{1}{2}}$ can be given a super-vertex algebra structure, as is known from e.g. [FFR91], [KW94], [Kac98].

In [ACJ13] and [Ang14] we introduced a $\mathbb{Z}$ grading dg on $F^{\otimes \frac{1}{2}}$ by assigning $dg(|0\rangle) = 0$ and

$$dg(\phi^D_{-n_k-\frac{1}{2}} \dots \phi^D_{-n_2-\frac{1}{2}} \phi^D_{-n_1-\frac{1}{2}}|0\rangle) = \#\{i = 1, 2, \dots, k | n_i = \text{odd}\} \ - \#\{i = 1, 2, \dots, k | n_i = \text{even}\}.$$

Denote the homogeneous component of degree $dg = n \in \mathbb{Z}$ by $F^{\otimes \frac{1}{2}}_{(n)}$, hence as vector spaces we have

$$F^{\otimes \frac{1}{2}} = \oplus_{n\in\mathbb{Z}} F^{\otimes \frac{1}{2}}_{(n)}. \tag{3.5}$$

We define the special vectors $v_n \in F^{\otimes \frac{1}{2}}_{(n)}$ by

$$v_0 = |0\rangle \in F^{\otimes \frac{1}{2}}_{(0)}; \tag{3.6}$$

$$v_n = \phi^D_{-2n+1-\frac{1}{2}} \dots \phi^D_{-3-\frac{1}{2}} \phi^D_{-1-\frac{1}{2}}|0\rangle \in F^{\otimes \frac{1}{2}}_{(n)}, \quad \text{for } n > 0; \tag{3.7}$$

$$v_{-n} = \phi^D_{-2n+2-\frac{1}{2}} \dots \phi^D_{-2-\frac{1}{2}} \phi^D_{-\frac{1}{2}}|0\rangle \in F^{\otimes \frac{1}{2}}_{(-n)}, \quad n > 0. \tag{3.8}$$

Note that the vectors $v_n \in F^{\otimes \frac{1}{2}}_{(n)}$ have minimal length $\tilde{L} = |n|$ among the vectors within $F^{\otimes \frac{1}{2}}_{(n)}$, and they are in fact the unique (up-to a scalar) vectors minimizing the length $\tilde{L}$, such that the index n_k is minimal too (n_k is identified from the smallest index appearing). One can view each of the vectors v_n as a vacuum-like vector in $F^{\otimes \frac{1}{2}}_{(n)}$, see below, and the dg grading as the analogue of the "charge" grading.

Note also that the super-grading derived from the second grading dg is compatible with the super-grading derived from the first grading $\tilde{L}$, as $dg \ mod\ 2 = \tilde{L} \ mod\ 2$.

Lastly, we recall the grading deg_h (from [Ang14])) on each of the components $F^{\otimes \frac{1}{2}}_{(n)}$ for each $n \in \mathbb{Z}$. Consider a monomial vector $v = \phi^D_{-n_k-\frac{1}{2}} \dots \phi^D_{-n_2-\frac{1}{2}} \phi^D_{-n_1-\frac{1}{2}}|0\rangle$ from $F^{\otimes \frac{1}{2}}_{(n)}$. One can view this vector as an "excitation" from the vacuum-like vector v_n, and count the n_i that should have been in v as compared to v_n, and also the n_i that should not have been in v as compared to v_n. Thus the grading deg_h (one can think of it as "energy") is defined as

$$\begin{aligned} deg_h(v) = &\sum \{ \lfloor \frac{n_l + 1}{2} \rfloor \mid n_l \text{ that should have been there but are not} \} \\ &+ \sum \{ \lfloor \frac{n_l + 1}{2} \rfloor \mid n_l \text{ that should not have been there but are} \}; \end{aligned}$$

here $\lfloor x \rfloor$ denotes the floor function, i.e., $\lfloor x \rfloor$ denotes the largest integer smaller or equal to x. Denote by $F^{\otimes \frac{1}{2}}_{(n,k)}$ the linear span of all vectors of grade $deg_h = k$ in $F^{\otimes \frac{1}{2}}_{(n)}$.

Recall the Heisenberg algebra $\mathcal{H}_\mathbb{Z}$ with relations $[h_m, h_n] = m\delta_{m+n,0}1, \ m, n$ integers. We proved in [Ang14] (and by the less traditional bicharacter construction in [Ang13]) that each $F^{\otimes \frac{1}{2}}_{(n)}$ is an irreducible highest weight module for the Heisenberg algebra via a 2-point-local field:

Proposition 3.1 *([Ang13], [Ang14]) The 2-point-local field $h^D(z)$ given by:*

$$h^D(z) = \frac{1}{2} : \phi^D(z)\phi^D(-z) : \tag{3.9}$$

has only odd-indexed modes ($h^D(z) = -h^D(-z)$), $h^D(z) = \sum_{n\in\mathbb{Z}} h_n z^{-2n-1}$, and has OPE with itself given by:

$$h^D(z)h^D(w) \sim \frac{zw}{(z^2-w^2)^2} \sim \frac{1}{4}\frac{1}{(z-w)^2} - \frac{1}{4}\frac{1}{(z+w)^2}. \tag{3.10}$$

Hence its modes, h_n, $n \in \mathbb{Z}$, generate the Heisenberg algebra $\mathcal{H}_\mathbb{Z}$. The neutral real fermion Fock space $F^{\otimes\frac{1}{2}}$ is thus a module for $\mathcal{H}_\mathbb{Z}$ via this 2-point-local field representation and decomposes into irreducible highest weight modules for $\mathcal{H}_\mathbb{Z}$ under this action as follows:

$$F^{\otimes\frac{1}{2}} = \oplus_{m\in\mathbb{Z}} F^{\otimes\frac{1}{2}}_{(m)} \cong \oplus_{m\in\mathbb{Z}} B_m, \quad \textit{where} \quad F^{\otimes\frac{1}{2}}_{(m)} \cong B_m, \quad B_m \cong \mathbb{C}[x_1, x_2, \dots, x_n, \dots], \quad \forall\, m \in \mathbb{Z}. \tag{3.11}$$

Recall the well-known Virasoro algebra Vir, the central extension of the complex polynomial vector fields on the circle. The Virasoro algebra Vir is the Lie algebra with generators L_n, $n \in \mathbb{Z}$, and central element C, with commutation relations

$$[L_m, L_n] = (m-n)L_{m+n} + \delta_{m,-n}\frac{(m^3-m)}{12}C; \quad [C, L_m] = 0, \ m, n \in \mathbb{Z}. \tag{3.12}$$

Equivalently, the 1-point-local Virasoro field $L(z) := \sum_{n\in\mathbb{Z}} L_n z^{-n-2}$ has OPE with itself given by:

$$L(z)L(w) \sim \frac{C/2}{(z-w)^4} + \frac{2L(w)}{(z-w)^2} + \frac{\partial_w L(w)}{(z-w)}. \tag{3.13}$$

We denote by $M(c,h)$ the irreducible highest weight Vir module with central charge $c \in \mathbb{C}$, where $h \in \mathbb{C}$ is the weight of the operator L_0 acting on the highest weight vector v_h, i.e., $M(c,h)$ is an irreducible Vir module generated by a vector $v_h \in M(c,h)$ with $Cv_h = cv_h$, $L_n v_h = 0$ for any $n > 0$, and $L_0 v_h = hv_h$. (Note that in some of the literature such Vir modules are called "lowest weight", instead of "highest weight", here we follow the convention of Kac, as in [Kac80], [KR87], [Kac90]). The seminal paper [FF83] (using the Kac determinant formula, [Kac80]) delineates the cases where $M(c,h)$ is a quotient by a nontrivial factor of the corresponding Verma module $VM(c,h)$, as opposed to the "generic" cases where $M(c,h) \cong VM(c,h)$.

Denote the formal character of a highest weight module V with highest weight (c,h) to be

$$ch_q^{Vir} V := tr_V q^{L_0} := \sum_{j\in\mathbb{Z}_+} (dim V_{h+j}) q^{h+j},$$

where V_{h+j} is the eigenspace of L_0 of weight $h+j$.

It is well known that $F^{\otimes\frac{1}{2}}$ is a module for the Virasoro algebra with central charge $c = \frac{1}{2}$ (see for example [FFR91], [KW94], [Wan99a]) via the 1-point local Virasoro field $L^{1/2}(z)$ given by

$$L^{1/2}(z) = \frac{1}{2} : \partial_z \phi^D(z)\phi^D(z) : . \tag{3.14}$$

Furthermore, it is well known that as Vir modules

$$F^{\otimes\frac{1}{2}}_{\bar{0}} \cong M\left(\frac{1}{2}, 0\right); \quad F^{\otimes\frac{1}{2}}_{\bar{1}} \cong M\left(\frac{1}{2}, \frac{1}{2}\right); \quad F^{\otimes\frac{1}{2}} \cong M\left(\frac{1}{2}, 0\right) \oplus M\left(\frac{1}{2}, \frac{1}{2}\right).$$

In [Ang14] we proved the following Jacobi identity holds:

Proposition 3.2 *(Corollary to Proposition 3.1) Define the graded dimension (character) of the Fock space* $F = F^{\otimes \frac{1}{2}}$ *as*

$$chF := tr_F q^{L_0^{1/2}} z^{h_0}.$$

We have

$$chF = \prod_{i=1}^{\infty}(1 + zq^{2i-1+\frac{1}{2}})(1 + z^{-1}q^{2i-2+\frac{1}{2}}) \tag{3.15}$$

$$= \frac{1}{\prod_{i=1}^{\infty}(1-q^{2i})} \sum_{n\in\mathbb{Z}} z^n q^{\frac{n}{2}} q^{n^2}. \tag{3.16}$$

By comparing the two identities we get the Jacobi identity

$$\prod_{i=1}^{\infty}(1-q^{2i})(1 + zq^{2i-\frac{1}{2}})(1 + z^{-1}q^{2i-\frac{3}{2}}) = \sum_{m\in\mathbb{Z}} z^m q^{\frac{m(2m+1)}{2}}. \tag{3.17}$$

Observe that we can specialize the graded dimension chF to the Virasoro character by evaluating at $z = 1$:

Corollary 3.3

$$ch_q^{Vir} F = ch_q M\left(\frac{1}{2}, 0\right) + ch_q M\left(\frac{1}{2}, \frac{1}{2}\right) = \prod_{i=0}^{\infty}(1 + q^{\frac{1}{2}} q^i) \tag{3.18}$$

$$= \frac{1}{\prod_{i=1}^{\infty}(1-q^{2i})} \sum_{n\in\mathbb{Z}} q^{n^2+\frac{n}{2}}. \tag{3.19}$$

If we use the q-Pochhammer symbol notation,

$$(a;q)_\infty := \prod_{i=0}^{\infty}(1-aq^i); \quad (a;q)_m := \prod_{i=0}^{m-1}(1-aq^i), \quad \text{with } (a;q)_0 := 1;$$

we can rewrite the first formula (3.18)

$$ch_q^{Vir} F = ch_q M\left(\frac{1}{2}, 0\right) + ch_q M\left(\frac{1}{2}, \frac{1}{2}\right) = (-q^{\frac{1}{2}}; q)_\infty.$$

The formula (3.18) is well known (going back to I. Frenkel and D. Friedan, as well as [FF83]), but here we obtain it as a direct evaluation at $z = 1$ of the formula (3.15). We can refine the proof of (3.16) from [Ang14] to obtain separate formulas for $ch_q M\left(\frac{1}{2}, 0\right)$ and $ch_q M\left(\frac{1}{2}, \frac{1}{2}\right)$:

Proposition 3.4

$$ch_q M\left(\frac{1}{2}, 0\right) = \frac{1}{\prod_{i=1}^{\infty}(1-q^{2i})} \sum_{\substack{n\in\mathbb{Z} \\ n\ even}} q^{n^2+\frac{n}{2}}; \tag{3.20}$$

$$ch_q M\left(\frac{1}{2}, \frac{1}{2}\right) = \frac{1}{\prod_{i=1}^{\infty}(1-q^{2i})} \sum_{\substack{n\in\mathbb{Z} \\ n\ odd}} q^{n^2+\frac{n}{2}} \tag{3.21}$$

These formulas have all positive coefficients in their sum, as

$$\frac{1}{\prod_{i=1}^{\infty}(1-q^i)} = \sum_{k\geq 0} p(k)q^k,$$

where $p(k)$ is the number of partitions $0 < k_1 \leq \cdots \leq k_l$ of $k = k_1 + \cdots + k_l$. Such formulas as (3.20) and (3.21) are called fermionic (see e.g. [KKMM93]), to distinguish them from the character sums with alternating sign coefficients (as in [RC85]), which are called bosonic.
Proof: We have, as vector spaces, that

$$M\left(\frac{1}{2},0\right) \cong F_{\bar{0}}^{\otimes\frac{1}{2}} = \oplus_{\substack{n\in\mathbb{Z} \\ n \text{ even}}} F_{(n)}^{\otimes\frac{1}{2}}; \qquad M\left(\frac{1}{2},\frac{1}{2}\right) \cong F_{\bar{1}}^{\otimes\frac{1}{2}} = \oplus_{\substack{n\in\mathbb{Z} \\ n \text{ odd}}} F_{(n)}^{\otimes\frac{1}{2}}.$$

We can now use the decomposition of each $F_{(n)}^{\otimes\frac{1}{2}}$ into $F_{(n,k)}^{\otimes\frac{1}{2}}$ by deg_h. From Proposition 3.1 it follows that a basis for $F_{(n,k)}^{\otimes\frac{1}{2}}$ is given by the elements $v_{h,n,\vec{k}} = h_{-k_l}\dots h_{-k_1}v_n$ with indecies varying with partitions $0 < k_1 \leq \cdots \leq k_l$ of $k = k_1 + \cdots + k_l$. First, we have

$$\begin{aligned}
L_0^{1/2}v_0 &= L_0^{1/2}|0\rangle = 0 \\
L_0^{1/2}v_n &= L_0^{1/2}\phi^D_{-2n+1-\frac{1}{2}}\cdots\phi^D_{-3-\frac{1}{2}}\phi^D_{-1-\frac{1}{2}}|0\rangle \\
&= \left((1+\frac{1}{2})+(3+\frac{1}{2})+\cdots+(2n-1+\frac{1}{2})\right)v_n = (n^2+\frac{n}{2})v_n, \quad \text{for } n>0; \\
L_0^{1/2}v_{-n} &= L_0^{1/2}\phi^D_{-2n+2-\frac{1}{2}}\cdots\phi^D_{-2-\frac{1}{2}}\phi^D_{-\frac{1}{2}}|0\rangle \\
&= \left((0+\frac{1}{2})+(2+\frac{1}{2})+(4+\frac{1}{2})+\cdots+(2n-2+\frac{1}{2})\right)v_{-n} = (n^2-\frac{n}{2})v_{-n}, \quad \text{for } n>0.
\end{aligned}$$

Now a direct calculation shows that ([Ang14]):

$$: h^D(w)^2 :=: h^D(w)h^D(w) := \frac{1}{4} : (\partial_{-w}\phi^D(-w))\phi^D(-w) : +\frac{1}{4} : (\partial_w\phi^D(w))\phi^D(w) : -\frac{1}{2w}h^D(w). \tag{3.22}$$

We can calculate by direct use of Wick's Theorem the OPE between $: h^D(z)^2 :$ and $h^D(w)$, and thus by use of the equation above the commutator of $L_0^{1/2}$ and $h^D(w)$; and we obtain that

$$[L_0^{1/2}, h_k^D] = -2kh_k^D. \tag{3.23}$$

Hence

$$\begin{aligned}
L_0^{1/2}v_{h,0,\vec{k}} &= L_0^{1/2}h_{-k_l}\dots h_{-k_1}v_n = (2k_1+\cdots+2k_l)\, v_{h,0,\vec{k}}; \\
L_0^{1/2}v_{h,n,\vec{k}} &= L_0^{1/2}h_{-k_l}\dots h_{-k_1}v_n = \left(2k_1+\cdots+2k_l+n^2+\frac{n}{2}\right)v_{h,n,\vec{k}} \quad \text{for } n>0; \\
L_0^{1/2}v_{h,-n,\vec{k}} &= L_0^{1/2}h_{-k_l}\dots h_{-k_1}v_{-n} = \left(2k_1+\cdots+2k_l+n^2-\frac{n}{2}\right)v_{h,-n,\vec{k}} \quad \text{for } n>0.
\end{aligned}$$

Now since there are $p(k)$ such elements for partitions $0 < k_1 \leq \cdots \leq k_l$ of $k = k_1 + \cdots + k_l$, we have for $F_{\bar{0}}^{\otimes\frac{1}{2}} \cong M(\frac{1}{2},0)$

$$\begin{aligned} ch_q M(\frac{1}{2},0) = tr_{F_{\bar{0}}^{\otimes\frac{1}{2}}} q^{L_0^{1/2}} &= \sum_{k\geq 0} p(k)q^{2k} + \sum_{\substack{n,k\in\mathbb{Z}_+\\ n\ even}} p(k)\left(q^{2k+n^2+\frac{n}{2}} + q^{2k+n^2-\frac{n}{2}}\right) \\ &= \sum_{k\in\mathbb{Z},k\geq 0} p(k)q^{2k}\cdot\left(1+\sum_{\substack{n\in\mathbb{Z}_+\\ n\ even}}\left(q^{n^2+\frac{n}{2}}+q^{n^2-\frac{n}{2}}\right)\right) \\ &= \frac{1}{\prod_{i=1}^{\infty}(1-q^{2i})}\sum_{\substack{n\in\mathbb{Z}\\ n\ even}} q^{n^2+\frac{n}{2}}. \end{aligned}$$

The calculation for $F_{\bar{1}}^{\otimes\frac{1}{2}} \cong M(\frac{1}{2},\frac{1}{2})$ is even simpler. □

We now turn to the new representations in this paper, the 2-point local Virasoro field representations on the Fock space $F^{\otimes\frac{1}{2}}$. In [Ang14] we proved that besides $L^{1/2}(z)$ there is a 2-parameter family of 2-point-local Virasoro fields on $F^{\otimes\frac{1}{2}}$:

Proposition 3.5 *([Ang14]) The 2-point-local field*

$$L^1(z^2) := \frac{1}{2z^2} : h^D(z)h^D(z) : \tag{3.24}$$

has only even-indexed modes, $L^1(z^2) := \sum_{n\in\mathbb{Z}} L_n^1(z^2)^{-n-2}$ and its modes L_n satisfy the Virasoro algebra commutation relations with central charge $c = 1$:

$$[L_m, L_n] = (m-n)L_{m+n} + \delta_{m,-n}\frac{(m^3-m)}{12}.$$

Equivalently, the 2-point-local field $L^1(z^2)$ has OPE with itself given by:

$$L^1(z^2)L^1(w^2) \sim \frac{1/2}{(z^2-w^2)^4} + \frac{2L^1(w^2)}{(z^2-w^2)^2} + \frac{\partial_{w^2}L^1(w^2)}{(z^2-w^2)}. \tag{3.25}$$

Furthermore, the 2-point-local field

$$L^{\lambda,b}(z^2) := L^1(z^2) + \frac{1-2\lambda}{4z^2}\partial_z h^D(z) - \frac{b}{z^3}h^D(z) + \frac{\left(b+\frac{1-2\lambda}{4}\right)^2 - 4\left(\frac{1-2\lambda}{4}\right)^2}{2z^4} = \sum_{n\in\mathbb{Z}} L_n^{\lambda,b}(z^2)^{-n-2} \tag{3.26}$$

is a Virasoro field for every $\lambda, b \in \mathbb{C}$ with central charge $-12\lambda^2+12\lambda-2$. If $\lambda = \frac{1}{2}$, $b = 0$, $L^{\frac{1}{2},0}(z^2) = L^1(z^2)$.

We now turn to the types of Vir representations these 2-point-local Virasoro fields generate on $F^{\otimes\frac{1}{2}}$, depending on the choices of the parameters (λ, b). First, observe that these representations of Vir, although of course expressible as 2-point local fermionic oscillator representations through the generating field $\phi^D(z)$ (see (3.22)), are in fact only dependent on the descendent Heisenberg field $h^D(z)$ from Proposition 3.1. It is then immediate that each $F_{(n)}^{\otimes\frac{1}{2}}$ for $n \in \mathbb{Z}$ is also a submodule for these 2-point-local Vir representations, and is in fact a highest weight module:

$$L_m^{\lambda,b}v_n = 0, \quad \text{for any } m,n\in\mathbb{Z}, \quad \text{where } m>0, \tag{3.27}$$

and we have for the highest weight vectors v_n $(n \in \mathbb{Z})$

$$\begin{aligned} L_0^{\lambda,b} v_n &= \left(\frac{n^2}{2} - bn - (\frac{1-2\lambda}{4})n + \frac{\left(b + \frac{1-2\lambda}{4}\right)^2 - 4(\frac{1-2\lambda}{4})^2}{2} \right) v_n \\ &= \left(\frac{\left(b + \frac{1-2\lambda}{4} - n\right)^2}{2} - 2\left(\frac{1-2\lambda}{4}\right)^2 \right) v_n. \end{aligned}$$

Observe that

$$c = 1 \quad \text{only for} \quad \lambda = \frac{1}{2}; \quad c = -12\lambda^2 + 12\lambda - 2 < 1 \quad \text{for} \quad \lambda \in \mathbb{R} \setminus \{\frac{1}{2}\}.$$

For "generic" cases of real $0 \le c \le 1$ and h the Verma modules $VM(c,h)$ are irreducible representations ([Kac80], [FF83]), and for those "generic" (c,h) we have $F_{(n)}^{\otimes \frac{1}{2}} \cong VM(c,h) = M(c,h)$. Furthermore, one is of course interested in the cases where the representation is unitary, and that leaves only the discrete series (see [FQS84], and e.g. [KR87]), or the case of $c = 1$ (see e.g. [Kac80], [FF83], [FQS84], [KR87]). For $F_{(n)}^{\otimes \frac{1}{2}}$ to be of discrete series type we need to have ([Kac80], [FF83], [FQS84], [KR87]):

$$\begin{aligned} c &= -12\lambda^2 + 12\lambda - 2 = 1 - \frac{6}{(m+2)(m+3)}, \quad \text{for } m \in \mathbb{Z}_{\ge 0}, \\ h &= h_{r,s} = \frac{((m+3)r - (m+2)s)^2 - 1}{4(m+2)(m+3)} \quad \text{for } r, s \in \mathbb{Z}_+,\ 1 \le s \le r \le m+1; \end{aligned}$$

which here gives us

$$(1 - 2\lambda)^2 = \frac{2}{(m+2)(m+3)}.$$

Now we have to consider for such λ also the highest weight h, which gives us

$$\frac{\left(b + \frac{1-2\lambda}{4} - n\right)^2}{2} = \frac{((m+3)r - (m+2)s)^2}{4(m+2)(m+3)} \quad \text{for some } r, s \in \mathbb{Z}_+,\ 1 \le s \le r \le m+1;$$

and thus

$$\left(\frac{b + \frac{1-2\lambda}{4} - n}{\frac{1-2\lambda}{4}} \right)^2 = 4\left((m+3)r - (m+2)s\right)^2, \quad \text{for } r, s \in \mathbb{Z}_+,\ 1 \le s \le r \le m+1.$$

Hence we can find $b \in \mathbb{R}$ which will produce a discrete-series-type representation on $F_{(n)}^{\otimes \frac{1}{2}}$. To summarize, if the parameters (λ, b) satisfy

$$\begin{aligned} (2\lambda - 1) &= \pm\sqrt{\frac{2}{(m+2)(m+3)}}, \quad \text{for } m \in \mathbb{Z}_+; \\ (b - n) &= \pm\frac{2\left((m+3)r - (m+2)s\right) \pm 1}{2\sqrt{2(m+2)(m+3)}}, \quad \text{for } r, s \in \mathbb{Z}_+,\ 1 \le s \le r \le m+1; \end{aligned}$$

then the submodule $F_{(n)}^{\otimes \frac{1}{2}}$ belongs to the discrete series. Thus we can choose values of the parameters (λ, b) to produce a 2-point local fermionic oscillator field representation of each of the discrete series Virasoro modules.

Next we consider the case of $c = 1$, i.e., $\lambda = \frac{1}{2}$. In this case we either have

$$L_0^{\frac{1}{2},b} v_n = \frac{(b-n)^2}{2} v_n = \frac{m^2}{4} v_n, \quad \text{for some } m \in \mathbb{Z};$$

i.e.,

$$(b-n)^2 = \frac{m^2}{2}, \quad \text{for some } m \in \mathbb{Z}; \tag{3.28}$$

or the submodule $F_{(n)}^{\otimes\frac{1}{2}}$ is irreducible. It is clear that if for some $n_1 \in \mathbb{Z}$ the parameter b satisfies

$$(b-n_1)^2 = \frac{m^2}{2}, \quad \text{for some } m \in \mathbb{Z}; \tag{3.29}$$

then for all other $n \in \mathbb{Z}$, $n \neq n_1$, we have $(b-n)^2 \neq \frac{m^2}{2}$, for any $m \in \mathbb{Z}$. Hence if the module $F_{(n_1)}^{\otimes\frac{1}{2}}$ is indeed reducible, then for all other $n \in \mathbb{Z}$, $n \neq n_1$, $F_{(n)}^{\otimes\frac{1}{2}}$ will be irreducible. Thus, when $\lambda = \frac{1}{2}$ we either have $(b-n)^2 \neq \frac{m^2}{2}$, for any $n, m \in \mathbb{Z}$ and thus all modules $F_{(n)}^{\otimes\frac{1}{2}}$ are irreducible (we call such b "generic"). Or exactly one the modules $F_{(n_1)}^{\otimes\frac{1}{2}}$ is completely reducible, the others are irreducible. The general structure of these reducible highest Vir modules with central charge 1 is well known from [FF83] (see also [KR87]), but here we can actually describe explicitly the singular vectors generating the submodules. We can without loss of generality assume that for $\lambda = \frac{1}{2}$ the one reducible submodule is $F_{(0)}^{\otimes\frac{1}{2}}$ (i.e., $b^2 = \frac{m^2}{2}$, for some $m \in \mathbb{Z}$), and thus the other $F_{(n)}^{\otimes\frac{1}{2}}$, $n \neq 0$, are irreducible.

Lemma 3.6 *Let $\lambda = \frac{1}{2}$, $b = \frac{m}{\sqrt{2}}$, $m \in \mathbb{Z}$ is fixed.*

Case I. *For $m \geq 0$, the following vectors in $F_{(0)}^{\otimes\frac{1}{2}}$, indexed by $k \in \mathbb{Z}$, $k \geq 0$, exhaust all singular vectors for the two-point local Vir representation on $F_{(0)}^{\otimes\frac{1}{2}}$:*

$\tilde{v}_{m,0} = v_0 = |0\rangle, \quad$ *for* $k = 0$;

$\tilde{v}_{m,k} = \phi^D_{-2(k+m)+1-\frac{1}{2}} \phi^D_{-2(k+m-1)+1-\frac{1}{2}} \cdots \phi^D_{-2m-1-\frac{1}{2}} \phi^D_{-2(k-1)-\frac{1}{2}} \phi^D_{-2(k-2)-\frac{1}{2}} \cdots \phi^D_{-\frac{1}{2}} |0\rangle, \quad$ *for* $k > 0$.

We have

$$L_j^{\frac{1}{2},\frac{m^2}{2}} \tilde{v}_{m,k} = 0, \quad \textit{for any } j > 0, k \geq 0;$$
$$L_0^{\frac{1}{2},\frac{m^2}{2}} \tilde{v}_{m,k} = \frac{1}{4}(m+2k)^2 \tilde{v}_{0,k}.$$

Case II. *For $m < 0$, the following vectors in $F_{(0)}^{\otimes\frac{1}{2}}$, indexed by $k \in \mathbb{Z}$, $k \geq -m$, exhaust all singular vectors for the two-point local Vir representation on $F_{(0)}^{\otimes\frac{1}{2}}$:*

$\tilde{v}_{m,-m} = v_0 = |0\rangle, \quad$ *for* $k = -m$;

$\tilde{v}_{m,k} = \phi^D_{-2(k+m)+1-\frac{1}{2}} \phi^D_{-2(k+m-1)+1-\frac{1}{2}} \cdots \phi^D_{-1-\frac{1}{2}} \phi^D_{-2(k-1)-\frac{1}{2}} \cdots \phi^D_{2(m-1)-\frac{1}{2}} \phi^D_{2m-\frac{1}{2}} |0\rangle, \quad$ *for* $k > -m$.

We have

$$L_j^{\frac{1}{2},\frac{m^2}{2}} \tilde{v}_{m,k} = 0, \quad \textit{for any } j > 0, k \geq -m;$$
$$L_0^{\frac{1}{2},\frac{m^2}{2}} \tilde{v}_{m,k} = \frac{1}{4}(m+2k)^2 \tilde{v}_{0,k}.$$

The proof is by direct calculation and we omit it.

Proposition 3.7 ***Case I.*** *Let $\lambda = \frac{1}{2}$, b is generic, i.e., $b \neq n_1 + \frac{m}{\sqrt{2}}$ for any $m, n_1 \in \mathbb{Z}$. Then as Virasoro modules with central charge $c = 1$*

$$F_{\bar{0}}^{\otimes\frac{1}{2}} = \oplus_{\substack{n\in\mathbb{Z}\\ n\ even}} M\big(1, \frac{(b-n)^2}{2}\big); \quad F_{\bar{1}}^{\otimes\frac{1}{2}} = \oplus_{\substack{n\in\mathbb{Z}\\ n\ odd}} M\big(1, \frac{(b-n)^2}{2}\big).$$

Case II. *Let $\lambda = \frac{1}{2}$, $b = n_1 + \frac{m}{\sqrt{2}}$ for some unique $m, n_1 \in \mathbb{Z}$ with n_1* ***even****. Then as Virasoro modules with central charge $c = 1$*

$$F_{\bar{0}}^{\otimes\frac{1}{2}} = \left(\oplus_{\substack{n\in\mathbb{Z}\\ n\ even\\ n\neq n_1}} M\big(1, \frac{(m-\sqrt{2}n)^2}{4}\big)\right) \bigoplus \left(\oplus_{\substack{n\geq 0\\ n\geq -m}} M\big(1, \frac{(m+2n)^2}{4}\big)\right);$$

$$F_{\bar{1}}^{\otimes\frac{1}{2}} = \oplus_{\substack{n\in\mathbb{Z}\\ n\ odd}} M\big(1, \frac{(m-\sqrt{2}n)^2}{4}\big).$$

Case III. *Let $\lambda = \frac{1}{2}$, $b = n_1 + \frac{m}{\sqrt{2}}$ for some unique $m, n_1 \in \mathbb{Z}$ with n_1* ***odd****. Then as Virasoro modules with central charge $c = 1$*

$$F_{\bar{0}}^{\otimes\frac{1}{2}} = \oplus_{\substack{n\in\mathbb{Z}\\ n\ even}} M\big(1, \frac{(m-\sqrt{2}n)^2}{4}\big);$$

$$F_{\bar{1}}^{\otimes\frac{1}{2}} = \left(\oplus_{\substack{n\in\mathbb{Z}\\ n\ odd\\ n\neq n_1}} M\big(1, \frac{(m-\sqrt{2}n)^2}{4}\big)\right) \bigoplus \left(\oplus_{\substack{n\geq 0\\ n\geq -m}} M\big(1, \frac{(m+2n)^2}{4}\big)\right).$$

We want to finish by showing an application of the above decomposition to calculating directly a positive sum (fermionic) representation of the characters of the Virasoro modules $M\left(\frac{1}{2}, \frac{1}{2}\right)$ and $M\left(\frac{1}{2}, 0\right)$. We can pick any b and use the Proposition above, together with the observation that we have a relation between the highest weights for $L_0^{1/2}$ and the highest weights for $L_0^{\lambda,b}$, via the connection represented by (3.22). In particular for $\lambda = \frac{1}{2}$, $b = 0$ we have $L^{\frac{1}{2},0}(z^2) = L^1(z^2)$; and from (3.22) we have that

$$L_0^{1/2} = 2L_0^1 + \frac{1}{2}h_0^D. \tag{3.30}$$

Hence we have from Proposition 3.7, Case II with $b = 0$:

$$ch_q M\big(\frac{1}{2}, \frac{1}{2}\big) = tr_{F_{\bar{1}}^{\otimes\frac{1}{2}}} q^{L_0^{1/2}} = tr_{F_{\bar{1}}^{\otimes\frac{1}{2}}} (q^2)^{L_0^1} q^{\frac{1}{2}h_0^D} = \sum_{\substack{n\in\mathbb{Z}\\ n\ \mathrm{odd}}} ch_{q^2} M\big(1, \frac{n^2}{2}\big) q^{\frac{n}{2}} = \sum_{\substack{n\in\mathbb{Z}\\ n\ \mathrm{odd}}} \frac{q^{n^2+\frac{n}{2}}}{\prod_{i=1}^\infty (1-q^{2i})}.$$

Here we have used the well known character formula $ch_q M\big(1, \frac{n^2}{2}\big) = \frac{q^{n^2/2}}{\prod_{i=1}^\infty (1-q^i)}$, for $n \in \mathbb{Z}, n \neq 0$, and we recover the formula (3.21) we obtained earlier.

Similarly, using Case II, $b = 0$ and the formula for $ch_q M\big(1, \frac{m^2}{4}\big)$ (see e.g. [RC85]):

$$ch_q M\big(1, \frac{m^2}{4}\big) = \frac{1}{\prod_{i=1}^\infty (1-q^i)} \big(q^{m^2/4} - q^{(m+2)^2/4}\big), \quad \text{for } m \in \mathbb{Z},$$

we get

$$ch_qM(\frac{1}{2},0) = tr_{F_0^{\otimes\frac{1}{2}}} q^{L_0^{1/2}} = tr_{F_0^{\otimes\frac{1}{2}}} (q^2)^{L_0^1} q^{\frac{1}{2}h_0^D} = \sum_{\substack{n\in\mathbb{Z}\\ n \text{ even}\\ n\neq 0}} ch_{q^2}M\big(1,\frac{n^2}{2}\big)q^{\frac{n}{2}} + \sum_{n\geq 0} ch_{q^2}M\big(1,n^2\big)$$

$$= \sum_{\substack{n\in\mathbb{Z}\\ n \text{ even}\\ n\neq 0}} \frac{q^{n^2+\frac{n}{2}}}{\prod_{i=1}^{\infty}(1-q^{2i})} + \sum_{n\geq 0} \frac{q^{2n^2}-q^{2(n+1)^2}}{\prod_{i=1}^{\infty}(1-q^{2i})} = \sum_{\substack{n\in\mathbb{Z}\\ n \text{ even}\\ n\neq 0}} \frac{q^{n^2+\frac{n}{2}}}{\prod_{i=1}^{\infty}(1-q^{2i})} + \frac{1}{\prod_{i=1}^{\infty}(1-q^{2i})}$$

$$= \sum_{\substack{n\in\mathbb{Z}\\ n \text{ even}}} \frac{q^{n^2+\frac{n}{2}}}{\prod_{i=1}^{\infty}(1-q^{2i})}.$$

Note that we could have used a generic b, for instance we have

$$L^{\frac{1}{2},-\frac{1}{4}}(z^2) = L^1(z^2) + \frac{1}{4z^3}h^D(z) + \frac{1}{32z^4}, \qquad L_n^{\frac{1}{2},-\frac{1}{4}} = \frac{1}{2}L_{2n}^{1/2} + \frac{1}{32}\delta_{n,0},$$

but the resulting character formulas would have been the same.

We want to remark that this application of multi-local Virasoro field representations to calculating character formulas can be extended more generally to the discrete Virasoro series, not only to the Ising case of $c = \frac{1}{2}$. The main new idea is to obtain a decomposition of (the vector space of) the irreducible modules represented by a one-point local Virasoro field of discrete central charge $c = 1 - \frac{6}{(m+2)(m+3)}$ into irreducible modules represented by a multi-local Virasoro field of charge 1 (or higher).

References

[ACJ13] Iana I. Anguelova, Ben Cox, and Elizabeth Jurisich. Representations of a_∞ and d_∞ with central charge 1 on the Fock space $F^{\otimes\frac{1}{2}}$. *Journal of Physics: Conference Series, 474(1)*, 474(1):20, 2013.

[ACJ14] Iana I. Anguelova, Ben Cox, and Elizabeth Jurisich. N-point locality for vertex operators: normal ordered products, operator product expansions, twisted vertex algebras. *J. Pure Appl. Algebra*, 218(12):2165–2203, 2014.

[Ang13] Iana I. Anguelova. Twisted vertex algebras, bicharacter construction and boson-fermion correspondences. *J. Math. Phys.*, 54(12):38, 2013.

[Ang14] Iana I. Anguelova. Boson-fermion correspondence of type D-A and multi-local Virasoro representations on the Fock space $F^{\otimes\frac{1}{2}}$. submitted, arXiv:1406.5158 [math-ph], 2014.

[BF00] Andrei G. Bytsko and Andreas Fring. Factorized combinations of Virasoro characters. *Comm. Math. Phys.*, 209(1):179–205, 2000.

[BS83] N. N. Bogoliubov and D. V. Shirkov. *Quantum fields.* Benjamin/Cummings Publishing Co. Inc. Advanced Book Program, Reading, MA, 1983. Translated from the Russian by D. B. Pontecorvo.

[DJKM81a] Etsurō Date, Michio Jimbo, Masaki Kashiwara, and Tetsuji Miwa. Transformation groups for soliton equations. III. Operator approach to the Kadomtsev-Petviashvili equation. *J. Phys. Soc. Japan*, 50(11):3806–3812, 1981.

[DJKM81b] Etsurō Date, Michio Jimbo, Masaki Kashiwara, and Tetsuji Miwa. Transformation groups for soliton equations. VI. KP hierarchies of orthogonal and symplectic type. *J. Phys. Soc. Japan*, 50(11):3813–3818, 1981.

[FF83] B. L. Feĭgin and D. B. Fuks. Verma modules over a Virasoro algebra. *Funktsional. Anal. i Prilozhen.*, 17(3):91–92, 1983.

[FFR91] Alex J. Feingold, Igor B. Frenkel, and John F. X. Ries. *Spinor construction of vertex operator algebras, triality, and $E_8^{(1)}$*, volume 121 of *Contemporary Mathematics*. American Mathematical Society, Providence, RI, 1991.

[FFW08] Boris Feigin, Omar Foda, and Trevor Welsh. Andrews-Gordon type identities from combinations of Virasoro characters. *Ramanujan J.*, 17(1):33–52, 2008.

[FQ97] Omar Foda and Yas-Hiro Quano. Virasoro character identities from the Andrews-Bailey construction. *Internat. J. Modern Phys. A*, 12(9):1651–1675, 1997.

[FQS84] Daniel Friedan, Zongan Qiu, and Stephen Shenker. Conformal invariance, unitarity, and critical exponents in two dimensions. *Phys. Rev. Lett.*, 52(18):1575–1578, 1984.

[Fre81] Igor B. Frenkel. Two constructions of affine Lie algebra representations and boson-fermion correspondence in quantum field theory. *J. Funct. Anal.*, 44(3):259–327, 1981.

[GKO86] P. Goddard, A. Kent, and D. Olive. Unitary representations of the Virasoro and super-Virasoro algebras. *Comm. Math. Phys.*, 103(1):105–119, 1986.

[GNO85] P. Goddard, W. Nahm, and D. Olive. Symmetric spaces, Sugawara's energy momentum tensor in two dimensions and free fermions. *Phys. Lett. B*, 160(1-3):111–116, 1985.

[GOS85] P. Goddard, D. I. Olive, and A. Schwimmer. The heterotic string and a fermionic construction of the E_8 Kac-Moody algebra. *Phys. Lett. B*, 157(5-6):393–399, 1985.

[Hua98] Kerson Huang. *Quantum field theory.* John Wiley & Sons Inc., New York, 1998. From operators to path integrals.

[Kac80] Victor G. Kac. Highest weight representations of infinite-dimensional Lie algebras. In *Proceedings of the International Congress of Mathematicians (Helsinki, 1978)*, pages 299–304. Acad. Sci. Fennica, Helsinki, 1980.

[Kac90] Victor G. Kac. *Infinite-dimensional Lie algebras.* Cambridge University Press, Cambridge, third edition, 1990.

[Kac98] Victor G. Kac. *Vertex algebras for beginners*, volume 10 of *University Lecture Series.* American Mathematical Society, Providence, RI, second edition, 1998.

[KKMM93] R. Kedem, T. R. Klassen, B. M. McCoy, and E. Melzer. Fermionic sum representations for conformal field theory characters. *Phys. Lett. B*, 307(1-2):68–76, 1993.

[KR87] Victor G. Kac and A. K. Raina. *Bombay lectures on highest weight representations of infinite-dimensional Lie algebras*, volume 2 of *Advanced Series in Mathematical Physics.* World Scientific Publishing Co. Inc., Teaneck, NJ, 1987.

[KW94] Victor G. Kac and Weiqiang Wang. Vertex operator superalgebras and their representations. In *Mathematical aspects of conformal and topological field theories and quantum groups (South Hadley, MA, 1992)*, volume 175 of *Contemp. Math.*, pages 161–191. Amer. Math. Soc., Providence, RI, 1994.

[KWY98] Victor G. Kac, Weiqiang Wang, and Catherine H. Yan. Quasifinite representations of classical Lie subalgebras of $W_{1+\infty}$. *Adv. Math.*, 139(1):56–140, 1998.

[LL04] James Lepowsky and Haisheng Li. *Introduction to vertex operator algebras and their representations*, volume 227 of *Progress in Mathematics.* Birkhäuser Boston Inc., Boston, MA, 2004.

[RC85] A. Rocha-Caridi. Vacuum vector representations of the Virasoro algebra. In *Vertex operators in mathematics and physics (Berkeley, Calif., 1983)*, volume 3 of *Math. Sci. Res. Inst. Publ.*, pages 451–473. Springer, New York, 1985.

[RT12] Karl-Henning Rehren and Gennaro Tedesco. Multilocal fermionization. *Letters in Mathematical Physics*, pages 1–18, September 2012.

[Wan99a] Weiqiang Wang. Dual pairs and infinite dimensional Lie algebras. In *Recent developments in quantum affine algebras and related topics (Raleigh, NC, 1998)*, volume 248 of *Contemp. Math.*, pages 453–469. Amer. Math. Soc., Providence, RI, 1999.

[Wan99b] Weiqiang Wang. Duality in infinite-dimensional Fock representations. *Commun. Contemp. Math.*, 1(2):155–199, 1999.

[Wel05] Trevor A. Welsh. Fermionic expressions for minimal model Virasoro characters. *Mem. Amer. Math. Soc.*, 175(827):viii+160, 2005.

Classical-quantum semigroups

Paolo Aniello

Dipartimento di Fisica dell'Università di Napoli "Federico II" and INFN – Sezione di Napoli, Complesso Universitario di Monte S. Angelo, via Cintia, 80126 Napoli, Italy

E-mail: paolo.aniello@na.infn.it

Abstract. An intriguing and deep analogy between classical and quantum states is revealed using the notion of positive definite function. By Bochner's theorem, (classical) positive definite functions on phase space can be obtained by taking the Fourier transform of probability measures; similarly, *quantum* positive definite functions are the image via the Fourier-Plancherel operator of Wigner quasi-probability distributions. Considering the basic properties of positive definite functions — classical and quantum — one is led to define a class of semigroups of operators, the so-called *classical-quantum semigroups*. It is then natural to wonder whether they have any physical meaning. It turns out that the classical-quantum semigroups can also be obtained by dequantizing a certain class of quantum dynamical semigroups, namely, the *classical noise semigroups*. This correspondence fits in a more general group-theoretical framework in which a larger class of quantum dynamical semigroups, the *twirling semigroups*, can be suitably dequantized. Connections with quantum information science will be briefly discussed.

1. Introduction and outline

The evolution of an *open* quantum system — a quantum system interacting with an environment system or reservoir — is well described by a *quantum dynamical semigroup*, provided that suitable assumptions on the nature of such interaction are satisfied (suitable regimes, e.g., weak coupling, singular coupling or low density; various approximations, e.g., the Markovian approximation) [1, 2]. Like any other semigroup of operators, a quantum dynamical semigroup is completely determined by its infinitesimal generator [3]. In the case of a finite-dimensional open system, this operator has a canonical form, independently derived by Gorini, Kossakowski and Sudarshan [4], and by Lindblad [5], around the mid-1970s.

Physicists usually prefer to express the fundamental laws and properties ruling the behaviour of a physical system in 'infinitesimal form' (rather than in 'integrated form') — e.g., Schrödinger equation (versus evolution operator), canonical commutation relations (versus commutation relations in the Weyl form [6]), etc. — and the integrated form is often just a formal expression. In the case we are considering, the usual approach amounts to fixing the infinitesimal generator of the evolution of a given open quantum system by writing a suitable *master equation*. In this regard, it turns out that the Gorini-Kossakowski-Lindblad-Sudarshan canonical form of the master equation appears not only when studying the dynamics of a finite-dimensional system, but in a more general setting — e.g., it is associated with every norm-continuous quantum dynamical semigroup [5] — and deriving the canonical master equation describing a given open quantum system is a fundamental task [2]. However, in some noteworthy cases this 'master equation approach' can be reversed by considering directly the integrated form of the evolution: the semigroup of operators itself is given — in some more or less explicit form — whereas the infinitesimal generator (thus, the master equation) can be derived. We will call this point of view the *semigroup approach*.

Undertaking the semigroup approach, we will consider a class of semigroups of operators whose definition stems in a natural and simple way from the notion of *positive definite function*. As briefly recalled in sect. 2, here one should actually distinguish two different cases. On one hand, we have the standard positive definite functions on $\mathbb{R}^n$, $n \geq 1$ — with $\mathbb{R}^n$ regarded as an (additive) abelian group — arising in harmonic analysis [7] and probability theory [8, 9], and directly related to the notion of *classical* state. On the other hand, associated with quantum states we have the *quantum* positive definite functions on $\mathbb{R}^{2n}$, a notion arising in the context of the phase-space formulation of quantum mechanics *à la* Weyl-Wigner-Groenewold-Moyal [10–14] (which is almost as old as the standard formulation itself), but seems to have been introduced only in the mid-1960s by Kastler [15], and by Loupias and Miracle-Sole [16,17]. It turns out that there is a nice interplay between classical and quantum positive definite functions [18], and this interplay is at the root of the definition of the aforementioned class of semigroups of operators, that will be therefore called *classical-quantum* semigroups.

The classical-quantum semigroups, whose mere definition — see sect. 3 — may be regarded as sort of mathematical curiosity, can actually be considered as quantum dynamical semigroups 'in disguise'. Otherwise stated, they can be regarded as a nonconventional representation of the dynamics of certain open quantum systems. The main aim of the present contribution is to give a precise sense to this claim, and in the rest of this section we will briefly outline the basic ideas and try to give a sketch of the general picture.

Indeed, a standard quantum dynamical semigroup acts in a Banach space of trace class operators, where the quantum states are realized by *density operators* (normalized, positive trace class operators). However, as already observed, quantum mechanics admits a phase-space formulation, which ultimately relies on group-theoretical methods; see [19–21] and references therein. In this formulation, a density operator is replaced with an ordinary function, living on a (symmetry) group or, more generally, on a homogeneous space of that group. These functions are often called *generalized Wigner functions* or *quantum tomograms* in the literature [19–24]. Of course, the archetypical approach is based on the standard Wigner functions, which amounts to considering phase-space translations as the relevant symmetry group. It is then natural to wonder what is the expression of the master equation or of the semigroup of operators, associated with an open quantum system, in terms of tomograms. Clearly, how unwieldy these expressions are strongly depends on the class of quantum dynamical semigroups one is considering.

There is a class of semigroups of operators whose definition involves two basic ingredients: a representation of a locally compact group in a Banach space and a convolution semigroup of probability measures on that group; see sect. 4 and references therein. We call the semigroups in this class *randomly generated semigroups* (RGSs). The class of RGSs contains, in particular, both 'classical' and 'quantum' objects. E.g., it contains the semigroups of operators describing the statistical properties of classical Brownian motion on Lie groups, i.e., the *probability semigroups* [25–27].

The mentioned group-theoretical framework of RGSs turns out to be a natural bridge between quantum mechanics on phase space — formulation that is based on group representation theory, as recalled above — and the theory of open quantum systems. In fact, for a suitable choice of the relevant (group, i.e. phase-space translations, and) group representation, by varying the convolution semigroup of measures involved in the construction one spans the (sub-)class of the *Wigner (quasi-probability) semigroups*. The Wigner semigroups act on phase-space functions, the standard Wigner functions, so by applying the symplectic Fourier-Plancherel transform one obtains a further class of semigroups of operators. The semigroups in this class turn out to coincide with the classical-quantum semigroups that are the main topic of the paper.

This way of re-deriving the classical-quantum semigroups simultaneously confirms a profound

Journal of Physics: Conference Series **563** (2014) 012002 doi:10.1088/1742-6596/563/1/012002

'group-theoretical nature' of these semigroups of operators and shows how to generalize their definition; see sect. 4. This generalization leads to the notion of *tomographic semigroup*. At this point, one is ready to recognize that the tomographic semigroups (more precisely, the *proper* tomographic semigroups) — hence, in particular, the classical-quantum semigroups (or the strictly related Wigner semigroups) — can be obtained by transforming via a linear isometry quantum dynamical semigroups of a certain type, the so-called *twirling semigroups*, introduced by Kossakowski in a seminal paper [28]; see sect. 5 and further references therein.

This last step closes our circle of ideas: the classical-quantum semigroups, despite their extremely simple definition, are quantum dynamical semigroups in disguise, where the disguise is obtained by a linear transformation mapping operators into phase-space functions. Precisely, the classical-quantum semigroups are the disguised counterpart of certain twirling semigroups, the so-called *classical-noise* semigroups. The origin of such a peculiar denomination for this class of quantum dynamical semigroups will become clear in sect. 6, where connections with quantum information science will be established.

Interestingly, not only the probability and the Wigner quasi-probability semigroups, but also all other semigroups of operators considered here — the tomographic semigroups (in particular, the classical-quantum semigroups) and the twirling semigroups (in particular, the classical-noise semigroups) — belong to the class of RGSs.

2. Positive definite functions: classical and quantum

In classical (statistical) mechanics, *states* are usually realized as probability measures on phase space — for the sake of notational simplicity, here and in the following we will consider the (1+1)-dimensional phase space $\mathbb{R}\times\mathbb{R}$ — and the expectation value of an *observable* $f\colon \mathbb{R}\times\mathbb{R}\to\mathbb{R}$ in the state μ is provided by the expression

$$\langle f\rangle_\mu = \int_{\mathbb{R}\times\mathbb{R}} f(q,p)\,\mathrm{d}\mu(q,p), \quad f\in \mathsf{C}_0(\mathbb{R}\times\mathbb{R}), \tag{1}$$

where $\mathsf{C}_0(\mathbb{R}\times\mathbb{R})$ is the space of continuous *real* functions on $\mathbb{R}\times\mathbb{R}$ vanishing at infinity. Here we are thinking of μ as a normalized, positive bounded functional on the C*-algebra of classical observables (whose selfadjoint part is $\mathsf{C}_0(\mathbb{R}\times\mathbb{R})$) but, clearly, formula (1) makes sense for every μ-integrable function f.

If one wants to deal with ordinary functions, rather than with probability measures, the state (associated with) μ can be replaced with its symplectic Fourier transform

$$\widetilde{\mu}(q,p) = \int_{\mathbb{R}\times\mathbb{R}} \mathrm{e}^{\mathrm{i}(qp'-pq')}\,\mathrm{d}\mu(q',p') = \big\langle (q',p')\mapsto \mathrm{e}^{\mathrm{i}(qp'-pq')}\big\rangle_\mu, \tag{2}$$

which is a bounded continuous function usually called the *characteristic function* associated with μ, in the context of probability theory [8, 9]. By *Bochner's theorem* [7–9], characteristic functions admit an *intrinsic* characterization: the convex set that they form (in the vector space of bounded, continuous $\mathbb{C}$-valued functions on $\mathbb{R}\times\mathbb{R}$) *coincides* with the convex set of normalized, continuous *positive definite functions*; i.e., with the convex set containing every continuous complex function $\widetilde{\mu}$ on $\mathbb{R}\times\mathbb{R}$ satisfying

$$\sum_{j,k} \widetilde{\mu}(z_j - z_k)\, c_j c_k^* \geq 0, \quad \text{(positivity)} \tag{3}$$

— for any finite set

$$\{z_1 \equiv (q_1,p_1),\ldots,z_n\equiv(q_n,p_n)\}\subset \mathbb{R}\times\mathbb{R} \tag{4}$$

and arbitrary complex numbers $c_1, \ldots, c_n$ — and normalized in such a way that

$$\widetilde{\mu}(0) = 1, \tag{5}$$

with $0 \equiv (0,0)$ denoting the origin in $\mathbb{R}\times\mathbb{R}$. It is worth stressing that in the first of conditions (3) the difference $z_j - z_k = (q_j - q_k, p_j - p_k)$ should be regarded as a group operation, where obviously the (additive) group $\mathbb{R}\times\mathbb{R}$ is involved. Indeed, the notion of positive definite function on a vector group extends in a natural way to abelian groups [7] and, more generally, to locally compact groups [29]. In the following, we will denote by C (alternatively, by $\check{\mathsf{C}}$) the convex cone of positive definite functions (respectively, the convex set of *normalized* positive definite functions) on $\mathbb{R}\times\mathbb{R}$.

On the other hand, in quantum mechanics (normal) states are usually realized as *density operators*, i.e., as normalized, positive trace class operators in a given Hilbert space. In this formalism, there is no direct analogue of the notion positive definite function. However, if one wishes to deal with ordinary functions — as in the classical case, where a state can be represented by a characteristic function, see (2) — 'phase-space formulations' of quantum theory are possible. The archetype — and, under certain respects, the most remarkable — of these approaches is mainly due to the pioneering work of Weyl, Wigner, Groenewold and Moyal [10–13], so that we will call it the *WWGM formulation.*

Setting $\hbar = 1$, in the WWGM formulation of quantum mechanics a *pure* state $\hat{\rho}_\psi = |\psi\rangle\langle\psi|$, $\psi \in \mathrm{L}^2(\mathbb{R})$ ($\|\psi\| = 1$), is (injectively) replaced with a function ϱ_ψ on phase space according to Wigner's prescription

$$\varrho_\psi(q,p) := \frac{1}{2\pi}\int_{\mathbb{R}} e^{-\mathrm{i}px}\, \psi\Big(q-\frac{x}{2}\Big)^{*}\, \psi\Big(q+\frac{x}{2}\Big)\ \mathrm{d}x\,. \tag{6}$$

This recipe, exploiting the spectral decomposition of a positive trace class operator, immediately extends to any mixed state in $\mathrm{L}^2(\mathbb{R})$ and, hence, to every trace class operator (by taking linear superpositions). The (complex Banach) space of functions that one obtains with such construction will be denoted by LW. The linear space LW contains a convex cone W, formed by those functions that correspond to *positive* trace class operators, and W contains the convex set $\check{\mathsf{W}}$ formed by the *Wigner functions*, i.e., the functions associated with density operators. Within the convex cone W, the Wigner functions are characterized by the normalization condition

$$\lim_{r\to+\infty}\int_{q^2+p^2<r} \varrho(q,p)\ \mathrm{d}q\,\mathrm{d}p = \mathrm{tr}(\hat{\rho}) = 1, \tag{7}$$

see e.g. [30], where $\varrho \in \check{\mathsf{W}}$ is the phase-space function associated with a certain state $\hat{\rho}$.

The function ϱ is real and, although not (in general) a genuine probability distribution — it may also assume negative values — it allows us to express the expectation value of an observable $\hat{A}$ in the state $\hat{\rho}$ (i.e., $\langle\hat{A}\rangle_{\hat{\rho}} = \mathrm{tr}(\hat{A}\,\hat{\rho})$) as a phase-space integral,

$$\langle\hat{A}\rangle_{\hat{\rho}} = \int_{\mathbb{R}\times\mathbb{R}} \mathcal{A}(q,p)\,\varrho(q,p)\ \mathrm{d}q\,\mathrm{d}p, \tag{8}$$

where $\mathcal{A}$ is another real function suitably associated with the selfadjoint operator $\hat{A}$. Due to the integral formulae (7) and (8), the Wigner function ϱ is often called a *quasi-probability distribution.*

Remark 1 To be precise, not all the observables can be realized as ordinary functions in the WWGM formulation. More generally, they will be suitable distributions (generalized functions), see [31]. However, such technicalities are not relevant for our purposes.

It is natural to wonder whether there is any *intrinsic* characterization of Wigner quasi-probability distributions — apart from the mere normalization condition (7), which makes sense within W. Only if such a characterization is possible, the WWGM approach can be regarded as a fully self-consistent formulation of quantum theory. A simple answer to this question is provided by a quantum version of Bochner's theorem (Kastler 1965 [15], Loupias and Miracle-Sole 1966 [16,17]; also see [32,33]). As in the case of classical states, a Wigner quasi-probability distribution ϱ admits a remarkable characterization in terms of its symplectic Fourier transform $\widetilde{\varrho}$. Precisely, a function $\widetilde{\varrho}\colon \mathbb{R}\times\mathbb{R}\to\mathbb{C}$ is the symplectic Fourier transform of a Wigner function,

$$\widetilde{\varrho} = 2\pi\,\mathcal{F}_{\mathrm{sp}}\,\varrho, \tag{9}$$

if and only if it is continuous, it satisfies the condition

$$\sum_{j,k} \widetilde{\varrho}(z_j - z_k)\, \mathrm{e}^{\mathrm{i}\omega(z_k,z_j)/2}\, c_j\, c_k^* \geq 0, \quad (\omega\text{-positivity}) \tag{10}$$

for every finite set $\{z_1 \equiv (q_1,p_1),\ldots,z_n \equiv (q_n,p_n)\} \subset \mathbb{R}\times\mathbb{R}$ and complex numbers $c_1,\ldots,c_n$ — where $\omega\colon (\mathbb{R}\times\mathbb{R})\times(\mathbb{R}\times\mathbb{R})\to\mathbb{R}$ is the standard *symplectic form* — and it is suitably normalized:

$$\widetilde{\varrho}(0) = 1. \tag{11}$$

Remark 2 The symplectic Fourier transform denoted above by $\mathcal{F}_{\mathrm{sp}}$ should be regarded as a unitary operator in the Hilbert space $\mathrm{L}^2(\mathbb{R}\times\mathbb{R})$ — a symplectic Fourier-Plancherel operator — rather than an integral expression like formula (2). Indeed, as shown in [30], a Wigner function is square integrable but, in general, it is *not* integrable (fact that, as the reader may have noted, is taken into account in the lhs of formula (7)). In order to obtain a unitary operator, we fix the normalization of $\mathcal{F}_{\mathrm{sp}}$ as follows:

$$(\mathcal{F}_{\mathrm{sp}} f)(q,p) = \frac{1}{2\pi}\int_{\mathbb{R}\times\mathbb{R}} f(q',p')\, \mathrm{e}^{\mathrm{i}(qp'-pq')}\, \mathrm{d}q'\mathrm{d}p', \quad f\in \mathrm{L}^1(\mathbb{R}\times\mathbb{R})\cap\mathrm{L}^2(\mathbb{R}\times\mathbb{R}). \tag{12}$$

This explains the factor 2π appearing on the rhs of relation (9).

Remark 3 By the previous facts, a real function ϱ on phase space is a Wigner distribution if and only if it belongs to $\mathrm{L}^2(\mathbb{R}\times\mathbb{R})$ and $\widetilde{\varrho} := 2\pi\,\mathcal{F}_{\mathrm{sp}}\,\varrho$ is a normalized *quantum positive definite function*, i.e., it is a continuous function satisfying the ω-positivity condition (10) and normalized according to (11) (the requirement that $\widetilde{\varrho}$ be continuous can be slightly relaxed — see [33] — but, again, this technical aspect is not relevant here). By contrast, the characteristic function of a probability measure will be sometimes referred to as a *classical positive definite function*.

In the following, we will denote by Q (alternatively, by $\check{\mathsf{Q}}$) the convex cone of quantum positive definite functions (respectively, the convex set of *normalized* quantum positive definite functions) on $\mathbb{R}\times\mathbb{R}$; namely, $\mathsf{Q} = \mathcal{F}_{\mathrm{sp}}\mathsf{W}$ and $\check{\mathsf{Q}} = 2\pi\,\mathcal{F}_{\mathrm{sp}}\check{\mathsf{W}}$.

Comparing conditions (3) and (10), the reader should note something more than a formal similarity. In the latter, the function

$$(\mathbb{R}\times\mathbb{R})\times(\mathbb{R}\times\mathbb{R}) \ni (z_1,z_2) \mapsto \mathrm{e}^{\mathrm{i}\omega(z_1,z_2)/2} \tag{13}$$

should be regarded as a (nontrivial) multiplier for the group $\mathbb{R}\times\mathbb{R}$ [34], the multiplier associated with the Weyl system and with the integrated form of canonical commutation relations [6], whereas in the former the trivial multiplier is implicitly involved. Otherwise stated, relation (3) is a clue of classical commutativity as opposed to quantum non-commutativity which is implicit in relation (10).

3. Playing with positive definite functions: classical-quantum semigroups

The *convolution* $\mu_1 \circledast \mu_2$ of two probability measures μ_1 and μ_2 (say, on a locally compact abelian group [29]) is a probability measure too; hence, taking the Fourier transform, by Bochner's theorem the point-wise product $\mathcal{C}_1\, \mathcal{C}_2$ of two positive definite functions $\mathcal{C}_1$ and $\mathcal{C}_2$ in C is again a positive definite function. To the same conclusion one is led exploiting relation (3) and Schur's product theorem [35], according to which the *Hadamard product* (i.e., the entrywise product) of two positive (semi-definite) matrices is positive too. What happens if we now take the point-wise product of a (classical) positive definite function $\mathcal{C} \in \mathsf{C}$ by a quantum positive definite function $\mathcal{Q} \in \mathsf{Q}$? It is immediately clear, by Schur's product theorem, that $\mathcal{C}\, \mathcal{Q}$ is once again a *quantum* positive definite function.

Remark 4 Clearly, the point-wise product of two classical positive definite functions, or of a classical positive definite function by a quantum one, preserves normalization: $\mathcal{C}, \mathcal{C}' \in \breve{\mathsf{C}}$, $\mathcal{Q} \in \breve{\mathsf{Q}} \Rightarrow \mathcal{C}\, \mathcal{C}' \in \breve{\mathsf{C}}$, $\mathcal{C}\, \mathcal{Q} \in \breve{\mathsf{Q}}$.

Comforted by the previous results, we can now feel free to play with the point-wise product of positive definite functions (classical or quantum), without paying attention, for the moment, to the physical meaning of our mathematical manipulations.

Consider then a *semigroup of positive definite functions*, i.e., a set $\{\mathcal{C}_t\}_{t\in\mathbb{R}^+}$ of normalized (classical) positive definite functions on $\mathbb{R} \times \mathbb{R}$ such that

$$\mathcal{C}_t\, \mathcal{C}_s = \mathcal{C}_{t+s}\,, \ t, s \geq 0, \quad \mathcal{C}_0 \equiv 1, \tag{14}$$

where $\mathcal{C}_t\, \mathcal{C}_s$ is a point-wise product. Such semigroups can be completely classified. Indeed, the (symplectic) Fourier transform of a semigroup of positive definite functions is a convolution semigroup of probability measures, and convolution semigroups admit a well known classification related to the Lévy-Kintchine formula [25–27]. Since, for every $t \geq 0$, $\mathcal{C}_t$ is a bounded continuous function, we can define in $\mathrm{L}^2(\mathbb{R} \times \mathbb{R})$ a bounded operator $\hat{\mathfrak{C}}_t$ by setting

$$(\hat{\mathfrak{C}}_t\, f)(q,p) := \mathcal{C}_t\,(q,p)\, f(q,p), \quad f \in \mathrm{L}^2(\mathbb{R} \times \mathbb{R}). \tag{15}$$

The set $\{\hat{\mathfrak{C}}_t\}_{t\in\mathbb{R}^+}$ is a semigroup of operators:

(i) $\hat{\mathfrak{C}}_t\, \hat{\mathfrak{C}}_s = \hat{\mathfrak{C}}_{t+s}$, $t, s \geq 0$ (one-parameter semigroup property);

(ii) $\hat{\mathfrak{C}}_0 = I$ (I denoting the identity operator).

It is now natural to consider the restriction of the semigroup of operators $\{\hat{\mathfrak{C}}_t\}_{t\in\mathbb{R}^+}$ to a linear subspace of $\mathrm{L}^2(\mathbb{R} \times \mathbb{R})$. Indeed, by linear superpositions, one can extend in a natural way the convex cone Q of *quantum* positive definite functions on $\mathbb{R} \times \mathbb{R}$ to a *complex* vector space LQ which turns out to be a *dense* linear subspace of $\mathrm{L}^2(\mathbb{R} \times \mathbb{R})$. A semigroup of operators $\{\mathfrak{C}_t\}_{t\in\mathbb{R}^+}$ in LQ is defined as follows. Since the point-wise product of a classical positive definite function by a quantum positive definite function is again quantum positive definite, we can set

$$(\mathfrak{C}_t\, \mathcal{Q})(q,p) := \mathcal{C}_t\,(q,p)\, \mathcal{Q}(q,p), \quad \mathcal{Q} \in \mathsf{LQ}, \tag{16}$$

where, with a slight abuse of notation, $\mathcal{Q}$ here is a linear superposition of quantum positive definite functions:

$$\mathcal{Q} = \mathcal{Q}_1 - \mathcal{Q}_2 + \mathrm{i}(\mathcal{Q}_3 - \mathcal{Q}_4), \quad \mathcal{Q}_1, \ldots, \mathcal{Q}_4 \in \mathsf{Q}\,. \tag{17}$$

It is clear that we have:

$$\mathfrak{C}_t\, \mathsf{Q} \subset \mathsf{Q}, \quad \mathfrak{C}_t\, \breve{\mathsf{Q}} \subset \breve{\mathsf{Q}}. \tag{18}$$

Remark 5 It can be shown that the semigroup of operators $\{\hat{\mathfrak{C}}_t\}_{t\in\mathbb{R}^+}$ — regarded as the map $t \mapsto \hat{\mathfrak{C}}_t$ — is continuous; moreover, since $|\mathcal{C}_t(q,p)| \le \mathcal{C}_t(0) = 1$, it is a *contraction semigroup*, i.e., $\|\hat{\mathfrak{C}}_t\| \le 1$. Similarly, endowing the sets C and LQ with suitable topologies, the semigroups $\{\mathcal{C}_t\}_{t\in\mathbb{R}^+}$ and $\{\mathfrak{C}_t\}_{t\in\mathbb{R}^+}$ turn out to be continuous too. In particular, the linear space LQ can be regarded as the image, via the Weyl (dequantization) map, of the Banach space of trace class operators, as it will be briefly recalled in sect. 4. However, as already stressed, we will focus on the main ideas rather than on technical details.

We will call the semigroups of operators $\{\hat{\mathfrak{C}}_t\}_{t\in\mathbb{R}^+}$ and $\{\mathfrak{C}_t\}_{t\in\mathbb{R}^+}$ a *classical-quantum semigroup* and a *proper classical-quantum semigroup*, respectively.

4. Classical-quantum semigroups as tomographic semigroups

The mere definition of a classical-quantum semigroup in terms of positive definite functions is so simple and abstract that the reader may hardly suspect, at first sight, that it has anything to do with the description of an open quantum system.

To clarify this point, consider the unitary operator $\mathcal{T}(q,p)$ in $\mathrm{L}^2(\mathbb{R}\times\mathbb{R})$ defined by

$$\big(\mathcal{T}(q,p)\, f\big)(\tilde{q},\tilde{p}) = \mathrm{e}^{-\mathrm{i}(q\tilde{p}-p\tilde{q})}\, f(\tilde{q},\tilde{p}). \tag{19}$$

The map $\mathbb{R}\times\mathbb{R} \ni (q,p) \mapsto \mathcal{T}(q,p)$ is a unitary representation. Assume that $\{\mu_t\}_{t\in\mathbb{R}^+}$ is the convolution semigroup of probability measures associated, via the symplectic Fourier transform, with the semigroup $\{\mathcal{C}_t\}_{t\in\mathbb{R}^+}$ of (normalized) positive definite functions which generates the classical-quantum semigroup (15). Observe now that

$$\hat{\mathfrak{C}}_t\, f = \int_{\mathbb{R}\times\mathbb{R}} \mathcal{T}(q,p)\, f \;\mathrm{d}\mu_t(q,p), \quad \forall f \in \mathrm{L}^2(\mathbb{R}\times\mathbb{R}), \tag{20}$$

where the rhs of (20) should be regarded as an integral of vector-valued functions.

Formula (20) shows that a classical-quantum semigroup can be regarded as a particular case of a class of semigroups of operators constructed in the following way. Given a *locally compact group* G, let $\mathfrak{V}$ be a (weakly continuous) representation of G in a real or complex Banach space $\mathcal{J}$, and let $\{\mu_t\}_{t\in\mathbb{R}^+}$ a convolution semigroup on G. The set $\{\mu_t[\mathfrak{V}]\}_{t\in\mathbb{R}^+}$, with $\mu_t[\mathfrak{V}]\colon \mathcal{J}\to\mathcal{J}$ denoting the bounded linear operator defined by

$$\mu_t[\mathfrak{V}]\,\Psi := \int_G \mathfrak{V}(g)\,\Psi\;\mathrm{d}\mu_t(g), \quad \forall \Psi \in \mathcal{J}, \tag{21}$$

is a semigroup of operators, a so-called *randomly generated semigroup* [36, 37].

Remark 6 The representation $\mathfrak{V}$ is also assumed to be uniformly bounded and, if the Banach space $\mathcal{J}$ is separable, the integral on the rhs of (21) can be regarded as a Bochner integral (in particular, this is the case of the integral on the rhs of (20)); see [36].

The (selfadjoint) unitary operator $\mathcal{F}_{\mathrm{sp}}$ determined by (12) intertwines the representation $\mathcal{T}$ with a unitary representation $\mathcal{S}$ of the group $\mathbb{R}\times\mathbb{R}$ in $\mathrm{L}^2(\mathbb{R}\times\mathbb{R})$,

$$\mathcal{S}(q,p) := \mathcal{F}_{\mathrm{sp}}\, \mathcal{T}(q,p)\, \mathcal{F}_{\mathrm{sp}}, \quad \forall (q,p) \in \mathbb{R}\times\mathbb{R}, \tag{22}$$

and obviously the same construction as above can be applied to this new representation. Taking into account the fact that

$$\big(\mathcal{S}(q,p)\, f\big)(\tilde{q},\tilde{p}) = f(\tilde{q}-q,\tilde{p}-p), \quad f \in \mathrm{L}^2(\mathbb{R}\times\mathbb{R}) \tag{23}$$

— i.e., $\mathcal{S}$ is the regular representation of the group $\mathbb{R}\times\mathbb{R}$ — we obtain the semigroup of operators

$$\{\hat{\mathfrak{W}}_t \equiv \mu_t[\mathcal{S}] \colon \mathrm{L}^2(\mathbb{R}\times\mathbb{R}) \to \mathrm{L}^2(\mathbb{R}\times\mathbb{R})\}_{t\in\mathbb{R}^+} \tag{24}$$

defined by

$$(\hat{\mathfrak{W}}_t f)(\tilde{q},\tilde{p}) = \int_{\mathbb{R}\times\mathbb{R}} (\mathcal{S}(q,p) f)(\tilde{q},\tilde{p})\, \mathrm{d}\mu_t(q,p) = \int_{\mathbb{R}\times\mathbb{R}} f(\tilde{q}-q, \tilde{p}-p)\, \mathrm{d}\mu_t(q,p). \tag{25}$$

One can show, moreover, that the representation $\mathcal{T}$ satisfies

$$\mathcal{T}(q,p)\,\breve{\mathsf{Q}} = \breve{\mathsf{Q}}\,. \tag{26}$$

From this relation it follows that

$$\mathcal{T}(q,p)\,\mathsf{LQ} = \mathsf{LQ} \quad \text{and} \quad \mathcal{T}(q,p)\,\mathsf{Q} = \mathsf{Q}\,. \tag{27}$$

The first of relations (27) is coherent with the fact that, as already observed, $\hat{\mathfrak{C}}_t\,\mathsf{LQ} \equiv \mu_t[\mathcal{T}]\,\mathsf{LQ} \subset \mathsf{LQ}$, and hence one can define a further semigroup of operators — the proper classical-quantum semigroup $\{\mathfrak{C}_t\}_{t\in\mathbb{R}^+}$ — just by restriction to the Banach space LQ. The second of relations (27) and relation (26) are then coherent with relations (18).

Relation (26) also implies that

$$\mathcal{S}(q,p)\,\mathsf{LW} = \mathsf{LW}\,, \quad \mathcal{S}(q,p)\,\mathsf{W} = \mathsf{W} \quad \text{and} \quad \mathcal{S}(q,p)\,\breve{\mathsf{W}} = \breve{\mathsf{W}}\,. \tag{28}$$

Accordingly, we can restrict the semigroup of operators $\{\hat{\mathfrak{W}}_t\}_{t\in\mathbb{R}^+}$ to the Banach space LW, so obtaining a new semigroup of operators $\{\mathfrak{W}_t \colon \mathsf{LW} \to \mathsf{LW}\}_{t\in\mathbb{R}^+}$, which is *positive* — namely, $\mathfrak{W}_t\,\mathsf{W} \subset \mathsf{W}$ — and preserves the normalization condition (7), i.e., $\mathfrak{W}_t\,\breve{\mathsf{W}} \subset \breve{\mathsf{W}}$ (thus, we may say that it is *trace-preserving*). The semigroups of operators $\{\hat{\mathfrak{W}}_t\}_{t\in\mathbb{R}^+}$ and $\{\mathfrak{W}_t\}_{t\in\mathbb{R}^+}$ will be called a *Wigner semigroup* and a *proper Wigner semigroup*, respectively.

Apart from *complete* positivity [1], whose role in the theory of open quantum systems was first recognized by Lindblad [5] (for the sake of simplicity, we will not consider this property here), positivity and preservation of the trace are the basic features of a quantum dynamical semigroup. It is then clear that the (proper) classical-quantum semigroups, and the strictly related (proper) Wigner semigroups, describe the evolution of open quantum systems in terms of phase-space functions.

It is now worth observing that, in their integral formulation (20), the classical-quantum semigroups can be suitably generalized. This generalization, which gives rise to a special class of randomly generated semigroups, the tomographic semigroups [18,36,37], is based on considering a certain representation of a locally compact group G.

Precisely, let Δ_G be the modular function on G, and let $\mathsf{m} : G\times G \to \mathbb{T}$ — with $\mathbb{T}$ denoting the circle group — be a multiplier for G [34], i.e.,

$$\mathsf{m}(g,e) = \mathsf{m}(e,g) = 1, \quad \forall g \in G, \tag{29}$$

and

$$\mathsf{m}(g_1, g_2 g_3)\,\mathsf{m}(g_2,g_3) = \mathsf{m}(g_1 g_2, g_3)\,\mathsf{m}(g_1,g_2), \quad \forall g_1,g_2,g_3 \in G. \tag{30}$$

Associated with the multiplier m, there is a function $\breve{\mathsf{m}} : G\times G \to \mathbb{T}$ defined by

$$\breve{\mathsf{m}}(g,h) := \mathsf{m}(g, g^{-1}h)^*\,\mathsf{m}(g^{-1}h, g), \quad \forall g,h \in G. \tag{31}$$

Consider then the map

$$\mathcal{T}_{\mathsf{m}}\colon G \to \mathcal{U}(\mathrm{L}^2(G)) \tag{32}$$

— with $\mathcal{U}(\mathrm{L}^2(G))$ denoting the unitary group of the Hilbert space $\mathrm{L}^2(G)$, of square integrable functions on G with respect to the left Haar measure — defined by

$$\big(\mathcal{T}_{\mathsf{m}}(g)f\big)(h) := \Delta_G(g)^{\frac{1}{2}}\ \check{\mathsf{m}}(g,h)\, f(g^{-1}hg), \quad f \in \mathrm{L}^2(G), \tag{33}$$

It turns out that $\mathcal{T}_{\mathsf{m}}$ is a strongly continuous unitary representation [21].

Therefore, for every convolution semigroup $\{\mu_t\}_{t\in\mathbb{R}^+}$ of probability measures on G we have a randomly generated semigroup associated with the pair $(\mathcal{T}_{\mathsf{m}}, \{\mu_t\}_{t\in\mathbb{R}^+})$, namely,

$$\mathfrak{T}_t^{\mathsf{m}} f := \mu_t[\mathcal{T}_{\mathsf{m}}]\, f = \int_G \mathcal{T}_{\mathsf{m}}(g) f \,\mathrm{d}\mu_t(g), \quad \forall f \in \mathrm{L}^2(G). \tag{34}$$

As anticipated, we call the semigroup of operators $\{\mathfrak{T}_t^{\mathsf{m}}\}_{t\in\mathbb{R}^+}$ a *tomographic semigroup*. The origin of this term becomes clear when one considers the case where there exists a *square integrable* projective representation U of the group G, with multiplier m [38,39]. In this case, one can define a map $\mathscr{D}$ that associates injectively a function in $\mathrm{L}^2(G)$ with every Hilbert-Schmidt operator — in particular, with every trace class operator — in the carrier Hilbert space of the representation U [21]. This map should be thought of as a *dequantization map*. If the group G is unimodular (e.g., a compact group) — i.e., $\Delta_G \equiv 1$ — the function ϱ associated with a density operator $\hat{\rho}$ is of the form

$$\varrho(g) \equiv (\mathscr{D}\hat{\rho})(g) = d_U^{-1}\,\mathrm{tr}(U(g)^*\hat{\rho}), \tag{35}$$

where $d_U > 0$ is a normalization constant depending on the representation U and on the normalization of the Haar measure on G. As mentioned in sect. 1, ϱ is sometimes called the quantum *tomogram* associated with $\hat{\rho}$. The range $\mathrm{Ran}(\mathscr{D})$ of the dequantization map is a closed subspace of $\mathrm{L}^2(G)$, invariant with respect to the representation $\mathcal{T}_{\mathsf{m}}$ [21]. Also invariant for $\mathcal{T}_{\mathsf{m}}$ is a dense linear subspace of $\mathrm{Ran}(\mathscr{D})$, i.e., the image via $\mathscr{D}$ of the trace class operators, and we will denote this space with $\mathscr{T\!C}(U)$. As a consequence, we obtain two further semigroups of operators by restricting $\{\mathfrak{T}_t^{\mathsf{m}}\}_{t\in\mathbb{R}^+}$ to $\mathrm{Ran}(\mathscr{D})$ and to $\mathscr{T\!C}(U)$. We will call the latter semigroup a *proper tomographic semigroup* since it acts on quantum tomograms.

The classical-quantum semigroups fit in this scheme with the following identifications. The group G has obviously to be identified with the vector group $\mathbb{R}\times\mathbb{R}$. The representation U is the *Weyl system* [6,21], i.e.,

$$U(q,p) = \exp(\mathrm{i}(p\hat{q} - q\hat{p}))\,, \tag{36}$$

with $\hat{q}$, $\hat{p}$ denoting the position and momentum operators in $\mathrm{L}^2(\mathbb{R})$. This is a (square integrable) projective representation,

$$U(q+\tilde{q}, p+\tilde{p}) = \mathsf{m}(q,p;\tilde{q},\tilde{p})\, U(q,p)\, U(\tilde{q},\tilde{p}), \tag{37}$$

and the multiplier m is given by

$$\mathsf{m}(q,p;\tilde{q},\tilde{p}) := \exp\Big(\frac{\mathrm{i}}{2}(q\tilde{p} - p\tilde{q})\Big). \tag{38}$$

Hence, the associated representation $\mathcal{T}_{\mathsf{m}}$ coincides precisely with $\mathcal{T}$, see (19). Although not immediately clear, it turns out that applying the symplectic Fourier-Plancherel transform to the expression on the rhs of (35) on gets precisely the Wigner function associated with $\hat{\rho}$ (up to a normalization factor) [20,21]. Moreover, in this case $\mathrm{Ran}(\mathscr{D}) = \mathrm{L}^2(\mathbb{R}\times\mathbb{R})$ [21,40], so that the proper tomographic semigroup is obtained by restricting the tomographic semigroup $\{\hat{\mathfrak{C}}_t\}_{t\in\mathbb{R}^+}$ to the (Banach) space $\mathscr{T\!C}(U) = \mathsf{LQ}$.

Journal of Physics: Conference Series **563** (2014) 012002 doi:10.1088/1742-6596/563/1/012002

5. Classical-quantum semigroups as quantum dynamical (twirling) semigroups

In the picture emerging in the previous section, there are two points that still need to be clarified:

- the meaning of the representation $\mathcal{T}_{\mathsf{m}}$ characterizing the tomographic semigroups within the full class of randomly generated semigroups;
- the form that a proper tomographic semigroup assumes once expressed in terms of standard Hilbert space operators.

These two issues are actually non unrelated as it will become soon evident.

Let μ be a probability measure on G. Denoting by $\mathcal{H}$ the carrier Hilbert space of the projective representation U and by $\mathcal{B}_1(\mathcal{H})$ the Banach space of trace class operators in $\mathcal{H}$, we can define a bounded linear operator

$$\mu[U \vee U] \colon \mathcal{B}_1(\mathcal{H}) \to \mathcal{B}_1(\mathcal{H}), \tag{39}$$

where

$$\mu[U \vee U]\, \hat{\rho} := \int_G U(g)\, \hat{\rho}\, U(g)^*\, \mathrm{d}\mu(g). \tag{40}$$

We call $\mu[U \vee U]$ a *twirling operator* [27]. This bounded operator is a (completely) positive and trace-preserving map, namely, a *quantum channel*. The notation $U \vee U$ has to do with the fact that the map $G \ni g \mapsto U \vee U(g)$, with

$$U \vee U(g) \colon \mathcal{B}_1(\mathcal{H}) \ni \hat{\rho} \mapsto U(g)\, \hat{\rho}\, U(g)^* \in \mathcal{B}_1(\mathcal{H}), \tag{41}$$

is an isometric representation, i.e., the standard symmetry action of G, on trace class operators, associated with U. Having once again a group representation at our disposal, we can consider the corresponding randomly generated semigroups; i.e., for every convolution semigroup $\{\mu_t\}_{t\in\mathbb{R}^+}$ on G we obtain a semigroup of operators $\{\mu_t[U \vee U]\}_{t\in\mathbb{R}^+}$ by setting

$$\mu_t[U \vee U]\, \hat{\rho} = \int_G U \vee U(g)\, \hat{\rho}\, \mathrm{d}\mu_t(g). \tag{42}$$

Since every twirling operator is a quantum channel, the *twirling semigroup* $\{\mu_t[U \vee U]\}_{t\in\mathbb{R}^+}$ is a quantum dynamical semigroup [27, 36, 37]. Clearly, if a probability measure on G is a member of a convolution semigroup, then a twirling operator associated with this measure is member of a twirling semigroup.

In order to illustrate the role that the twirling semigroups play in our framework — assuming that the representation U is square integrable — let us *dequantize* them. To this aim, it is fundamental to observe that the dequantization map $\mathscr{D}$ associated with U, satisfies the intertwining relation

$$\mathscr{D}\, U \vee U(g)\, \hat{\rho} = \mathcal{T}_{\mathsf{m}}(g)\, \mathscr{D}\, \hat{\rho}, \quad \forall \hat{\rho} \in \mathcal{B}_1(\mathcal{H}). \tag{43}$$

It follows that, for every $t \geq 0$, we have:

$$\mathscr{D}\, \mu_t[U \vee U]\, \hat{\rho} = \mathfrak{T}_t^{\mathsf{m}}\, \mathscr{D}\, \hat{\rho}. \tag{44}$$

Therefore, the representation $\mathcal{T}_{\mathsf{m}}$ is (a natural extension to $L^2(G)$ of) the quantum symmetry action of G on the tomograms living in $\mathscr{T}\!\mathscr{C}(U) \subset \mathrm{Ran}(\mathscr{D})$, and the proper tomographic semigroup is nothing but a twirling semigroup expressed in terms of such tomograms.

6. Conclusions and connections with quantum information science

In classical statistical mechanics, there is a natural one-to-one correspondence between *states*, realized as probability measures on phase space, and positive definite functions. The latter are usually more convenient to deal with, since they are ordinary $\mathbb{C}$-valued functions, rather than measures that are in general quite abstract objects. A similar correspondence arises in the context of the WWGM formulation of quantum mechanics, where the quantum positive definite functions can be obtained as symplectic Fourier-Plancherel transforms of Wigner quasi-probability distributions on phase space. We have shown that there is a natural notion of semigroup of (classical) positive definite functions, and by this notion one is led to define a class of semigroups of operators, the classical-quantum semigroups. The physical content of these abstract mathematical objects becomes clear as soon as one realizes that they can be regarded as the dequantized counterpart of a certain class of quantum dynamical semigroups, the *classical-noise* semigroups.

This term is borrowed from quantum information science, where a *classical-noise channel* is a completely positive, trace-preserving map $\mathscr{C}$ of the form

$$\mathscr{C}\hat{\rho} = \int_{\mathbb{C}} D(z)\,\hat{\rho}\,D(z)^*\,\gamma(z)\,\mathrm{d}^2z, \tag{45}$$

where γ is a Gaussian probability distribution and $D(z) := \exp(za^\dagger - \bar{z}a)$ — with $a^\dagger$, a denoting the creation and annihilation operators — is known in the context of quantum optics as the *displacement operator* [41, 42]; i.e., with the standard identifications the map $\mathbb{C} \ni z \mapsto D(z)$ is nothing but the Weyl system (36). This class of Gaussian quantum channels has been studied extensively; in particular, estimates of the channel capacity and of the minimum Rényi and Wehrl output entropies have been investigated, see [43, 44] and references therein.

Therefore, assuming that the twirling semigroup $\{\mu_t[U \vee U]\}_{t\in\mathbb{R}^+}$ is defined identifying the representation U with the Weyl system and $\{\mu_t\}_{t\in\mathbb{R}^+}$ with a *Gaussian* convolution semigroup [27, 37] on $\mathbb{R} \times \mathbb{R}$, we obtain a semigroup of operators which consists of classical noise channels.

The proper classical-quantum semigroups are contained in a natural way in the class of proper tomographic semigroups. Also this larger class of semigroups of operators may be thought of as the dequantized version of certain quantum dynamical semigroups, namely, the twirling semigroups associated with square integrable representations. It is worth mentioning that the twirling semigroups acting in finite-dimensional spaces are of particular interest in quantum information science because these semigroups of operators are formed by random unitary maps [27, 37], i.e., by quantum channels that are *perfectly corrigible*, according to the definition given by Gregoratti and Werner [45].

References

[1] Holevo A S 2001 *Statistical Structure of Quantum Theory* (Berlin: Springer)
[2] Breuer H-P and Petruccione F 2002 *The Theory of Open Quantum Systems* (Oxford: Oxford University Press)
[3] Yosida K 1968 *Functional Analysis* (Berlin: Springer)
[4] Gorini V, Kossakowski A and Sudarshan E C G 1976 Completely positive dynamical semigroups of N-level systems *J. Math. Phys.* **17** (1976) 821
[5] Lindblad G 1976 On the generators of quantum dynamical semigroups *Commun. Math. Phys.* **48** 119
[6] Aniello P 2010 On the notion of Weyl system *Journal of Russian Laser Research* **31** 102
[7] Rudin W 1962 *Fourier Analysis on Groups* (New York: Interscience)
[8] Lucaks E 1970 *Characteristic Functions* 2nd edn (London: Griffin)
[9] Klenke A 2014 *Probability theory* 2nd edn (London: Springer)
[10] Weyl H 1927 Quantenmechanik und Gruppentheorie *Z. Phys.* **46** 1
[11] Wigner E 1932 On the quantum correction for thermodinamic equilibrium *Phys. Rev.* **40** 749
[12] Groenewold H J 1946 On the principles of elementary quantum mechanics *Physica* **12** 405

[13] Moyal J E 1949 Quantum mechanics as a statistical theory *Proc. Camb. Phil. Soc.* **45** 99
[14] Hillery M, O'Connell R F, Scully M O and Wigner E P 1984 Distribution functions in physics: fundamentals *Phys. Rep.* **106** 121
[15] Kastler D 1965 The C*-algebras of free boson field *Commun. Math. Phys.* **1** 14
[16] Loupias G and Miracle-Sole S 1966 C*-algèbre des systèmes canoniques. I *Commun. Math. Phys.* **2** 31
[17] Loupias G and Miracle-Sole S 1967 C*-algèbre des systèmes canoniques. II *Ann. Inst. Henri Poincaré Sect. A* **6** 39
[18] Aniello P 2013 Operators versus functions: from quantum dynamical semigroups to tomographic semigroups *J. Phys.: Conf. Series* **474** 012005
[19] Ali S T, Antoine J P and Gazeau J P 2000 *Coherent States, Wavelets and Their Generalizations* (New York: Springer)
[20] Aniello P, Man'ko V I and Marmo G 2008 Frame transforms, star products and quantum mechanics on phase space *J. Phys. A: Math. Theor.* **41** 285304
[21] Aniello P 2009 Star products: a group-theoretical point of view *J. Phys. A: Math. Theor.* **42** 475210
[22] Man'ko O V, Man'ko V I and Marmo G 2002 Alternative commutation relations, star products and tomography *J. Phys. A: Math. Gen.* **35** 699
[23] Man'ko V I, Marmo G and Vitale P 2005 Phase space distributions and a duality symmetry for star products *Phys. Lett. A* **334** 1
[24] Aniello P, Ibort A, Man'ko V I and Marmo G 2009 Remarks on the star product of functions on finite and compact groups *Phys. Lett. A* **373** 401
[25] Grenander W 1963 *Probabilities on Algebraic Structures* (New York: Wiley)
[26] Heyer H 1977 *Probability Measures on Locally Compact Groups* (Berlin: Springer)
[27] Aniello P, Kossakowski A, Marmo G and Ventriglia F 2010 Brownian motion on Lie groups and open quantum systems *J. Phys. A: Math. Theor.* **43** 265301
[28] Kossakowski A 1972 On quantum statistical mechanics of non-Hamiltonian systems *Rep. Math. Phys.* **3** 247
[29] Folland G B 1995 *A Course in Abstract Harmonic Analysis* (Boca Raton, FL: CRC)
[30] Daubechies I 1983 Continuity statements and counterintuitive examples in connection with Weyl quantization *J. Math. Phys.* **24** 1453
[31] Daubechies I 1980 On the distributions corresponding to bounded operators in the Weyl quantization *Commun. Math. Phys.* **75** 229
[32] Narcowich F J and O'Connell F R 1986 Necessary and sufficient conditions for a phase-space function to be a Wigner dstribution *Phys. Rev. A* **34** 1
[33] Narcowich F J 1989 Distributions of $\hbar$-positive type and applications *J. Math. Phys.* **30** 2565
[34] Varadarajan V S 1985 *Geometry of Quantum Theory* 2nd edn (New York: Springer)
[35] Horn R A and Johnson C R 2013 *Matrix Analysis* 2nd edn (Cambridge: Cambridge University Press)
[36] Aniello P 2011 On a certain class of semigroups of operators *Open Syst. & Inf. Dynamics* **18** 129
[37] Aniello P 2013 Quantum dynamical semigroups, group representations and convolution semigroups *Physica Scripta* **T153** 014003
[38] Aniello P, Cassinelli G, De Vito E and Levrero A 1998 Square-integrability of induced representations of semidirect products *Rev. Math. Phys.* **10** 301
[39] Aniello P 2006 Square integrable projective representations and square integrable representations modulo a relatively central subgroup *Int. J. Geom. Meth. Mod. Phys.* **3** 233
[40] Pool J C T 1966 Mathematical aspects of Weyl correspondence *J. Math. Phys.* **7** 66
[41] Aniello P, Man'ko V, Marmo G, Solimeno S and Zaccaria F 2000 On the coherent states, displacement operators and quasidistributions associated with deformed quantum oscillators *J. Opt. B: Quantum Semiclass. Opt.* **2** 718
[42] Schleich W P 2001 *Quantum Optics in Phase Space* (Berlin: Wiley-VCH)
[43] Harrington J and Preskill J 2001 Achievable rates for the Gaussian quantum channel *Phys. Rev. A* **64** 062301
[44] Giovannetti V, Lloyd S, Maccone L, Shapiro J H and Yen B J 2004 Minimum Rényi and Wehrl entropies at the output of bosonic channels *Phys. Rev. A* **70** 022328
[45] Gregoratti M and Werner R F 2004 On quantum error-correction by classical feedback in discrete time *J. Math. Phys.* **45** 2600

Clusters of bound particles in a quantum integrable many-body system and number theory

B Basu-Mallick[1], T Bhattacharyya[2] and D Sen[3]

[1] Theory Division, Saha Institute of Nuclear Physics, 1/AF Bidhan Nagar, Kolkata 700 064, India

[2] Department of Physics, St. Xavier's College, 30 Park Street, Kolkata 700 016, India

[3] Centre for High Energy Physics, Indian Institute of Science, Bangalore 560 012, India

E-mail: bireswar.basumallick@saha.ac.in

Abstract. We construct clusters of bound particles for a quantum integrable derivative δ-function Bose gas in one dimension. It is found that clusters of bound particles can be constructed for this Bose gas for some special values of the coupling constant, by taking the quasi-momenta associated with the corresponding Bethe state to be equidistant points on a *single* circle in the complex momentum plane. Interestingly, there exists a connection between the above mentioned special values of the coupling constant and some fractions belonging to the Farey sequences in number theory. This connection leads to a classification of the clusters of bound particles for the derivative δ-function Bose gas and the determination of various properties of these clusters like their size and their stability under a variation of the coupling constant.

1. Introduction

One-dimensional (1D) quantum integrable many-body systems with short range interactions have emerged as an active area of research [1]-[20], due to their effectiveness in describing recent experiments using strongly interacting ultracold atomic gases [21]-[27]. Indeed, many results of these experiments have been understood within the framework of the 1D quantum integrable Lieb-Liniger model or the δ-function Bose gas, which can be solved exactly through the Bethe ansatz.

In this context it may be recalled that, for a large class of quantum integrable systems, the method of coordinate Bethe ansatz directly yields exact eigenfunctions in the coordinate representation. One can study the asymptotic form of these eigenfunctions in the limit of infinite length of the system (i.e., when all x_i's are allowed to take value in the range $-\infty < x_i < \infty$). If the probability density associated with an eigenfunction decays sufficiently fast when the relative distance between any two particle coordinates tends towards infinity (for a translationally invariant system), a bound state is formed. It is well known, for the case of the δ-function Bose gas with $N \geq 2$, that bound states exist for all negative values of the coupling constant [2, 3, 5, 6, 7, 8, 9]. The quasi-momenta associated with such a bound state are represented by equidistant points lying on a straight line or 'string' parallel to the imaginary axis in the complex momentum plane. For the case of the δ-function Bose gas with negative values of the coupling constant, one can also construct Bethe eigenfunctions corresponding to more complex structures like clusters of bound particles. The quasi-momenta corresponding to such clusters of bound particles are represented through discrete points lying on several 'strings', all of which are parallel to the imaginary axis in the complex momentum plane [3, 9, 12, 13, 14, 15].

Similar to the case of the δ-function Bose gas mentioned above, there exists another exactly solvable and quantum integrable bosonic system with a Hamiltonian given by

$$\mathcal{H}_N = -\hbar^2 \sum_{j=1}^{N} \frac{\partial^2}{\partial x_j^2} + 2i\hbar^2\eta \sum_{l<m} \delta(x_l - x_m) \left(\frac{\partial}{\partial x_l} + \frac{\partial}{\partial x_m}\right), \tag{1.1}$$

where we have chosen $2m = 1$ and η is a real (nonzero) dimensionless coupling constant for this choice of mass [28, 29, 30, 31, 32]. The Hamiltonian (1.1) of this derivative δ-function Bose gas can be obtained by projecting that of an integrable derivative nonlinear Schrödinger (DNLS) quantum field model on the N-particle subspace. Classical and quantum versions of such DNLS field models have found applications in different areas of physics like circularly polarized nonlinear Alfven waves in plasma, quantum properties of optical solitons in fibers, and in some chiral Tomonaga-Luttinger liquids obtained from the Chern-Simons model defined in two dimensions [33, 34, 35, 36, 37, 38, 39]. The scattering and bound states of the derivative δ-function Bose gas (1.1) have been studied extensively by using the methods of coordinate as well as algebraic Bethe ansatz [28, 29, 30, 31, 32, 40, 41, 42]. It turns out that for the cases $N = 2$ and $N = 3$, bound states of this model can be constructed for any value of η within its full range: $0 < |\eta| < \infty$. However, for any given value of $N \geq 4$, the derivative δ-function Bose gas allows bound states in only certain non-overlapping ranges of the coupling constant η (the union of these ranges yields a proper subset of the full range of η), and such non-overlapping ranges of η can be determined by using the Farey sequences in number theory [40, 41, 42].

In analogy with the case of the δ-function Bose gas, one may think that clusters of bound particles can only be constructed for the case of derivative δ-function Bose gas by properly assigning the corresponding quasi-momenta on several concentric circles or circular 'strings' in the complex momentum plane. However, we have recently found that, for the Hamiltonian (1.1) with some special values of the coupling constant η, clusters of bound particles can be constructed in a much simpler way by assigning the corresponding quasi-momenta as equidistant points on a *single* circle in the complex momentum plane [43]. The purpose of the present article is mainly to review the key results of Ref. [43] and make some additional comment. The arrangement of this article is as follows. In Sec. 2, we first discuss the general form of Bethe eigenstates for the derivative δ-function Bose gas. Then we identify a sufficient condition for which such a Bethe eigenstate would represent clusters of bound particles. In Sec. 3, we classify all possible solutions of this sufficient condition and obtain different types of clusters of bound particles for the derivative δ-function Bose gas. In Sec. 4 we discuss various properties of such clusters of bound particles, such as the sizes of the clusters, their stability under the variation of the coupling constant. We end with some concluding remarks in Sec. 5.

2. Construction of clusters of bound particles

In the coordinate representation, the eigenvalue equation for the Hamiltonian (1.1) may be written as

$$\mathcal{H}_N \, \tau_N(x_1, x_2, \cdots, x_N) = E \, \tau_N(x_1, x_2, \cdots, x_N) , \tag{2.1}$$

where $\tau_N(x_1, x_2, \cdots, x_N)$ denotes a completely symmetric N-particle wave function. Since $\mathcal{H}_N$ commutes with the total momentum operator given by

$$\mathcal{P}_N = -i\hbar \sum_{j=1}^{N} \frac{\partial}{\partial x_j} , \tag{2.2}$$

$\tau_N(x_1, x_2, \cdots, x_N)$ can be chosen as a simultaneous eigenfunction of these two commuting operators. Note that $\mathcal{H}_N$ remains invariant while $\mathcal{P}_N$ changes sign if we change the sign of η and transform all the $x_i \to -x_i$ at the same time; such a transformation may be called as 'parity transformation'. Due to the invariance of $\mathcal{H}_N$ under this parity transformation, it is sufficient to study the eigenvalue problem (2.1) for one particular sign of η, say, $\eta > 0$. The eigenfunctions for $\eta < 0$ case can then be constructed from those for $\eta > 0$ case by simply changing $x_i \to -x_i$; this leaves all energy eigenvalues invariant but reverses the sign of the corresponding momentum eigenvalues.

For the purpose of solving the eigenvalue problem (2.1) through the coordinate Bethe ansatz, it is convenient to divide the coordinate space $R^N \equiv \{x_1, x_2, \cdots x_N\}$ into various N-dimensional sectors defined through inequalities like $x_{\omega(1)} < x_{\omega(2)} < \cdots < x_{\omega(N)}$, where $\{\omega(1), \omega(2), \cdots, \omega(N)\}$ represents a permutation of the integers $\{1, 2, \cdots, N\}$. Since the interaction part of the Hamiltonian (1.1) vanishes within each such sector, the resulting eigenfunction can be expressed as a superposition of free particle wave functions. The coefficients associated with these free particle wave functions can be computed by using the interaction part of the Hamiltonian (1.1), which is nontrivial only at the boundary of two adjacent sectors. In the region $x_1 < x_2 < \cdots < x_N$, such eigenfunctions can be written in the form [28, 30]

$$\tau_N(x_1, x_2, \cdots, x_N) = \frac{1}{\sqrt{N!}} \sum_{\omega} \left(\prod_{l<m} \frac{A(k_{\omega(m)}, k_{\omega(l)})}{A(k_m, k_l)} \right) \rho_{\omega(1),\omega(2),\cdots,\omega(N)}(x_1, x_2, \cdots, x_N) , \quad (2.3)$$

where

$$\rho_{\omega(1),\omega(2),\cdots,\omega(N)}(x_1, x_2, \cdots, x_N) = \exp \{i(k_{\omega(1)}x_1 + \cdots + k_{\omega(N)}x_N)\} , \quad (2.4)$$

k_n's are all distinct quasi-momenta, ω represents an element of permutation group for the integers $\{1, 2, ...N\}$ and $\sum_\omega$ implies summing over all such permutations. The coefficient $A(k_l, k_m)$ in Eq. (2.3) is obtained by solving the two-particle problem related to the derivative δ-function Bose gas and this coefficient is given by

$$A(k_l, k_m) = \frac{k_l - k_m + i\eta(k_l + k_m)}{k_l - k_m} . \quad (2.5)$$

The eigenvalues of the momentum (2.2) and Hamiltonian (1.1) operators, corresponding to the eigenfunctions $\tau_N(x_1, x_2, \cdots, x_N)$ of the form (2.3), are easily obtained as

$$\mathcal{P}_N \, \tau_N(x_1, x_2, \cdots, x_N) = \hbar\Big(\sum_{j=1}^{N} k_j\Big) \, \tau_N(x_1, x_2, \cdots, x_N) , \quad (2.6a)$$

$$\mathcal{H}_N \, \tau_N(x_1, x_2, \cdots, x_N) = \hbar^2\Big(\sum_{j=1}^{N} k_j^2\Big) \, \tau_N(x_1, x_2, \cdots, x_N) . \quad (2.6b)$$

Next, we shall discuss how Bethe states of the form in (2.3) lead to the bound states of the derivative δ-function Bose gas, by allowing k_j's to take complex values in an appropriate way. As mentioned earlier, for a translationally invariant system, a wave function represents a localized bound state if the corresponding probability density decays sufficiently fast when any of the relative coordinates measuring the distance between a pair of particles tends towards infinity. To obtain the condition for which the Bethe state (2.3) would represent such a localized bound state, let us first consider the following wave function in the region $x_1 < x_2 < \cdots < x_N$:

$$\rho_{1,2,\cdots,N}(\, x_1, x_2, \cdots, x_N \,) = \exp \, (i \sum_{j=1}^{N} k_j x_j) , \quad (2.7)$$

where k_j's in general are complex valued wave numbers. Since the corresponding momentum eigenvalue given by $\hbar \sum_{j=1}^{N} k_j$ must be a real quantity, one obtains the condition

$$\sum_{j=1}^{N} q_j \; = \; 0 \, , \tag{2.8}$$

where q_j denotes the imaginary part of k_j. By using (2.8), the probability density for the wave function $\rho_{1,2,\cdots,N}(\, x_1, x_2, \cdots, x_N \,)$ in (2.7) can be expressed as

$$|\rho_{1,2,\cdots,N}(\, x_1, x_2, \cdots, x_N \,)|^2 \; = \; \exp \left\{ \, 2 \sum_{r=1}^{N-1} \Big(\sum_{j=1}^{r} q_j \Big) \, y_r \, \right\} , \tag{2.9}$$

where the y_r's are the $N-1$ relative coordinates: $y_r \equiv x_{r+1} - x_r$. Hence, the probability density in (2.9) decays exponentially in the limit $y_r \to \infty$ for one or more values of r, provided that all the following conditions are satisfied:

$$q_1 < 0 \, , \quad q_1 + q_2 < 0 \, , \quad \cdots\cdots \, , \; \sum_{j=1}^{N-1} q_j < 0 \, . \tag{2.10}$$

It should be observed that the wave function (2.7) is obtained by taking ω as the identity permutation in (2.4). However, the Bethe state (2.3) also contains terms like (2.4) with ω representing all possible nontrivial permutations. The conditions which ensure the decay of such a term, associated with any nontrivial permutation ω, are evidently given by

$$q_{\omega(1)} < 0 \, , \quad q_{\omega(1)} + q_{\omega(2)} < 0 \, , \quad \cdots\cdots \, , \; \sum_{j=1}^{N-1} q_{\omega(j)} < 0 \, . \tag{2.11}$$

It is easy to check that above conditions, in general, contradict the conditions given in Eq. (2.10). To bypass this problem and ensure an overall decaying wave function (2.3), it is sufficient to assume that the coefficients of all terms $\rho_{\omega(1),\omega(2),\cdots,\omega(N)}(x_1, x_2, \cdots, x_N)$ with nontrivial permutations take the zero value. This leads to a set of relations given by

$$A(k_r, k_{r+1}) \; = \; 0, \quad \text{for} \quad r \in \Omega_N \, , \tag{2.12}$$

where $\Omega_N \equiv \{1,\, 2,\, \cdots,\, N-1\}$. Consequently, the simultaneous validity of the conditions (2.8), (2.10) and (2.12) ensures that the Bethe state $\tau_N(x_1, x_2, \cdots, x_N)$ (2.3) would represent a bound state.

Let us now analyse the conditions (2.8), (2.10) and (2.12) for the case of the derivative δ-function Bose gas. Using the conditions (2.8) and (2.12) along with Eq. (2.5), one can easily derive an expression for all the quasi-momenta as

$$k_n \; = \; \chi \, e^{-i(N+1-2n)\phi} \, , \tag{2.13}$$

where χ is a real, non-zero parameter, and ϕ is related to the coupling constant η as

$$\phi \; = \; \tan^{-1}(\eta) \;\; \Longrightarrow \;\; \eta = \tan\phi \, . \tag{2.14}$$

To obtain an unique value of ϕ from the above equation, it may be restricted to the fundamental region $-\frac{\pi}{2} < \phi (\neq 0) < \frac{\pi}{2}$. Furthermore, since we have seen that $\mathcal{H}_N$ (1.1) remains invariant under the 'parity transformation', it is enough to study the corresponding eigenvalue problem only within the range $0 < \phi < \frac{\pi}{2}$. Next, we consider the remaining conditions (2.10) for the existence of a localized bound state. Since summation over the imaginary parts of k_n's in (2.13) yields

$$\sum_{j=1}^{l} q_j = -\chi \, \frac{\sin(l\phi)}{\sin\phi} \, \sin[(N-l)\phi] \,, \tag{2.15}$$

Eq. (2.10) can be expressed as

$$\chi \, \frac{\sin(l\phi)}{\sin\phi} \, \sin[(N-l)\phi] \; > \; 0 \,, \quad \text{for} \quad l \in \Omega_N \,, \tag{2.16}$$

where Ω_N denotes the set of integers $\{1, 2, \cdots, N-1\}$. Consequently, for any given values of ϕ and N, a bound state would exist when all the inequalities in Eq. (2.16) are simultaneously satisfied for some real non-zero value of χ. In our earlier works it was shown that, for any given value of $N \geq 4$, the derivative δ-function Bose gas allows bound states in only certain non-overlapping ranges of the coupling constant ϕ called 'bands' and the location of these bands can be determined exactly [40, 42].

We would now like to find the conditions for constructing clusters of bound particles in the case of the derivative δ-function Bose gas. To this end, we shall first discuss the concept of a 'clustered state' for any translationally invariant system, and then give a prescription for finding the Bethe states representing clusters of bound particles. Let us consider a system of N particles which are divided into some groups or clusters — with at least one group containing more than one particle. It is assumed that particles within the same group behave like the constituents of a bound state, but particles corresponding to different groups behave like the constituents of a scattering state. More precisely, a wave function corresponding to such an N-particle system satisfies the following two conditions. If the relative distance between any two particles belonging to the same group goes to infinity, the probability density corresponding to the N-particle wave function decays in the same way as a bound state. On the other hand, if the relative distance between any two particles belonging to different groups tends towards infinity (keeping the relative distances among all particles belonging to the same group unchanged), the probability density remains finite similar to a scattering state. If any wave function corresponding to a N-particle system satisfies these two conditions, we define it as a clustered state.

Next, let us discuss how the conditions for constructing a clustered state can be implemented for the case of the plane wave function (2.7). Since any eigenvalue of the momentum operator $\mathcal{P}_N$ must be a real quantity, Eq. (2.8) is also obeyed for this case. To proceed further, let us choose a specific value of N given by $N = 4$. For this case, the probability density (2.9) may be explicitly written as

$$|\rho_{1,2,\cdots,4}(\, x_1, x_2, \cdots, x_4 \,)|^2 \;=\; \exp\,\{2q_1y_1 + 2(q_1+q_2)y_2 + 2(q_1+q_2+q_3)y_3\}\,. \tag{2.17}$$

Suppose, the conditions (2.10) for a bound state formation are slightly modified for this case as

$$q_1 < 0, \qquad q_1 + q_2 = 0, \qquad q_1 + q_2 + q_3 < 0\,.$$

Taking into account this new condition, it is easy to see that when $y_1 = x_2 - x_1$ or $y_3 = x_4 - x_3$ tends towards infinity, the probability density in Eq. (2.17) still decays like a bound state. On the other hand, when $y_2 = x_3 - x_2$ tends towards infinity, the probability density in Eq. (2.17) remains finite. Hence, the clusters of particles given by $\{1, 2\}$ and $\{3, 4\}$ satisfy all the criteria of a clustered state. Generalizing this specific example in a straightforward way for any given values of N and ϕ, we replace some of the inequalities in Eq. (2.10) by equalities. In this way, we find out the conditions for obtaining a clustered state from the plane wave function (2.7) as

$$\sum_{i=1}^{l} q_i = 0\,, \quad \text{for } l \in \Omega_{N,\phi}\,, \tag{2.18a}$$

$$\sum_{i=1}^{l} q_i < 0\,, \quad \text{for } l \in (\Omega_N - \Omega_{N,\phi})\,, \tag{2.18b}$$

where $\Omega_{N,\phi}$ denotes any non-empty proper subset of Ω_N and $(\Omega_N - \Omega_{N,\phi})$ is the complementary set of $\Omega_{N,\phi}$.

Next, we try to find the simplest possible condition for which any Bethe state of the form (2.3) would represent clusters of bound particles. Let us assume that the quasi-momenta corresponding to this Bethe state satisfy the relations (2.12). As a result, the coefficients of all plane waves except (2.7) take the zero value within the Bethe state (2.3). Thus, due to the relations (2.12), the Bethe state (2.3) would reduce to the plane wave function (2.7). Next, we assume that the quasi-momenta corresponding to this Bethe state also satisfy the relations (2.8) and (2.18a,b). Hence, the plane wave function (2.7) represents a clustered state. Consequently, Eqs. (2.8), (2.12) and (2.18a,b) together yield a sufficient and simplest possible condition for which the Bethe state (2.3) would represent clusters of bound particles. Let us now analyse this condition for the case of the derivative δ-function Bose gas. Using Eqs. (2.8) and (2.12) along with the form of $A(k_l, k_m)$ given in (2.5) it is easy to see that, similar to the case of a localized bound state, the quasi-momenta associated with clusters of bound particles can be written in the form (2.13) and the imaginary parts of these quasi-momenta satisfy the relation (2.15). With the help of Eq. (2.15), we can recast Eqs. (2.18a,b) as

$$\chi\,\frac{\sin(l\phi)}{\sin\phi}\,\sin[(N-l)\phi] = 0\,, \quad \text{for} \quad l \in \Omega_{N,\phi}\,, \tag{2.19a}$$

$$\chi\,\frac{\sin(l\phi)}{\sin\phi}\,\sin[(N-l)\phi] > 0\,, \quad \text{for} \quad l \in (\Omega_N - \Omega_{N,\phi})\,. \tag{2.19b}$$

For any given values of N and ϕ, the above equations clearly give the simplest possible condition for which the Bethe state (2.3) would represent clusters of bound particles. Let us assume that these equations are satisfied for some values of ϕ and N, where $\Omega_{N,\phi}$ is given by

$$\Omega_{N,\phi} = \{l_1, l_2, \cdots, l_p\}\,, \tag{2.20}$$

with $1 \leq p < N-1$. Then from Eq. (2.9) it follows that the sets of particles given by $\{1,\cdots,l_1\}, \{l_1+1,\cdots,l_2\}, \cdots\cdots, \{l_{p-1}+1,\cdots,l_p\}, \{l_p+1,\cdots,N\}$ represent $(p+1)$ number of clusters of bound particles. Moreover, the numbers of particles present within each of these clusters, i.e., the size of the clusters, may be written in the form

$$\{\{\, l_1, l_2 - l_1, \cdots, l_p - l_{p-1}, N - l_p \,\}\}\,. \tag{2.21}$$

Since the quasi-momenta associated with both bound states and clusters of bound particles are given by Eq. (2.13), the momentum and energy eigenvalues for clusters of bound particles can be derived in exactly the same way as has been done earlier [40] for the case of a bound state. Inserting the quasi-momenta given in Eq. (2.13) to Eqs. (2.6a,b), we obtain the momentum eigenvalue as

$$P = \hbar\chi\,\frac{\sin(N\phi)}{\sin\phi}\,, \tag{2.22}$$

and the energy eigenvalue as

$$E = \frac{\hbar^2 \chi^2 \sin(2N\phi)}{\sin(2\phi)} . \tag{2.23}$$

Let us make a comment at this place. Using Eq. (2.22) we find that, for any given values of N and ϕ, the parameter χ is proportional to the total momentum P. Hence, due to Eqs. (2.9) and (2.15) it follows that, in the limit of large relative distances between particle coordinates, the probability density for a bound state wave function (or for a cluster within a clustered state) decays exponentially over a length which is inversely proportional to the total momentum P. This surprising result is a consequence of the fact that since we have set $2m = 1$ and the parameter η is dimensionless in the Hamiltonian (1.1), the only length scale appearing in an eigenstate of the Hamiltonian is $\hbar/P$. Hence the decay length must be proportional to $\hbar/P$.

3. Farey sequences and clusters of bound particles

For the purpose of constructing clusters of bound particles in the case of the derivative δ-function Bose gas, here our aim is to find out all possible solutions of Eqs. (2.19a,b). Some properties of the Farey sequences [44] in number theory will play a crucial role in our analysis. Due to the existence of the parity transformation, as mentioned in the earlier section, it is sufficient to concentrate on values of ϕ lying in the range $0 < \phi < \frac{\pi}{2}$. Within this range of ϕ, $\sin\phi > 0$ and hence the conditions (2.19a,b) for forming clusters of bound particles reduce to

$$\chi \, \sin(l\phi) \, \sin[(N-l)\phi] = 0 , \qquad \text{for} \;\; l \in \Omega_{N,\phi} , \tag{3.1a}$$

$$\chi \, \sin(l\phi) \, \sin[(N-l)\phi] > 0 , \qquad \text{for} \;\; l \in (\Omega_N - \Omega_{N,\phi}) . \tag{3.1b}$$

It is easy to see that the condition (3.1a) would be satisfied if and only if ϕ/π can be expressed in the form

$$\frac{\phi}{\pi} = \frac{a}{b} , \tag{3.2}$$

where $\{a, b\}$ are relatively prime integers (i.e, the greatest common divisor of a and b is 1), taking values within the ranges

$$0 < a < \frac{b}{2} , \qquad 2 < b \leq N - 1 . \tag{3.3a, b}$$

Due to Eq. (3.3b) it is evident that, clusters of bound particles can exist only for $N \geq 4$. In the following, we shall establish a connection of the fractions ϕ/π, given by Eqs. (3.2) and (3.3a,b), with the elements of Farey sequences in number theory.

For a positive integer N, the Farey sequence is defined to be the set of all the fractions a/b in increasing order such that (i) $0 \leq a \leq b \leq N$, and (ii) $\{a, b\}$ are relatively prime integers [44]. The Farey sequences for the first few values of N are given by

$$\begin{array}{ll}
F_1 : & \frac{0}{1} \;\; \frac{1}{1} \\
F_2 : & \frac{0}{1} \;\; \frac{1}{2} \;\; \frac{1}{1} \\
F_3 : & \frac{0}{1} \;\; \frac{1}{3} \;\; \frac{1}{2} \;\; \frac{2}{3} \;\; \frac{1}{1} \\
F_4 : & \frac{0}{1} \;\; \frac{1}{4} \;\; \frac{1}{3} \;\; \frac{1}{2} \;\; \frac{2}{3} \;\; \frac{3}{4} \;\; \frac{1}{1} \\
F_5 : & \frac{0}{1} \;\; \frac{1}{5} \;\; \frac{1}{4} \;\; \frac{1}{3} \;\; \frac{2}{5} \;\; \frac{1}{2} \;\; \frac{3}{5} \;\; \frac{2}{3} \;\; \frac{3}{4} \;\; \frac{4}{5} \;\; \frac{1}{1}
\end{array} \tag{3.4}$$

These sequences enjoy several properties, of which we list the relevant ones below.
(i) Let $a/b, a'/b'$ are two fractions appearing in the Farey sequence F_N. Then $a/b < a'/b'$ ($a'/b' < a/b$) are two successive fractions in F_N, if and only if the following two conditions are satisfied:

$$a'b \; - \; ab' \; = \; 1\;(-1)\,, \tag{3.5a}$$
$$b + b' \; > \; N\,. \tag{3.5b}$$

It then follows that both a and b' are relatively prime to a' and b.
(ii) For $N \geq 2$, if n/N is a fraction appearing somewhere in the sequence F_N (this implies that N and n are relatively prime according to the definition of F_N), then the fractions a_1/b_1 and a_2/b_2 appearing immediately to the left and to the right respectively of n/N satisfy

$$a_1\,,\; a_2 \; \leq \; n\,, \quad \text{and} \quad a_1 \,+\, a_2 \; = \; n\,,$$
$$b_1\,,\; b_2 \; < \; N\,, \quad \text{and} \quad b_1 \,+\, b_2 \; = \; N\,. \tag{3.6}$$

To apply the above mentioned Farey sequence in the present context, let us define a subset of F_N as

$$F'_N = \left\{ \frac{a}{b} \;\middle|\; \frac{a}{b} \in F_N, \;\; \frac{0}{1} < \frac{a}{b} < \frac{1}{2} \right\}, \tag{3.7}$$

and a subset of F'_N as

$$F''_N = \left\{ \frac{n}{N} \;\middle|\; \frac{n}{N} \in F'_N \right\}. \tag{3.8}$$

Using these definitions of various subsets of a Farey sequence, it is easy to show that

$$F'_N = F'_{N-1} \cup F''_N\,. \tag{3.9}$$

Furthermore, it is worth noting that, Eqs. (3.2) and (3.3a,b) can equivalently be expressed as

$$\frac{\phi}{\pi} \in F'_{N-1}\,. \tag{3.10}$$

Consequently, it follows that the condition (3.1a) for cluster formation is obeyed if and only if $\phi/\pi \in F'_{N-1}$. In this context it may be observed that, due to Eq. (3.9), all the elements of F'_{N-1} are also present in F'_N. By using such an embedding of F'_{N-1} into F'_N, we find that any fraction $a/b \in F'_{N-1}$ belongs to one of the four distinct classes, which are defined in the following:

I. At least one of the fractions nearest to a/b (from either the left or the right side) in the sequence F'_N lies in the set F''_N. Then, from a property of the Farey sequences, it follows that $\{b, N\}$ are relatively prime integers in this case.
II. None of the nearest fractions of a/b (from the left or right side) in the sequence F'_N lies in the set F''_N, and $\{b, N\}$ are relatively prime integers.
III. N is divisible by b. Clearly, $\{b, N\}$ are not relatively prime integers in this case.
IV. N is not divisible by b, and $\{b, N\}$ are not relatively prime integers.

To demonstrate the above mentioned classification through an example, let us choose $N = 6$. For this case, the sets F'_5, F'_6 and F''_6 are given by

$$F'_5: \quad \frac{1}{5} \quad \frac{1}{4} \quad \frac{1}{3} \quad \frac{2}{5}$$
$$F'_6: \quad \frac{1}{6} \quad \frac{1}{5} \quad \frac{1}{4} \quad \frac{1}{3} \quad \frac{2}{5}\,; \qquad F''_6: \;\; \frac{1}{6}\,.$$

Using the embedding of F'_5 into F'_6, it is easy to verify that each fraction in F'_5 falls under one of the four classes discussed above. More precisely, the fractions 1/5, 1/4, 1/3 and 2/5 belong to type I, type IV, type III and type II respectively. Returning back to the general case we note that, for any fraction $a/b \in F'_{N-1}$, $\{b, N\}$ are either relatively prime integers or not relatively prime integers. If $\{b, N\}$ are relatively prime integers, then it is obvious that a/b must be an element of either type I or type II. On the other hand, if $\{b, N\}$ are not relatively prime integers, then a/b must be an element of either type III or type IV. In this way, one can show that any fraction $a/b \in F'_{N-1}$ belongs to one of these four distinct classes.

Through a lengthy analysis which uses the properties (3.5a,b) and (3.6) of a Farey sequence, we have shown that fractions of types I and III belonging to F'_{N-1} satisfy the relation (3.1b), while type II and type IV fractions do not satisfy (3.1b) [43]. Hence, clusters of bound particles are formed for the case of derivative δ-function Bose gas, if and only if the corresponding coupling constant $\phi/\pi = a/b \in F'_{N-1}$ is a fraction of type I or type III.

4. Some properties of clusters of bound particles

In this section, we shall compute the number of clusters present within a Bethe state representing clusters of bound particles and the sizes of these clusters (i.e., number of bound particles present in each of these clusters). Subsequently, we shall analyse the behavior of these clusters of bound particles under small variations of the coupling constant.

4.1. Sizes of the clusters of bound particles

Let us first consider clusters of bound particles when $\phi/\pi = a/b$ is taken as any fraction of type I within the set F'_{N-1}. In section 2 we have seen that, to find the number and sizes of the clusters within a Bethe state, we have to determine the set $\Omega_{N,\phi}$ for which Eq. (3.1a) is satisfied. Since $\{N, b\}$ are relatively prime integers for any fraction of type I, we can express N as

$$N = pb + r\,, \tag{4.1}$$

where $1 \leq r \leq b-1$. Hence, for the discrete variable l taking values within the set Ω_N, the zero points of the functions $\sin l\phi$ and $\sin(N-l)\phi$ are respectively given by the sets

$$S_1 \equiv \{b, 2b, \cdots\cdots, pb\}, \quad S_2 \equiv \{N-b, N-2b, \cdots\cdots, N-pb\}\,. \tag{4.2a, b}$$

Combining the sets S_1 and S_2 by using Eqs. (4.1) and (4.2a,b), we obtain $\Omega_{N,\phi}$ as

$$\Omega_{N,\phi} = S_1 \cup S_2 = \{r,\ b,\ r+b,\ 2b, \cdots\cdots,\ r+(p-1)b,\ pb\}\,. \tag{4.3}$$

Comparing (4.3) with (2.20), and also using (2.21), it is easy to see that the sizes of the clusters are given by

$$\{\{r,\ b-r,\ r,\ b-r,\ \cdots\cdots,\ r,\ b-r,\ r\}\}\,. \tag{4.4}$$

Hence, for any fraction of type I, the corresponding Bethe state contains $(p+1)$ number of clusters of size r and p number of clusters of size $(b-r)$. Next, by using the method of contradiction, we would like to show that these two possible sizes of the clusters, i.e., r and $(b-r)$, must be relatively prime integers. To this end, let us first assume that b and r are not relatively prime integers. Therefore, we can write b and r as $b = \alpha b'$ and $r = \alpha r'$, where $\alpha > 1$. Substituting these values of b and r in Eq. (4.1), we find that

$$N = \alpha(pb' + r')\,.$$

Thus α is a common factor of N and b. However, this result contradicts the fact that $\{N, b\}$ must be relatively prime integers for any fraction of type I. Hence it is established that $\{b, r\}$ are relatively prime integers. From this relation, it trivially follows that $\{r, b-r\}$ are relatively prime integers and, in particular, $r \neq b - r$. Consequently from Eq. (4.4) we find that, for any fraction of type I, the corresponding Bethe state contains heterogeneous clusters of two different sizes. As a special case, let us consider any fraction of type I with denominator satisfying the relation $b > N/2$. Due to Eq. (4.1) it follows that, $p = 1$, $r = N - b$ and $b - r = 2b - N$ for this case. Hence, the corresponding Bethe state contains two clusters of the size $(N - b)$ and one cluster of the size $(2b - N)$.

Next, we consider the clusters of bound particles corresponding to any fraction of type III. In this case N can be written as $N = pb$, where p is an integer greater than one. Consequently, for the variable l taking value within the set Ω_N, the zero points of the functions $\sin l\phi$ and $\sin(N-l)\phi$ coincide with each other and yield $\Omega_{N,\phi}$ as

$$\Omega_{N,\phi} = \{b,\ 2b,\ 3b, \cdots\cdots, (p-1)b\}\,. \tag{4.5}$$

Hence p number of clusters are formed in this case. Comparing (4.5) with (2.20), and also using (2.21), it is easy to see that each cluster of this type has the size b. In other words, the corresponding Bethe state (2.3) contains N/b of homogeneous clusters, each of which is made of b number of bound particles. In Table 1 we show all the fractional values of ϕ/π for which clusters of bound particles exist within the range of N given by $4 \leq N \leq 10$, the types of these fractions and the sizes of the corresponding clusters using the notation of Eq. (2.21).

4.2. Stability of clusters of bound particles and binding energies

Here, we shall discuss about the stability of clusters of bound particles under infinitesimal variations of the coupling constant. For $N \geq 4$, let us choose any specific value of the coupling constant ϕ within the range $0 < \phi < \frac{\pi}{2}$ such that clusters of bound particles are formed. If one increases or decreases this value of ϕ by an infinitesimal amount, it is obvious that all inequalities in Eq. (3.1b) would continue to be satisfied and all equalities in Eq. (3.1a) would be transformed into some inequalities. Consequently, clusters of bound particles cease to exist even for a very small change of the coupling constant. One of the following two different cases can occur in such a situation. In the first case, at least one of the equalities in Eq. (3.1a) is transformed into an inequality of the form

$$\chi\ \sin(l\phi)\ \sin[(N-l)\phi]\ <\ 0\,. \tag{4.6}$$

It is evident that, the probability density of the corresponding Bethe state (2.3) would diverge if the relative distance between at least one pair of particle coordinates tends towards infinity. As a result, this Bethe state becomes ill-defined and disappears from the Hilbert space of the Hamiltonian (1.1) of derivative the δ-function Bose gas. So we may say that clusters of bound particles become unstable in this case. Let us now consider the second case, for which all of equalities in Eq. (3.1a) are transformed into inequalities of the form

$$\chi\ \sin(l\phi)\ \sin[(N-l)\phi]\ >\ 0\,, \tag{4.7}$$

due to an infinitesimal change of the coupling constant ϕ. It is evident that, for this case, Eqs. (3.1a,b) are transformed to Eq. (2.16) within the range of ϕ given by $0 < \phi < \frac{\pi}{2}$. As a result, clusters of bound particles merge with each other and produce a localized bound state containing only one cluster of particles. Therefore, in this second case, we may say that clusters of bound particles turn into a localized bound state with only one cluster.

Table 1. The fractional values of ϕ/π for which clusters of bound particles exist for $4 \leq N \leq 10$, the types of these fractions and the sizes of the corresponding clusters are shown.

N	Value of ϕ/π	Type	Size of the clusters
4	1/3	I	$\{\{1,2,1\}\}$
5	1/4	I	$\{\{1,3,1\}\}$
5	1/3	I	$\{\{2,1,2\}\}$
6	1/5	I	$\{\{1,4,1\}\}$
6	1/3	III	$\{\{3,3\}\}$
7	1/6	I	$\{\{1,5,1\}\}$
7	1/4	I	$\{\{3,1,3\}\}$
7	1/3	I	$\{\{1,2,1,2,1\}\}$
7	2/5	I	$\{\{2,3,2\}\}$
8	1/7	I	$\{\{1,6,1\}\}$
8	1/4	III	$\{\{4,4\}\}$
8	1/3	I	$\{\{2,1,2,1,2\}\}$
8	2/5	I	$\{\{3,2,3\}\}$
9	1/8	I	$\{\{1,7,1\}\}$
9	1/5	I	$\{\{4,1,4\}\}$
9	1/4	I	$\{\{1,3,1,3,1\}\}$
9	1/3	III	$\{\{3,3,3\}\}$
9	3/7	I	$\{\{2,5,2\}\}$
10	1/9	I	$\{\{1,8,1\}\}$
10	1/5	III	$\{\{5,5\}\}$
10	2/7	I	$\{\{3,4,3\}\}$
10	1/3	I	$\{\{1,2,1,2,1,2,1\}\}$
10	2/5	III	$\{\{5,5\}\}$

Using the above mentioned procedure, we have found that [43] clusters of bound particles associated with fractions of type III become unstable and cease to exist for any small change of the coupling constant. On the other hand, for the case of fractions of type I, clusters of bound particles transmute to a localized bound state containing only one cluster if the value of ϕ is slightly changed towards the direction of the nearest fraction n/N. However, such clusters of bound particles become unstable if the value of ϕ is slightly changed towards the opposite direction. Consequently, fractions of type I lie at the end points of the bands containing localized bound states.

5. Conclusion

In this article, we have explored how clusters of bound particles can be constructed in the simplest possible way for the case of an exactly solvable derivative δ-function Bose gas. To this end, we consider a sufficient condition for which Bethe states of the form in (2.3) would lead to clusters of bound particles. It is found that this sufficient condition can be satisfied by taking the quasi-momenta of the corresponding Bethe state to be equidistant points on a *single* circle having its centre at the origin of the complex momentum plane. Furthermore, the coupling constant (ϕ) and the total number of particles (N) of this derivative δ-function Bose gas must satisfy the relations in (3.1a,b). For any given $N \geq 4$, it is found that Eq. (3.1a) is satisfied if ϕ/π takes any value within the set F'_{N-1}, which is a subset of the Farey sequence F_{N-1}. Then we classify all fractions belonging to the set F'_{N-1} into four types. It turns out that fractions of types I and III belonging to F'_{N-1} satisfy the remaining relation (3.1b), while type II and type IV fractions do not satisfy (3.1b). Consequently, clusters of bound particles can be constructed for the derivative δ-function Bose gas only for special values of ϕ/π given by the fractions of types I and III within the set F'_{N-1}.

We have also computed the sizes of the above mentioned clusters of bound particles, i.e., the number of particles present within each of these clusters. It is found that any fraction of type I within the set F'_{N-1} leads to heterogeneous clusters of bound particles having two different sizes. On the other hand, any fraction of type III within the set F'_{N-1} leads to homogeneous clusters of bound particles having only one size. Interestingly, clusters of bound particles associated with fractions of type I and type III transform in rather different ways under a small variation of the coupling constant. For example, it is found that clusters of bound particles associated with fractions of type III cease to exist for any small change of the coupling constant. On the other hand, clusters of bound particles corresponding to fractions of type I turn into localized bound states consisting of a single cluster if the value of the coupling constant is slightly increased or decreased.

In this paper we have analyzed a particular type of sufficient condition for constructing clusters of bound particles in the case of the derivative δ-function Bose gas in the finite volume limit. However, in future, it might be interesting to explore other possible ways of constructing clusters of bound particles for this exactly solvable system. As is well known, many properties of the complex quasi-momenta for the case of the δ-function Bose gas can be derived relatively easily by using the solutions of the corresponding finite volume Bethe ansatz equations. A similar analysis of the finite volume Bethe ansatz equations may be helpful in finding all possible clusters of bound particles for the derivative δ-function Bose gas in the infinite volume limit.

Acknowledgments

B.B.M. thanks the organizers of 'The XXIInd International Conference on Integrable Systems and quantum symmetries (ISQS-22)' for inviting to present this work and J.S. Caux for fruitful discussions. B.B.M. also thanks the Abdus Salam International Centre for Theoretical Physics for a Senior Associateship, which partially supported this work. D.S. thanks the Department of Science and Technology, India for financial support through the grant SR/S2/JCB-44/2010.

References

[1] Lieb E H and Liniger W 1963 *Phys. Rev.* **130** 1605
[2] McGuire J B 1964 *J. Math. Phys.* **5** 622
[3] Yang C N 1968 *Phys. Rev.* **168** 1920
[4] Yang C N and Yang C P 1969 *J. Math. Phys.* **10** 1115
[5] Thacker H B 1981 *Rev. Mod. Phys.* **53** 253
[6] Faddeev L D 1980 *Sov. Sci. Rev.* C **1** 107
[7] Sklyanin E K 1990 *Yang-Baxter equation in integrable systems* (Advanced Series in Mathematical Physics Vol. 10) ed M Jimbo (Singapore: World Scientific) p. 121
[8] Korepin V E, Bogoliubov N M and Izergin A G 1993 *Quantum Inverse Scattering Method and Correlation Functions* (England: Cambridge University Press)
[9] Takahashi M 1999 *Thermodynamics of One-Dimensional Solvable Models* (England: Cambridge University Press)
[10] Olshanii M 1998 *Phys. Rev. Lett.* **81** 938
[11] Gangardt D M and Shlyapnikov G V 2003 *Phys. Rev. Lett.* **90** 010401
[12] Calabrese P and Caux J 2007 *Phys. Rev. Lett.* **98** 150403
[13] Calabrese P and Caux J 2007 *J. Stat. Mech.* P08032
[14] Dotsenko V 2010 *J. Stat. Mech.* P07010
[15] Prolhac S and Spohn H 2011 *J. Math. Phys.* **52** 122106
[16] Kormos M, Mussardo G and Trombettoni A 2009 *Phys. Rev. Lett.* **103** 210404
[17] Kormos M, Mussardo G and Trombettoni A 2010 *Phys. Rev.* A **81** 043606

Journal of Physics: Conference Series **563** (2014) 012003 doi:10.1088/1742-6596/563/1/012003

[18] Muth D and Fleischhauer M 2010 *Phys. Rev. Lett.* **105** 150403.

[19] Kormos M, Chou Y and Imambekov A 2011 *Phys. Rev. Lett.* **107** 230405

[20] Cazalilla M A, Citro R, Giamarchi T, Orignac E and Rigol M 2011 *Rev. Mod. Phys.* **83** 1405

[21] Khaykovich L, Schreck F, Ferrari G, Bourdel T, Cubizolles J, Carr L D, Castin Y and Salomon C 2002 *Science* **296** 1290

[22] Strecker K E, Partridge G B, Truscott A G and Hulet R G 2002 *Nature (London)* **417** 150

[23] Kheruntsyan K V, Gangardt D M, Drummond P D and Shlyapnikov G V 2003 *Phys. Rev. Lett* **91** 040403

[24] Kinoshita T, Wenger T and Weiss D S 2004 *Science* **305** 1125

[25] Paredes B, Widera A, Murg V, Mandel O, Fölling S, Cirac I, Shlyapnikov G V, Hänsch T W and Bloch I 2004 *Nature (London)* **429** 277

[26] Kinoshita T, Wenger T and Weiss D S 2005 *Phys. Rev. Lett.* 190406

[27] Kinoshita T, Wenger T and Weiss D S 2006 *Nature (London)* **440** 900

[28] Gutkin E 1987 *Ann. Phys.* **176** 22

[29] Kundu A and Basu-Mallick B 1993 *J. Math. Phys.* **34** 1052

[30] Shnirman A G, Malomed B A and Ben-Jacob E 1994 *Phys. Rev.* A **50** 3453

[31] Basu-Mallick B and Bhattacharyya T 2002 *Nucl. Phys.* B **634** 611

[32] Basu-Mallick B and Bhattacharyya T 2003 *Nucl. Phys.* B **668** 415

[33] Chen H H, Lee Y C and Liu C S 1979 *Phys. Scr.* **20** 490

[34] Kaup D J and Newell A C 1978 *J. Math. Phys.* **19** 798

[35] Min H and Park Q H 1996 *Phys. Lett.* B **388** 621

[36] Wadati M, Sanuki H, Konno K and Ichikawa Y H 1978 *Rocky Mountain J. Math.* **8** 323

[37] Clarkson P A 1992 *Nonlinearity* **5** 453

[38] Rosenbluh M and Shelby R M 1991 *Phys. Rev. Lett.* **66** 153

[39] Aglietti U, Griguolo L, Jackiw R, Pi S Y and Seminara D 1996 *Phys. Rev. Lett.* **77** 4406

[40] Basu-Mallick B, Bhattacharyya T and Sen D 2003 *Nucl. Phys.* B **675** 516

[41] Basu-Mallick B, Bhattacharyya T and Sen D 2004 *Phys. Lett.* A **325** 375

[42] Basu-Mallick B, Bhattacharyya T and Sen D 2004 *Mod. Phys. Lett.* A **19** 2697

[43] Basu-Mallick B, Bhattacharyya T and Sen D 2013 *Nucl. Phys.* B **871** 362

[44] Niven I, Zuckerman H S and Montgomery H L 2000 *An Introduction to the Theory of Numbers* (New York: John Wiley)

Yang–Mills gauge fields conserving the symmetry algebra of the Dirac equation in a homogeneous space

A I Breev[1,2] **and A V Shapovalov**[1,2]

[1] Department of Theoretical Physics, Tomsk State University, 634050 Tomsk, Russia
[2] Department of Higher Mathematics and Mathematical Physics, Tomsk Polytechnic University, 634050 Tomsk, Russia

E-mail: breev@mail.tsu.ru, shpv@phys.tsu.ru

Abstract. We consider the Dirac equation with an external Yang–Mills gauge field in a homogeneous space with an invariant metric. The Yang–Mills fields for which the motion group of the space serves as the symmetry group for the Dirac equation are found by comparison of the Dirac equation with an invariant matrix differential operator of the first order. General constructions are illustrated by the example of de Sitter space. The eigenfunctions and the corresponding eigenvalues for the Dirac equation are obtained in the space $\mathbb{R}^2 \times \mathbb{S}^2$ by a noncommutative integration method.

Introduction

Exact integration of relativistic wave equations for strong external fields is the topical problem in studying various effects in quantum field theory and cosmology where the standard S-matrix method does not work [1].

The main technique for exact integration of the equations is based on the classical method of separation of variables (SoV) [2, 3, 4]. There are a large number of works dealing with classification of external fields admitting separation of variables in relativistic quantum equations (see, e.g., [5] and references therein). In this connection, the integration of relativistic wave equations with external fields by means of methods other than the SoV method can provide new possibilities to study the relativistic quantum wave equations for classical and quantum fields and their interactions.

A new method of exact integration of linear partial differential equations was proposed [6] and applied to quantum equations [7, 8, 9]. This method differs from the classical SoV method and uses non-commutative algebras to describe the symmetry operators of the equation under consideration.

In this work, we consider the non-commutative symmetries of the Dirac equation with the potential of an external gauge Yang-Mills field on some homogeneous space.

The generators of a Lie transformation group acting on the homogeneous space form a non-commutative symmetry algebra for the Dirac equation. So a Lie transformation group of the homogeneous space will be the symmetry group of the Dirac equation. In general, the symmetry of the Dirac equation breaks down in the presence of an external gauge field. Our aim is to answer the question: what are the Yang-Mills gauge fields which do not destroy the symmetry group of the Dirac equation in the homogeneous space? Note that this problem was solved for the Klein-Gordon equation [10]. The symmetry of the Dirac equation in the presence of external gauge fields was also studied by the authors of Refs. [11] and [12]. We adopt natural units $\hbar = c = G = 1$, unless stated otherwise.

The paper is organized as follows: In Section 1 we briefly describe the necessary concepts and notations related to homogeneous spaces [13, 14]. The construction of an invariant differential operator with matrix coefficients on a homogeneous space is introduced following Refs. [10, 15], in Section 2.

In Section 3 we present the Dirac equation on a homogeneous space with an invariant metric tensor in terms of a first-order invariant matrix operator. The spinor connection and the symmetry operators of the Dirac equation are shown to define the isotropy representation of a spinor space. The generators of the spinor representation are found explicitly.

Next, in Section 4, we find the gauge potentials of an external Yang–Mills field that do not change the symmetry algebra of the Dirac equation. For these potentials, the Dirac equation is presented as a system of equations on a motion group.

An illustration of the general results obtained is given in Section 5 by the example of the homogeneous space $\mathbb{R}^2 \otimes \mathbb{S}^2$ with an invariant metric. The homogeneous space is shown to admit the external magnetic field preserving the symmetry algebra of the Dirac equation. The spectrum of the Dirac equation and the corresponding eigenstates are found by a noncommutative integration method.

In the final Section 6, we find the Yang–Mills gauge fields preserving the symmetry of the Dirac equation in de Sitter space.

In Section 7 we give our conclusion remarks.

1. An invariant metric on a homogeneous space

Here we provide some basic concepts and notations relevant to the theory of homogeneous spaces.

Let G be a simply connected real Lie group with a Lie algebra $\mathfrak{g}$ and let M be a homogeneous space with right action of the group G. For any $x \in M$ there exists an isotropy subgroup $H_x \in G$. Denote by H a stationary subgroup of a point $x_0 \in M$, and let $\mathfrak{h}$ be the Lie algebra of H. The homogeneous space M is diffeomorphic to the manifold G/H of right cosets Hg, where H is the isotropy subgroup. The group of transformations G can be regarded as a principal bundle (G, π, M, H) with the structure group H, the base M, and the projection map $\pi : G \to M$. The Lie algebra $\mathfrak{g}$ is decomposed into a direct sum of subspaces $\mathfrak{g} = \mathfrak{h} \oplus \mathfrak{m}$, where $\mathfrak{m} \simeq T_{x_0}M$ is complement to $\mathfrak{h}$.

The coordinates of an arbitrary element of $g \in G$ can be written as $g = hs(x)$, $h \in H$, i.e., $g^A = (x^a, h^\alpha)$, where $A, B, C, \ldots = 1, \ldots, \dim \mathfrak{g}; a, b, c, \ldots = 1, \ldots, \dim M; \alpha, \beta, \gamma, \ldots = 1, \ldots, \dim \mathfrak{h}$, and $s : M \to G$ is a local and smooth section of G.

Let us introduce an invariant metric on the homogeneous space M. Suppose that $\mathbf{G}$ is a non-degenerate Ad_H-*invariant* quadratic form on a subspace $\mathfrak{m} \subset \mathfrak{g}$,

$$\mathbf{G}(\overline{[X, Y]}, \overline{Z}) + \mathbf{G}(\overline{Y}, \overline{[X, Z]}) = 0, \quad X \in \mathfrak{h}, \quad Y, Z \in \mathfrak{g}, \tag{1}$$

where the bar denotes the projection map of the Lie algebra $\mathfrak{g}$ onto the subspace $\mathfrak{m}$. The quadratic form $\mathbf{G}$ defines an invariant inner product Ad_H on the tangent space $T_{x_0}M \simeq \mathfrak{m}$. By the action of a Lie group G with right-hand shifts R_g ($g \in G$) in the homogeneous space M, we define the inner product throughout the space M as

$$g_M(\tau, \tau')(x) = \mathbf{G}((R_{g^{-1}})_* \tau, (R_{g^{-1}})_* \tau'), \quad \tau, \tau' \in T_x M, \quad x = \pi(g). \tag{2}$$

The Ad_H-invariance (1) is the necessary and sufficient condition for the inner product (2) to be invariant with respect to the Lie group G action on the homogeneous space M. The inner product (2) defines an *invariant* metric g_M on the homogeneous space M [14]. From

(2) we can write down the covariant components of the metric tensor in local coordinates as $(i, j, k, l, \ldots = 1, \ldots, \dim M)$

$$g_{ij}(x) = G_{ab}\sigma^a_i(x, e_H)\sigma^b_j(x, e_H), \quad G_{ab} \equiv \mathbf{G}(e_a, e_b), \quad a, b - 1, \ldots, \dim M. \tag{3}$$

Here e_a are the fixed basis vectors of the space $\mathfrak{m}$, $\sigma^b(g) \equiv -(R_g)^* e^b$ is the basis of right-invariant 1-forms, and e^b are the basis vectors in the dual space $\mathfrak{m}^*$: $\langle e_a, e^b\rangle = \delta^b_a$, and e_H is the identity element of H. The contravariant components of the metric tensor (3) can be represented as

$$g^{ij}(x) = G^{ab}\eta^i_a(x, e_H)\eta^j_b(x, e_H), \quad G^{ab} = (G_{ab})^{-1}, \quad \eta^i_a = (\sigma^a_i)^{-1},$$

where $\eta_a(g) = -(R_g)_* e_a$ are right-invariant vector fields on the Lie group G. On the basis of the algebra $\mathfrak{g}$, the Ad_H-invariance condition (1) takes the form

$$G_{ab}C^a_{c\alpha} + G_{ac}C^a_{b\alpha} = 0, \tag{4}$$

where $C^A_{AB} = [e_A, e_B]^C$ are the structure constants of $\mathfrak{g}$.

Let us remark that any non-degenerate symmetric 2-form G_{ab} satisfying condition (4) defines an invariant metric g_M on the homogeneous space M.

The metric tensor (3) defines the Christoffel symbols of the Levi-Civita connection as [8, 14]

$$\begin{aligned}\Gamma^i_{jk}(x) =&\Gamma^a_{bc}\sigma^b_j(x, e_H)\sigma^c_k(x, e_H)\eta^i_a(x, e_H)- \\ &- \sigma^b_j(x, e_H)\eta^i_{b,k}(x, e_H) - C^a_{b\alpha}\sigma^b_j(x, e_H)\sigma^\alpha_k(x, e_H)\eta^i_a(x, e_H).\end{aligned} \tag{5}$$

The coefficients Γ^a_{bc} are determined by the components G^{ab} of the quadratic form $\mathbf{G}$ and by the structure constants of the Lie algebra:

$$\Gamma^a_{bc} = -\frac{1}{2}C^a_{bc} - \frac{1}{2}G^{ad}\left[G_{ec}C^e_{bd} + G_{eb}C^e_{cd}\right]. \tag{6}$$

Thus, the Levi-Civita connection is defined by the algebraic properties of a homogeneous space with an invariant metric.

2. Invariant matrix differential operator of the first order

In this section, we consider algebraic conditions for a first-order linear differential operator with matrix coefficients invariant on the homogeneous space M. We follow Ref. [10], which presents a study of a more general invariant linear matrix differential operator of the second order was studied.

Denote by $C^\infty(M, V)$ and $C^\infty(G, V)$ two spaces of functions taking values on a linear space V and defined on a homogeneous space M and on a transformation group G, respectively.

The functions on the homogeneous space M can be considered as functions defined on the Lie group G and invariant along the fibers H of the bundle G [14]. In our case, if functions take values on the vector space V, the space $C^\infty(M, V)$ is isomorphic to the function space

$$\hat{\mathcal{F}} - \{\varphi \in C^\infty(G, V) \mid \varphi(hg) = U(h)\varphi(g), \quad h \in H\},$$

where $U(h)$ is an exact representation of the isotropy group H on V. For an arbitrary function $\varphi \in \hat{\mathcal{F}}$, we have

$$\varphi(g) = \varphi(hs(x)) = U(h)\varphi(s(x)), \quad g = (x, h). \tag{7}$$

Then we can identify $\varphi(s(x))$ with a function $\varphi \in C^\infty(M,V)$. Formula (7) gives an explicit form of the isomorphism $\hat{\mathcal{F}} \simeq C^\infty(M,V)$. Differentiating (7) with respect to h^α and assuming $h = e_H$, we obtain:

$$(\eta_\alpha + \Lambda_\alpha)\,\varphi(g) = 0, \quad \Lambda_\alpha = \frac{\partial U(h)}{\partial h^\alpha}|_{h=e_H}. \tag{8}$$

Here Λ_α are generators of the group H in the linear space V. Formula (8) is the infinitesimal consequence of (7). The isotropy subgroup H is assumed to be connected. Then conditions (7) and (8) are equivalent.

From (8) we can see that a linear differential operator R leaves invariant the function space $\hat{\mathcal{F}}$ if

$$(\eta_\alpha + \Lambda_\alpha)R\varphi(g) = [\eta_\alpha + \Lambda_\alpha, R]\varphi(g) = 0, \quad \varphi \in \hat{\mathcal{F}}. \tag{9}$$

Denote by $L(\hat{\mathcal{F}})$ a space of linear differential operators on $C^\infty(G,V)$ satisfying the condition

$$[\eta_\alpha + \Lambda_\alpha, R]|_{\hat{\mathcal{F}}} = 0, \quad \alpha = 1,\ldots,\dim\mathfrak{h}. \tag{10}$$

Then, given relation (7), the action of $R \in L(\hat{\mathcal{F}})$ on a function $\varphi(g)$ from the space $\hat{\mathcal{F}}$ can be written as

$$R\varphi(g) = U(h)\left(U^{-1}(h)RU(h)\right)\varphi(s(x)). \tag{11}$$

Multiplying equation (9) by $U^{-1}(h)$ and taking into account that $\eta_\alpha U(h) = -\Lambda_\alpha U(h)$, we obtain

$$U^{-1}(h)[\eta_\alpha + \Lambda_\alpha, R]U(h)\varphi(s(x)) = [\eta_\alpha, U^{-1}(h)RU(h)]\varphi(s(x)) = \\ \eta_\alpha\left(U^{-1}(h)RU(h)\varphi(s(x))\right) = 0.$$

Hence, the operator $U^{-1}(h)RU(h)$ is independent of h and (11) can be presented as

$$R\varphi(g) = U(h)R_M\varphi(s(x)), \quad R_M \equiv \left(U^{-1}(h)RU(h)\right)|_{h=e_H} = RU(h)|_{h=e_H}. \tag{12}$$

Thus, for any operator R from $L(\hat{\mathcal{F}})$ there exists an operator R_M in a homogeneous space M which acts on the acts on the functions of the space $C^\infty(M,V)$. We call R_M *the projection operator*, $R_M = \pi_* R$. For example, for a linear differential operator of the first order

$$R_1 = B^a(x,h)\partial_{x^a} + B^\alpha(x,h)\partial_{h^\alpha} + B(x,h),$$

the projection is:

$$R_M^{(1)} = \pi_* R_1 = B^a(x,e_H)\partial_{x^a} + B^\alpha(x,e_H)\Lambda_\alpha + B(x,e_H). \tag{13}$$

On the other hand, any linear differential operator R_M defined on $C^\infty(M,V)$ corresponds to an operator $R = U(h)R_M U^{-1}(h) \in L(\hat{\mathcal{F}})$. Thus we have the isomorphism $L(\hat{\mathcal{F}}) \simeq L(C^\infty(M,V))$ given by (12).

Let $\xi_X(g)$ be a left-invariant vector field on a Lie group G, $X \in \mathfrak{g}$. Since the left-invariant vector fields commute with the right-invariant ones, condition (10) is fulfilled. Using (13), we find the corresponding operator on the homogeneous space in the form:

$$X' = \pi_*\xi_X = \xi_X^a(x)\partial_{x^a} + \xi_X^\alpha(x,e_H)\Lambda_\alpha, \quad X \in \mathfrak{g}. \tag{14}$$

It is easy to verify the following commutation relations:

$$[X',Y'] = [U^{-1}(h)\xi_X U(h), U^{-1}\xi_Y U(h)]|_{h=e_H} = [U^{-1}(h)[\xi_X,\xi_Y]U(h)]|_{h=e_H} = \\ [U^{-1}(h)\xi_{[X,Y]}U(h)]|_{h=e_H} = [X,Y]'$$

Journal of Physics: Conference Series **563** (2014) 012004 doi:10.1088/1742-6596/563/1/012004

for all X, Y, of the form (14), $X, Y \in \mathfrak{g}$. Consequently, the operators X' corresponding to the left-invariant vector fields ξ_X are the *generators* of a transformation group acting on $C^\infty(M, V)$.

An operator $R_M \in L(C^\infty(M, V))$ is invariant under the action of a Lie group of transformations, if R_M commutes with X':

$$[R_M, X'] = [U^{-1}(h)RU(h), U^{-1}(h)\xi_X U(h)]|_{h=e_H} = U^{-1}(h)[R, \xi_X]U(h)|_{h=e_H} = 0. \tag{15}$$

From (15) it follows that the operator R_M is invariant with respect to the transformation group if and only if the corresponding operator $R \in L(\hat{\mathcal{F}})$ commutes with the left-invariant vector fields:

$$[R, \xi_X] = 0, \quad X \in \mathfrak{g}. \tag{16}$$

Suppose that $R_M^{(1)} \in C^\infty(M, V)$ is a first-order linear differential operator invariant with respect to the group action. By virue of (16), this operator corresponds to a first-order polynomial of the right-invariant vector fields:

$$R_{(1)} = B^a \eta_a(x, h) + B^\alpha \eta_\alpha(h) + \tilde{B}.$$

The projection map of $B^\alpha \eta_\alpha(h)$ is a constant $B^\alpha \Lambda_\alpha$, which can be removed from the operator $R_M^{(1)}$ by changing the variable $B = \tilde{B} + B^\alpha \Lambda_\alpha$. Therefore, we can put $B^\alpha = 0$ without loss of generality. Substituting the operator $R_{(1)}$ in condition (10), we get

$$\begin{aligned}[][\eta_\alpha + \Lambda_\alpha, R_{(1)}]|_{\hat{\mathcal{F}}} &= ([b^a, \Lambda_\alpha]\eta_a + b^a[\eta_a, \eta_\alpha] + [B, \Lambda_\alpha])_{\hat{\mathcal{F}}} = \\ &\left([B^a, \Lambda_\alpha] + B^b C^a_{b\alpha}\right)\eta_a|_{\hat{\mathcal{F}}} + [B, \Lambda_\alpha] - B^a C^\beta_{a\alpha}\Lambda_\beta = 0.\end{aligned}$$

Also, we have the following system of algebraic equations for the coefficients B^a and B:

$$[B^a, \Lambda_\alpha] + B^b C^a_{b\alpha} = 0, \tag{17}$$

$$[B, \Lambda_\alpha] - B^a C^\beta_{a\alpha}\Lambda_\beta = 0. \tag{18}$$

Under the conditions (17)–(18), the projection map of $R_{(1)}$ on the homogeneous space yields the desired form of the first-order invariant linear differential operator:

$$R_M^{(1)} = \pi_* R_{(1)} = B^a \eta_a^i(x, e_H)\partial_{x^i} + B^a \eta_a^\alpha(x, e_H)\Lambda_\alpha + B. \tag{19}$$

Thus, any linear first-order differential operator acting on functions from $C^\infty(M, V)$ and invariant with respect to the action of the transformation group has the form (19), where the matrix coefficients B^a and B satisfy the algebraic system of equations (17)–(18).

3. The Dirac equation on a homogeneous space

In this section, we consider the Dirac equation on a four-dimensional homogeneous space M. We assume that in the four-dimensional homogeneous space M, an invariant metric g_M of signature $(+, -, -, -)$ and a Levi-Civita connection are given. Denote by V_Ψ a space of spinor fields on M.

Let us write down the Dirac equation on the space M as an equation on a four-dimensional Lorentzian manifold M [12] as follows:

$$\left(i\gamma^k(x)[\nabla_k + \Gamma_k(x)] - m\right)\psi(x) = 0. \tag{20}$$

Here ∇_k is the covariant derivative corresponding to the Levi-Civita connection on M and m is the mass of the field $\psi \in C^\infty(M, V_\Psi)$. The Dirac gamma matrices, $\gamma^k(x)$, satisfy the condition

$$\{\gamma_i(x), \gamma_j(x)\} = 2g_{ij}(x)E_4, \tag{21}$$

where E_4 denotes an identity matrix. The spinor connection $\Gamma_k(x)$ satisfies the conditions $[\nabla_k + \Gamma_k(x), \gamma_j(x)] = 0$, $\operatorname{Tr}\Gamma_k(x) = 0$ and can be presented in explicit form as [12]:

$$\Gamma_k(x) = -1/4(\nabla_k\gamma_j(x))\gamma^j(x).$$

We will seek a solution to (21) in terms of a tetrad decomposition:

$$\gamma^k(x) = \hat\gamma^a\eta^k_a(x, e_H), \quad \hat\gamma^b = \gamma^k(x)\sigma^b_k(x, e_H). \tag{22}$$

The constant matrices $\hat\gamma^a$ are the tetrad components of $\gamma^k(x)$ and satisfy the system of algebraic equations

$$\{\hat\gamma^a, \hat\gamma^b\} = 2G^{ab}E_4. \tag{23}$$

It follows from (3) that the gamma matrices with subscripts are of the form:

$$\gamma_i(x) = g_{ij}(x)\gamma^j(x) = \hat\gamma_a\sigma^a_i(x, e_H), \quad \hat\gamma_a \equiv G_{ab}\hat\gamma^b. \tag{24}$$

The spinor connection is defined by the following theorem.

Theorem 1 *Let*

$$\Gamma(x) = \gamma^k(x)\Gamma_k(x)$$

be the spinor connection on a four-dimensional homogeneous space M with an invariant metric g_M. Then we have

$$\Gamma(x) = \hat\gamma^a\left(\Gamma_a + \eta^\alpha_a(x, e_H)\Lambda^s_\alpha\right), \quad \Gamma_a = -\frac{1}{4}\Gamma^d_{ba}\hat\gamma^b\hat\gamma_d, \quad \Lambda^s_\alpha = -\frac{1}{8}G_{ac}C^a_{\alpha b}[\hat\gamma^b, \hat\gamma^c]. \tag{25}$$

Proof. Let us write $\Gamma(x)$ so that the covariant derivative $\nabla_j\gamma_k$ be expressed in terms of the Christoffel symbols $\Gamma^l_{kj}(x)$:

$$\Gamma(x) = \frac{1}{4}\gamma^j(x)\gamma^k(x)\left(\partial_{x^j}\gamma_k(x) - \Gamma^l_{kj}(x)\gamma_l(x)\right). \tag{26}$$

Substituting the Christoffel symbols (5) and expressions (22) and (24) to (26), we obtain

$$\Gamma(x) = \hat\gamma^a\Gamma_a + \frac{1}{4}C^d_{b\alpha}\hat\gamma^a\hat\gamma^b\hat\gamma_d\sigma^\alpha_j(x, e_H)\eta^j_a(x, e_H).$$

Using the property (4) of the invariant metric, we reduce the expression $C^d_{b\alpha}\hat\gamma^b\hat\gamma_d$ to

$$C^d_{b\alpha}\hat\gamma^b\hat\gamma_d = C^d_{b\alpha}G_{dc}\hat\gamma^b\hat\gamma^c = -C^d_{c\alpha}G_{db}\hat\gamma^b\hat\gamma^c = \frac{1}{2}C^d_{b\alpha}G_{dc}[\hat\gamma^b, \hat\gamma^c] = 4\Lambda^s_\alpha.$$

From the chain of equalities

$$\sigma^\alpha_j(x, e_H)\eta^j_a(x, e_H) = \sigma^\alpha_A(x, e_H)\eta^A_a(x, e_H) - \sigma^\alpha_\beta(e_H)\eta^\beta_a(x, e_H) =$$
$$\delta^\alpha_a - (-\delta^\alpha_\beta)\eta^\beta_a(x, e_H) = \eta^\alpha_a(x, e_H),$$

we easily obtain the required expression (25) for the spinor connection. ∎

Thus we can write down the Dirac equation (20) on the homogeneous space M with the invariant metric g_M and the gamma matrices of the form (22):

$$\mathcal{D}^0_M\psi = m\psi, \quad \mathcal{D}^0_M = i\hat\gamma^a\left[\eta^j_a(x, e_H)\partial_{x^j} + \Gamma_a + \eta^\alpha_a(x, e_H)\Lambda^s_\alpha\right].$$

The set of matrices Λ^s_α determines the spinor representation of the group H in the space V_Ψ.

Journal of Physics: Conference Series **563** (2014) 012004 doi:10.1088/1742-6596/563/1/012004

Theorem 2 *The matrices Λ^s_α are the generators of the group H representation on the space V_Ψ.*

Proof. We clime that the matrices Λ^s_α satisfy the commutation relations

$$[\Lambda^s_\alpha, \Lambda^s_\beta] = C^\gamma_{\alpha\beta}\Lambda^s_\gamma. \tag{27}$$

Indeed, the commutator of two matrices Λ^s_α and Λ^s_β can be written as:

$$[\Lambda^s_\alpha, \Lambda^s_\beta] = -\frac{1}{4}C^d_{\beta b}[\Lambda^s_\alpha, \hat\gamma^b\hat\gamma_d] = -\frac{1}{4}C^d_{\beta b}\left([\Lambda^s_\alpha, \hat\gamma^b]\hat\gamma_d + \hat\gamma^b[\Lambda^s_\alpha, \hat\gamma_d]\right). \tag{28}$$

Using (4), (23), and (24), we find the commutator of Λ_α with the gamma matrices $\hat\gamma^a$:

$$[\Lambda^s_\alpha, \hat\gamma^a] = \frac{1}{4}C^d_{d\alpha}[\hat\gamma^b\hat\gamma_d, \hat\gamma^a] = \frac{1}{2}C^d_{b\alpha}\left(\delta^a_d\hat\gamma^b - G^{ab}\hat\gamma_d\right) = \frac{1}{2}\left(G^{bd}C^a_{b\alpha} - G^{ab}C^d_{b\alpha}\right)\hat\gamma_d = C^a_{b\alpha}\hat\gamma^b. \tag{29}$$

In the same way, we obtain for the gamma matrices with subscript indices:

$$[\Lambda^s_\alpha, \hat\gamma_a] = C^b_{\alpha a}\hat\gamma_b. \tag{30}$$

Substitution of (29)–(30) in (28) yields

$$[\Lambda^s_\alpha, \Lambda^s_\beta] = \frac{1}{4}\left(C^d_{\beta e}C^e_{\beta b} - C^e_{\beta b}C^d_{\alpha e}\right)\hat\gamma^b\hat\gamma_d. \tag{31}$$

The expression enclosed in the parentheses takes the form

$$C^d_{\beta e}C^e_{\beta b} - C^e_{\beta b}C^d_{\alpha e} = \left[C^A_{\alpha b}C^d_{\beta A} + C^A_{b\beta}C^d_{\alpha A} + C^A_{\beta\alpha}C^d_{bA}\right] + C^\gamma_{\alpha\beta}C^d_{b\gamma}. \tag{32}$$

Applying the Jacobi identity for structure constants to the expression in parentheses, we see that it vanishes. Substituting (32) in (31), we obtain (27). ∎

Let us associate the Dirac operator $\mathcal{D}^0_M$ with an operator $\mathcal{D}^0_G$ using a projection map π_* similar to (19).

Theorem 3 *The Dirac operator $\mathcal{D}^0_M$ in the homogeneous space M with the invariant metric g_M can be presented as:*

$$\mathcal{D}^0_M = \pi_*\mathcal{D}^0_G, \quad \mathcal{D}^0_G \equiv i\hat\gamma^a[\eta_a(g) + \Gamma_a] \in L(\hat{\mathcal{F}}_\Psi) \tag{33}$$

Proof. Comparing the Dirac operator $\mathcal{D}^0_M$ with the first-order invariant matrix differential operator (19) on the homogeneous space M, we obtain

$$B^a = i\hat\gamma^a, \quad B = i\hat\gamma^a\Gamma_a. \tag{34}$$

The Dirac operator $\mathcal{D}^0_M$ in (33) is defined if the coefficients B^a and B of the form (34) satisfy equations (17)–(18). From (29) it follows that the commutator of Λ^s_α and $\hat\gamma^a$ satisfies the first condition in (17). In this case, condition (18) is reduced to the expression

$$[\Gamma, \Lambda_\alpha] = C^\beta_{a\alpha}\hat\gamma^a\Lambda_\beta, \quad \Gamma = \hat\gamma^a\Gamma_a. \tag{35}$$

The commutator of Γ and Λ^s_α can be presented in terms of the commutator $[\Lambda^s_\alpha, \Gamma_a]$:

$$[\Lambda^s_\alpha, \Gamma] = [\Lambda^s_\alpha, \hat\gamma^a]\Gamma_a + \hat\gamma^a[\Lambda^s_\alpha, \Gamma_a]. \tag{36}$$

Using (25), in view of properties (29)–(30), we obtain:

$$[\Lambda^s_\alpha, \Gamma_a] = -\frac{1}{4}\Gamma^d_{ba}\left([\Lambda^s_\alpha, \hat\gamma^b]\hat\gamma_d + \hat\gamma^b[\Lambda^s_\alpha, \hat\gamma_d]\right) = -\frac{1}{4}\Gamma^d_{ba}\left(C^c_{\alpha d}\hat\gamma^b\hat\gamma_c - C^b_{\alpha c}\hat\gamma^c\hat\gamma_d\right). \quad (37)$$

Substituting (37) in (36), we have

$$[\Lambda^s_\alpha, \Gamma] = \frac{1}{4}\left(C^e_{\alpha a}\Gamma^c_{be} - C^c_{\alpha c}\Gamma^e_{ba} + C^e_{\alpha b}\Gamma^c_{ea}\right)\hat\gamma^a\hat\gamma^b\hat\gamma_c. \quad (38)$$

From (6) and the Jacobi identity for the structure constants of the Lie algebra $\mathfrak{g}$, it follows that

$$C^c_{\alpha e}\Gamma^e_{ba} = C^e_{\alpha a}\Gamma^c_{be} + C^e_{\alpha b}\Gamma^c_{ea} + C^\beta_{\alpha a}C^c_{\beta b}. \quad (39)$$

Substituting (39) in (38), we obtain (35).

Thus, relations (17) and (18) are satisfied. Then the Dirac operator $\mathcal{D}^0_M$ can be obtained as the projection of the operator $B^a\eta_a + B$, where B^a and B are determined by (34), onto M and we come to the projection map (33). ∎

From this theorem we immediately obtain

Corollary 1 *The generators*

$$X' = \xi^a_X(x)\partial_{x^a} + \xi^\alpha_X(x, e_H)\Lambda^s_\alpha, \quad X \in \mathfrak{g},$$

of a representation of the Lie algebra $\mathfrak{g}$ in the space V_Ψ are the symmetry operators of the Dirac operator $\mathcal{D}^0_M$ on the homogeneous space M.

4. The Dirac equation with an external gauge field admitting the motion group of a homogeneous space as a symmetry group

Here we consider the Dirac equation with an external gauge Yang–Mills field on a homogeneous space M. Our aim is to find the Yang–Mills potentials for which the Dirac equation admits the motion group of the homogeneous space as a symmetry group. Let V_K be a set of vector fields on M transforming according to the fundamental representation of an N-dimensional gauge Lie group K.

A multiplet of N spinor fields on M can be considered as a space $C^\infty(M, V)$ of functions on M which take values on a linear space $V = V_K \otimes V_\Psi$.

The potential A_i, $i = 1, \dots, \dim M = 4$, of the gauge Yang–Mills field takes values in the Lie algebra $\mathfrak{k}$ of the gauge group K, $A_i = gA^{\bar a}_i(x)T_{\bar a}$, where g is the coupling constant and the generators $T_{\bar a}$ of the gauge group take values in V_K:

$$[T_{\bar a}, T_{\bar b}] = f^{\bar c}_{\bar a\bar b}T_{\bar c}, \quad \bar a, \bar b, \bar c = 1, \dots, N.$$

Here $f^{\bar c}_{\bar a\bar b}$ are the structure constants of the gauge group K.

As a result, the Dirac equation on M with an external non-Abelian gauge field can be written as

$$\mathcal{D}_M\psi = m\psi, \quad \mathcal{D}_M = \mathcal{D}^0_M\mathbb{E}_N + i\hat\gamma^a\eta^j_a(x, e_H)A_j,$$

where $\mathbb{E}_N$ is an identity matrix on the space V_K.

Let the Dirac equation possess the group of motions G of the homogeneous space M as a symmetry group. Then, the inclusion of an external non-Abelian gauge potential in the Dirac operator breaks its the symmetry, in the general case, since the Dirac operator with the external field, in contrast to $\mathcal{D}^0_M$, is no longer an invariant operator:

$$\mathcal{D}_M = i\mathbb{E}_4\hat\gamma^a\eta^j_a(x, e_H)\partial_{x^j} + i\hat\gamma^a\left(\eta^\alpha_a(x, e_H)\Lambda^s_\alpha\mathbb{E}_4 + \eta^j_a(x, e_H)A_j\right) + i\Gamma\mathbb{E}_4. \quad (40)$$

We will seek the Yang-Mills potentials A_i for which the symmetry group of the Dirac equation is the group of motions G. In this case, the operator $\mathcal{D}_M$ must be invariant under the group of motions and can be presented in the form of (19). Comparing (40) with (19), we obtain

$$B^a = i\hat{\gamma}^a \mathbb{E}_4, \quad B = i\Gamma\mathbb{E}_4, \quad \eta_a^\alpha(x, e_H)\Lambda_\alpha = \eta_a^\alpha \Lambda_\alpha^s \mathbb{E}_4 + \eta_a^j(x, e_H) A_j, \tag{41}$$

and, hence,

$$\eta_a^j(x, e_H) A_j = \eta_a^\alpha(x, e_H)\Lambda_\alpha^k, \tag{42}$$

where Λ_α^k take values on the space V_K. Let us now multiply (42) by $\sigma_j^a(x, e_H)$ and perform summation over a. Here $\sigma_j^a(x, e_H)$ is the inverse matrix to $\eta_a^j(x, e_H)$. Considering that

$$\sigma_j^a(x, e_H)\eta_a^\alpha(x, e_H) = \sigma_j^A(x, e_H)\eta_A^\alpha(x, e_H) - \sigma_j^\beta(x, e_H)\eta_\beta^\alpha(e_H) =$$
$$\delta_j^\alpha - (-\delta_\beta^\alpha)\sigma_j^\beta(x, e_H) = \sigma_j^\alpha(x, e_H),$$

we finally obtain

$$A_j = \sigma_j^\alpha(x, e_H)\Lambda_\alpha^k. \tag{43}$$

From (43) it follows that the gauge group K of the potentials A_j is isomorphic to the isotropy subgroup H of the homogeneous space M.

Also, when condition (43) is fulfilled, the generators Λ_α of the representation on the space $V_K \otimes V_\Psi$ can be written as:

$$\Lambda_\alpha = \Lambda_\alpha^s \mathbb{E}_4 + \Lambda_\alpha^k. \tag{44}$$

Substituting (44) in the commutation relations for Λ_α, we see that

$$[\Lambda_\alpha, \Lambda_\beta] - C_{\alpha\beta}^\gamma \Lambda_\gamma = \left([\Lambda_\alpha^s, \Lambda_\beta^s] - C_{\alpha\beta}^\gamma \Lambda_\gamma^s\right)\mathbb{E}_4 + [\Lambda_\alpha^k, \Lambda_\beta^k] - C_{\alpha\beta}^\gamma \Lambda_\gamma^k = 0.$$

The expression in parentheses vanishes according to Theorem 1. Consequently, the generators Λ_α^k satisfy the commutation relations $[\Lambda_\alpha^k, \Lambda_\beta^k] = C_{\alpha\beta}^\gamma \Lambda_\gamma^k$. Thus, Λ_α^k are the generators of the isotropy subgroup H on the representation space V_K of the gauge group K. Then we have:

$$U(h) = \exp(h^\alpha \Lambda_\alpha^s \mathbb{E}_4)\exp(h^\beta \Lambda_\alpha^k) = U^s(h)U^k(h).$$

For gauge fields of the form (43), the Dirac operator $\mathcal{D}_M$ takes the form of (19):

$$\mathcal{D}_M = i\hat{\gamma}^a\left(\mathbb{E}_4 \eta_a^i(x, e_H)\partial_i + \eta_a^\alpha(x, e_H)\Lambda_\alpha\right) + i\Gamma\mathbb{E}_4.$$

Relations (17)–(18) for the coefficients B^a and B are the necessary and sufficient conditions for the operators (19) to be invariant. Substituting (41) in (17)–(18) and taking into account relations (35), we obtain the condition:

$$\hat{\gamma}^a C_{a\alpha}^\beta \Lambda_\beta^k = 0, \quad \alpha, \beta = 1, \ldots, \dim \mathfrak{h}. \tag{45}$$

By virtue of the linear independence of the gamma matrices $\hat{\gamma}^a$, this condition is satisfied if and only if the structure constants $C_{a\alpha}^\beta$ are zero. In other words, condition (45) is equivalent to *reductivity* of the homogeneous space:

$$[\mathfrak{m}, \mathfrak{h}] \subset \mathfrak{m}. \tag{46}$$

Let us say that a Dirac operator $\mathcal{D}_M$ on homogeneous space M with an invariant metric g_M admits an external gauge field if the motion group of M is the symmetry group of the Dirac equation. Thus, we have

Theorem 4 *The Dirac operator $\mathcal{D}_M$ admits an external gauge field if and only if the homogeneous space is reductive.*

In this case, the isotropy subgroup H is the gauge group. The potential of the gauge field is determined by the representation generators Λ^k_α of the subgroup H on V_K (see (43)). Under the conditions of Theorem 4, the Dirac operator $\mathcal{D}_M$ can be redefined as the projection map of the operator

$$\mathcal{D}_G = i\hat{\gamma}^a\left(\eta_a(g) + \Gamma\right) \in L(\hat{\mathcal{F}}). \tag{47}$$

Comparing (47) and (33), we conclude that the external gauge potential (43) in the Dirac operator does not change the operator $\mathcal{D}_G$ and leads to the replacement

$$U(h) = U^s(h) \to U(h) = U^k(h)U^s(h).$$

We are now in a position to give the following theorem:

Theorem 5 *Let the Dirac equation on a homogeneous space M with an invariant metric g_M and an external gauge field admits a group of motions G as the symmetry group. Then the Dirac equation is equivalent to the system of equations*

$$\mathcal{D}_G\psi(g) = m\psi(g), \quad (\eta_\alpha + \Lambda^s_\alpha \mathbb{E}_N + \Lambda^k_\alpha)\psi(g) = 0, \tag{48}$$

where $\mathcal{D}_G$ is the Dirac operator without an external gauge field on the Lie group G.

5. The Dirac equation in an $\mathbb{R}^2 \otimes \mathbb{S}^2$ space

The homogeneous space M with the transformation group $G = \mathbb{R}^2 \otimes SO(3)$ and the isotropy subgroup $SO(2)$ is topologically isomorphic to the Cartesian product of a two-dimensional plane $\mathbb{R}^2$ and a sphere $\mathbb{S}^2$. Let us introduce a local coordinate system (t, x, ϕ, θ) on M, where $(t, x) \in \mathbb{R}^2$, $\phi \in (0, 2\pi)$, and $\theta \in (0, \pi)$. The invariant metric in the local coordinates is given as

$$ds^2 = dt^2 - c_1^{-1}dx^2 - c_2^{-1}d\Omega_2, \quad d\Omega_2 = d\theta^2 + \sin\theta^2 d\phi^2, \quad c_1 > 0, c_2 > 0. \tag{49}$$

Denote by $\{e_1, e_2, e_3, e_4, e_5\}$ a fixed basis of the Lie algebra $\mathfrak{g} = \mathbb{R}^2 \times \mathfrak{so}(3)$ of the Lie group G, where $\mathbb{R}^2 = \{e_1, e_2\}$, $\mathfrak{so}(3) = \{e_3, e_4, e_5\}$. The non-zero commutation relations are

$$[e_3, e_4] = e_5, \quad [e_5, e_3] = e_4, \quad [e_4, e_5] = e_3.$$

The isotropy subalgebra $\mathfrak{h}$ is generated by the basis element e_5: $H = \exp(he_5)$, $h \in (0, 2\pi)$. The invariant metric (49) is defined by a non-degenerate 2-form $G^{ab} = \mathrm{diag}(1, -c_1, -c_2, -c_2)$ on the basis $\{e_a\}$, where $c_1 > 0, c_2 > 0$.

The local coordinates (t, x, ϕ, θ, h) are defined by canonical coordinates of the second kind on the Lie group G:

$$g(t, x, \phi, \theta, h) = e^{he_5}e^{(\theta - \frac{\pi}{2})e_4}e^{\phi e_3}e^{xe_2}e^{te_1}, \quad (t, x, \phi, \theta) \in M, \quad h \in H. \tag{50}$$

In the coordinates (50), the right-invariant basis of the 1-forms $\sigma^A(x, h)$ reads

$$\sigma^1 = -dt, \quad \sigma^2 = -dx, \quad \sigma^3 = -\sin\theta\cos h d\phi + \sin h d\theta,$$
$$\sigma^4 = -\sin\theta\sin h d\phi - \cos h d\theta, \quad \sigma^5 = -\cos\theta d\phi - dh.$$

Since the isotropy subgroup is one-dimensional, the gauge group that does not change the symmetry of the equation is Abelian. For $K = U(1)$, we take the potential (43) of an external electromagnetic field:

$$A_1 = A_2 = A_4 = 0, \quad A_3 = -\varepsilon\cos\theta, \quad \varepsilon \in \mathbb{R}. \tag{51}$$

The potential (51) describes a stationary magnetic field depending on the variable θ with the strength tensor

$$F = F_{ij}dx^i \wedge dx^j = -\varepsilon \sin\theta d\phi \wedge d\theta.$$

From Theorem 5 it follows that the Dirac equation on the homogeneous space M with the magnetic field (51) is equivalent to the system of equations

$$(i\hat{\gamma}^a\eta_a(g) - m)\psi(g) = 0, \quad (-\partial_h + \frac{1}{2c_2}\hat{\gamma}^3\hat{\gamma}^4 + \varepsilon)\psi(g) = 0, \tag{52}$$

where $\eta_a(g)$ are given by

$$\eta_1 = -\partial_t, \quad \eta_2 = -\partial_x, \quad \eta_5 = -\partial_h,$$
$$\eta_3 = -\frac{\cos h}{\sin\theta}\partial_\phi + \sin h\partial_\theta + \cos h \cot\theta\partial_h, \quad \eta_4 = -\frac{\sin h}{\sin\theta}\partial_\phi - \cos h\partial_\theta + \sin h\cot\theta\partial_h.$$

The left invariant vector fields

$$\xi_1 = \partial_t, \quad \xi_2 = \partial_x, \quad \xi_3 = \partial_\phi,$$
$$\xi_4 = -\cot\theta\sin\phi\partial_\phi + \cos\phi\partial_\theta + \frac{\sin\phi}{\sin\theta}\partial_h, \quad \xi_5 = -\cot\theta\cos\phi\partial_\phi - \sin\phi\partial_\theta + \frac{\cos\phi}{\sin\theta}\partial_h$$

on the Lie group G provide symmetry operators for the Dirac equation (52).

Let us find the spectrum of the Dirac equation (52). The operators ξ_1 and ξ_2 form an Abelian algebra $\mathbb{R}^2$ and allow one to separate the variables t and x:

$$\psi(g) = e^{-i(\omega t + j_1 x)}f(\phi,\theta,h), \quad i\xi_1\psi(g) = \omega\psi(g), \quad i\xi_2\psi(g) = j_1\psi(g). \tag{53}$$

The operators ξ_3, ξ_4, and ξ_5 form a Lie algebra $\mathfrak{so}(3)$. We take into account this symmetry algebra using the noncommutative integration method [6].

According to this method, we will solve equations (52) together with the system

$$(\xi_A(g) + l_A(q,\lambda))\psi(g) = 0, \quad \left(\eta_A(g) + \overline{l_A(q',\lambda)}\right)\psi(g) = 0, \quad (A = 3,4,5). \tag{54}$$

We call $l_A(q,\lambda)$ *the operators of λ-representation* for the Lie algebra $\mathfrak{so}(3)$ on the space $L_2(Q, d\mu(q))$ of functions determined on the Lagrangian submanifold Q of a coadjoint orbit (K-orbit) of the Lie algebra $\mathfrak{so}(3)$ [7]. The submanifold Q has the topology of a cylinder: $q = \alpha + i\beta \in Q$, $\alpha \in (0, 2\pi)$, $\beta \in \mathbb{R}^1$. The covector $\lambda = (j_2, 0, 0) \in \mathfrak{so}^*(3)$, $j_2 > 0$, parameterizes a non-degenerate K-orbit. The operators $-il_A(q,\lambda)$ are hermitian in $L_2(Q, d\mu(q))$ with respect to the inner product

$$(\psi_1(q), \psi_2(q)) = \int_Q \overline{\psi_1(q)}\psi_2(q)d\mu(q), \quad d\mu(q) = \frac{(2j_2+1)!}{2^{j_2}(j_2!)^2}\frac{dq\wedge d\overline{q}}{(1+\cos(q-\overline{q}))^{j_2+1}}$$

and are given by the following equations [7]:

$$l_3(q,\lambda) = -i(\sin(q)\partial_q - j_2\cos(q)), \quad l_4(q,\lambda) = -i(\cos(q)\partial_q + j_2\sin(q)), \quad l_5(q,\lambda) = \partial_q.$$

In the framework of the non-commutative integration method, the system of equations (54) is represented as a generalized Fourier transform [8, 9]:

$$f(\phi,\theta,h) = \sum_{j_2=0}^{\infty}(2j_2+1)\int_Q \psi(q,q',j_2)D^{j_2}_{\overline{q}q'}(\phi,\theta,h)d\mu(q')d\mu(q),$$

where $D^{j_2}_{q\overline{q}'}(\phi,\theta,h)$ is given by the formula

$$D^{j_2}_{q\overline{q}'}(\phi,\theta,h)=\frac{2^{j_2}(j_2!)^2}{(2j_2)!}\Big(\sin\theta\cos\phi+\cos\left(h+\overline{q}'\right)(\cos q\cos\phi-i\sin\phi)-$$
$$i\cos\theta\sin q+\sin\left(h+\overline{q}'\right)(-i\cos\theta\cos\phi-\cos\theta\cos q\sin\phi+\sin\theta\sin q)-i\sin\theta\cos q\sin\phi\Big)^{j2}.$$

Substituting (53) in the Dirac equation (52) and taking into account relation (54), we obtain a system of equations for $\psi(q,q',j_2)$:

$$\left(-\omega\hat{\gamma}^1-j_1\hat{\gamma}^2+i\hat{\gamma}^3 l_3(q',j_2)+i\hat{\gamma}^4 l_4(q',j_2)-m\right)\psi(q,q',j_2)=0, \tag{55}$$
$$\left(l_5(q',j_2)+\frac{1}{2c_2}\hat{\gamma}^3\hat{\gamma}^4+\varepsilon\right)\psi(q,q',j_2)=0 \tag{56}$$

From (56) it follows that

$$\psi(q,q',j_2)=\exp\left(-\left[\frac{1}{2c_2}\hat{\gamma}^3\hat{\gamma}^4+\varepsilon\right]q'\right)R(q,\omega,j_1,j_2). \tag{57}$$

Substituting (57) in the reduced equation (55), we find the linear system of equations

$$\left(\omega\hat{\gamma}^1+j_1\hat{\gamma}^2+\left(j_2+\frac{1}{2}\right)\hat{\gamma}^3+\varepsilon\hat{\gamma}^4+m\right)R(q,\omega,j_1,j_2)=0. \tag{58}$$

Finally, from the compatibility conditions for this system, we obtain the following relation determining the spectrum of the Dirac equation:

$$\omega^2=j_1^2c_1+\left[\varepsilon^2+\frac{1}{4}+j_2(j_2+1)\right]c_2+m^2.$$

Note that the spectral parameter j_1 is continuous because the variable x takes values on the non-compact space $\mathbb{R}^1$. The parameter j_2 has the physical meaning of an orbital quantum number and it takes only a discrete set of values. The expression for $j_2(j_2+1)$ corresponds to the spectrum of the Laplace operator on the sphere $\mathbb{S}^2$.

In the context of Kirillov's coadjoint orbit method [16], the quantization of j_2 is prescribed by the condition of *integer-valued orbits.* Note that this condition is equivalent to the Kostant-Souriau condition in the context of the geometric quantization method [17].

The basis of solutions for the linear system of equations (58) reads

$$R^{s=1}(\omega,j_1,j_2)=\begin{pmatrix}\frac{i\sqrt{c_2}(j_2+1/2)-\sqrt{c_1}j_1}{\omega+m}\\ \frac{\varepsilon\sqrt{c_2}}{\omega+m}\\ 0\\ 1\end{pmatrix},\quad R^{s=-1}(\omega,j_1,j_2)=\begin{pmatrix}-\frac{\varepsilon\sqrt{c_2}}{\omega+m}\\ -\frac{i\sqrt{c_2}(j_2+1/2)+\sqrt{c_1}j_1}{\omega+m}\\ 1\\ 0\end{pmatrix},$$

where the spin index s enumerates the basis vectors. The basis of the original Dirac equation (52) can be written in the framework of the noncommutative integration method as:

$$\psi_\tau(t,x,\phi,\theta)=e^{-i(\omega t+j_1t)}D^{j_2}_q(\phi,\theta)R_\tau,\quad \tau=(j_0,j_1,j_2,q,s),$$
$$D^{j_2}_q(\phi,\theta)=\int_Q e^{-(\frac{1}{2c_2}\hat{\gamma}^3\hat{\gamma}^4+\varepsilon)q'}D^{\lambda}_{q\overline{q}'}(\phi,\theta,e_H)d\mu(q').$$

Here τ denotes the set of quantum numbers for the basis of solutions.

6. The Dirac equation in de-Sitter space

Consider de Sitter space M as a homogeneous space with de Sitter isometry group of transformations $G = SO(1,4)$ and an isotropy Lorentz subgroup $H = SO(1,3)$. The space M is topologically isomorphic to $R^1 \times S^3$ and has constant positive curvature.

The de Sitter group $SO(1,4)$ is a rotation group of the 5-dimensional pseudo-Euclidean space with the metric $G_{AB} = \mathrm{diag}(1,-1,-1,-1,-1)$. The algebra $\mathfrak{g} = \mathfrak{so}(1,4)$ of the de Sitter group can be defined in terms of the basis $\{E_{AB} \mid A < B\}$ by the following commutation relations:

$$[E_{AB}, E_{CD}] = G_{AD}E_{BC} - G_{AC}E_{BD} + G_{BC}E_{AD} - G_{BD}E_{AC},$$

where $A, B, C, D = 1, \dots, 5$. The basis E_{AB} can be written as

$$E_{ab} = e_{ab}, \ (a < b), \quad E_{a5} = e_a/\varepsilon, \quad [e_a, e_b] = \varepsilon^2 e_{ab}, \quad a, b = 1, \dots, 4.$$

Here the basis e_{ab} forms an isotropy subalgebra $\mathfrak{h} = \mathfrak{so}(1,3)$, and ε is a parameter defining the curvature of de Sitter space, $R = 12\varepsilon^2$. Define canonical coordinates of the second kind for a Lie group G by the formula

$$g(t, x, y, z, h_1, h_2, h_3, h_4, h_5, h_6) = e^{h_6 e_{34}} e^{h_5 e_{24}} e^{h_4 e_{23}} e^{h_3 e_{14}} e^{h_2 e_{13}} e^{h_1 e_{12}} e^{z e_4} e^{y e_3} e^{x e_2} e^{t e_1}. \tag{59}$$

Here (t, x, y, z) are local coordinates on the de Sitter space M and h_{ab} are local coordinates on the isotropy subgroup H. The right-invariant 1-forms $\sigma^\alpha(x, e_H)$ in the canonical coordinates (59) are written as

$$\begin{gathered}
\sigma^{12} = \varepsilon \sin \varepsilon x dt - dh_1, \quad \sigma^{13} = \varepsilon \cos \varepsilon x \sin \varepsilon y dt - dh_2, \\
\sigma^{14} = \varepsilon \cos \varepsilon x \cos \varepsilon y \sin \varepsilon z dt - dh_3, \quad \sigma^{23} = \varepsilon \sin \varepsilon y dx - dh_4, \\
\sigma^{24} = \varepsilon \cos \varepsilon y \sin \varepsilon z dx - dh_5, \quad \sigma^{34} = \varepsilon \sin \varepsilon z dy - dh_6.
\end{gathered}$$

For the isotropy subgroup H, we choose generators of a representation on a space V, $\dim V = 4$, in the form

$$\Lambda_{ab} = \frac{1}{4}[\gamma^a, \gamma^b], \quad \{\gamma^a, \gamma^b\} = 2\eta^{ab}\mathbb{E}_4, \quad \eta^{ab} = \mathrm{diag}(1,-1,-1,-1), \quad a < b.$$

The potentials of the external gauge field for which the de Sitter group is the symmetry group of the Dirac equation are given by

$$\begin{gathered}
A_1 = \varepsilon \sin \varepsilon x \Lambda_{12} + \varepsilon \cos \varepsilon x \sin \varepsilon y \Lambda_{13} + \varepsilon \cos \varepsilon x \cos \varepsilon y \sin \varepsilon z \Lambda_{14}, \\
A_2 = \varepsilon \sin \varepsilon y \Lambda_{23} + \varepsilon \cos \varepsilon y \sin \varepsilon z \Lambda_{24}, \\
A_3 = \varepsilon \sin \varepsilon z \Lambda_{34}, \quad A_4 = 0.
\end{gathered}$$

These potentials define the gauge field strength tensor

$$\begin{gathered}
F_{12} = -\varepsilon^2 \cos \varepsilon x \cos^2 \varepsilon y \cos^2 \varepsilon z \Lambda_{12}, \quad F_{13} = -\varepsilon^2 \cos \varepsilon x \cos \varepsilon y \cos^2 \varepsilon z \Lambda_{13}, \\
F_{14} = -\varepsilon^2 \cos \varepsilon x \cos \varepsilon y \cos \varepsilon z \Lambda_{14}, \quad F_{23} = -\varepsilon^2 \cos \varepsilon y \cos^2 \varepsilon z \Lambda_{23}, \\
F_{24} = -\varepsilon^2 \cos \varepsilon y \cos \varepsilon z \Lambda_{24}, \quad F_{34} = -\varepsilon^2 \cos \varepsilon z \Lambda_{34}.
\end{gathered} \tag{60}$$

The solutions of the Dirac equation with the external gauge field (60) can be obtained using the noncommutative integration method [6]. The structure of the K-orbits for the de Sitter group is rather complicated [16]. Detail consideration of the noncommutative integration method and exact solution construction for the Dirac equation in the case under consideration is beyond the scope of the present work and will be the subject of a separate study.

7. Conclusion remarks

We have obtained the Yang–Mills potentials for which the symmetry group of the Dirac equation is a group of transformations of a homogeneous space M. Such gauge fields are found by representation of the Dirac equation in terms of an invariant matrix differential operator of the first order. The matrix coefficients of this operator satisfy an algebraic system of equations that is equivalent to the condition that the gauge fields do not change the symmetry of the Dirac equation, and this is possible if the homogeneous space M is reductive. For the reductive space M, these gauge fields are determined by the Lie symmetry algebra of the equation. The gauge group in this case is isomorphic to a subgroup H of the isotropic homogeneous space M.

The noncommutative integration method [6] can be effectively used to construct solutions of the Dirac equation with the gauge fields (43).

Following this method, we have found the spectrum and the basis of solutions of the Dirac equation in an $\mathbb{R}^2 \times \mathbb{S}^2$ space with an invariant metric. This is a reductive homogeneous space, and the potentials (43) describe the external magnetic field. The Dirac equation spectrum is shown to be represented as a continuous part corresponding to a non-compact subgroup of $\mathbb{R}$ and a part corresponding to the spectrum of the Laplace operator on a two-dimensional sphere $\mathbb{S}^2$. Also, we have obtained explicit formulae for the Yang–Mills potentials conserving the symmetry group of the Dirac equation in de Sitter space.

Exact solutions of the Dirac equation with the gauge fields of the form (43) can be used to investigate the effects of vacuum polarization and particle creation in homogeneous cosmological models [1, 9]. Note that seaking exact solutions to the Dirac equation on homogeneous spaces is a mathematically attractive problem, but it is far from trivial even when external fields are absent [1]. On the other hand, the results obtained contribute to the study of mathematical properties of external fields with internal symmetry.

Acknowledgments

The work was partially supported by Tomsk State University Competitiveness Improvement Program and by program 'Nauka' under contract No. 1.676.2014/ K.

References

[1] Birrell N D and Davies P C 1982 *Quantum Fields in Curved Space* (Cambrigde: Cambridge Univ. Press)
[2] Kalnins E G 1986 *Separation of Variables in Riemannian Spaces of Constant Curvature* (New York : Wiley)
[3] Kalnins E G, Miller W, Williams G C 1992 Recent advances in the use of separation of variables methods in general relativity *Phil. Trans. R. Soc. Lond. A* **340** 337
[4] Miller W 1977 *Symmetry and Separation of Variables* (MA: Addison-Wesley)
[5] Bagrov V G and Gitman D M 1990 *Exact solutions of relativistic wave equations* (Dordrecht: Kluwer);
[6] Shapovalov A V and Shirokov I V 1995 Noncommutative integration of linear differential equations *Theor. Math. Phys.* **104** 921
[7] Baranovsky S P, Mikheev V V and Shirokov I V 2001 Quantum Hamiltonian Systems on K-Orbits: Semiclassical Spectrum of the Asymmetric Top *Theor. Math. Phys.* **129** 1311
[8] Breev A I, Magazev A A and Shirokov I V 2011 Vacuum polarization of a scalar field on Lie groups and homogeneous spaces *Theor. Math. Phys.* **167** 468
[9] A I Breev 2014 Scalar field vacuum polarization on homogeneous spaces with an invariant metric *Theor. Math. Phys.* **178** 59
[10] Kurnyavko O L and Shirokov I V 2008 Construction of invariant scalar particle wave equations on Riemannian manifolds with external gauge fields *Theor. Math. Phys.* **156** 1169
[11] Bagrov V G and Shapovalov A V 1986 Symmetry of the Dirac equation with on external nonabelian gauge field *Soviet Physics Journal* **18** 95
[12] Bagrov V G, Shapovalov A V and Yevseyevich A A 1991 Separation of variables in the Dirac equation in Stackel spaces: II. External gauge fields *Class. Quantum Grav.* **8** 163
[13] Dubrovin B, Fomenko A and Novikov S 1990 *Modern geometry. Part 2. The geometry and topology of manifolds* (New York: Springer)
[14] Kobayashi S and Nomizu K 1963 *Foundations of Differential Geometry I* (New York: Interscience)

[15] Baranovskiii S P and Shirokov I V 2003 Prolongations of Vector Fields on Lie Groups and Homogeneous Spaces *Theor. Math. Phys.* **135** 510
[16] Kirillov A A 1976 *Elements of the Theory of Representation* (New York: Springer)
[17] Kostant B 1970 Quantization and unitary representations *Lecture Notes in Math.* **170** 87

Journal of Physics: Conference Series **563** (2014) 012005 doi:10.1088/1742-6596/563/1/012005

An integrable Hierarchy on the Perfect Schrödinger Lie Algebra

Paolo Casati

Dipartimento di Matematica e applicazioni,Università di Milano-Bicocca, Via Cozzi 53, I-20125 Milano, Italy

E-mail: paolo.casati@unimib.it

Abstract. We construct an integrable hierarchy of coupled integrable solitons equations on the non semisimple perfect Schrödinger Lie algebra $\mathfrak{sl}(2) \ltimes V(1)$.

1. Introduction

Since the celebrated discovery of Gelfand Dickey hierachies [16] and their generalizations to any affine Kac–Moody Lie algebra made by Drinfeld and Sokolov in their famous paper [18], a deep and far reaching theory on the integrable systems defined on simple Lie algebras has been developed by many authors [12] [2] [3] [9] [10] [22] [13] [23] [20] [14] [24] [8] [37] [17] [25].

However in the last years have been discovered examples of integrable hierarchies of partial differential equations which live on non semisimple Lie algebras. Most of these new examples are due to Ma and his collaborators (see for example [26] [27] [28] [29] [31] [32], [33]) who have studied hierarchies on semidirect product of Lie algebras, but there exists also hierarchies defined on truncated current Lie algebras [6] [7] [4] or on Frobenius valued Lie algebras [39].

As far as we know the hierarchies of PDE which arise from such Lie algebras are systems of coupled equations. In particular if the Lie algebra is a semidirect product between a simple Lie algebra and an abelian subalgebra or a truncated current Lie algebra, the corresponding equations are systems where a (usually) known soliton equation is coupled with other ones, which can be sometimes regarded as perturbation of the first one. Such type of coupled PDE's as been for the first time considered in an important work by Ma and Fuchssteiner [30]. However also coupled equations which do not enter in this scheme have been studied in the literature [36].

The aim of this paper is to proceed further into this line of research presenting a new integrable hierarchy of PDE's living on the perfect but not semisimple Schrödinger Lie algebra $\mathfrak{sl}(2) \ltimes V(1)$. Beyond its intrinsic interest, these are new coupled soliton equations which contains the KdV equation as "first equation", we hope that this case can be a first step to construct a theory of integrable systems defined on perfect Lie algebra at least in the case when the perfect Lie algebra $\mathfrak{g}$ has a Levi decomposition: $\mathfrak{g} = \mathfrak{s} \ltimes \mathfrak{r}$ where the nilradical part $\mathfrak{r}$ is a irreducible module of the simple part $\mathfrak{s}$.

In the present work we investigate the Lax equation theory of such system which it is likely to be extended to more general cases of perfect Lie algebras.

The paper is organized as follows. In the second section, we briefly describe the Schrödinger Lie algebra and its properties. In the third Section we construct the integrable hierarchy using the lax operator formalism. In the fourth and last Section we determine the reduced coupled equations.

The author wishes to thank professor Laszlo Feher, for his interest in these new hierarchies defined on perfect Lie algebras and for sending me a deep and very general description of "generalized Drinfeld–Sokolov" hierachies [21], which will followed here. The author whises also to thank the organizers and partecipants of the Integrable systems and quantum symmetries meeting for the fruitful and wonderful time spent in Prague.

2. The Schrödinger Lie Algebra

In this section we shall describe the Schrödinger Lie algebra $\mathfrak{s} = \mathfrak{sl}(2) \ltimes V(1)$ [5]. The Lie algebra $\mathfrak{s}$ is a perfect Lie algebra i.e., it equals its commutator ideal $[\mathfrak{s}, \mathfrak{s}] = \mathfrak{s}$, of dimension 5 spanned by the elements H, E, F, P, Q with Lie brackets

$$\begin{array}{lll} [H,E] = 2E, & [H,F] = -2F, & [E,F] = H \\ [H,P] = P & [H,Q] = -Q & \\ [E,P] = 0 & [E,Q] = P & \\ [F,P] = Q & [F,Q] = 0 & \\ [P,Q] = 0. & & \end{array} \tag{1}$$

However The Lie algebra $\mathfrak{s}$ is not semisimple, because it has a non trivial ideal spanned by the elements P, Q. In fact the Schrödinger Lie algebra is the easiest perfect Lie algebra (other than semisimple). Its Levi decomposition is

$$\mathfrak{s} \ltimes V(1)$$

with semisimple part $\mathfrak{s} = \mathfrak{sl}(2)$ and nilradical $V(1) = \{P, Q\}$. Observe that $V(1)$ is nothing else but the irreducible two dimensional $\mathfrak{sl}(2)$–module with highest weight 1, which explains our notation. It may be worth to note here then in the literature the name Schrödinger Lie algebra is given also to the non trivial central extension of $\mathfrak{sl}(2) \ltimes V(1)$ [19].

The Lie algebra $\mathfrak{s}$ can be embedded in $\mathfrak{sl}(3)$ as follows [35]

$$(H, E, F, P, Q) \mapsto \begin{pmatrix} H & E & 0 \\ F & -H & 0 \\ -Q & P & 0 \end{pmatrix}. \tag{2}$$

There exist non semisimple Lie algebras, which own an ad–invariant non degenerate bilinear form [6] [4], this is unfortunately not the case of $\mathfrak{s}$.

Proposition 2.1 *For any ad–invariant bilinear form* $\langle\cdot,\cdot\rangle$ *on* $\mathfrak{s}$ *we have* $\langle P, \mathfrak{s}\rangle = 0$.

Proof Indeed since $\langle\cdot,\cdot\rangle$ is ad–invariant we have

$$\begin{aligned}
&\langle P, H\rangle = \langle [H, P], H\rangle = -\langle P, [H, H]\rangle = 0\\
&\langle P, E\rangle = \langle [E, Q], E\rangle = -\langle Q, [E, E]\rangle = 0\\
&\langle P, F\rangle = -\tfrac{1}{2}\langle P, [H, F]\rangle = \tfrac{1}{2}\langle [H, P], F\rangle = \tfrac{1}{2}\langle P, F\rangle \Longrightarrow \langle P, F\rangle = 0\\
&\langle P, P\rangle = \langle [H, P], P\rangle = \langle H, [P, P]\rangle = 0\\
&\langle P, Q\rangle = \langle P, [F, P]\rangle = \langle [P, P], F\rangle = 0.
\end{aligned}$$

□

This last fact does not really affect the construction of the integrable hierarchy trought the Lax operator but makes the bihamiltonian approach rather "uncomfortable", as we shall discuss briefly later.

Let $\mathcal{L}(\mathfrak{s})$ be the loop algebra given by all the formal Laurent series in a parameter λ with values in $\mathfrak{s}$:

$$\mathcal{L}(\mathfrak{s}) = \mathfrak{s}\otimes C[\lambda] = \left\{\sum_{k\leq N} X_k\lambda^k |\, X_k \in \mathfrak{s}\right\} \tag{3}$$

Now let Λ be the element of $\mathcal{L}(\mathfrak{s})$:

$$\Lambda = \begin{pmatrix} 0 & 1 & 0\\ \lambda & 0 & 0\\ \lambda & 0 & 0\end{pmatrix}. \tag{4}$$

It is easy to check that

$$\mathrm{Ker}(\Lambda) = \begin{pmatrix} 0 & v & 0\\ v & 0 & 0\\ v & 0 & 0\end{pmatrix}, \tag{5}$$

while

$$\mathrm{Im}(\Lambda) = \begin{pmatrix} w_1 & w_2 & 0\\ -w_2 & -w_1 & 0\\ w_3 & w_4 & 0\end{pmatrix}. \tag{6}$$

Hence

$$\mathrm{Ker}(\Lambda) \cap \mathrm{Im}\,(\Lambda) = \{0\} \qquad \mathrm{Ker}(\Lambda) \oplus \mathrm{Im}(\Lambda) = \mathcal{L}(\mathfrak{s}). \tag{7}$$

From equation (5) we have that that the annihilator $\mathcal{L}(\mathfrak{s})_\Lambda$ of Λ in $\mathcal{L}(\mathfrak{s})$, $\mathcal{L}(\mathfrak{s})_\Lambda = \{X \in \mathcal{L}|\ [\Lambda, X] = 0\} = \mathrm{Ker}(\Lambda)$ is

$$\mathcal{L}(\mathfrak{s})_\Lambda = \left\{c_k\Lambda^k |\, c_k \in \mathbb{C}\ k \in \mathbb{Z}\right\}.$$

This corresponds to the fact that in the case of the Drinfled Sokolov hierarchies the annihilator of Λ is a Cartan subalgebra. On $\mathcal{L}(\mathfrak{s})$ we can introduce a $\mathbb{Z}$–gradation by setting:

$$\deg(H) = 0, \quad \deg(E) = \deg(P) = 1, \quad \deg(F) = \deg(Q) = -1 \quad \deg(\lambda) = 2.$$

With respect to this gradiation Λ is a homogeneous element with grade 1. We denote by $\mathcal{L}(\mathfrak{s})^k$ the subset given by of all elements of $\mathcal{L}(\mathfrak{s})$ of degree k. If $X = \sum_{k\leq p} X^k \in \mathcal{L}(\mathfrak{s})$ we denote by $(X)_+ = \sum_{k\geq 0} X^k$ its projection on the space spanned by the positive homogeneous elements.

3. The Integrable Hierarchy

Let us construct commuting differential polynomial vector fields on the "phase space"

$$\mathcal{M} = C^\infty\left(S^1, \mathcal{L}(\mathfrak{s})\right) \tag{8}$$

We begin by defining the Lax operator L of our hierachy: the differential:

$$L = \partial_x + \Lambda + Q(x) = \partial_x + \begin{pmatrix} 0 & 1 & 0 \\ \lambda & 0 & 0 \\ \lambda & 0 & 0 \end{pmatrix} + \begin{pmatrix} p_1 & 0 & 0 \\ q_1 & -p_1 & 0 \\ q_2 & p_2 & 0 \end{pmatrix} \tag{9}$$

where with ∂_x is denoted the derivative along the coordinate x on S^1. Observe that the elements of Q have degree between -1 and 0 i,e, less then $\deg(\Lambda) = 1$.

Following [21] the next step is

Lemma 3.1 *(dressing lemma). For $Q(x)$ defined as in (9) consider the equation*

$$L := \partial_x + \Lambda + J(x) \mapsto e^{ad(F)}(L) = \partial_x + \Lambda + H(x) \tag{10}$$

where $F(x) = \sum_{k<0} F^k, H(x) = \sum_{k<0} H^k \in \oplus_{k<0}\mathcal{L}(\mathfrak{s})^k$ and $H(x) \in Ker(ad\Lambda)$. Then there exists a solution $F(x)$, $H(x)$ of (10) for any given $J(x)$. In terms of a particular solution, given by $F_0(x)$ $H(x)$ the general solution $F(x)$ is determined by

$$e^{adF} = e^{adK} e^{adF_0}, \tag{11}$$

where $K(x) \in \oplus_{j\leq 0}\mathcal{L}(\mathfrak{s})^j_\Lambda$ is arbitrary. One has a unique solution $F_J(x)$ satisfying $F_J(x) \in Im(ad\Lambda))^{<0}$, and the components of this solution are differential polynomials in $J(x)$.

Proof Following the lines of Drinfeld-Sokolov [18], the existence is proved by induction based on the gradation and the use of the decomposition in (7) in each step. The uniqueness property is straightforward to see. The proof also shows that the components of the unique solution $F_J(x) \in \mathrm{Im}(\mathrm{ad}\Lambda))$ are differential polynomials in $J(x)$.

□

Set now

$$\mathcal{H}^+_\Lambda = \mathcal{L}(\mathfrak{s})_\Lambda \cap \bigoplus_{k\geq 0} \mathcal{L}(\mathfrak{s})^k.$$

For any $b \in \mathcal{H}^+_\Lambda$ define

$$B_b(J) = e^{\mathrm{ad}F_J}(b). \tag{12}$$

Then we can give

Journal of Physics: Conference Series **563** (2014) 012005 doi:10.1088/1742-6596/563/1/012005

Definition 3.2 *For any $b \in \mathcal{H}_\Lambda^+$ let $\frac{\partial}{\partial t_b}$ the vector field*

$$\frac{\partial}{\partial t_b} J(x) = [B_b(J), L] .$$

With the same arguments used in [18] [2] [21] one can prove

Theorem 3.3 *(i) The vector fields on $\mathcal{M}$ defined by (12) pairwise commute with each other with respect to the natural Lie derivative,*

$$\left[\frac{\partial}{\partial t_a}, \frac{\partial}{\partial t_b}\right] \qquad a, b \in \mathcal{H}_\Lambda^+$$

(ii) The quantities H^k of Lemma 3.1 are the conserved densities for the hierarchy.

Actually for the still easy case of the Schrödinger Lie algebra it is possible to construct directly the equations of the hierarchy without reffering to the dressing method, which however seems to be very useful to prove Therem 3.3.

We have indeed

Theorem 3.4 *For L given as in (9) there exist a solution $V(x) = \sum_{k\leq 1} V_{-k}\lambda^k$ $V_k \in C^\infty(S^1, \mathfrak{s})$ of the "Casimir" equation*

$$[L, V] = 0. \tag{13}$$

Proof Explicitly equation (13) is

$$\left[\partial_x + \Lambda + Q(x), \sum_{k\leq 1} V_{-k}\lambda^k\right] = 0$$

equating the coefficient with the same power in λ we obtain

$$\begin{aligned} &[A, V_{-1}] = 0 \\ &(V_k)_x + [Q(x) + B, V_k] = [V_{k+1}, A] \quad k \geq -1 \end{aligned} \tag{14}$$

where

$$A = F - Q = \begin{pmatrix} 0 & 0 & 0 \\ 1 & 0 & 0 \\ 1 & 0 & 0 \end{pmatrix} \quad B = E = \begin{pmatrix} 0 & 1 & 0 \\ 0 & 0 & 0 \\ 0 & 0 & 0 \end{pmatrix}. \tag{15}$$

Let us first observe that

$$\mathrm{Im}(\mathrm{ad}A) = \begin{pmatrix} v_1 & 0 & 0 \\ v_2 & -v_1 & 0 \\ v_3 & v_1 & 0 \end{pmatrix}. \tag{16}$$

The first equation in (14); $[A, V_{-1}] = 0$ implies that

$$V_{-1} = \begin{pmatrix} 0 & 0 & 0 \\ \gamma_{-1} & 0 & 0 \\ \delta_{-1} & 0 & 0 \end{pmatrix}$$

with $\gamma_{-1}(x)$ and $\delta_{-1}(x)$ arbitrary functions. Now since

$$(V_{-1})_x + [Q(x) + B, V_{-1}] = \begin{pmatrix} \gamma_{-1} & 0 & 0 \\ (\gamma_{-1})_x - 2p_1\gamma_{-1} & -\gamma_{-1} & 0 \\ (\delta_{-1})_x + p_2\gamma_{-1} - p_1\delta_{-1} & \delta_{-1} & 0 \end{pmatrix} \in \mathrm{Im}(\mathrm{ad}A)$$

we can choose at this step $\gamma_{-1} = \delta_{-1} = 1$. The equation $(V_{-1})_x + [Q(x) + B, V_{-1}] = [V_0, A]$ has then solution

$$V_0 = \begin{pmatrix} p_1 & 1 & 0 \\ \gamma_0 & -p_1 & 0 \\ \delta_0 & p_2 & 0 \end{pmatrix}.$$

with $\gamma_0(x)$ and $\delta_0(x)$ arbitrary functions. Again we have

$$(V_0)_x + [Q(x) + B, V_0] = \begin{pmatrix} p_{1x} - q_1 + \gamma_0 & 0 & 0 \\ \gamma_{0x} + 2q_1p_1 - 2p_1\gamma_0 & -p_{1x} + q_1 - \gamma_0 & 0 \\ \delta_{0x} + q_2p_1 + p_2q_1 + p_1\delta_0 + p_2\gamma_0 & p_{2x} + q_2 - \delta_0 & 0 \end{pmatrix}.$$

Therefore in order to have this latter expression in Im(ad$A$) (i.e., of the form (16)) and solve equation (14) at the next step we must impose $\delta_0 = p_{2x} + p_{1x} + q_2 - q_1 + \gamma_0$, while γ_0 remains arbitrary.

It should now be clear how to proceed by induction. Observe indeed that this still arbitrary function γ_0 will appears as linear factor in the entries of position $(1, 1)$ and $(2, 2)$ of V_1. Suppose by induction that we have already determine the elements V_k until the index $k = n$ with

$$V_n = \begin{pmatrix} a + \gamma_{n-1} & b & 0 \\ \gamma_n & -a - \gamma_{n-1} & 0 \\ \delta_n & c & 0 \end{pmatrix}$$

where γ_{n-1} γ_n and δ_n still arbitrary. Since we have

$$(V_n)_x + [Q(x) + B, V_n]$$
$$= \begin{pmatrix} a_x + \gamma_{(n-1)x} - bq_1 + \gamma_n & b_x + 2p_1b - 2a - 2\gamma_{n_1} & 0 \\ \gamma_{nx} + 2q_1a + 2q - 1\gamma_{n-1} - 2p_1\gamma_n & -a_x - \gamma_{(n-1)x} + bq_1 - \gamma_n & 0 \\ \delta_{nx} + q_2a + q_2\gamma_{n-1} - p_1\delta_n + p_2\gamma_n - cq_1 & c_x + q_2b - p_1a - p_2\gamma_{n-1} + cp_1 - \delta_n & 0 \end{pmatrix}$$

Journal of Physics: Conference Series **563** (2014) 012005 doi:10.1088/1742-6596/563/1/012005

to solve now equation (14) for V_{n+1} we have to impose

$$\begin{aligned}
\gamma_{n-1} &= a - p_1 b - \tfrac{1}{2} b_x \\
\delta_n &= -\tfrac{1}{2} b_{xx} + c_x + 2a - x + p_{1x} b + p_1 b_x + \tfrac{1}{2} p_2 b_x + b(q_2 - q_1) \\
&\quad - p_1 a_1 - p_2 a + p_2 p_1 b = c p_2 + \gamma_n
\end{aligned}$$

while γ_n still remains arbitrary and will be determined at the next step. Hence V_{n+1} has the form required by the induction Hypothesis and this closes the proof.

□

From the last theorem follows almost immediately

Theorem 3.5 *The equations of the hierarchy (3.2) are linear combination of*

$$\frac{\partial}{\partial t_k} L = \left[(\lambda^k V_k)_+, L \right]. \tag{17}$$

Proof The above equations (17) follow directly from equation (14). Now since V commutes with the Lax operator L from the dressing Lemma 10 and the property of Λ follows that $e^{\mathrm{ad}(-F)} V$ must belong to $\mathcal{L}(\mathfrak{s})_\Lambda$.

□

The first non trivial equations written in components are

$$\begin{aligned}
\frac{\partial p_1}{\partial t_3} &= -\frac{\partial p_2}{t_3} = \frac{1}{2}\left(p_{1x} + q_1 + p_1^2 \right) \\
\frac{\partial q_1}{\partial t_3} &= \frac{1}{2} p_{1xx} + \frac{1}{2} q_{1x} - q_1 p_1 - p_1^3 \\
\frac{\partial q_2}{\partial t_3} &= p_{2xx} + \frac{1}{2} p_{1xx} + q_{2x} + q_{1x} + 2 p_1 p_{1x} + p_{2x} p_2 - \frac{1}{2} p_{2x} p_1 + 2 q_2 p_1 + 2 q_1 p_1 \\
&\quad + 2 q_2 p_2 + p_1^2 p_2 + p_1^3.
\end{aligned}$$

Let us here remark that usually along with the Lax theory of integrable systems one can develop a bi–Hamiltonian theory. See for example for detailed descriptions of these theory the papers [11] [12] [10] [17] [34] or the books [16] [1].

A bi–Hamiltonian manifold $\mathcal{M}$ is a manifold equipped with two compatible Poisson structures, i.e., two Poisson tensors P_0 and P_1 such that the pencil $P_\lambda = P_1 - \lambda P_0$ is a Poisson tensor for any $\lambda \in \mathbb{C}$. A bihamiltonian vector field X on $\mathcal{M}$ is a vector field which is Hamiltonian with respect to both Poisson tensors (and therefore with respect to any Poisson tensors in the pencil).

The central idea of the bihamiltonian theory of the hierarchies of PDE's is to view them as collections of bihamiltonian vector fields on a (usually infinite dimensional) bihamiltonian manifold $\mathcal{M}$.

In our case where $\mathcal{M} = C^\infty(S^1, \mathcal{L}(\mathfrak{s}))$ it is still possible to define the canonical Lie Poisson tensor as

$$P_S(V) = -\partial_x V + [S, V] \qquad S, V \in C^\infty(S^1, \mathcal{L}(\mathfrak{s})), \tag{18}$$

and it can be easily shown [38] that this Poisson tensor is compatible with the constant Poisson tensors obtained by freezing it in any point, giving to $\mathcal{M}$ the structure of a bi–Hamiltonian manifold. But, since we do not have a non degenerate ad–invariant bilinear form on $\mathfrak{s}$, we are left with to choices either we take an ad–invariant degenerated linear form but the corresponding Poisson bracket $\{\cdot,\cdot\}$:

$$\{F,G\}(m) = \langle dF, P_m dG\rangle_m \qquad \forall F,G \in C^\infty(\mathcal{M},\mathbb{C}) \quad m\in\mathcal{M}$$

would be degenerate, or we use a non degenerate but not ad–invariant form and then the Poisson tensor would be not anymore skew symmetric.

4. The reduced Hierachy

Viewed on $\mathcal{M}$ the equations of the hierarchy seem to be rather complicated. However we can show that there exist a gauge transformation which brings them to a hierarchy of a system of two coupled equations, where the first ones are those of the KdV hierarchy.

Let T be the abelian group

Theorem 4.1 *The gauge group*

$$T = \left\{ \begin{pmatrix} 1 & 0 & 0 \\ t_1 & 1 & 0 \\ t_2 & 0 & 1 \end{pmatrix} \mid t_i \in C^\infty(S^1,\mathbb{C}) \right\}.$$

bring the Lax operator in the form:

$$L' = TLT^{-1} = \partial_x + TT_x^{-1} + T(Q+B)T^{-1} = \partial_x + \begin{pmatrix} 0 & 1 & 0 \\ u_1 & 0 & 0 \\ u_2 & 0 & 0 \end{pmatrix}.$$

with

$$\begin{aligned} u_1 &= p_{1x} + 2p_1^2 + q_1^2 \\ u_2 &= -p_{2x} - p_1p_2 + q_2 \end{aligned} \tag{19}$$

Proof by direct computation.

□

We can compute using the transformation (19) the first non trivial gauge fixed equations in term of the variable u_1 u_2 obtaining

$$\begin{aligned} \frac{\partial u_1}{\partial t_3} &= \frac{1}{4}u_{1xxx} + \frac{3}{2}u_1u_{1x} \\ \frac{\partial u_2}{\partial t_3} &= u_{2xx} - \frac{3}{4}u_{1xx} + \frac{3}{4}u_2u_{1x} + \frac{1}{2}u_{2x}u_2 - \frac{3}{4}u_1^2 + u_1u_2. \end{aligned} \tag{20}$$

which is a sistem of coupled equation which the KdV as first one. As far as we know this hierarchy of coupled KdV equation has been not yet considered in the literature.

[1] M. Blaszak *Multi-Hamiltonian theory of dynamical systems.* Texts and Monographs in Physics. Springer-Verlag, Berlin, 1998. x+350 pp. ISBN: 3-540-64251-X

[2] M. F. de Groot, T. Hollowood, J. L. Miramontes, *Generalized Drinfel'd-Sokolov hierarchies.* Comm. Math. Phys. **145** (1992), no. 1, 57–84.

[3] N. J. Burroughs, M. F. de Groot, T. Hollowood, J. L. Miramontes, *Generalized Drinfel'd-Sokolov hierarchies. II. The Hamiltonian structures.* Comm. Math. Phys. **153** (1993), no. 1, 187–215.

[4] *Drinfeld-Sokolov hierarchies on truncated current Lie algebras* Banach center Publication **94** (2011) 163-?171.

[5] A. O. Barut R. Raczka *Theory of Group Representations and Applications*, PWN Polish Scientific Publishers Warszawa 1977 xix+717 ISBN: 83-01-02716-9

[6] P. Casati G. Ortenzi *New Integrable Hierachies from Vertex Operator Representations of Polynomial Lie Algebras*, J. Geom. Phys. Vol 56 (2006), **3**, 418–449.

[7] P. Casati G. Ortenzi *Bihamiltonian equations on polynomial Virasoro algebras*, Journal of Nonlinear Mathematical Physics Vol 13 (2006), **3**, 352–364.

[8] P. Casati A. Della Vedova, G. Ortenzi *The soliton equations associated with the affine Kac–Moody Lie algebra* $G_2^{(1)}$ J. Geom. Phys. **58** (2008), 377–386.

[9] P. Casati, G. Falqui, F. Magri, M. Pedroni *Bi–Hamiltonian reductions and* w_n*-algebras* J. Geom. Phys. 26 (1998), no. 3-4, 291–310.

[10] P. Casati, G. Falqui, F. Magri, M. Pedroni *A note on fractional KdV hierarchies* J. Math. Phys. 38 (1997), no. 9, 4606–4628.

[11] P. Casati, F. Magri, M. Pedroni *Bi-Hamiltonian manifolds and* τ*-function.* Mathematical aspects of classical field theory (Seattle, WA, 1991), 213–234, Contemp. Math., **132**, Amer. Math. Soc., Providence, RI, 1992.

[12] P. Casati, M. Pedroni *Drinfeld-Sokolov reduction on a simple Lie algebra from the bi-Hamiltonian point of view* , Lett. Math. Phys. 25 (1992), no. 2, 89–101. Letters in Mathematical Physics 16, no. 2 (1988): 12532

[13] F. Delduc, L. Feher, *Regular Conjugacy Classes in the Weyl Group and Integrable Hierarchies* J.Phys.A **28**, 5843-5882

[14] F. Delduc, L. Feher, L. Gallot, *Nonstandard Drinfeld-Sokolov reduction* J. Phys. A 31 (1998), no. 25, 5545-5563.

[15] E. Date, M. Jimbo, M. Kashiwara, T. Miwa, "Transformation Groups for Soliton Equations." In: *Proceedings of R.I.M.S. Symposium on Nonlinear Integrable Systems–Classical Theory and Quantum Theory,* M. Jimbo, T. Miwa eds., pp. 39–119, World Scientific, Singapore, 1983.

[16] L. A. Dickey *Soliton equations and Hamiltonian systems.* Second edition. Advanced Series in Mathematical Physics, 26. World Scientific Publishing Co., Inc., River Edge, NJ, 2003. xii+408 pp. ISBN: 981-238-173-2

[17] B. Dubrovin, S. Liu, Y. Zhang *Frobenius manifolds and central invariants for the Drinfeld–Sokolov bihamiltonian structures* Advances in Mathematics 219 (2008) 780–837.

[18] V. G. Drinfeld, V. V. Sokolov, *Lie Algebras and Equations of Korteweg–de Vries Type.* J. Sov. Math. **30** (1985), 1975–2036

[19] B *Classification of simple weight modules with finite-dimensional weight spaces over the Schrodinger algebra* arXiv:1309.1346 [math.RT]

[20] L. Feher, *KdV type systems and W-algebras in the Drinfeld-Sokolov approach* (1997) arXiv:hep-th/9510001v2

[21] L. Feher, *Integrable hierarchies in the Drinfeld-Sokolov approach* private communication

[22] L. Feher, John Harnad, I. Marshall, *Generalized Drinfeld-Sokolov Reductions and KdV Type Hierarchies* Commun.Math.Phys. **154** (1992) 181-214

[23] L. Feher, I. Marshall, *Extensions of the matrix Gelfand-Dickey hierarchy from generalized Drinfeld-Sokolov reduction* Commun.Math.Phys. **183** (1997) 423-461

[24] B. Khesin, F. Malikov *Universal Drinfel'd-Sokolov reduction and matrices of complex size.* Comm. Math. Phys. 175 (1996), no. 1, 113–134.

[25] S. Liu, C. Wu, Y. Zhang *On the DrinfeldSokolov Hierarchies of D Type* Int. ernational Mathematics Research Notices Advance Int. Math. Res. Not. (2010), Notices, Article ID rnq138, 45 pages doi:10.1093/imrn/rnq138

[26] W.X. Ma, *Integrable couplings of soliton equations by perturbations I: a general theory and application to the KdV hierarchy*, Methods Appl. Anal. **7** (2000) 21?55.

[27] W.X. Ma, *A bi-Hamiltonian formulation for triangular systems by perturbations*, J. Math. Phys. textbf43 (2002) 1408–1421.

[28] W.X. Ma, *Enlarging spectral problems to construct integrable couplings of soliton equations*, Phys. Lett. A **316** (2003) 72–76.

[29] W.X. Ma, M. Chen, *Hamiltonian and quasi-Hamiltonian structures associated with semi-direct sums of Lie algebras*, J. Phys. A: Math. Gen. **39** (2006) 10787–10801.

[30] W.X. Ma, B. Fuchssteiner, *Integrable theory of the perturbation equations*, Chaos Solitons Fractals **7** (1996) 1227–1250.

[31] W.X. Ma, X.X. Xu, Y.F. Zhang, *Semi-direct sums of Lie algebras and continuous integrable couplings*, Phys. Lett. A **351** (2006) 125–130.

[32] W.X. Ma, X.X. Xu, Y.F. Zhang, *Semidirect sums of Lie algebras and discrete integrable couplings*, J. Math. Phys. **47** (2006) 053501. 16 pp.

[33] W. X. Ma, Z. Zhu *Constructing nonlinear discrete integrable Hamiltonian couplings* Computers and Mathematics with Applications **60** (2010) 2601–2608

[34] A. Panasyuk *Bi–Hamiltonian structures with symmetries, Lie pencils and integrable systems*. J. Phys. A 42 (2009), no. 16, 165205, 20 pp.

[35] A. Piard, *Sur des représentations indècomposables de dimension finie de* SL(2).R^2, J. Geom. Phys. 3 (**1**) (1986) pag. 1-53.

[36] A. Restuccia, A. Sotomayor, *The Hamiltonian structure of a coupled system derived from a supersymmetric breaking of super Korteweg-de Vries equations*,Journal of Mathematical Physics **54** 113510 (2013)

[37] C. Z. Wu, *A remark on KacWakimoto hierarchies of D-type*. Journal of Physics A: Mathematical and Theoretical 43 (2010): 035201

[38] I. Zakharevich, *The second Gel'fand-Dickey bracket as a bracket on a Poisson-Lie Grassmannian*. Comm. Math. Phys. 159 (1994), no. 1, 93–119.

[39] *The Frobenius–Virasoro algebra and Euler equations*, J. Geom. Phys. **86** (2014), 203–210.

Journal of Physics: Conference Series **563** (2014) 012006 doi:10.1088/1742-6596/563/1/012006

ALGEBRAIC FORMULATION OF HADRONIC SUPERSYMMETRY BASED ON OCTONIONS: NEW MASS FORMULAS AND FURTHER APPLICATIONS

Sultan Catto[1,2,†], Yasemin Gürcan[3], Amish Khalfan[4] and Levent Kurt[3,†]

[1]Physics Department, The Graduate School, City University of New York, New York, NY 10016-4309
[2]Theoretical Physics Group, Rockefeller University, 1230 York Avenue, New York, NY 10021-6399
[3] Department of Science, Borough of Manhattan CC, The City University of NY, New York, NY 10007
[4] Physics Department, LaGuardia CC, The City University of New York, LIC, NY 11101

† Work supported in part by the PSC-CUNY Research Awards, and DOE contracts No. DE-AC-0276 ER 03074 and 03075; and NSF Grant No. DMS-8917754.

Abstract.

A special treatment based on the highest division algebra, that of octonions and their split algebraic formulation is developed for the description of diquark states made up of two quark pairs. We describe symmetry properties of mesons and baryons through such formulation and derive mass formulae relating π, ρ, N and Δ trajectories showing an incredible agreement with experiments. We also comment on formation of diquark-antidiquark as well as pentaquark states and point the way toward applications into multiquark formulations expected to be seen at upcoming CERN experiments. A discussion on relationship of our work to flux bag models, string pictures and to string-like configurations in hadrons based on spectrum generating algebras will be given.

1. Introduction

In mid sixties Miyazawa, in a series of papers[1], extended the $SU(6)$ group to the supergroup $SU(6/21)$ that could be generated by constituent quarks and diquarks that could be transformed to each other. In particular, he found the following: (a) A general definition of $SU(m/n)$ superalgebras, expressing the symmetry between m bosons and n fermions, with Grassman-valued parameters. (b) A derivation of the super-Jacobi identity. (c) The relation of the baryon mass splitting to the meson mass splitting through the new mass formulae.

This work contained the first classification of superalgebras (later rediscovered by mathematicians in the seventies). Because of the field-theoretic prejudice against $SU(6)$, Miyazawa's work was generally ignored. Supersymmetry was, of course, rediscovered in the seventies within the dual resonance model by Ramond[2], and Neveu and Schwarz[3]. Golfand and Likhtman[4], and independently Volkov and Akulov[5], proposed the extension of the Poincaré group to the super-Poincaré group. Examples of supersymmetric field theories were given and the general method based on the super-Poincaré group was discovered by Wess and Zumino[6]. The super-Poincaré group allowed transformations between fields

Journal of Physics: Conference Series **563** (2014) 012006 doi:10.1088/1742-6596/563/1/012006

associated with different spins 0, $\frac{1}{2}$ and 1. The Coleman-Mandula theorem was amended in 1975 by Haag, Lopuszanski and Sohnius to allow super-Poincaré e group $\times G_{int}$ as the maximum symmetry of the S-matrix. Unfortunately, $SU(6)$ symmetry was still forbidden.

2. SU(6) and Hadronic Supersymmetry

How do we interpret the symmetries of the QCD spectrum in this light? In the ultraviolet, the running coupling constant tends to zero and quarks behave like free point particles. Thus an approximate conformal symmetry exists, allowing spin to be conserved separately from orbital angular momentum. Thus spin behaves as an internal quantum number; this makes a $SU(6)$ symmetry possible, since the quarks are almost free Dirac particles. Single vector-gluon exchange breaks this symmetry; thus, as shown by Glashow, Georgi and deRujula[7], the mass-degeneracy of hadrons of different spins is lifted by a hyperfine-interaction term.

Here is the main point. In the infrared we expect confinement to set in. The quark-antiquark potential becomes proportional to the distance. Careful studies of quarkonium spectra and lattice-gauge calculations show that at large separation the quark forces become spin-independent. QCD is also flavor independent. We therefore find approximate spin- and flavor-independent quark binding forces; these are completely consistent with $SU(6)$ symmetry. This is not an exact symmetry, but is a good starting point, before spin and flavor effects are included.

There is good phenomenological evidence that in a rotationally excited baryon a quark-diquark $(q - D)$ structure is favored over a three-quark (qqq) structure[8],[9],[10]. Eguchi[11] had shown that it is energetically favorable for the three quarks in a baryon to form a linear structure with a quark on one end and bilocal structure qq at the other end. Similarly, Bars and Hanson[12], and independently Johnson and Thorn[13] had shown that the string-like hadrons may be pictured as vortices of color flux lines which terminate on concentration of color at the end points. A baryon with three valence quarks would be arranged as a linear chain of molecule where the largest angular momentum for a state of a given mass is expected when two quarks are at one end, and the third is at the other: At large spin, two of the quarks form a diquark at one end of the string, the remaining quark being at the other. Regge trajectories for mesons and baryons are closely parallel; both have a slope of about $0.9(GeV)^{-2}$. If the quarks are light, the underlying quark-diquark symmetry leads to a Miyazawa symmetry between mesons and baryons. Thus we studied QCD with a weakly broken supergroup $SU(6/21)$. Note that the fundamental theory is not supersymmetric. For quarks, the generators of the Poincaré group and those of the color group $SU(3)^c$ commute. It is only the effective Hamiltonian which exhibits an approximate supersymmetry among the bound states $q\bar{q}$ and qD.

Under the color group $SU(3)^c$, meson $q\bar{q}$ and diquark $(D = qq)$ states transform as[10],[14]

$$qq:\ \mathbf{3}\times\mathbf{3} = \bar{\mathbf{3}} + \mathbf{6}\ ;\quad q\bar{q}:\ \mathbf{3}\times\bar{\mathbf{3}} = \mathbf{1} + \mathbf{8} \tag{1}$$

and under the spin flavor $SU(6)$ they transform as

$$qq:\ \mathbf{6}\times\mathbf{6} = \mathbf{15} + \mathbf{21}\ ;\quad q\bar{q}:\ \mathbf{6}\times\bar{\mathbf{6}} = \mathbf{1} + \mathbf{35} \tag{2}$$

Dimensions of internal degrees of quarks and diquarks are shown in the following table:

	$SU_f(3)$	$SU_s(2)$	$dim.$
q	□	$s = 1/2$	$3 \times 2 = 6$
D	□ □	$s = 1$	$6 \times 3 = 18$
	□ □	$s = 0$	$3 \times 1 = 3$

If one writes qqq as qD, then the quantum numbers of D are $\bar{\mathbf{3}}$ for color since when combined with q must give a color singlet, and $\mathbf{21}$ for spin-flavor since combined with color must give antisymmetric wavefunctions. The quantum numbers for $\bar{q}$ are for color, $\bar{\mathbf{3}}$, and for spin-flavor, $\bar{\mathbf{6}}$. Thus $\bar{q}$ and D have the same quantum numbers (color forces can not distinguish between $\bar{q}$ and D). Therefore there is a dynamic supersymmetry in hadrons with supersymmetric partners

$$\psi = \begin{pmatrix} \bar{q} \\ D \end{pmatrix}, \quad \bar{\psi} = (q \ \ \bar{D}) \tag{3}$$

We can obtain all hadrons by combining ψ and $\bar{\psi}$: mesons are $q\bar{q}$, baryons are qD, and exotics are $D\bar{D}$ states. Inside rotationally excited baryons, QCD leads to the formation of diquarks well separated from the remaining quark. At this separation the scalar, spin-independent, confining part of the effective QCD potential is dominant. Since QCD forces are also flavor-independent, the force between the quark q and the diquark D inside an excited baryon is essentially the same as the one between q and the antiquark $\bar{q}$ inside an excited meson. Thus the approximate spin-flavor independence of hadronic physics expressed by $SU(6)$ symmetry is extended to $SU(6/21)$ supersymmetry[10],[14] through a symmetry between $\bar{q}$ and D, resulting in parallelism of mesonic and baryonic Regge trajectories.

3. Color Algebra and Octonions

We shall now give an algebraic justification to our remarks above. We will find all the answers in an algebra we build in terms of octonions and their split basis. The exact, unbroken color group $SU(3)^c$ is the backbone of the strong interaction. It is worthwhile to understand its role in the diquark picture more clearly.

In what follows we first give a simple description of octonion algebra (also known as Cayley algebra). Later we'll show how to build split octonion algebra that will close into a fermionic Heisenberg algebra. Split octonion algebra will then be shown to produce algebra of color forces in QCD in application to hadronic supersymmetry when the split units and their conjugates become associated with quark and antiquark fields, respectively.

An octonion x is a set of eight real numbers

$$x = (x_0, x_1, \ldots, x_7) = x_0e_0 + x_1e_1 + \ldots + x_7e_7 \tag{4}$$

that are added like vectors and multiplied according to the rules

$$e_0 = 1, \quad e_0e_i = e_ie_0 = e_i, \quad i = 0, 1, \ldots, 7 \tag{5}$$

$$e_\alpha e_\beta = -\delta_{\alpha\beta} + \epsilon_{\alpha\beta\gamma}e_\gamma. \quad \alpha, \beta, \gamma = 1, 2, \ldots, 7 \tag{6}$$

where e_0 is the multiplicative unit element and e_α's are the imaginary octonion units. The structure constants $\epsilon_{\alpha\beta\gamma}$ are completely antisymmetric and take the value 1 for combinations

$$\epsilon_{\alpha\beta\gamma} = (165), (257), (312), (471), (543), (624), (736) \tag{7}$$

Note that summation convention is used for repeated indices.

The octonion algebra $\mathcal{C}$ is an algebra defined over the field $\mathbf{Q}$ of rational numbers, which as a vector space over $\mathbf{Q}$ has dimension 8.

We shall now give reasons for incorporation of the octonion algebra for hadronic physics, showing only they through their split octonionic parts one can provide the correct description of the color algebra in hadrons. Later in another publication we shall show[15] a previously unknown multiplication rules for octonions by producing a wheel that allows generalized multiplication rules for doublets and triplets of octonionic units.

First, the reasons: Two of the colored quarks in the baryon combine into an anti-triplet $\mathbf{3} \times \mathbf{3} = \bar{\mathbf{3}} + (\mathbf{6})$, and in a nucleon $\mathbf{3} \times \bar{\mathbf{3}} = \mathbf{1} + (\mathbf{8})$. The $(\mathbf{6})$ partner of the diquark and the $(\mathbf{8})$ partner of the nucleon do not exist. In hadron dynamics the only color combinations to consider are $\mathbf{3} \times \mathbf{3} \to \bar{\mathbf{3}}$ and $\bar{\mathbf{3}} \times \mathbf{3} \to \mathbf{1}$. These relations imply the existence of split octonion units u_i defined below through a representation of the Grassmann algebra $\{u_i, u_j\} = 0$, $i = 1, 2, 3$. What is a bit strange is that operators u_i, unlike ordinary fermionic operators, are not associative. We also have $\frac{1}{2}[u_i, u_j] = \epsilon_{ijk}\, u_k^*$. The Jacobi identity does not hold since $[u_i, [u_j, u_k]] = -ie_7 \neq 0$, where e_7, anticommute with u_i and u_i^*.

The behavior of various states under the color group are best seen if we use split octonion units defined by[16]

$$u_0 = \frac{1}{2}(1 + ie_7), \qquad u_0^* = \frac{1}{2}(1 - ie_7) \tag{8}$$

$$u_j = \frac{1}{2}(e_j + ie_{j+3}), \quad u_j^* = \frac{1}{2}(e_j - ie_{j+3}), \quad j = 1, 2, 3 \tag{9}$$

The automorphism group of the octonion algebra is the 14-parameter exceptional group G_2. The imaginary octonion units $e_\alpha (\alpha = 1, ..., 7)$ fall into its 7-dimensional representation.

Under the $SU(3)^c$ subgroup of G_2 that leaves e_7 invariant, u_0 and u_0^* are singlets, while u_j and u_j^* correspond, respectively, to the representations $\mathbf{3}$ and $\bar{\mathbf{3}}$. The multiplication table can now be written in a manifestly $SU(3)^c$ invariant manner (together with the complex conjugate equations):

$$u_0^2 = u_0, \qquad u_0 u_0^* = 0 \tag{10}$$

$$u_0 u_j = u_j u_0^* = u_j, \qquad u_0^* u_j = u_j u_0 = 0 \tag{11}$$

$$u_i u_j = -u_j u_i = \epsilon_{ijk} u_k^* \tag{12}$$

$$u_i u_j^* = -\delta_{ij} u_0 \tag{13}$$

where ϵ_{ijk} is completely antisymmetric with $\epsilon_{ijk} = 1$ for ijk = 123, 246, 435, 651, 572, 714, 367; and zero otherwise. Here, one sees the virtue of octonion multiplication. If we consider the direct products

$$C: \quad \mathbf{3} \otimes \bar{\mathbf{3}} = \mathbf{1} + \mathbf{8} \tag{14}$$

$$G: \quad \mathbf{3} \otimes \mathbf{3} = \bar{\mathbf{3}} + \mathbf{6} \tag{15}$$

for $SU(3)^c$, then these equations show that octonion multiplication gets rid of $\mathbf{8}$ in $\mathbf{3} \otimes \bar{\mathbf{3}}$, while it gets rid of $\mathbf{6}$ in $\mathbf{3} \otimes \mathbf{3}$. Combining Eq.(12) and Eq.(13) we find

$$(u_i u_j) u_k = -\epsilon_{ijk} u_0^* \tag{16}$$

Thus the octonion product leaves only the color part in $\mathbf{3} \otimes \bar{\mathbf{3}}$ and $\mathbf{3} \otimes \mathbf{3} \otimes \mathbf{3}$, so that it is a natural algebra for colored quarks.

For convenience we now produce the following multiplication table for the split octonion units:

	u_0	u_0^*	u_k	u_k^*
u_0	u_0	0	u_k	0
u_0^*	0	u_0^*	0	u_k^*
u_j	0	u_j	$\epsilon_{jki} u_i^*$	$-\delta_{jk} u_0$
u_j^*	u_j^*	0	$-\delta_{jk} u_0^*$	$\epsilon_{jki} u_i$

It is worth noting that u_i and u_j^* behave like fermionic annihilation and creation operators:

$$\{u_i, u_j\} = \{u_i^*, u_j^*\} = 0, \quad \{u_i, u_k^*\} = -\delta_{ik} \tag{17}$$

For more recent reviews on octonions and nonassociative algebras we refer to papers by Okubo[17], Baez[18] and Catto[19].

The quarks, being in the triplet representation of the color group $SU(3)^c$, are represented by the local fields $q_\alpha^i(x)$, where $i = 1, 2, 3$ is the color index and α the combined spin-flavor index. Antiquarks at point y are color antitriplets $q_\beta^i(y)$. Consider the two-body systems

$$C_{\alpha j}^{\beta i} = q_\alpha^i(x_1)\bar{q}_\beta^j(x_2) \tag{18}$$

$$G_{\alpha\beta}^{ij} = q_\alpha^i(x_1)q_\beta^j(x_2) \tag{19}$$

so that C is either a color singlet or color octet, while G is a color antitriplet or a color sextet. Now C contains meson states that are color singlets and hence observable. The octet $q-\bar{q}$ state is confined and not observed as a scattering state. In the case of two-body G states, the antitriplets are diquarks which, inside a hadron can be combined with another triplet quark to give observable, color singlet, three-quark baryon states. The color sextet part of G can only combine with a third quark to give unobservable color octet and color decuplet three-quark states. Hence the hadron dynamics is such that the **8** part of C and the **6** part of G are suppressed. This can best be achieved by the use of the above octonion algebra[20]. The dynamical suppression of the octet and sextet states in Eq.(18) and Eq.(19) is , therefore, automatically achieved. The split octonion units can be contracted with color indices of triplet or antitriplet fields. For quarks and antiquarks we can define the "transverse" octonions (calling u_0 and u_0^* longitidunal units)

$$q_\alpha = u_i q_\alpha^i = \mathbf{u} \cdot \mathbf{q}_\alpha, \quad \bar{q}_\beta = u_i^\dagger \bar{q}_\beta^j = -\mathbf{u}^* \cdot \bar{\mathbf{q}}_\beta \tag{20}$$

We find

$$q_\alpha(1)\bar{q}_\beta(2) = u_0 \mathbf{q}_\alpha(1) \cdot \bar{\mathbf{q}}_\beta(2) \tag{21}$$

$$\bar{q}_\alpha(1)q_\beta(2) = u_0^* \bar{\mathbf{q}}_\alpha(1) \cdot \mathbf{q}_\beta(2) \tag{22}$$

$$G_{\alpha\beta}(12) = q_\alpha(1)q_\beta(2) = \mathbf{u}^* \cdot \mathbf{q}_\alpha(1) \times \mathbf{q}_\beta(2) \tag{23}$$

$$G_{\beta\alpha}(21) = q_\beta(2)q_\alpha(1) = \mathbf{u}^* \cdot \mathbf{q}_\beta(2) \times \mathbf{q}_\alpha(1) \tag{24}$$

Because of the anticomutativity of the quark fields, we have

$$G_{\alpha\beta}(12) = G_{\beta\alpha}(21) = \frac{1}{2}\{q_\alpha(1), q_\beta(2)\} \tag{25}$$

If the diquark forms a bound state represented by a field $D_{\alpha\beta}(x)$ at the center-of-mass location x

$$x = \frac{1}{2}(x_1 + x_2) \tag{26}$$

when x_2 tends to x_1 we can replace the argument by x, and we obtain

$$D_{\alpha\beta}(x) = D_{\beta\alpha}(x) \tag{27}$$

so that the local diquark field must be in a symmetric representation of the spin-flavor group. If the latter is taken to be $SU(6)$, then $D_{\alpha\beta}(x)$ is in the 21-dimensional symmetric representation, given by

$$(\mathbf{6} \otimes \mathbf{6})_s = \mathbf{21} \tag{28}$$

If we denote the antisymmetric 15 representation by $\Delta_{\alpha\beta}$, we see that the octonionic fields single out the 21 diquark representation at the expense of $\Delta_{\alpha\beta}$. We note that without this color algebra supersymmetry would give antisymmetric configurations as noted by Salam and Strathdee[21] in their possible supersymmetric generalization of hadronic supersymmetry. Using the nonsingular part of the operator product expansion we can write

$$\tilde{G}_{\alpha\beta}(\mathbf{x}_1, \mathbf{x}_2) = D_{\alpha\beta}(x) + \mathbf{r} \cdot \mathbf{\Delta}_{\alpha\beta}(x) \tag{29}$$

The fields $\Delta_{\alpha\beta}$ have opposite parity to $D_{\alpha\beta}$; $\mathbf{r}$ is the relative coordinate at time t if we take $t = t_1 = t_2$. They play no role in the excited baryon which becomes a bilocal system with the 21- dimensional diquark as one of its constituents.

Now consider a three-quark system at time t. The c.m. and relative coordinates are

$$\mathbf{R} = \frac{1}{\sqrt{3}}(\mathbf{r}_1 + \mathbf{r}_2 + \mathbf{r}_3) \tag{30}$$

$$\vec{\rho} = \frac{1}{\sqrt{6}}(2\mathbf{r}_3 - \mathbf{r}_1 - \mathbf{r}_2) \tag{31}$$

$$\mathbf{r} = \frac{1}{\sqrt{2}}(\mathbf{r}_1 - \mathbf{r}_2) \tag{32}$$

giving

$$\mathbf{r}_1 = \frac{1}{\sqrt{3}}\mathbf{R} - \frac{1}{\sqrt{6}}\vec{\rho} + \frac{1}{\sqrt{2}}\mathbf{r} \tag{33}$$

$$\mathbf{r}_2 = \frac{1}{\sqrt{3}}\mathbf{R} - \frac{1}{\sqrt{6}}\vec{\rho} - \frac{1}{\sqrt{2}}\mathbf{r} \tag{34}$$

$$\mathbf{r}_3 = \frac{1}{\sqrt{3}}\mathbf{R} + \frac{2}{\sqrt{6}}\vec{\rho} \tag{35}$$

The baryon state must be a color singlet, symmetric in the three pairs (α, x_1), (β, x_2), (γ, x_3). We find

$$(q_\alpha(1)q_\beta(2))q_\gamma(3) = -u_0^* F_{\alpha\beta\gamma}(123) \tag{36}$$

$$q_\gamma(3)(q_\alpha(1)q_\beta(2)) = -u_0 F_{\alpha\beta\gamma}(123) \tag{37}$$

so that

$$-\frac{1}{2}\{\{q_\alpha(1), q_\beta(2)\}, q_\gamma(3)\} = F_{\alpha\beta\gamma}(123) \tag{38}$$

The operator $F_{\alpha\beta\gamma}(123)$ is a color singlet and is symmetrical in the three pairs of coordinates. We have

$$F_{\alpha\beta\gamma}(123) = B_{\alpha\beta\gamma}(\mathbf{R}) + \vec{\rho} \cdot \mathbf{B}'(\mathbf{R}) + \mathbf{r} \cdot \mathbf{B}''(\mathbf{R}) + C \tag{39}$$

where C is of order two and higher in $\vec{\rho}$ and $\mathbf{r}$. Because $\mathbf{R}$ is symmetric in $\mathbf{r}_1$, $\mathbf{r}_2$ and $\mathbf{r}_3$, the operator $B_{\alpha\beta\gamma}$ that creates a baryon at $\mathbf{R}$ is totally symmetrical in its flavor-spin indices. In the $SU(6)$ scheme it belongs to the (56) representation. In the bilocal $q - D$ approximation we have $\mathbf{r} = 0$ so that $F_{\alpha\beta\gamma}$

is a function only of $\mathbf{R}$ and $\vec{\rho}$ which are both symmetrical in $\mathbf{r}_1$ and $\mathbf{r}_2$. As before, $\mathbf{B}'$ belongs to the orbitally excited 70^- representation of $SU(6)$. The totally antisymmetrical (20) is absent in the bilocal approximation. It would only appear in the trilocal treatment that would involve the 15-dimensional diquarks. Hence, if we use local fields, any product of two octonionic quark fields gives a (21) diquark

$$q_\alpha(\mathbf{R})q_\beta(\mathbf{R}) = D_{\alpha\beta}(\mathbf{R}) \tag{40}$$

and any nonassociative combination of three quarks, or a diquark and a quark at the same point give a baryon in the 56^+ representation:

$$(q_\alpha(\mathbf{R})q_\beta(\mathbf{R}))q_\gamma(\mathbf{R}) = -u_0^* B_{\alpha\beta\gamma}(\mathbf{R}) \tag{41}$$

$$q_\alpha(\mathbf{R})(q_\beta(\mathbf{R})q_\gamma(\mathbf{R})) = -u_0 B_{\alpha\beta\gamma}(\mathbf{R}) \tag{42}$$

$$q_\gamma(\mathbf{R})(q_\alpha(\mathbf{R})q_\beta(\mathbf{R})) = -u_0 B_{\alpha\beta\gamma}(\mathbf{R}) \tag{43}$$

$$(q_\gamma(\mathbf{R})q_\alpha(\mathbf{R}))q_\beta(\mathbf{R}) = -u_0^* B_{\alpha\beta\gamma}(\mathbf{R}) \tag{44}$$

The bilocal approximation gives the $(35 + 1)$ mesons and the 70^- baryons with $\ell = 1$ orbital excitation.

If we consider a (28×28) octonionic matrix

$$Z = \begin{pmatrix} u_0 M & u_o B & \mathbf{u}\cdot\mathbf{Q} \\ u_0 B^\dagger & u_0 N & \mathbf{u}\cdot\mathbf{D}^* \\ \epsilon\mathbf{u}^*\cdot\mathbf{Q}^\dagger & \epsilon\mathbf{u}^*\cdot\mathbf{D}^\dagger & u_0^* L \end{pmatrix} \tag{45}$$

here ϵ can be 1, -1 or 0. M and N are respectively 6×6 and 21×21 hermitian matrices, B a regular 6×21 matrix, $\mathbf{u}\cdot\mathbf{Q}$ a 6×1 column matrix, $\mathbf{u}\cdot\mathbf{D}^*$ a 21×1 column matrix, and L a 1×1 scalar. Such matrices close under anticommutator operations for $\epsilon = 1$. Matrices iZ close under commutator operations. In either case, they don't satisfy the Jacobi identity. But for $\epsilon = 0$, when the algebra is no longer semi-simple, the Jacobi identity is satisfied and we obtain a hadronic superalgebra which is an extension of the algebra $SU(6/21)$. Its automorphism group includes $SU(6) \times SU(21) \times SU(3)^c$. Thus color is automatically incorporated.

4. QCD Justification of $U(6/21)$ Supersymmetry and Its Breaking

We shall first discuss the validity domain of $SU(6/21)$ supersymmetry. The diquark structure with spin $s = 0$ and $s = 1$ emerges in inelastic inclusive lepton-baryon collisions with high impact parameters that excite the baryon rotationally, resulting in inelastic structure functions based on point-like quarks and diquarks instead of three point-like quarks, In this case both mesons and baryons are bilocal with large separation of constituents.

Also, there is a symmetry between color antitriplet diquarks with $s = 0$ and $s = 1$ and color antitriplet antiquarks with $s = \frac{1}{2}$. This is only possible if the force between quark q and antiquark $\bar{q}$, and also between q and diquark D is mediated by a zero spin object that sees no difference between the spins of $\bar{q}$ and D. The object can be in color states that are either singlet or octet since q and D are both triplets. Such an object is provided by scalar flux tubes of gluons that dominate over the one gluon exchange at large distances. Various strong coupling approximations to QCD, like lattice gauge theory[22],[23], 't Hooft's $\frac{1}{N}$ approximation[24] when N, the number of colors, is very large, or the elongated bag model[13] all give a linear potential between widely separated quarks and an effective string that approximates the gluon flux tube. In such a theory it is energetically favorable for the three quarks in a baryon to form a linear structure with a quark in the middle and two at the ends, or, for high rotational excitation, a bilocal linear structure (diquark) at one end and a quark at the other end. In order to illustrate these points we start with the suggestion of Johnson and Thorn[13] that the string-like hadrons may be pictured as vortices of color flux lines which terminate on concentration of color at the end points. The color flux connecting opposite ends is the same for mesons and baryons giving an explanation for the same slope of meson and baryon trajectories[10].

To construct a solution which yields a maximal angular momentum for a fixed mass we consider a bag with elongated shape rotating about the center of mass with an angular frequency ω. Its ends have the maximal velocity allowed, which is the speed of light ($c = 1$). Thus, a given point inside the bag, at a distance r from the axis of rotation moves with a velocity

$$v = \vec{\omega} \cdot \vec{r} = \frac{2r}{L} \tag{46}$$

where L is the length of the string. In this picture the bag surface will be fixed by balancing the gluon field pressure against the confining vacuum pressure B, which (in analogy to electrodynamics) can be written in the form

$$\frac{1}{2}\sum_{\alpha=1}^{8}(E_\alpha^2 - B_\alpha^2) = B \tag{47}$$

Using Gauss' law, the color electric field E through the flux tube connecting the color charges at the ends of the string is given by

$$\int \vec{E}_\alpha \cdot d\vec{S} = E_\alpha A = g\frac{1}{2}\lambda_\alpha \tag{48}$$

where$A(r)$ is the cross-section of the flux tube at distance r from the center and $g\frac{1}{2}$ is the color electric charge which is the source of E_α. By analogy with classical electrodynamics the color magnetic field $\vec{B}_\alpha(r)$ associated with the rotation of the color electric field is

$$\vec{B}_\alpha(r) = \vec{v}(r) \times \vec{E}_\alpha(r) \tag{49}$$

at a point moving with a velocity $\vec{v}(r)$. For the absolute values this implies

$$B_\alpha = v\, E_\alpha \tag{50}$$

because $\vec{v}(r)$ is perpendicular to $\vec{E}_\alpha(r)$. Using last three equations together with

$$< \sum_{\alpha=1}^{8}(\frac{1}{2}\lambda_\alpha)^2 >= \frac{4}{3} \tag{51}$$

for the $SU(3)^c$ triplet in Eq.(47) we obtain for the crosssection of the bag

$$A(r) = \sqrt{\frac{2}{3B}}\, g\, \sqrt{1 - v^2} \tag{52}$$

which shows the expected Lorentz contraction.

The total energy E of the bag

$$E = E_q + E_G + BV \tag{53}$$

is the sum of the quark energy E_q, the gluon field energy E_G and the volume energy of the bag, BV. Because the quarks at the ends move with the a speed close to speed of light, their energy is simply given by

$$E_q = 2p \tag{54}$$

where p is the momentum of a quark, a diquark or an antiquark, respectively. For the gluon energy, by analogy with electrodynamics, one obtains from Eqs.(48-50) the result

$$\begin{aligned} E_G &= \frac{1}{2}\int d^3x \sum_{\alpha=1}^{8}(E_\alpha^2 + B_\alpha^2) \\ &= \sqrt{\frac{2}{3}} g\sqrt{B}L \int_0^1 dv \frac{1+v^2}{\sqrt{1-v^2}} = \sqrt{\frac{2}{3}} g\sqrt{B}L\frac{3\pi}{4} \end{aligned} \tag{55}$$

and for the volume energy

$$\begin{aligned} BV &= 2B\int_0^{\frac{L}{2}} A(r)\,dr = 2B\int_0^1 \sqrt{\frac{2}{3B}} g\sqrt{1-v^2}\frac{L}{2}dv \\ &= \sqrt{\frac{2}{3}} g\sqrt{B}L\frac{\pi}{4} = \frac{BA(0)L\pi}{4} \end{aligned} \tag{56}$$

It is obvious from Eq.(55) that the gluon field energy is proportional to the length L of the bag. The gluon field energy and the volume energy of the bag together correspond to a linear rising potential of the form

$$V(L) = E_G + BV = bL \tag{57}$$

where

$$b = \sqrt{\frac{2B}{3}} g\pi \tag{58}$$

The total angular momentum J of this classical bag is the sum of the angular momenta of the quarks at the two ends

$$J_q = pL \tag{59}$$

and the angular momentum J_G of the gluon field. From Eq.(49) we get

$$\vec{E}_\alpha \times \vec{B}_\alpha = \vec{v}E_\alpha^2 \tag{60}$$

for the momentum of the gluon field, and hence

$$\begin{aligned} J_G &= |\int_{bag} d^3\vec{r}\sum_{\alpha=1}^{8} \vec{r}\times(\vec{E}_\alpha \times \vec{B}_\alpha)| = 2\int_0^{\frac{L}{2}} dr A(r) r v E_\alpha^2 \\ &= \frac{16}{3L}g^2\int_0^{\frac{L}{2}} \frac{r^2 dr}{A(r)} = \sqrt{\frac{2}{3}} g\sqrt{B}L^2\frac{\pi}{4} \end{aligned} \tag{61}$$

where we have used Eq.(46) and Eq.(48) in the third step. We can now express the total energy of the bag in terms of angular momenta. Putting these results back into expressions for E_q and E_G, we arrive at

$$E_q = \frac{2J_q}{L}, \qquad E_G = \frac{3J_G}{L} \tag{62}$$

so that the bag energy now becomes

$$\begin{aligned} E &= \frac{2J_q}{L} + \frac{3J_G}{L} + \sqrt{\frac{2B}{3}} Lg\frac{\pi}{L} = \frac{2J_q + 4J_G}{L} \\ &= \frac{2(J + J_G)}{L} = \frac{1}{L}(2J + \sqrt{\frac{2}{3}} g\sqrt{B}L^2\frac{\pi}{2}) \end{aligned} \tag{63}$$

Minimizing the total energy for a fixed angular momentum with respect to the length of the bag, $\frac{dE}{dt} = 0$ gives the relation

$$-\frac{2J}{L^2} + \sqrt{\frac{2}{3}} g\sqrt{B}\frac{\pi}{2} = 0 \tag{64}$$

so that

$$L^2 = \frac{4J}{g\pi}\sqrt{\frac{3}{2B}} \tag{65}$$

Re-inserting this into Eq.(63) we arrive at

$$E = 2\sqrt{Jg\pi}(\frac{2B}{3})^{\frac{1}{4}} \tag{66}$$

or

$$J = (\sqrt{\frac{3}{2B}}\frac{1}{4\pi g})E^2 = (\sqrt{\frac{3}{2B}}\frac{1}{8\pi^{\frac{3}{2}}}\frac{1}{\sqrt{\alpha_s}})E^2 = \alpha^{'}(0)M^2 \tag{67}$$

where $M = E$ and we used $\alpha_s = \frac{g^2}{4\pi}$, the unrationalized color gluon coupling constant. We can now let $\alpha^{'}(0)$ defined by the last equation which is the slope of the Regge trajectory as

$$\alpha^{'}(0) = \sqrt{\frac{3}{2B}}\frac{1}{8\pi^{\frac{3}{2}}}\frac{1}{\sqrt{\alpha_s}} = \frac{1}{4b} \tag{68}$$

where b was defined in Eq.(57).

The parameters B and α_s have been determined[25],[26] using the experimental information from the low lying hadron states: $B^{\frac{1}{4}} = 0.146\ GeV$ and $\alpha_s = 0.55\ GeV$. If we use these values in Eq.(68) we find

$$\alpha^{'}(0) = 0.88\ (GeV)^{-2} \tag{69}$$

in remarkable agreement with the slope determined from experimental data which is about $0.9\ (GeV)^{-2}$.

The total phenomenological non-relativistic potential then is the well known superposition of the Coulomb-like and confining potentials $V(r) = \frac{a}{r} + br$ where $r = |\vec{r}_1 - \vec{r}_2|$ is the distance between q and $\bar{q}$ in a meson, or betwee q and D in a baryon with high angular momentum. This is verified in lattice QCD to a high degree of accuracy[27] ($a = \frac{-c\alpha_c}{r}$, where c is the color factor and α_c the strong coupling strength).

It is interesting to know that all this is related very closely to the dual strings. Indeed we can show that the slope given in Eq.(68) is equivalent to the dual string model formula for the slope if we associate the "proper tension" in the string with the proper energy per unit length of the color flux tube and the volume.

By proper energy per unit length we mean the energy per unit length at a point in the bag evaluated in the rest system of that point. This will be

$$T_0 = \frac{1}{2}\sum_{\alpha} E_\alpha^2 A_0 + BA_0 \tag{70}$$

The fact that $\frac{1}{2}\sum_\alpha E_\alpha^2 = B$ in the rest sytem gives

$$T_0 = 2BA_0 \tag{71}$$

where A_0 is the cross-sectional area of the bag. If in Eq.(52) we let $A = A_0$ when $v = 0$, then using

$$A_0 = \sqrt{\frac{2}{3B}}g \tag{72}$$

we find

$$T_0 = 2\sqrt{\frac{2}{3}}g\sqrt{B} = 4\sqrt{\frac{2\pi}{3}}\sqrt{\alpha_s}\sqrt{B} \tag{73}$$

for the proper tension. In the dual string the slope and proper tension are related by the formula[28]

$$T_0 = \frac{1}{2\pi\alpha'} \tag{74}$$

so the slope is

$$\alpha' = \frac{1}{8}\sqrt{\frac{3}{2}}\frac{1}{\pi^{\frac{3}{2}}}\frac{1}{\sqrt{\alpha_s}}\frac{1}{\sqrt{B}} \tag{75}$$

which is identical to the earlier formula we produced in Eq.(68).

It would appear from Eq.(73) that the ratio of volume to field energy would be one-to-one in one space dimension in contrast to the result one-to-three which holds for a three dimensional bag[29]. However, the ratio one-to-one is true only in the rest system at a point in the bag, and each position along the x-axis is of course moving with a different velocity. Indeed we see from Eq.(55) and Eq.(56) that the ratio of the total volume energy to the total field energy is given by one-to three in conformity with the virial theorem[29].

In the string model of hadrons we have $E^2 \sim J$ between the energy and the angular momentum of the rotating string. If we denote by $\rho(r)$ the mass density of the string, and by v and ω its linear and angular velocities, respectively, the energy and the angular momentum of the rotating string are given by

$$E = 2\int \frac{\rho(r)}{\sqrt{1-\omega^2 r^2}}dr = \frac{2}{\omega}\int_0^1 \frac{\rho(v)}{\sqrt{1-v^2}}dv \tag{76}$$

and

$$J = 2\int \frac{\rho(r)}{\sqrt{1-\omega^2 r^2}}r^2\omega dr = \frac{2}{\omega^2}\int_0^1 \frac{\rho(v)}{\sqrt{1-v^2}}v^2 dv \tag{77}$$

and hence the relation

$$E^2 \propto J. \tag{78}$$

If the string is loaded with mass points at its ends, they no longer move with speed of light, however, the above relation still holds approximately for the total energy and angular momentum of the loaded string.

We now look at various ways of partitioning of the total angular momentum into two subsystems. Figures (a), (b) and (c) show the possible configurations of three quarks in a baryon.

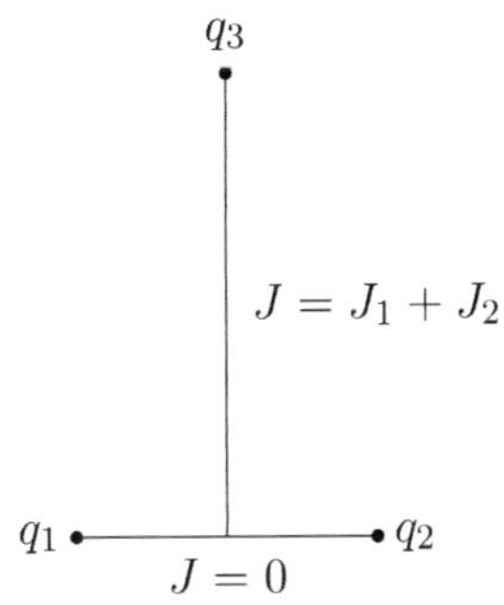

Fig.(a)

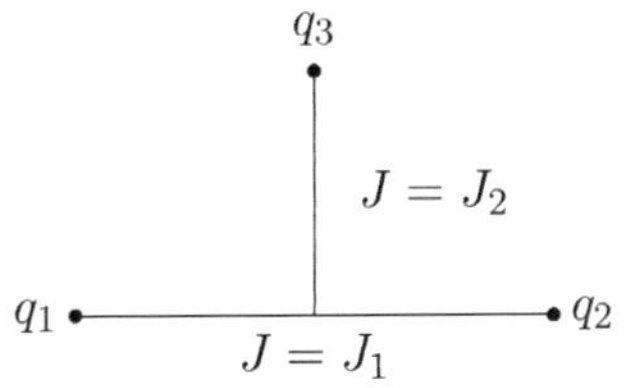

Fig.(b)

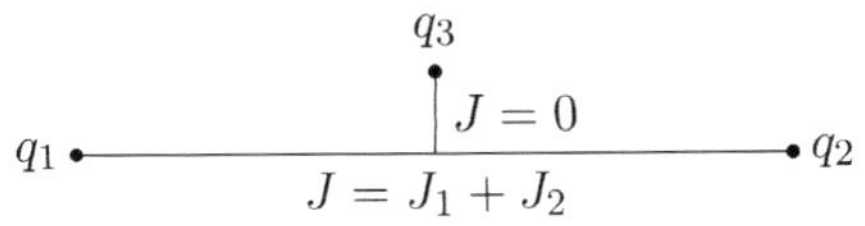

Fig.(c)

If we put the proportionality constant in Eq.(78) equal to unity, then the naive evaluation of energies yield

$$E^2 = J_1 + J_2 = E_1^2 + E_2^2 \leq (E_1 + E_2)^2 = E'^2 \tag{79}$$

where E and E' denote the energies corresponding to figures (a) or (c). In the case of figure (b), J_1, J_2 are the angular momenta corresponding to the energies E_1 and E_2 of the subsystems. The equality in Eq.(79) holds only if E_1 or E_2 is zero. Therefore for each fixed total angular momentum its most unfair partition into two subsystems gives us the lowest energy levels, and its more or less fair partition gives rise to energy levels on daughter trajectories. Hence on the leading baryonic trajectory we have a quark-diquark structure (Fig.(a)), or a linear molecule structure (Fig.(c)). On the other hand on low-lying trajectories we have more or less symmetric ($J_1 \sim J_2$) configuration of quarks.

In Table 3, we give the flavor-spin content for three quark and quark-diquark baryons that correspond to $Fig.(a)$ and $Fig.(c)$ shown above.

Since high J hadronic states on leading Regge trajectories tend to be bilocal with large separation of their constituents, they fulfill all the conditions for supersymmetry between $\bar{q}$ and D. Then the only difference between the energies of $(q\bar{q})$ mesons and (qD) baryons comes from the different masses of their constituents, namely $m_q = m_{\bar{q}} = m$, and $m_D \sim 2m$. For high J this is the main source of symmetry breaking which is spin independent. We will show how we can obtain sum rules from this breaking. The part of the mass operator that gives rise to this splitting is a diagonal element of $U(6/21)$ that commutes with $SU(6)$.

Let us now consider the spin dependent breaking of $U(6/21)$. For low J states the (qD) system becomes trilocal(qqq), the flux tube degenerates to a single gluon propagator that gives spin-dependent forces in addition to the Coulomb term $\frac{a}{r}$. In this case we have the regime studied by de Rujula, Georgi and Glashow, where the breaking is due to hyperfine splitting caused by the exchange of single gluons that have spin 1. These mass splittings give rise to different intercept of the Regge trajectories are given by

$$\Delta m_{12} = k\frac{\vec{\mathbf{S}}_1 \cdot \vec{\mathbf{S}}_2}{m_1 m_2}, \quad k = |\psi(0)|^2 \tag{80}$$

both for baryons and mesons at high energies. But at low energies the baryon becomes a trilocal object (with three quarks) and the mass splitting is given by

$$\Delta m_{123} = \frac{1}{2}k(\frac{\vec{\mathbf{S}}_1 \cdot \vec{\mathbf{S}}_2}{m_1 m_2} + \frac{\vec{\mathbf{S}}_2 \cdot \vec{\mathbf{S}}_3}{m_2 m_3} + \frac{\vec{\mathbf{S}}_3 \cdot \vec{\mathbf{S}}_1}{m_3 m_1}) \tag{81}$$

where m_1, m_2 and m_3 are the masses of the three different quark constituents.

The element of $U(6/21)$ that give rise to such splittings is a diagonal element of its $U(21)$ subgroup and gives rise to $s(s+1)$ terms that behave like an element of the (405) representation of $SU(6)$ in the $SU(6)$ mass formulae. The splitting of isospin multiplets is due to a symmetry breaking element in the (35) representation of $SU(6)$. Hence all symmetry breaking terms are in the adjoint representation of $U(6/21)$. If we restrict ourselves to the non-strange sector of hadrons with approximate $SU(4)$ symmetry, effective supersymmetry will relate the splitting in m^2 between the Δ ($s = \frac{3}{2}, I = \frac{3}{2}$) and N ($s = \frac{1}{2}, I = \frac{1}{2}$) to the splitting between ω ($s = 1, I = 0$) and π ($s = 0, I = 1$) so that

$$m_\Delta^2 - m_N^2 \simeq m_\omega^2 - m_\pi^2 \tag{82}$$

which is satisfied to within 5%.

For classification of supergroups including $SU(m/n)$ we refer to a paper by Viktor Kac[30].

5. Linear Mass Formulae

Based on the flavor $SU(3)$ and its breaking into its $SU(2) \times U(1)$ maximal subgroup of isospin and hypercharge, in 1962 the Gell-Mann-Okubo[31],[32] mass formula illuminated the low lying hadronic spectrum. It led to the pseudoscalar mass formula. The mass formula for the vector mesons presented a more delicate problem since the isospin singlet members of the nine vector mesons, namely the physical ω and ϕ were mixtures of octet and singlet states, involving a mixing angle θ_V as a new parameter. A year later Okubo proposed[33] a model for the determination of this mixing angle by requiring the nine vector mesons to fit into a 3×3 matrix. The group theoretic interpretation of ideal mixing followed soon after[34] with the enlargement of $SU(3)$ to $SU(3)_q \times SU(3)_{\bar{q}}$, one $SU(3)$ being associated with the quarks and the other $SU(3)$ with the antiquarks that are constituents of the vector mesons. Then the nonet corresponds to the representation $(3,3)$ of this group. Since the u and d quarks are much lighter than the strange quark s, the $SU(2) \times SU(2)$ subgroup is not badly broken, so that we must decompose with respect to the subgroup $SU(2)_q \times U(1)_q \times SU(2)_{\bar{q}} \times U(1)_{\bar{q}}$ by using the (I, Y) labels for each $SU(2) \times U(1)$. They are shown in the table 1 below.

With respect to the diagonal $SU(3)$ subgroup the hypercharge Y is the sum of Y_q and $Y_{\bar{q}}$ while the isospin I is zero for ω and ϕ, one for ρ, and $\frac{1}{2}$ for K^* and $\bar{K}^*$. Now, Okubo's octet breaking hypothesis involves the octet-singlet mixture given by

$$K = I(I+1) - \frac{Y^2}{4}. \tag{83}$$

For the nonet the energy breaking requires the combination

$$E = E_0 + a(K_q + K_{\bar{q}}) + bK. \tag{84}$$

We find the following assignments shown in the table below:

Particle	I_q	$I_{\bar{q}}$	I	Y_q	$Y_{\bar{q}}$	Y	K_q	$K_{\bar{q}}$	K_q+$K_{\bar{q}}$	K
ω	1/2	1/2	0	1/3	-1/3	0	13/18	13/18	13/9	0
ρ	1/2	1/2	1	1/3	-1/3	0	13/18	13/18	13/9	2
K^*	1/2	0	1/2	1/3	2/3	1	13/18	-1/9	11/18	1/2
$\bar{K}^*$	0	1/2	1/2	-2/3	-1/3	-1	-1/9	13/18	11/18	1/2
ϕ	0	0	0	-2/3	2/3	0	-1/9	-1/9	-2/9	0

Table 1. Particle assignments

Note that the sum $K_q + K_{\bar{q}}$ of the two octet breakings gives equal spacing for the energy levels and degeneracy for ω and ρ. It was shown in Okubo's paper that the rest energy breaking formula leads to a quadratic mass formula when the energy differences are large with respect to the mean energy as in the case of the pseudoscalar mesons and to a linear mass formula when the ratio of the energy splittings to the mean energy is small as in the case of baryons. The vector mesons being nearer in mass to the baryons than the pseudoscalar mesons we can use the linear mass formula as also suggested by the value of the mixing angle being nearer perfect mixing in this case. Then we get

$$m_\omega = \mu + \frac{13}{9}a, \tag{85}$$

$$m_\rho = \mu + \frac{13}{9}a + 2b, \tag{86}$$

$$m_{K^*} = m_{\bar{K}^*} = \mu + \frac{11}{18}a + \frac{1}{2}b, \tag{87}$$

$$m_\phi = \mu - \frac{2}{9}a, \tag{88}$$

leading to the mass sum rule:

$$m_\omega + m_\rho = 2(2m_{K^*} - m_\phi). \tag{89}$$

With the choice

$$\mu = 987.9\ MeV, \quad a = -142.1\ MeV, \quad b = -3.6\ MeV \tag{90}$$

we find the following masses as compared to experimental values given in Particle Data Table:[35]

Particle	our result	experiment
m_ω	782.74	782.65
m_ρ	775.54	775.49
$m_{K^*} = m_{\bar{K}^*}$	899.36	891.66
m_ϕ	1020	1019.46

Table 2. Calculated masses vs. experiment

and the mass formula Eq.(89) gives deviations of less than one percent with above choices. The Eq.(89) which we have derived here using purely the $SU(3) \times SU(3)$ group theoretical assignments shown in Table1 is usually written as the squared mass formula

$$4m_K^2 = 2m_\phi^2 + m_\omega^2 + m_\rho^2 \tag{91}$$

for which the deviations are much larger than the linear formula.

It is also important to note that the quantum numbers of Table 1 which forbid the the decay of the ϕ into pions, or more generally of the $s\bar{s}$ system into systems involving u and d quarks is consistent with the OZI rule. This rule must be violated in QCD through gluonic intermediate states and yet it is surprisingly well verified, reinforcing the symmetry breaking chain that gives Eq.(84).

The diquark structure in the baryon will cause the form factors in lepton-baryon scattering to deviate from the model with three point-quarks in the bag. Donachie et al.[36] and Fredrickson et al.[37] have shown that the inclusion of both spin one and spin zero diquark states in the nucleon can explain some of the deviations from the quark-parton model without including higher order gluon corrections. According to our analysis such diquark states should occur in deep inelastic scattering only at high impact parameter when the nucleon is thrown into a rotationally excited state.

The exotic meson states should occur whether one statrts from a string model[38], a relativistic oscillator model[39], or a bag model[40] within the mass range 1.2 - 2 GeV. Their experimental absence is a real puzzle. However recently some possible evidence for the formation of $s = 2,\ l = 2$ meson resonances that would fit nicely in the (405) representation of $SU(6)$ has been found in the analysis of $\gamma\gamma \to \rho\rho$ reactions by Li and Liu[41]. More experimental confirmation must be forthcoming before the issue is settled.

We have also derived the Hamiltonian of the relativistic quark model and have found new exact solutions[42] using confluent hypergeometric functions. In the process we have built generating functions for the hypergeometric series from the point of view of solutions of a differential equation and have obtained quadratic mass formulae in a remarkable agreement with experiments.

At this point we would also like to mention that Iachello and his collaborators [43] have obtained a similar mass formulae based on algebraic methods for the quadratic case. Iachello proposed that starting from a spectrum generating algebra G, one can write a chain of subalgebras

$$G \supset G' \supset G'' \supset \cdots \tag{92}$$

and that the Hamiltonian can be expanded in terms of invariants of the chain of subalgebras

$$H = \alpha C(G) + \beta C(G') + \gamma C(G'') + \cdots \tag{93}$$

where $C(G)$ denotes one of the invariants of G. They have also[44],[45] successfully applied their analysis into the study of triatomic molecules. Connections and comparisons between their formalism and ours will be exploited in another publication[46].

Table 3. Flavor-Spin Content for three quark and quark-diquark baryons

Configuration	$SU_c(3)$		$SU_f(3)$	$SU_s(2)$
q	$\square \equiv \underline{3}$		$\square \equiv \underline{3}$	$S = 1/2$
$D \equiv q^2$	$\begin{smallmatrix}\square\\\square\end{smallmatrix} \equiv \underline{\bar{3}}$		$\square\square \equiv \underline{6}$	$S = 0$ not allowed; $S = 1$
			$\begin{smallmatrix}\square\\\square\end{smallmatrix} \equiv \underline{\bar{3}}$	$S = 0$; $S = 1$ not allowed
q^3	$\begin{smallmatrix}\square\\\square\\\square\end{smallmatrix} \equiv \underline{1}$		$\square\square\square \equiv \underline{10}$	$S = 3/2$
			$\begin{smallmatrix}\square\square\\\square\end{smallmatrix} \equiv \underline{8}$	$S = 1/2$
$q - D$	$\begin{smallmatrix}\square\\\square\\\square\end{smallmatrix} \equiv \underline{1}$	$\underline{3} \times \underline{\bar{3}}$	$\begin{smallmatrix}\square\square\\\square\end{smallmatrix} \equiv \underline{8}$	$S = 1/2$
			$\begin{smallmatrix}\square\\\square\\\square\end{smallmatrix} \equiv \underline{1}$	$S = 1/2$
		$\underline{3} \times \underline{6}$	$\square\square\square \equiv \underline{10}$	$S = 1/2$
			$\square\square\square \equiv \underline{10}$	$S = 3/2$
			$\begin{smallmatrix}\square\square\\\square\end{smallmatrix} \equiv \underline{8}$	$S = 1/2$
			$\begin{smallmatrix}\square\square\\\square\end{smallmatrix} \equiv \underline{8}$	$S = 3/2$

References

[1] H. Miyazawa, *Prog. Theor. Phys.* **36** *(1966) 1266; Phys. Rev.* **170** *(1968) 1586.*
[2] P. Ramond, *Phys. Rev.* **D3** *(1971)2415.*
[3] A. Neveu and J. Schwarz, *Nucl.Phys.* **B31** *(1971) 86.*
[4] T.A. Golfand and E.P. Likhtman, *JETP Lett.***13** *(1971) 323.*
[5] D.V. Volkov and V.P. Akulov, *Phys. Lett.* **46b***(1973) 109.*
[6] J. Wess and B. Zumino, *Nucl.Phys.* **B70** *(1974) 39; Phys. Lett.* **49b** *(1974) 52.*
[7] A. de Rujula, H. Georgi and S.L. Glashow, *Phys. Rev.* **D12** *(1975) 147.*
[8] D.B. Lichtenberg, W. Namgung, E. Predazzi and J.G. Wills, *Phys. Lett.* **48** *(1982) 1653*; W. Namgung and D.B. Lichtenberg, *Lett. Nuovo Cimento* **41** *(1984) 597*; A. Martin, *Z. Phys.* **C32** *(1986) 359* and in "Symmetry in Nature" vol.II, p. 523 (Scuola Normale Superiore, Pisa, 1989).
[9] M. Anselmino, E. Predazzi, S. Ekelin, S. Fredriksson and D.B. Lichtenberg, *Rev. Mod. Phys.* **65** *(1993) 1199.*
[10] S. Catto and F. Gürsey, *Nuovo Cimento* **A 86** *(1985) 201.*
[11] T. Eguchi, *Phys. Lett. B,* **59** *(1975) 457.*
[12] I. Bars and A.J. Hanson, *Phys. Rev.***D13** *(1976) 1744.*
[13] For strings as elongated bags see K. Johnson and C.B. Thorn, *Phys. Rev.* **D13** *(1976) 1934.*
[14] S. Catto and F. Gürsey, *Nuovo Cimento* **A99** *(1988) 685.*
[15] S. Catto, *To be published.*
[16] M. Günaydin and F. Gürsey, *J. Math. Phys.* **14** *(1973) 1651.*
[17] S. Okubo, *"Introduction to Octonion and other Nonassociative Algebras." Montroll Memorial Lecture Series in Mathematical Physics, 2. Cambridge Univ. Press (1995).*
[18] J. Baez, *Bull. Am. Math. Soc.* **39(2)** *(2002) 145-205.*
[19] S. Catto, *"Exceptional Projective Geometries and Internal Symmetries" ArXiv: Hep-th/0302079*
[20] Properties of this algebra is also discussed in G. Domokos and S. Kövesi-Domokos, *J. Math. Phys.* **19** *(1978) 1477.*
[21] A. Salam and J. Strathdee, *Nucl. Phys. B***87** *(1975) 85.*
[22] K.N. Wilson, *Phys. Rev.* **D10** *(1974) 2445.*
[23] R.L. Jaffe, *Phys. Rev,* **D15** *(1977) 281.*
[24] G. 't Hooft, *Nucl. Phys.* **B72** *(1974) 461.*
[25] A. Chodos, R.L. Jaffe, K. Johnson and C.B. Thorn, *Phys. Rev.* **D10** *(1974) 2599.*
[26] T. De Grand, R.L. Jaffe, K. Johnson and J. Kiskis, *Phys. Rev.* **D12** *(1975) 2060.*
[27] S.W. Otto and J.D. Stark, *Phys. Rev. Lett.* **52** *(1984) 2328.*
[28] P. Goddard, J. Goldstone, C. Rebbi and C.B. Thorn, *Nucl. Phys.* **B56** *(1973) 109.*
[29] A. Chodos, et.al., *Phys. Rev.* **D9** *(1974) 3471.*
[30] For the classification of supergroups including U(m,n), see V.G. Kac, *Comm. Math. Phys.* **53** *(1977) 31.*
[31] M. Gell-Mann, Phys. Rev. **125**, 1062 (1962)
[32] S. Okubo, Prog. Theo. Phys. **27**, 949 (1962)
[33] S. Okubo, Phys. Lett. 4, 14(1963).
[34] F. Gürsey, T.D. Lee and M. Nauenberg, Phys.Rev.135, B467(1964).
[35] Review of Particle Data Properties.
[36] A. Donachie and P.V. Landshoff, phys. Lett. **95B** (1980) 437.
[37] S. Fredrickson, M. Jandel and T. Larsson, *Z. Phys.* **C19** *(1983) 53*; *Phys. Rev. Lett.* **51** *(1983) 2179.*
[38] H. Harari, *Phys. Rev. Lett* **22** *(1969) 562*; J.L. Rosner, *Phys. Rev. Lett* **22** *(1969) 689.*
[39] P.J. Mulders, A.T. Aerts and J.J. de Swart, *Phys. Rev.* **D21** *(1980) 2653.*
[40] R.L. Jaffe, *Phys. Rev.* **D15** *(1977) 281.*
[41] B.A. Li and K.F. Liu, *Phys. Rev. Lett.* **51** *(1983) 1510.*
[42] S. Catto and Y. Choun, *To be published.*
[43] F. Iachello, *Nucl. Phys.* **A497** *(1989) 23c*; F. Iachello, N.C. Mukhopadhyay and L. Zhang, *Phys. Rev.* **B44** *(1991) 898.*
[44] F. Iachello and R.D. Levine, *J. Chem. Phys.***77** *(1982) 3046.*
[45] O.S. van Roosmalen, F. Iachello, R.D. Levine, and A.E.L. Dieperink, *J. Chem. Phys.* **79** *(1983) 2515.*
[46] S. Catto, Y. Choun and L. Kurt *To be published.*

Acknowledgments:

One of us (SC) would like to thank Professor Čestmir Burdik for invitation to Prague and for his kind hospitality. We would like to thank Itzhak Bars, Francesco Iachello, Ramzi Khuri and Susumu Okubo for enlightening conversations. This work is supported by DOE contracts No. DE-AC-0276 ER 03074 and 03075, and PSC-CUNY Research Awards.

Ground state configurations of charged particles in a disk at zero temperature

M. Cerkaski[1] and R.G.Nazmitdinov [2,3]

[1] Henryk Niewodniczanski Institute of Nuclear Physics, Department of Theoretical Physics, Krakow, Poland
[2] Departament de Física, Universitat de les Illes Balears, 07122 Palma de Mallorca, Spain
[3] BLTP, Joint Institute for Nuclear Research, 141980 Dubna, Russia

E-mail: Marcin.Cerkaski@ifj.edu.pl

Abstract. We discuss a novel theoretical approach which explains the self-organization of charged particles in a disk geometry. It allows to calculate readily equilibrium configurations for $n \leq 400$ with a remarkable accuracy, when compared with the molecular dynamics calculations.

1. Introduction

There is an enormous interest in mesoscopic systems consisting of a finite number of interacting particles in confined geometry [1, 2]. Progress in a modern technology allows to study various phenomena on the same scale, from Bose condensate with thousand atoms to quantum dots with a few electrons, providing a rich information about correlation effects in such systems. Nowadays many ideas and concepts introduced, for example, in condensed matter are realized and analysed with a high accuracy. In particular, long ago Wigner predicted for electrons interacting by means of the Coulomb forces in a three-dimensional space the onset of crystallization at low enough densities and temperatures [3]. At these conditions the potential energy dominates over the kinetic energy and defines the equilibrium configurations of electron systems. Indeed, signatures of the Wigner crystallization were observed in two-dimensional (2D) distributions of electrons on the surface of liquid helium [4].

Nowadays the question how a finite number of charged particles arrange themselves in a restricted planar geometry attracts many researches. J.J. Thompson was among first who suggested an instructive solution for interacting electrons, reducing the 3D harmonic oscillator confinement potential to a circular (2D) harmonic oscilator [5]. He developed an analytical approach which enables to one to trace a self-organization of a small number of electrons in a family of rings (shells) with specific numbers of electrons. Although the number of particles in outer and inner rings are changing as a function of the total number of electrons, each shell is characterised by a certain discrete symmetry. Namely, N point charges located on the ring create equidistant nodes on the ring, with the angle $\alpha = 2\pi/N$. Similar shell pattern have been found much later by means of Monte–Carlo (MC) calculations for charged particles (ions and electrons) confined by a 2D parabolic [6, 7] and by a hard wall potentials [8, 9] for small and large systems. Structures of polygonal patterns, similar to ones obtained in the effective harmonic oscillator confinement, have been observed in experimental measurements [10, 11]. Note, that in many cases, the polygonal pattern of equally charge particles is sufficiently regular.

The purpose of this contribution is to discuss a semi-analytical approach for analysis of ground state properties of charged particles in a disk as a function of particle number. We consider particles interacting by means of the Coulomb interaction at zero temperature in order to avoid the admixture of metastable states to the ground state. We will illuminate specific properties which distinguish the electronic configurations in the circular potential from the parabolic one.

2. Coulomb nteraction of rings

The Coulomb energy of n unit charges e forming a regular polygon [ring] has the following form [5]:

$$E_n(R) = \frac{n\, S_n\, e^2}{4\, R}\,, \qquad S_n = \sum_{k=1}^{n-1} \frac{1}{\sin \frac{\pi}{n} k}\,. \tag{1}$$

Evidently, one needs to know the interaction between rings. As it was shown in [12], the interaction between two rings with n and m charges is determined as:

$$E_{nm}(r_1, r_2, \psi) = \sum_{i=1}^{n}\sum_{j=1}^{m} \epsilon(r_1, r_2, \psi_{ij}^{nm} + \psi) = \mathrm{G} \times \sum_{k=1}^{\mathrm{L}} \epsilon(r_1, r_2, \psi_k + \psi)\,, \tag{2}$$

$$\epsilon(r_1, r_2, \theta) = (r_1^2 + {r_2}^2 - 2\, r_1\, r_2\, \cos\theta)^{-1/2}\,, \tag{3}$$

where $\psi_{ij}^{nm} = 2\,\pi(i/n - j/m)$, ψ is the relative angular offset between the two rings. Here, $\{\psi_k = \Delta \times k, k = 1, \ldots, L\}$; $L \equiv \mathrm{LCM}(n, m)$ and $G \equiv \mathrm{GCD}(n, m) = n \times m/L$ are the least common multiple and greatest common divisor of the of numbers (n, m), respectively. The energy is a periodic function with a $\Delta = 2\pi/L$ periodicity. In turn, this result shows that these kind of functions are invariant under angle transformations corresponding to the cyclic group of L elements.

Let us denote $r(R) = \min(\max)(r_1, r_2)$. Employing the expansion $\epsilon(1, x, \psi) = \Sigma_{l \geq 0} x^l\, P_l(\cos\psi)$, we obtain

$$E_{nm} = \langle E_{nm} \rangle + \Delta E_{mn}, \tag{4}$$

$$\Delta E_{mn} = R^{-1}\,(n\, m) \sum_{k=1}^{\infty} H_k(r/R)\,(r/R)^{k\times \mathrm{L}}\, \cos\,(k \times \mathrm{L}\,\psi) \tag{5}$$

Here, a zero order harmonic term is defined as

$$\langle E_{nm} \rangle \equiv \langle E_{nm}(r_1, r_2) \rangle = \frac{2\, n\, m\, K(4\, r_1\, r_2/(r_1 + r_2)^2)\, e^2}{\pi\, R} = \frac{2\, n\, m\, K((r/R)^2)\, e^2}{\pi R}\,, \tag{6}$$

and

$$H_k(x) = h_{k,\mathrm{L}}\; {}_2F_1(\tfrac{1}{2},\, \tfrac{1}{2} + k \times \mathrm{L},\, 1 + k \times \mathrm{L};\, x^2), \tag{7}$$

$$h_{k,\mathrm{L}} = [2^{k\times \mathrm{L}}\,(k \times \mathrm{L})!]^{-1}\, 2\,(2\, k\, \mathrm{L} - 1)!!\,. \tag{8}$$

Since $r/R < 1$, it is enough to consider only a leading term in the sum (5)

$$\Delta E_{mn} \approx \frac{nm}{R}\, H_1(r/R)\,(r/R)^L\, \cos(\mathrm{L}\,\psi), \qquad H_1(r/R) \approx h_{1,L} \tag{9}$$

3. Equilibrium configurations

The total energy is defined as

$$E_{tot}(\mathbf{n}, \mathbf{r}, \psi) = \sum_{i=1}^{p} E_{n_i} + \sum_{i=2}^{p} \sum_{j=1}^{i-1} E_{n_i n_j}(r_i, r_j, \psi_i - \psi_j)\,, \tag{10}$$

where a particle number $n = n(\mathbf{n}) = \Sigma_{i=1}^{p} n_i$, p is a number of rings for a configuration $\mathbf{c} = (\mathbf{n}, \mathbf{r}, \psi)$. The numerical analysis [12] demonstrates that

$$E_{n_i,n_j}(r_i, r_j, \psi_i - \psi_j) \approx \langle E_{n,m}(r_i, r_j)\rangle, \tag{11}$$

holds for $n \leq 2000$ with a high accuracy. The equilibrium configuration of particles can be obtained by minimizing Eq.(10) with respect to $(p, \mathbf{n}, \mathbf{r})$, i.e., finding the partition corresponding to the lowest total energy. For a given partition, the set of equations that determines the optimal radiuses $(r_i, i = 2, \ldots, p)$ is

$$r_i^2 \sum_{j=i+1}^{p} \frac{n_j\, \mathrm{E}((r_j/r_i)^2)}{{r_j}^2 - r_i^2} + r_i \sum_{j=1}^{i-1} n_j \left(\frac{r_j\, \mathrm{E}((r_i/r_j)^2)}{r_j^2 - r_i^2} - \frac{\mathrm{K}((r_i/r_j)^2)}{r_j} \right) = \frac{\pi}{8} S_{n_i}\,, \tag{12}$$

where $\mathrm{K} = \mathrm{X}_{-1}$, $\mathrm{E} = \mathrm{X}_1$ are complete elliptic integrals of first (second) kind:$\mathrm{X}_p(x) = \int_0^{\pi/2} dt\,(1 - x \sin^2 t)^{-p/2}$.

It is enough a few iterations of Eq.(12) to reach an optimal energy configuration (10) for a given partition. Making a grid of different partitions one can readily find the equilibrium configuration for a total number n.

4. Structure of magic configurations

The minimization of energy with respect to the ring's partition numbers $\mathbf{n}$ leads to the following "magic configurations":

$$\begin{array}{rllllllllll}
11: & 11 \\
29: & 6 & 23 \\
55: & 5 & 13 & 37 \\
90: & 5 & 12 & 20 & 53 \\
135: & 5 & 12 & 19 & 29 & 70 \\
186: & 5 & 12 & 19 & 26 & 37 & 87 \\
246: & 5 & 12 & 18 & 25 & 34 & 46 & 106 \\
316: & 5 & 11 & 18 & 25 & 33 & 42 & 56 & 126 \\
394: & 5 & 11 & 18 & 25 & 32 & 40 & 50 & 66 & 147
\end{array} \tag{13}$$

These configurations are characterized with complete p shells. For example, for $n = 394$ we obtain $p = 9$ filled shells. The addition of one electron creates a *centered core configurations* with a number of electrons $6k, k = 1, 2, 3 \ldots$, starting from the one particle in the center

$$\begin{array}{rlllllllllll}
12: & 1 & 11 \\
30: & 1 & 6 & 23 \\
56: & \mathbf{1} & \mathbf{6} & \mathbf{12} & 37 \\
92: & \mathbf{1} & \mathbf{6} & \mathbf{12} & 20 & 53 \\
136: & \mathbf{1} & \mathbf{6} & \mathbf{12} & 19 & 28 & 70 \\
187: & \mathbf{1} & \mathbf{6} & \mathbf{12} & \mathbf{18} & 26 & 37 & 87 \\
248: & \mathbf{1} & \mathbf{6} & \mathbf{12} & \mathbf{18} & 25 & 34 & 46 & 106 \\
317: & \mathbf{1} & \mathbf{6} & \mathbf{12} & \mathbf{18} & 25 & 32 & 42 & 55 & 126 \\
395: & \mathbf{1} & \mathbf{6} & \mathbf{12} & \mathbf{18} & \mathbf{24} & 32 & 40 & 50 & 65 & 147.
\end{array} \tag{14}$$

As a result, a new shell appears which is filling by electrons. The sequence of electrons in the filled shells of the *centered core configurations* is a characteristic property of the hexagonal lattice. It signals on the onset of the crystalliztion in the disk geometry for $n \geq 200$.

5. Asymptotic limit: $n \gg 1$

The density of the hexagonal lattice is defined as $\rho = c/a^2 e$, where $c = 2/\sqrt{3}$ and a is the distance between two nearest neighbor particles. At large n one can define this distance as $a = R/p$. As a result, one obtains

$$\rho = ne/\pi R^2 \Rightarrow p^2 \approx \frac{\sqrt{3}}{2\pi} \times n \tag{15}$$

To consider the density of the ring we employ the factorization of ϱ: $\varrho(r,\phi) = \rho(r) \times (2\pi)^{-1}$ which leads to a normalization of a radial integral: $Q = \int_0^R drr\, \rho_R(r)$.

The distribution $\rho_R(r)$ can be obtained from the Ritz problem of the minimization of the Coulomb energy on a two–dimensional disk of radius R. To this aim we consider the functional $G_{[\rho]} = E_{[\rho]} g(V^0_{[\rho]})$ where

$$E_{[\rho]}(r) = \int_0^R dr\, A_{[\rho]}(r)\, \rho(r) = \int_0^R dr\, B_{[\rho]}(r)\, \rho(r), \tag{16}$$

$$A_{[\varrho]}(r) = c_0 \int_0^r dy\, y\, \mathrm{K}((y/r)^2)\, \varrho(y), \qquad B_{[\varrho]}(r) = c_0\, r \int_r^{\bar{R}} dy\, \mathrm{K}((r/y)^2)\, \varrho(y), \tag{17}$$

$$V^0_{[\rho]} = \int_0^R dr\, r\, \rho(r). \tag{18}$$

Hence the variational principle $\delta E_{[\rho]} = 0$ gives

$$0 = (A_{[\varrho]} + B_{[\varrho]})(r) + (E/g)\, g'(V^0)\, r\,, \tag{19}$$

where $c_0 = 2/\pi$. Since the integration of r.h.s. over r gives $2\,E + (E/g)\, g'(V^0)\, V^0$, the solution exists only if $g(x) = x^{-2}$. Denoting $X_m(z) = R^{-2}\, X_{[\varrho_m]}(z)\, (X = A, B)$, where $\varrho_m(r) = (r/R)^m = z^m$, we obtain

$$\begin{pmatrix} A_m(z) \\ B_m(z) \end{pmatrix} = \sum_{n=0}^{\infty} K_n \begin{pmatrix} \frac{z^{m+2}}{2+2n+m} \\ \frac{z\, h_n^m(z)}{1-2n+m} \end{pmatrix} = \begin{pmatrix} 0 \\ \frac{z\, \tilde{B}_m(z^2)}{m+1} \end{pmatrix} + \begin{pmatrix} \frac{z^{2+m}\, \tilde{A}_m(1)}{2+m} \\ \frac{-z^{m+2}\, \tilde{B}_m(1)}{1+m} \end{pmatrix}. \tag{20}$$

Here

$$K_n = \frac{((2\,n-1)!!)^2}{2^{2\,n}\,((n)!)^2}, \qquad h_n^m(z) = z^{2\,n} - z^{1+m}, \tag{21}$$

$$\tilde{A}_m(z) = {}_3F_2(\tfrac{1}{2}, \tfrac{1}{2}, 1 + \tfrac{m}{2}; 1, 2 + \tfrac{m}{2}; z), \quad \tilde{B}_m(z) = {}_3F_2(\tfrac{1}{2}, \tfrac{1}{2}, -\tfrac{1+m}{2}; 1, \tfrac{1-m}{2}; z),$$

are considered for $m = 0, 2, \ldots$. Eq.(20) leads to the following condition

$$0 = (m+1)\, \tilde{A}_m - (m+2)\, \tilde{B}_m(1) \Longrightarrow A_m(z) + B_m(z) = (m+1)^{-1}\, z\, \tilde{B}_m(z^2). \tag{22}$$

The l.h.s. is a nontrivial identity fulfilled by generalized hypergeometric functions ${}_3\mathrm{F}_2$. Expanding $\varrho = \varrho(\mathbf{a})$ in series

$$\varrho(\mathbf{a}) = (Q/R^2) \times \sum_{n=0}^{\infty} a_n\, (r/R)^{2\,n} \tag{23}$$

and employing Eq.(22), we rewrite Eq.(19) in the form

$$z\left(\sum_{m=0}^{\infty} a_m\,(2\,m+1)^{-1}\,\tilde{B}_{2\,m}(z^2) - 2\,(R/Q^2)\,E_{[\varrho]}\right)\,Q = 0. \tag{24}$$

Evidently, $z \neq 0$, and, therefore, we obtain

$$E(\mathbf{a}) = \frac{Q^2}{2\,R} \times \sum_{m=0}^{\infty} \frac{a_m}{2\,m+1}, \tag{25}$$

$$1 = Q^{-1}\,Q_{[\varrho(\mathbf{a})]} = \sum_{m=0}^{\infty} \frac{a_m}{2\,m+2} = F_0(\mathbf{a}), \tag{26}$$

$$F_\alpha(\mathbf{a}) = R^{-2\,\alpha}\,\langle r^{2\,\alpha}\rangle_\rho = \sum_{m=0}^{\infty} \frac{a_m}{2\,m+2\,\alpha+2}. \tag{27}$$

We require that the higher order terms in the sums in Eqs.(25) are zeros, and obtain:

$$F_{-(2\,\alpha+1)/2}(\mathbf{a}) = 0, \qquad \alpha = 1, 2, \ldots, \qquad E(\mathbf{a}) = \frac{Q^2}{2\,R} \times F_{-1/2}(\mathbf{a})/F_0(\mathbf{a}), \tag{28}$$

As a result, we have

$$a_m = (-1)^m \binom{-1/2}{m}. \tag{29}$$

This result is consistent with the Taylor expansion of

$$\rho_R(r) = Q\,R^{-1}(R^2 - r^2)^{-1/2}. \tag{30}$$

For finite n, the minimal energy E_n contains a smooth part. The numerical results are well fitted by series

$$E_{(\mathbf{a},\mathbf{b})}(n) \approx \frac{(ne)^2}{R}\left(\frac{\pi}{4} + \sum_{k=1}^{p} a_k/n^k + \sum_{k=1}^{q} b_k\,n^{-k}\,\ln n\right). \tag{31}$$

Here, the coefficient $a_0 = \pi/4$ is found from Eq.(25) , takng into account that

$$\Sigma_{m=0}^{\infty}\,a_m/(2\,m+1) = \pi/2. \tag{32}$$

For finite (p, q), the "effective coefficients" $(\mathbf{a}, \mathbf{b})$ could be fitted by a minimization:

$$\Sigma_n {\delta_n}^2, \qquad \delta_n = (E_n - E_{(\mathbf{a},\mathbf{b})}(n))\,R/(en)^2 \tag{33}$$

In the interval $n \in [20, 400]$ a good estimation is obtained for $(p, q) = (4, 3)$, which is

$$\begin{pmatrix}\mathbf{a}\\ \mathbf{b}\end{pmatrix} = \begin{pmatrix}76.7756, 1266.32, -617.616, -745.263\\ -16.9075, -552.892, -1502.45\end{pmatrix} \tag{34}$$

6. Summary

The zero–harmonic limit, applied for the study of equilibrium states of finite number of charged particles in a disk, essentially reduces the optimization problem for the ground state configuration. The proposed method reproduces with a remarkable accuracy the results obtained with the aid of numerical simulation (see details in [12]).

Acknowledgments
This work was supported in part by Bogoliubov-Infeld program of BLTP and RFBR grant 14-02-00723.

References

[1] Saarikoski H, Reimann S M, Harju A and Manninen M 2010 *Rev. Mod. Phys.* **82** 2785
[2] Birman J L, Nazmitdinov R G and Yukalov V I 2013 *Phys. Rep.* **526** 1
[3] Wigner E P 1934 *Phys. Rev. B* **46** 1002
[4] Grimes C C and Adams G 1979 *Phys. Rev. Lett.* **42** 795
[5] Thomson J J 1904 *Phil. Mag.* **7** 237
[6] Lozovik Yu E and Mandelshtam V A 1992 *Phys. Lett. A* **165** 469
[7] Bolton F and Rössler U 1992 *Superlatt. Microstruct.* **13** 139
[8] Bedanov V and Peeters F M 1994 *Phys. Rev. B* **49** 2667
[9] Kong M, Partoens B, Matulis A and Peeters F M 2004 *Phys. Rev. E* **69** 036412
[10] Saint Jean, M, Even C and Guthmann C 2001 *Eur.Phys.Lett.* **55** 45
[11] Rousseau E, Ponarin D, Hristakos L, Avenel O, Varoquaaux E and Mukharsky Y 2009 *Phys. Rev. B* **79** 045406
[12] Cerkaski M, Nazmitdinov R G and Puente A 2014 *eprint arXiv:1403.4878*

Multiplet classification for SU(n,n)

V.K. Dobrev

Institute of Nuclear Research and Nuclear Energy, Bulgarian Academy of Sciences, 72 Tsarigradsko Chaussee, 1784 Sofia, Bulgaria

Abstract. In the present paper we review our project of systematic construction of invariant differential operators on the example of the non-compact algebras $su(n, n)$ for $n = 2, 3, 4$. We give explicitly the main multiplets of indecomposable elementary representations and some reduced multiplets. We give explicitly the minimal representations. Due to the recently established parabolic relations the multiplet classification results are valid also for the algebras $sl(2n, \mathbb{R})$ and when $n = 2k$ for the algebras $su^*(4k)$ with suitably chosen maximal parabolic subalgebras.

1. Introduction

Invariant differential operators play very important role in the description of physical symmetries - starting from the early occurrences in the Maxwell, d'Allembert, Dirac, equations, to the latest applications of (super-)differential operators in conformal field theory, supergravity and string theory. Thus, it is important for the applications in physics to study systematically such operators.

In a recent paper [1] we started the systematic explicit construction of invariant differential operators. We gave an explicit description of the building blocks, namely, the parabolic subgroups and subalgebras from which the necessary representations are induced. Thus we have set the stage for study of different non-compact groups.

Since the study and description of detailed classification should be done group by group we had to decide which groups to study. Since the most widely used algebras are the conformal algebras $so(n,2)$ in n-dimensional Minkowski space-time we concentrated on a class that shares most of their properties. This class consists of:

$$so(n,2), \quad sp(n,\mathbb{R}), \quad su(n,n), \quad so^*(4n), \quad E_{7(-25)}$$

the corresponding analogs of Minkowski space-time V being:

$$\mathbb{R}^{n-1,1}, \quad \mathrm{Sym}(n,\mathbb{R}), \quad \mathrm{Herm}(n,\mathbb{C}), \quad \mathrm{Herm}(n,\mathbb{Q}), \quad \mathrm{Herm}(3,\mathbb{O})$$

involving the four division algebras $\mathbb{R}, \mathbb{C}, \mathbb{Q}, \mathbb{O}$.

In view of applications to physics, we proposed to call these algebras 'conformal Lie algebras', (or groups) [2].

We have started the study of the above class in the framework of the present approach in the cases: $so(n,2)$, $su(n,n)$, $sp(n,\mathbb{R})$, $E_{7(-25)}$, cf. [2, 3, 4, 5].

Lately, we discovered an efficient way to extend our considerations beyond this class introducing the notion of 'parabolically related non-compact semisimple Lie algebras' [5].

• *Definition:* Let $\mathcal{G}, \mathcal{G}'$ be two non-compact semisimple Lie algebras with the same complexification $\mathcal{G}^{\mathbb{C}} \cong \mathcal{G}'^{\mathbb{C}}$. We call them parabolically related if they have parabolic subalgebras $\mathcal{P} = \mathcal{M} \oplus \mathcal{A} \oplus \mathcal{N}$, $\mathcal{P}' = \mathcal{M}' \oplus \mathcal{A}' \oplus \mathcal{N}'$, such that: $\mathcal{M}^{\mathbb{C}} \cong \mathcal{M}'^{\mathbb{C}}$ $(\Rightarrow \mathcal{P}^{\mathbb{C}} \cong \mathcal{P}'^{\mathbb{C}})$.$\diamondsuit$

Certainly, there are many such parabolic relationships for any given algebra $\mathcal{G}$. Furthermore, two algebras $\mathcal{G}, \mathcal{G}'$ may be parabolically related with different parabolic subalgebras.

In the present paper we review our results on the case of $su(n,n)$, cf. [3, 6, 7]. Due to the parabolic relationships these would be valid also for $sl(2n,\mathbb{R})$, and if $n = 2k$ also for $su^*(4k)$.

2. Preliminaries

Let G be a semisimple non-compact Lie group, and K a maximal compact subgroup of G. Then we have an *Iwasawa decomposition* $G = KA_0N_0$, where A_0 is Abelian simply connected vector subgroup of G, N_0 is a nilpotent simply connected subgroup of G preserved by the action of A_0. Further, let M_0 be the centralizer of A_0 in K. Then the subgroup $P_0 = M_0A_0N_0$ is a *minimal parabolic subgroup* of G. A *parabolic subgroup* $P = M'A'N'$ is any subgroup of G which contains a minimal parabolic subgroup.

Further, let $\mathcal{G}, \mathcal{K}, \mathcal{P}, \mathcal{M}, \mathcal{A}, \mathcal{N}$ denote the Lie algebras of G, K, P, M, A, N, resp.

For our purposes we need to restrict to *maximal parabolic subgroups* $P = MAN$, i.e. $\text{rank}A = 1$, resp. to *maximal parabolic subalgebras* $\mathcal{P} = \mathcal{M} \oplus \mathcal{A} \oplus \mathcal{N}$ with $\dim \mathcal{A} = 1$.

Let ν be a (non-unitary) character of A, $\nu \in \mathcal{A}^*$, parameterized by a real number d, called the *conformal weight* or energy.

Further, let μ fix a discrete series representation D^μ of M on the Hilbert space V_μ, or the finite-dimensional (non-unitary) representation of M with the same Casimirs.

We call the induced representation $\chi = \text{Ind}_P^G(\mu \otimes \nu \otimes 1)$ an *elementary representation* of G [8]. (These are called *generalized principal series representations* (or *limits thereof*) in [9].) Their spaces of functions are:

$$\mathcal{C}_\chi = \{\mathcal{F} \in C^\infty(G, V_\mu) \mid \mathcal{F}(gman) = e^{-\nu(H)} \cdot D^\mu(m^{-1})\,\mathcal{F}(g)\}$$

where $a = \exp(H) \in A'$, $H \in \mathcal{A}'$, $m \in M'$, $n \in N'$. The representation action is the left regular action:

$$(\mathcal{T}^\chi(g)\mathcal{F})(g') = \mathcal{F}(g^{-1}g') , \quad g, g' \in G .$$

ERs are important due to the following fundamental result:
Theorem [10, 11]: Every irreducible admissible representation of G is equivalent to a subrepresentation of an ER.

• An important ingredient in our considerations are the *highest/lowest weight representations* of $\mathcal{G}^{\mathbb{C}}$. These can be realized as (factor-modules of) Verma modules V^Λ over $\mathcal{G}^{\mathbb{C}}$, where $\Lambda \in (\mathcal{H}^{\mathbb{C}})^*$, $\mathcal{H}^{\mathbb{C}}$ is a Cartan subalgebra of $\mathcal{G}^{\mathbb{C}}$, weight $\Lambda = \Lambda(\chi)$ is determined uniquely from χ [12].

Actually, since our ERs may be induced from finite-dimensional representations of $\mathcal{M}$ (or their limits) the Verma modules are always reducible. Thus, it is more convenient to use *generalized Verma modules* $\tilde{V}^\Lambda$ such that the role of the highest/lowest weight vector v_0 is taken by the (finite-dimensional) space $V_\mu v_0$. For the generalized Verma modules (GVMs) the reducibility is controlled only by the value of the conformal weight d. Relatedly, for the intertwining differential operators only the reducibility w.r.t. non-compact roots is essential.

• One main ingredient of our approach is as follows. We group the (reducible) ERs with the same Casimirs in sets called *multiplets* [13]. The multiplet corresponding to fixed values of the Casimirs may be depicted as a connected graph, the vertices of which correspond to the reducible ERs and the lines (arrows) between the vertices correspond to intertwining operators. The explicit parametrization of the multiplets and of their ERs is important for understanding of the situation.

In fact, the multiplets contain explicitly all the data necessary to construct the intertwining differential operators. Actually, the data for each intertwining differential operator consists of the pair (β, m), where β is a (non-compact) positive root of $\mathcal{G}^{\mathbb{C}}$, $m \in \mathbb{N}$, such that the BGG Verma module reducibility condition (for highest weight modules) is fulfilled [14]:

$$(\Lambda + \rho, \beta^\vee) \ = \ m \ , \quad \beta^\vee \equiv 2\beta/(\beta, \beta)$$

ρ is half the sum of the positive roots of $\mathcal{G}^{\mathbb{C}}$. When the above holds then the Verma module with shifted weight $V^{\Lambda - m\beta}$ (or $\tilde{V}^{\Lambda - m\beta}$ for GVM and β non-compact) is embedded in the Verma module V^Λ (or $\tilde{V}^\Lambda$). This embedding is realized by a singular vector v_s determined by a polynomial $\mathcal{P}_{m,\beta}(\mathcal{G}^-)$ in the universal enveloping algebra $(U(\mathcal{G}_-))\, v_0$, $\mathcal{G}^-$ is the subalgebra of $\mathcal{G}^{\mathbb{C}}$ generated by the negative root generators [15]. More explicitly, [12], $v^s_{m,\beta} = \mathcal{P}_{m,\beta}\, v_0$ (or $v^s_{m,\beta} = \mathcal{P}_{m,\beta}\, V_\mu\, v_0$ for GVMs). Then there exists [12] an intertwining differential operator

$$\mathcal{D}_{m,\beta} \ : \ \mathcal{C}_{\chi(\Lambda)} \ \longrightarrow \ \mathcal{C}_{\chi(\Lambda - m\beta)}$$

given explicitly by:

$$\mathcal{D}_{m,\beta} \ = \ \mathcal{P}_{m,\beta}(\widehat{\mathcal{G}^-})$$

where $\widehat{\mathcal{G}^-}$ denotes the right action on the functions $\mathcal{F}$.

In most of these situations the invariant operator $\mathcal{D}_{m,\beta}$ has a non-trivial invariant kernel in which a subrepresentation of $\mathcal{G}$ is realized. Thus, studying the equations with trivial RHS:

$$\mathcal{D}_{m,\beta}\, f \ = \ 0 \ , \qquad f \in \mathcal{C}_{\chi(\Lambda)} \ ,$$

is also very important. For example, in many physical applications in the case of first order differential operators, i.e., for $m = m_\beta = 1$, these equations are called conservation laws, and the elements $f \in \ker \mathcal{D}_{m,\beta}$ are called conserved currents.

Below in our exposition we shall use the so-called Dynkin labels:

$$m_i \ \equiv \ (\Lambda + \rho, \alpha_i^\vee) \ , \quad i = 1, \ldots, n,$$

where $\Lambda = \Lambda(\chi)$, ρ is half the sum of the positive roots of $\mathcal{G}^{\mathbb{C}}$.

We shall use also the so-called Harish-Chandra parameters [16]:

$$m_\beta \equiv (\Lambda + \rho, \beta) \ ,$$

where β is any positive root of $\mathcal{G}^{\mathbb{C}}$. These parameters are redundant, since they are expressed in terms of the Dynkin labels, however, some statements are best formulated in their terms. (Clearly, both the Dynkin labels and Harish-Chandra parameters have their origin in the BGG reducibility condition.)

3. The Lie algebra $su(n,n)$ and parabolically related

Let $\mathcal{G} = su(n,n)$, $n \geq 2$. The maximal compact subgroup is $\mathcal{K} \cong u(1) \oplus su(n) \oplus su(n)$, while $\mathcal{M} = sl(n,\mathbb{C})_{\mathbb{R}}$. The number of ERs in the main multiplets is equal to [5]

$$|W(\mathcal{G}^{\mathbb{C}},\mathcal{H}^{\mathbb{C}})| \,/\, |W(\mathcal{M}^{\mathbb{C}},\mathcal{H}^{\mathbb{C}}_m)| = \binom{2n}{n}$$

The signature of the ERs of $\mathcal{G}$ is:

$$\chi = \{n_1,\ldots,n_{n-1},n_{n+1}\ldots,n_{2n-1};\ c\}\,, \quad n_j \in \mathbb{N}\,, \quad c = d - \tfrac{1}{2}n^2 \tag{1}$$

The restricted Weyl reflection is given by the Knapp–Stein integral operators [17]:

$$\begin{aligned} &G_{KS} : \mathcal{C}_\chi \longrightarrow \mathcal{C}_{\chi'}\,, \qquad \chi' = \{(n_1,\ldots,n_{n-1},n_{n+1},\ldots,n_{2n-1})^*;\ -c\}\,, \\ &(n_1,\ldots,n_{n-1},n_{n+1},\ldots,n_{2n-1})^* \doteq (n_{n+1},\ldots,n_{2n-1},n_1,\ldots,n_{n-1}) \end{aligned} \tag{2}$$

Multiplets

Below we give the multiplets for $su(n,n)$ for $n = 2,3,4$. They are valid also for $sl(2n,\mathbb{R})$ with $\mathcal{M}$-factor $sl(n,\mathbb{R}) \oplus sl(n,\mathbb{R})$, and when $n = 2k$ these are multiplets also for the parabolically related algebra $su^*(4k)$ with $\mathcal{M}$-factor $su^*(2k) \oplus su^*(2k)$,

There are several types of multiplets: the main type, which contains maximal number of ERs/GVMs, the finite-dimensional and the discrete series representations, and many reduced types of multiplets.

The multiplets of the main type are in 1-to-1 correspondence with the finite-dimensional irreps of $su(n,n)$, i.e., they will be labelled by the $2n-1$ positive Dynkin labels $m_i \in \mathbb{N}$.

4. Multiplets of SU(2,2), SL(4,$\mathbb{R}$) and SU*(4)

The main multiplet contains six ERs whose signatures can be given in the following pair-wise manner:

$$\begin{aligned} \chi_0^\pm &= \{(m_1,m_3)^\pm;\pm\frac{(m_1+2m_2+m_3)}{2}\} \\ \chi'^\pm &= \{(m_{12},m_{23})^\pm;\pm\tfrac{1}{2}(m_1+m_3)\} \\ \chi''^\pm &= \{(m_2,m_{13})^\pm;\pm\tfrac{1}{2}(m_3-m_1)\} \end{aligned} \tag{3}$$

where we have used for the numbers $m_\beta = (\Lambda(\chi)+\rho,\beta)$ the same compact notation as for the roots β, and

$$(n_1,n_3)^- = (n_1,n_3)\,, \qquad (n_1,n_3)^+ = (n_1,n_3)^* = (n_3,n_1) \tag{4}$$

These multiplets were given first for $su^*(4)$ [18].

Obviously, the pairs in (3) are related by Knapp-Stein integral operators, i.e.,

$$G_{KS} : \mathcal{C}_{\chi^\mp} \longrightarrow \mathcal{C}_{\chi^\pm}$$

The multiplets are given explicitly in Fig. 1, where we use the notation: $\Lambda^\pm = \Lambda(\chi^\pm)$. Each intertwining differential operator is represented by an arrow accompanied by a symbol i_{jk} encoding the root α_{jk} and the number $m_{\alpha_{jk}}$ which is involved in the BGG criterion. This notation is used to save space, but it can be used due to the fact that only intertwining differential operators which are non-composite are displayed, and that the data β, m_β, which is involved in the embedding $V^\Lambda \longrightarrow V^{\Lambda - m_\beta,\beta}$ turns out to involve only the m_i corresponding to simple roots, i.e., for each β, m_β there exists $i = i(\beta, m_\beta, \Lambda) \in \{1, \ldots, 2n-1\}$, such that $m_\beta = m_i$. Hence the data α_{jk}, $m_{\alpha_{jk}}$ is represented by i_{jk} on the arrows.

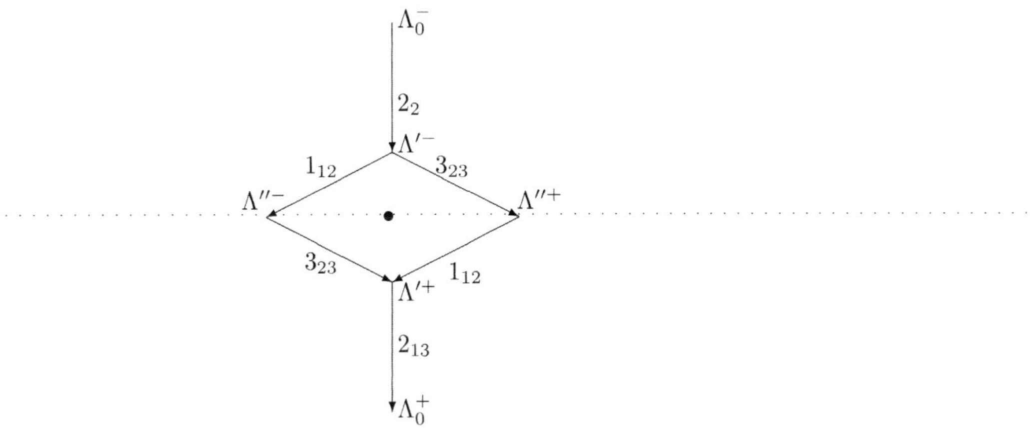

Fig. 1. Main multiplets for $su(2,2)$, $su^*(4)$ and $sl(4,\mathbb{R})$

The pairs $\Lambda^\pm$ are symmetric w.r.t. to the bullet in the middle of the figure - this represents the Weyl symmetry realized by the Knapp-Stein operators.

Matters are arranged so that in every multiplet only the ER with signature χ_0^- contains a finite-dimensional nonunitary subrepresentation in a finite-dimensional subspace $\mathcal{E}$. The latter corresponds to the finite-dimensional irrep of $\mathcal{G}$ with signature $\{m_1, m_2, m_3\}$ of dimension: $m_1 m_2 m_3 m_{12} m_{23} m_{13}/6$. The subspace $\mathcal{E}$ is annihilated by the operator G^+, and is the image of the operator G^-. The subspace $\mathcal{E}$ is annihilated also by the intertwining differential operator acting from χ^- to χ'^-. When all $m_i = 1$ then $\dim \mathcal{E} = 1$, and in that case $\mathcal{E}$ is also the trivial one-dimensional UIR of the whole algebra $\mathcal{G}$. Furthermore in that case the conformal weight is zero: $d = 2 + c = 2 - \frac{1}{2}(m_1 + 2m_2 + m_3)_{|m_i=1} = 0$.

In the conjugate ER χ_0^+ there is a unitary subrepresentation in an infinite-dimensional subspace $\mathcal{D}$. It is annihilated by the operator G^-, and is the image of the operator G^+.

All the above is valid also for the algebras $sl(4,\mathbb{R}) \cong so(3,3)$ and $su^*(4) \cong so(5,1)$. However, the latter two do not have discrete series representations. On the other hand the algebra $su(2,2) \cong so(4,2)$ had discrete series representations and furthermore highest/lowest weight series representations.

Thus, in the case of $su(2,2)$ the ER χ_0^+ contains both the holomorphic discrete series representation and the conjugate anti-holomorphic discrete series. The direct sum of the latter two is realized in the invariant subspace $\mathcal{D}$ of the ER χ_0^+.

Note that the corresponding lowest weight GVM is infinitesimally equivalent only to the holomorphic discrete series, while the conjugate highest weight GVM is infinitesimally equivalent to the anti-holomorphic discrete series.
The conformal weight of the ER χ_0^+ has the restriction $d = 2 + c = 2 + \frac{1}{2}(m_1 + 2m_2 + m_3) \geq 4$.

Remark on SU(1,1)

As we mentioned the case $su(1,1)$ is well known - it was studied 60 years ago in the isomorphic form $sl(2,\mathbb{R})$ by Gelfand et al [19] and by Bargmann [20]. In the current setting it was given in [21]. Here we shall only mention that the multiplets contain two ERS/GVMs (cf. $\binom{2n}{n}_{n=1} = 2$), and we can take as their representatives the pair $\Lambda^{\pm}$ and all statements that fit the setting are true. In fact, the old results are prototypical for these pairs, which appear once for each algebra of the conformal type.$\diamondsuit$

Reduced multiplets.

There are three types of reduced multiplets, R_1, R_2, R_3. Each of them contains two ERS/GVMs and may be obtained from the main multiplet by setting formally $m_1 = 0$, $m_2 = 0$, $m_3 = 0$, resp. The signatures are

$$\begin{aligned} {}_1\chi^{\pm} &= \{(m_2, m_{23})^{\pm}; \pm\tfrac{1}{2}m_3\} \\ {}_2\chi^{\pm} &= \{(m_1, m_3)^{\pm}; \pm\tfrac{1}{2}(m_1 + m_3)\} \\ {}_3\chi^{\pm} &= \{(m_{12}, m_2)^{\pm}; \pm\tfrac{1}{2}m_1\} \end{aligned} \tag{5}$$

The above is valid for the parabolically related algebras $su(2,2), su^*(4), sl(4,\mathbb{R})$. For $su(2,2)$ the ER ${}_2\chi^+$ contains the limits of the (anti)holomorphic discrete series representations. Its conformal weight has the restriction $d = 2 + \frac{1}{2}(m_1 + m_3) \geq 3$.

Actually, types R_1, R_3 are conjugated under the * operation (that is not the Weyl symmetry since the sign of c is not changed).

Finally, there is the reduced multiplet R_{13} containing a single representation

$$\chi^s = \{(m,m); 0\}$$

This multiplet may be omitted from this classification since it contains no operators, but its importance was understood in the framework of conformal supersymmetry, i.e., in the multiplet classification for the superconformal algebra $su(2,2/N)$ given in [22]. It turns out that the infinite multiplets of $su(2,2/N)$ have as building blocks all mentioned above multiplets of $su(2,2)$ - sextets, doublets and singlets.

5. Multiplets of SU(3,3) and SL(6,$\mathbb{R}$)

The main multiplet contains 20 ERs/GVMs whose signatures can be given in the following pair-wise manner:

$$\begin{aligned} \chi_0^{\pm} &= \{(m_1, m_2, m_4, m_5)^{\pm}; \pm m_\rho\} \\ \chi_a^{\pm} &= \{(m_1, m_{23}, m_{34}, m_5)^{\pm}; \pm(m_\rho - m_3)\} \\ \chi_b^{\pm} &= \{(m_{12}, m_3, m_{24}, m_5)^{\pm}; \pm(m_\rho - m_{23})\} \\ \chi_{b'}^{\pm} &= \{(m_1, m_{24}, m_3, m_{45})^{\pm}; \pm(m_\rho - m_{34})\} \\ \chi_c^{\pm} &= \{(m_2, m_3, m_{14}, m_5)^{\pm}; \pm(m_\rho - m_{13})\} \\ \chi_{c'}^{\pm} &= \{(m_{12}, m_{34}, m_{23}, m_{45})^{\pm}; \pm(m_\rho - m_{24})\} \\ \chi_{c''}^{\pm} &= \{(m_1, m_{25}, m_3, m_4)^{\pm}; \pm(m_\rho - m_{35})\} \\ \chi_d^{\pm} &= \{(m_2, m_{34}, m_{13}, m_{45})^{\pm}; \pm(m_\rho - m_{14})\} \\ \chi_{d'}^{\pm} &= \{(m_{12}, m_{35}, m_{23}, m_4)^{\pm}; \pm(m_\rho - m_{25})\} \\ \chi_e^{\pm} &= \{(m_2, m_{35}, m_{13}, m_4)^{\pm}; \pm(m_\rho - m_{15})\} \end{aligned} \tag{6}$$

where $m_\rho = \frac{1}{2}(m_1 + 2m_2 + 3m_3 + 2m_4 + m_5)$. They are given in Fig. 2.

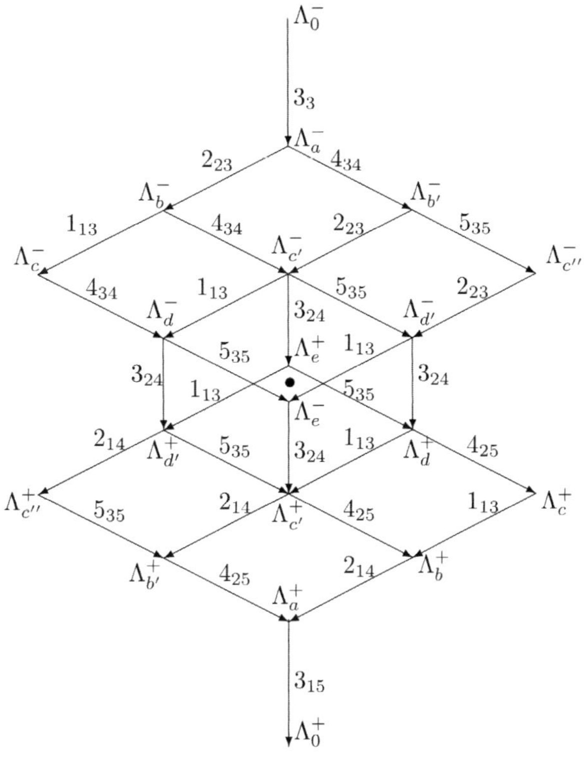

Fig. 2. Main multiplets for $su(3,3)$ and $sl(6,\mathbb{R})$

All general facts that were stated in the $SU(2,2)$ case are valid also here, in particular, the special role of the pair $\chi_0^\pm$. The finite-dimensional irreps $\mathcal{E}$ of $su(3,3)$ or $sl(6,\mathbb{R})$ are sitting in the ERs χ_0^- and have dimension as the UIRs of $SU(6)$.

Reduced multiplets.
There are five types of reduced multiplets, R_a^3, $a = 1,\dots,5$, which may be obtained from the main multiplet by setting formally $m_a = 0$. Multiplets of type R_4^3, R_5^3, are conjugate to the multiplets of type R_2^3, R_1^3, resp., and are not shown.

The reduced multiplets of type R_3^3 contain 14 ERs/GVMs with signatures:

$$\begin{aligned}
\chi_0^\pm &= \{\,(m_1,m_2,m_4,m_5)^\pm\,;\,\pm m_\rho\,\} \\
\chi_b^\pm &= \{\,(m_{12},0,m_{24},m_5)^\pm\,;\,\pm(m_\rho-m_2)\,\} \\
\chi_{b'}^\pm &= \{\,(m_1,m_{24},0,m_{45})^\pm\,;\,\pm(m_\rho-m_4)\,\} \\
\chi_c^\pm &= \{\,(m_2,0,m_{14},m_5)^\pm\,;\,\pm(m_\rho-m_{12})\,\} \\
\chi_{c''}^\pm &= \{\,(m_1,m_{25},0,m_4)^\pm\,;\,\pm(m_\rho-m_{45})\,\} \\
\chi_d^\pm &= \{\,(m_2,m_4,m_{12},m_{45})^\pm\,;\,\pm(m_\rho-m_{12,4})=\mp(m_\rho-m_{2,45})\,\} \\
\chi_e^\pm &= \{\,(m_2,m_{45},m_{12},m_4)^\pm\,;\,\pm(m_\rho-m_{2,4})=\pm\tfrac{1}{2}(m_1+m_5)\,\}\,,
\end{aligned} \tag{7}$$

here $m_\rho = \frac{1}{2}(m_1+2m_2+2m_4+m_5)$. These multiplets are given in Fig. 3. They may be called the main type of reduced multiplets since for $su(3,3)$ in χ_0^+ are contained the limits of the (anti)holomorphic discrete series.

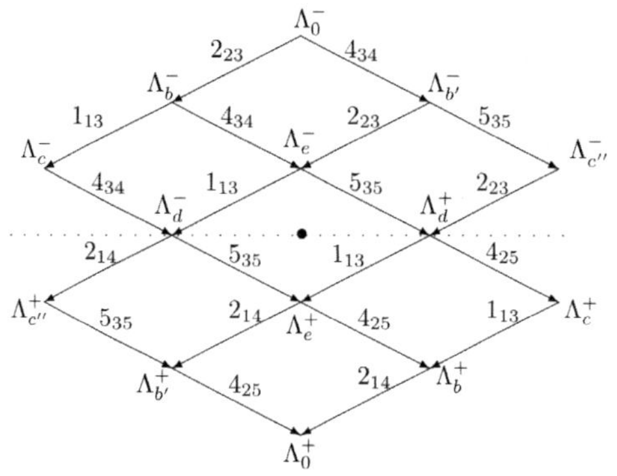

Fig. 3. Reduced multiplets of type R_3^3 for $su(3,3)$ and $sl(6,\mathbb{R})$

The reduced multiplets of type R_2^3, resp., R_1^3, contain 14 ERs/GVMs each. These multiplets are given in Fig. 4, resp., Fig. 5 :

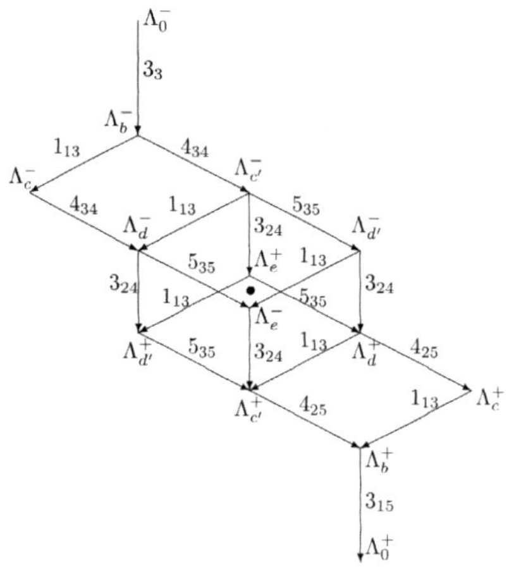

Fig. 4. Reduced multiplets of type R_2^3 for $su(3,3)$ and $sl(6,\mathbb{R})$

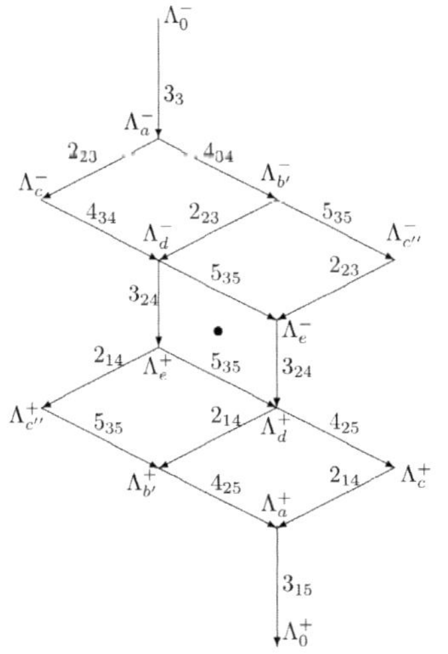

Fig. 5. Reduced multiplets of type R^3_1 for $su(3,3)$ and $sl(6,\mathbb{R})$

Further reduction of multiplets

There are further reductions of the multiplets denoted by R^3_{ab}, $a,b = 1,\ldots,5$, $a<b$, which may be obtained from the main multiplet by setting formally $m_a = m_b = 0$. From these ten reductions four (for $(a,b) = (1,2),(2,3),(3,4),(4,5)$) do not contain representations of physical interest, i.e., induced from finite-dimensional irreps of the $\mathcal{M}$ subalgebra. From the others R^3_{35} and R^3_{25} are conjugated to R^3_{13} and R^3_{14}, resp., as explained above. Thus, we present only four types of multiplets.

The reduced multiplets of type R^3_{13} contain 10 ERs/GVMs with signatures:

$$\begin{aligned}
\chi_a^\pm &= \{\,(0,m_2,m_4,m_5)^\pm\,;\ \pm m_\rho\,\} \\
\chi_b^\pm &= \{\,(m_2,0,m_{2,4},m_5)^\pm\,;\ \pm(m_\rho-m_2)\,\} \\
\chi_{b'}^\pm &= \{\,(0,m_{2,4},0,m_{45})^\pm\,;\ \pm(m_\rho-m_4)\,\} \\
\chi_c^\pm &= \{\,(0,m_{2,45},0,m_4)^\pm\,;\ \pm(m_\rho-m_{45})\,\} \\
\chi_d^\pm &= \{\,(m_2,m_4,m_2,m_{45})^\pm\,;\ \pm(m_\rho-m_{2,4}) = \pm\tfrac{1}{2}m_5\,\}\,,
\end{aligned} \tag{8}$$

here $m_\rho = m_2 + m_4 + \frac{1}{2}m_5$. The multiplets are given in Fig. 6.

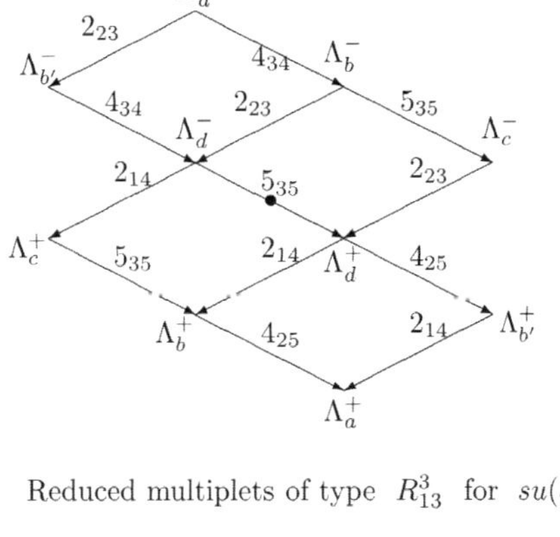

Fig. 6. Reduced multiplets of type R^3_{13} for $su(3,3)$ and $sl(6,\mathbb{R})$

Note that the differential operator (of order m_5) from χ_d^- to χ_d^+ is a degeneration of an integral Knapp-Stein operator.

The reduced multiplets of type R^3_{15} contain 10 ERs/GVMs with signatures:

$$\begin{aligned}
\chi_0^\pm &= \{\,(0, m_2, m_4, 0)^\pm\,;\, \pm m_\rho\,\} \\
\chi_a^\pm &= \{\,(0, m_{23}, m_{34}, 0)^\pm\,;\, \pm(m_\rho - m_3)\,\} \\
\chi_b^\pm &= \{\,(m_2, m_3, m_{24}, 0)^\pm\,;\, \pm(m_\rho - m_{23})\,\} \\
\chi_{b'}^\pm &= \{\,(0, m_{24}, m_3, m_4)^\pm\,;\, \pm(m_\rho - m_{34})\,\} \\
\chi_d^\pm &= \{\,(m_2, m_{34}, m_{23}, m_4)^\pm\,;\, \pm(m_\rho - m_{24}) = \pm\tfrac{1}{2}m_3\,\}\ ,
\end{aligned} \tag{9}$$

here $m_\rho = m_2 + \frac{3}{2}m_3 + m_4$. The multiplets are given in Fig. 7. Here the differential operator (of order m_3) from χ_d^- to χ_d^+ is a degeneration of an integral Knapp-Stein operator.

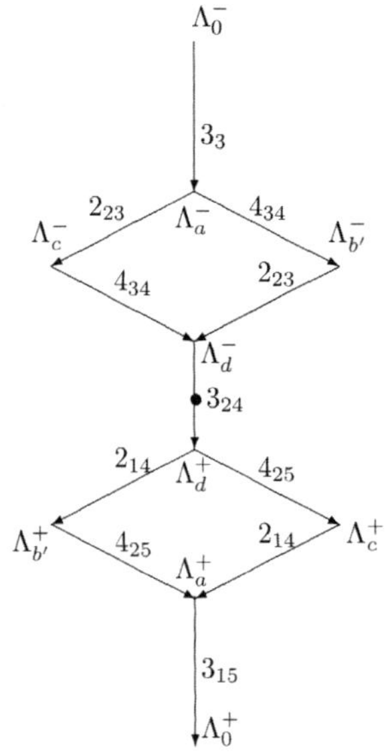

Fig. 7. Reduced multiplets of type R^3_{15} for $su(3,3)$ and $sl(6,\mathbb{R})$

The reduced multiplets of type R^3_{14}, R^3_{24}, contain 10 ERs/GVMs each, the corresponding multiplets being given below:

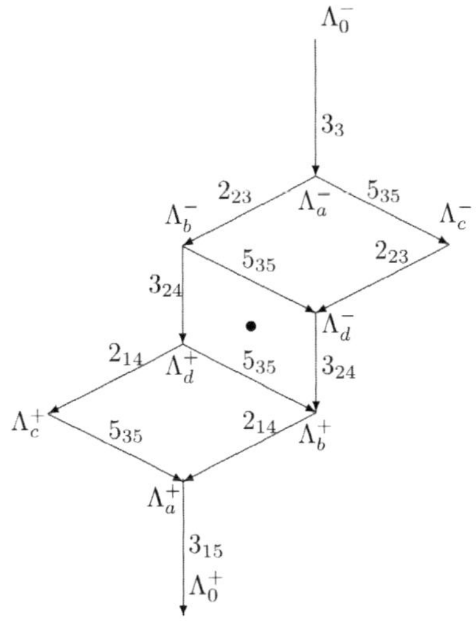

Fig. 8. Reduced multiplets of type R^3_{14} for $su(3,3)$ and $sl(6,\mathbb{R})$

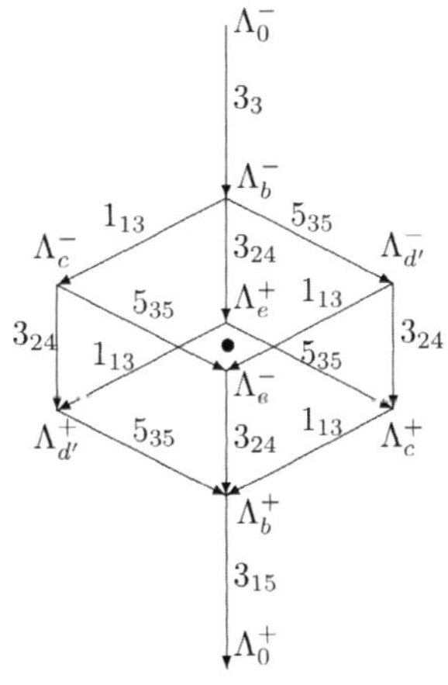

Fig. 9. Reduced multiplets of type R^3_{24} for $su(3,3)$ and $sl(6,\mathbb{R})$

Last reduction of multiplets

There are further reductions of the multiplets - triple and quadruple, but only one triple reduction contains representations of physical interest. Namely, this is the multiplet R^3_{135}, which may be obtained from the main multiplet by setting formally $m_1 = m_3 = m_5 = 0$. It contains 7 ERs/GVMs with signatures:

$$\begin{aligned}
\chi_a^{\pm} &= \{\, (0, m_2, m_4, 0)^{\pm}\,;\ \pm m_\rho = \pm m_{2,4} \,\} \\
\chi_b^{\pm} &= \{\, (0, m_{2,4}, 0, m_4)^{+}\,,\ \pm m_2 \,\} \\
\chi_{b'}^{\pm} &= \{\, (m_2, 0, m_{2,4}, 0)^{\pm}\,;\ \pm m_4 \,\} \\
\chi_d &= \{\, (m_2, m_4, m_2, m_4)\,;\ 0 \,\}
\end{aligned} \tag{10}$$

The multiplets are given below:

Journal of Physics: Conference Series **563** (2014) 012008 doi:10.1088/1742-6596/563/1/012008

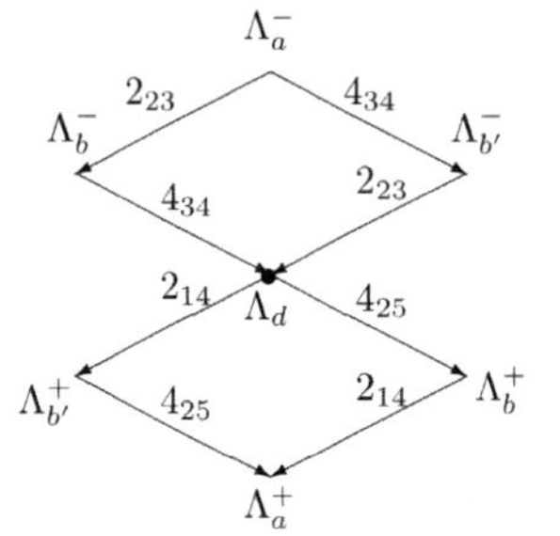

Fig. 10. Reduced multiplets of type R^3_{135} for $su(3,3)$ and $sl(6,\mathbb{R})$

The representation χ^d is a singlet, not in a pair, since it has zero weight c, and the $\mathcal{M}$ entries are self-conjugate. It is placed in the middle of the figure as the bullet. That ER contains the *minimal irreps* characterized by two positive integers which are denoted in this context as $m_2\,, m_4\,$. Each such irrep is the kernel of the two invariant differential operators $\mathcal{D}^{m_2}_{14}$ and $\mathcal{D}^{m_4}_{25}\,$, which are of order $m_2\,,\ m_4\,$, resp., corresponding to the noncompact roots $\alpha_{14}\,,\ \alpha_{25}\,$, resp.

6. Multiplets of SU(4,4), SL(8,$\mathbb{R}$) and SU*(8)

The main multiplet R^4 contains 70 ERs/GVMs whose signatures can be given in the following pair-wise manner:

$$\begin{array}{rcl}
\chi_0^\pm &=& \{\,(m_1,m_2,m_3,m_5,m_6,m_7)^\pm\,;\ \pm m_\rho\,\} \\
\chi_{00}^\pm &=& \{\,(m_1,m_2,m_{34},m_{45},m_6,m_7)^\pm\,;\ \pm(m_\rho-m_4)\,\} \\
\chi_{10}^\pm &=& \{\,(m_1,m_{23},m_4,m_{35},m_6,m_7)^\pm\,;\ \pm(m_\rho-m_{34})\,\} \\
\chi_{01}^\pm &=& \{\,(m_1,m_2,m_{35},m_4,m_{56},m_7)^\pm\,;\ \pm(m_\rho-m_{45})\,\} \\
\chi_{20}^\pm &=& \{\,(m_{12},m_3,m_4,m_{25},m_6,m_7)^\pm\,;\ \pm(m_\rho-m_{24})\,\} \\
\chi_{11}^\pm &=& \{\,(m_1,m_{23},m_{45},m_{34},m_{56},m_7)^\pm\,;\ \pm(m_\rho-m_{35})\,\} \\
\chi_{02}^\pm &=& \{\,(m_1,m_2,m_{36},m_4,m_5,m_{67})^\pm\,;\ \pm(m_\rho-m_{46})\,\} \\
\chi_{30}^\pm &=& \{\,(m_2,m_3,m_4,m_{15},m_6,m_7)^\pm\,;\ \pm(m_\rho-m_{14})\,\} \\
\chi_{21}^\pm &=& \{\,(m_{12},m_3,m_{45},m_{24},m_{56},m_7)^\pm\,;\ \pm(m_\rho-m_{25})\,\} \\
\chi_{12}^\pm &=& \{\,(m_1,m_{23},m_{46},m_{34},m_5,m_{67})^\pm\,;\ \pm(m_\rho-m_{36})\,\} \\
\chi_{03}^\pm &=& \{\,(m_1,m_2,m_{37},m_4,m_5,m_6)^\pm\,;\ \pm(m_\rho-m_{47})\,\} \\
\chi_{31}^\pm &=& \{\,(m_2,m_3,m_{45},m_{14},m_{56},m_7)^\pm\,;\ \pm(m_\rho-m_{15})\,\} \\
\chi_{22}^\pm &=& \{\,(m_{12},m_3,m_{46},m_{24},m_5,m_{67})^\pm\,;\ \pm(m_\rho-m_{26})\,\} \\
\chi_{13}^\pm &=& \{\,(m_1,m_{23},m_{47},m_{34},m_5,m_6)^\pm\,;\ \pm(m_\rho-m_{37})\,\} \\
\chi_{32}^\pm &=& \{\,(m_2,m_3,m_{46},m_{14},m_5,m_{67})^\pm\,;\ \pm(m_\rho-m_{16})\,\} \\
\chi_{23}^\pm &=& \{\,(m_{12},m_3,m_{47},m_{24},m_5,m_6)^\pm\,;\ \pm(m_\rho-m_{27})\,\} \\
\chi_{33}^\pm &=& \{\,(m_2,m_3,m_{47},m_{14},m_5,m_6)^\pm\,;\ \pm(m_\rho-m_{17})\,\} \\
\chi_{00}'^\pm &=& \{\,(m_1,m_{24},m_5,m_3,m_{46},m_7)^\pm\,;\ \pm(m_\rho-m_{35}-m_4)\,\} \\
\chi_{10}'^\pm &=& \{\,(m_{12},m_{34},m_5,m_{23},m_{46},m_7)^\pm\,;\ \pm(m_\rho-m_{25}-m_4)\,\} \\
\chi_{01}'^\pm &=& \{\,(m_1,m_{24},m_{56},m_3,m_{45},m_{67})^\pm\,;\ \pm(m_\rho-m_{36}-m_4)\,\} \\
\chi_{20}'^\pm &=& \{\,(m_2,m_{34},m_5,m_{13},m_{46},m_7)^\pm\,;\ \pm(m_\rho-m_{15}-m_4)\,\} \\
\chi_{11}'^\pm &=& \{\,(m_{12},m_{34},m_{56},m_{23},m_{45},m_{67})^\pm\,;\ \pm(m_\rho-m_{26}-m_4)\,\} \\
\chi_{02}'^\pm &=& \{\,(m_1,m_{24},m_{57},m_3,m_{45},m_6)^\pm\,;\ \pm(m_\rho-m_{37}-m_4)\,\} \\
\chi_{20}''^\pm &=& \{\,(m_{13},m_4,m_5,m_2,m_{36},m_7)^\pm\,;\ \pm(m_\rho-m_{25}-m_{34})\,\} \\
\chi_{21}''^\pm &=& \{\,(m_{13},m_4,m_{56},m_2,m_{35},m_{67})^\pm\,;\ \pm(m_\rho-m_{26}-m_{34})\,\} \\
\chi_{12}''^\pm &=& \{\,(m_{12},m_{35},m_6,m_{23},m_4,m_{57})^\pm\,;\ \pm(m_\rho-m_{26}-m_{45})\,\} \\
\chi_{02}''^\pm &=& \{\,(m_1,m_{25},m_6,m_3,m_4,m_{57})^\pm\,;\ \pm(m_\rho-m_{36}-m_{45})\,\}
\end{array} \qquad (11)$$

Journal of Physics: Conference Series **563** (2014) 012008 doi:10.1088/1742-6596/563/1/012008

$$
\begin{aligned}
\chi'^{\pm}_{30} &= \{\,(m_{23}, m_4, m_5, m_{12}, m_{36}, m_7)^{\pm}\,;\ \pm(m_\rho - m_{15} - m_{34})\,\}\\
\chi'^{\pm}_{21} &= \{\,(m_2, m_{34}, m_{56}, m_{13}, m_{45}, m_{67})^{\pm}\,;\ \pm(m_\rho - m_{16} - m_4)\,\}\\
\chi'^{\pm}_{12} &= \{\,(m_{12}, m_{34}, m_{57}, m_{23}, m_{45}, m_6)^{\pm}\,;\ \pm(m_\rho - m_{27} - m_4)\,\}\\
\chi'^{\pm}_{03} &= \{\,(m_1, m_{25}, m_{67}, m_3, m_4, m_{56})^{\pm}\,;\ \pm(m_\rho - m_{37} - m_{45})\,\}\\
\chi'^{\pm}_{40} &= \{\,(m_3, m_4, m_5, m_1, m_{26}, m_7)^{\pm}\,;\ \pm(m_\rho - m_{15} - m_{24})\,\}\\
\chi'^{\pm}_{31} &= \{\,(m_{23}, m_4, m_{56}, m_{12}, m_{35}, m_{67})^{\pm}\,;\ \pm(m_\rho - m_{16} - m_{34})\,\}\\
\chi'^{\pm}_{22} &= \{\,(m_2, m_{34}, m_{57}, m_{13}, m_{45}, m_6)^{\pm}\,;\ \pm(m_\rho - m_{17} - m_4)\,\}\\
\chi''^{\pm}_{22} &= \{\,(m_{13}, m_4, m_{57}, m_2, m_{35}, m_6)^{\pm}\,;\ \pm(m_\rho - m_{27} - m_{34})\,\}\,,
\end{aligned}
$$

where $m_\rho = \frac{1}{2}(m_1 + 2m_2 + 3m_3 + 4m_4 + 3m_5 + 2m_6 + m_7)$. The multiplets are given explicitly in Fig. 11 (first in [3]).

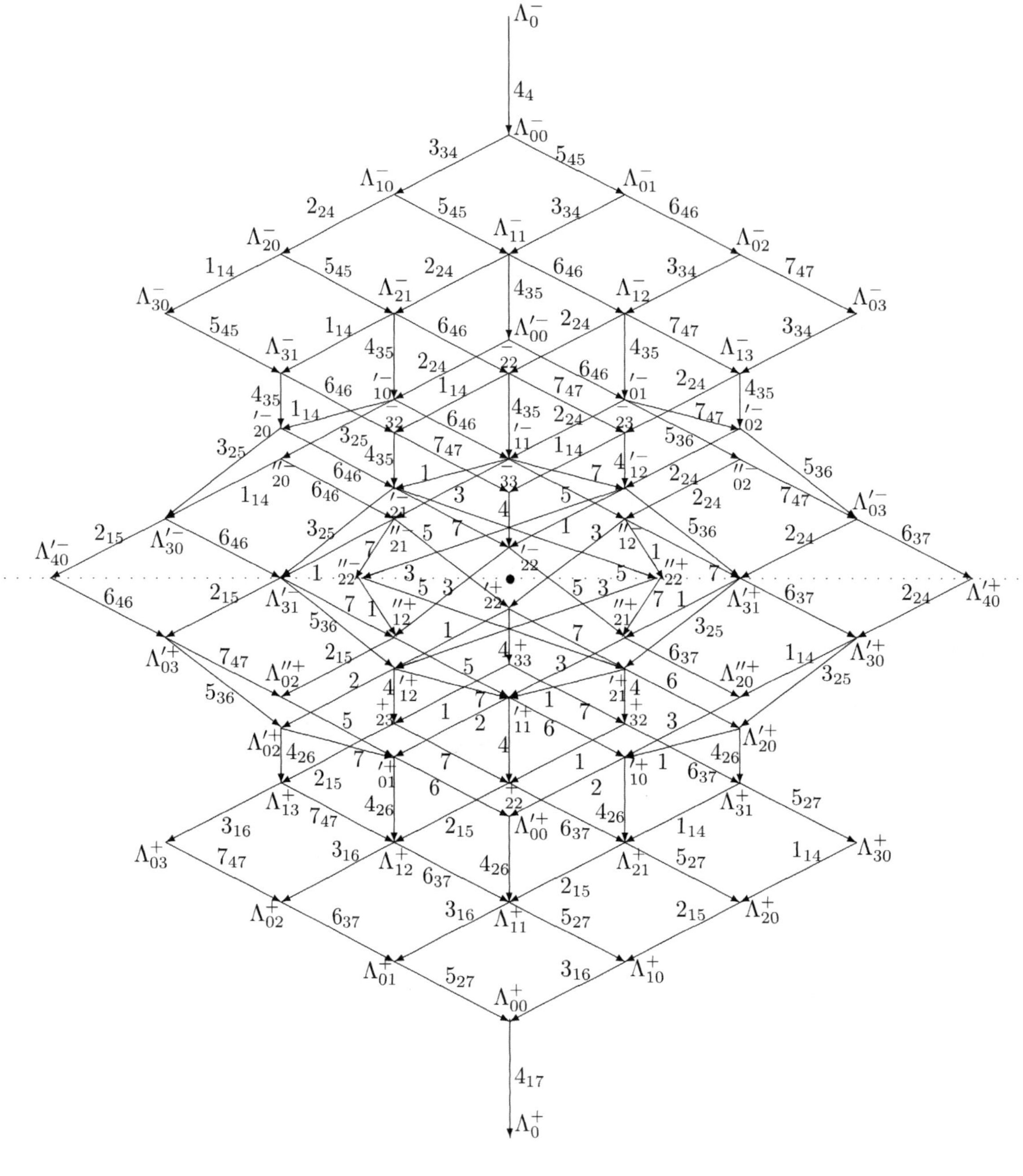

Fig. 11. Main multiplets R^4 for $su(4,4)$, $su^*(8)$ and $sl(8,\mathbb{R})$

Main reduced multiplets

There are nine physically relevant and essentially different reductions of multiplets denoted by R^4_3, R^4_3, R^4_2, R^4_1. Each of them contains 50 ERs/GVMs [7]. Here we present only the reduced multiplets R^4_4. Their 50 ERs/GVMs has signatures that can be given in the following pair-wise manner:

$$\begin{aligned}
\chi_0^{\pm} &= \{\,(m_1, m_2, m_3, m_5, m_6, m_7)^{\pm}\,;\, \pm m_\rho\,\} \\
\chi_{10}^{\pm} &= \{\,(m_1, m_{23}, 0, m_{3,5}, m_6, m_7)^{\pm}\,;\, \pm(m_\rho - m_3)\,\} \\
\chi_{01}^{\pm} &= \{\,(m_1, m_2, m_{3,5}, 0, m_{56}, m_7)^{\pm}\,;\, \pm(m_\rho - m_5)\,\} \\
\chi_{20}^{\pm} &= \{\,(m_{12}, m_3, 0, m_{23,5}, m_6, m_7)^{\pm}\,;\, \pm(m_\rho - m_{23})\,\} \\
\chi_{11}^{\pm} &= \{\,(m_1, m_{23}, m_5, m_3, m_{56}, m_7)^{\pm}\,;\, \pm(m_\rho - m_{3,5})\,\} \\
\chi_{02}^{\pm} &= \{\,(m_1, m_2, m_{3,56}, 0, m_5, m_{67})^{\pm}\,;\, \pm(m_\rho - m_{56})\,\} \\
\chi_{30}^{\pm} &= \{\,(m_2, m_3, 0, m_{13,5}, m_6, m_7)^{\pm}\,;\, \pm(m_\rho - m_{13})\,\} \\
\chi_{21}^{\pm} &= \{\,(m_{12}, m_3, m_5, m_{23}, m_{56}, m_7)^{\pm}\,;\, \pm(m_\rho - m_{23,5})\,\} \\
\chi_{12}^{\pm} &= \{\,(m_1, m_{23}, m_{56}, m_3, m_5, m_{67})^{\pm}\,;\, \pm(m_\rho - m_{3,56})\,\} \\
\chi_{03}^{\pm} &= \{\,(m_1, m_2, m_{3,57}, 0, m_5, m_6)^{\pm}\,;\, \pm(m_\rho - m_{57})\,\} \\
\chi_{31}^{\pm} &= \{\,(m_2, m_3, m_5, m_{13}, m_{56}, m_7)^{\pm}\,;\, \pm(m_\rho - m_{13,5})\,\} \\
\chi_{22}^{\pm} &= \{\,(m_{12}, m_3, m_{56}, m_{23}, m_5, m_{67})^{\pm}\,;\, \pm(m_\rho - m_{23,56})\,\} \\
\chi_{13}^{\pm} &= \{\,(m_1, m_{23}, m_{57}, m_3, m_5, m_6)^{\pm}\,;\, \pm(m_\rho - m_{3,57})\,\} \\
\chi_{32}^{\pm} &= \{\,(m_2, m_3, m_{56}, m_{13}, m_5, m_{67})^{\pm}\,;\, \pm(m_\rho - m_{13,56})\,\} \\
\chi_{23}^{\pm} &= \{\,(m_{12}, m_3, m_{57}, m_{23}, m_5, m_6)^{\pm}\,;\, \pm(m_\rho - m_{23,57})\,\} \\
\chi_{33}^{\pm} &= \{\,(m_2, m_3, m_{57}, m_{13}, m_5, m_6)^{\pm}\,;\, \pm(m_\rho - m_{13,57})\,\} \\
\chi_{20}^{\prime\prime\pm} &= \{\,(m_{13}, 0, m_5, m_2, m_{3,56}, m_7)^{\pm}\,;\, \pm(m_\rho - m_{23,5} - m_3)\,\} \\
\chi_{21}^{\prime\prime\pm} &= \{\,(m_{13}, 0, m_{56}, m_2, m_{3,5}, m_{67})^{\pm}\,;\, \pm(m_\rho - m_{23,56} - m_3)\,\} \\
\chi_{12}^{\prime\prime\pm} &= \{\,(m_{12}, m_{3,5}, m_6, m_{23}, 0, m_{57})^{\pm}\,;\, \pm(m_\rho - m_{23,56} - m_5)\,\} \\
\chi_{02}^{\prime\prime\pm} &= \{\,(m_1, m_{23,5}, m_6, m_3, 0, m_{57})^{\pm}\,;\, \pm(m_\rho - m_{3,56} - m_5)\,\} \\
\chi_{30}^{\prime\pm} &= \{\,(m_{23}, 0, m_5, m_{12}, m_{3,56}, m_7)^{\pm}\,;\, \pm(m_\rho - m_{13,5} - m_3)\,\} \\
\chi_{03}^{\prime\pm} &= \{\,(m_1, m_{23,5}, m_{67}, m_3, 0, m_{56})^{\pm}\,;\, \pm(m_\rho - m_{3,57} - m_5)\,\} \\
\chi_{40}^{\prime\pm} &= \{\,(m_3, 0, m_5, m_1, m_{23,56}, m_7)^{\pm}\,;\, \pm(m_\rho - m_{13,5} - m_{23})\,\} \\
\chi_{31}^{\prime\pm} &= \{\,(m_{23}, 0, m_{56}, m_{12}, m_{3,5}, m_{67})^{\pm}\,;\, \pm(m_\rho - m_{13,56} - m_3)\,\} \\
\chi_{22}^{\prime\prime\pm} &= \{\,(m_{13}, 0, m_{57}, m_2, m_{3,5}, m_6)^{\pm}\,;\, \pm(m_\rho - m_{23,57} - m_3)\,\}\,,
\end{aligned} \tag{12}$$

here $m_\rho = \frac{1}{2}(m_1 + 2m_2 + 3m_3 + 3m_5 + 2m_6 + m_7)$. This is a very important type of reduced multiplets since for $su(4,4)$ in χ_0^+ are contained the limits of the (anti)holomorphic discrete series. The multiplets are given in Fig. 12.

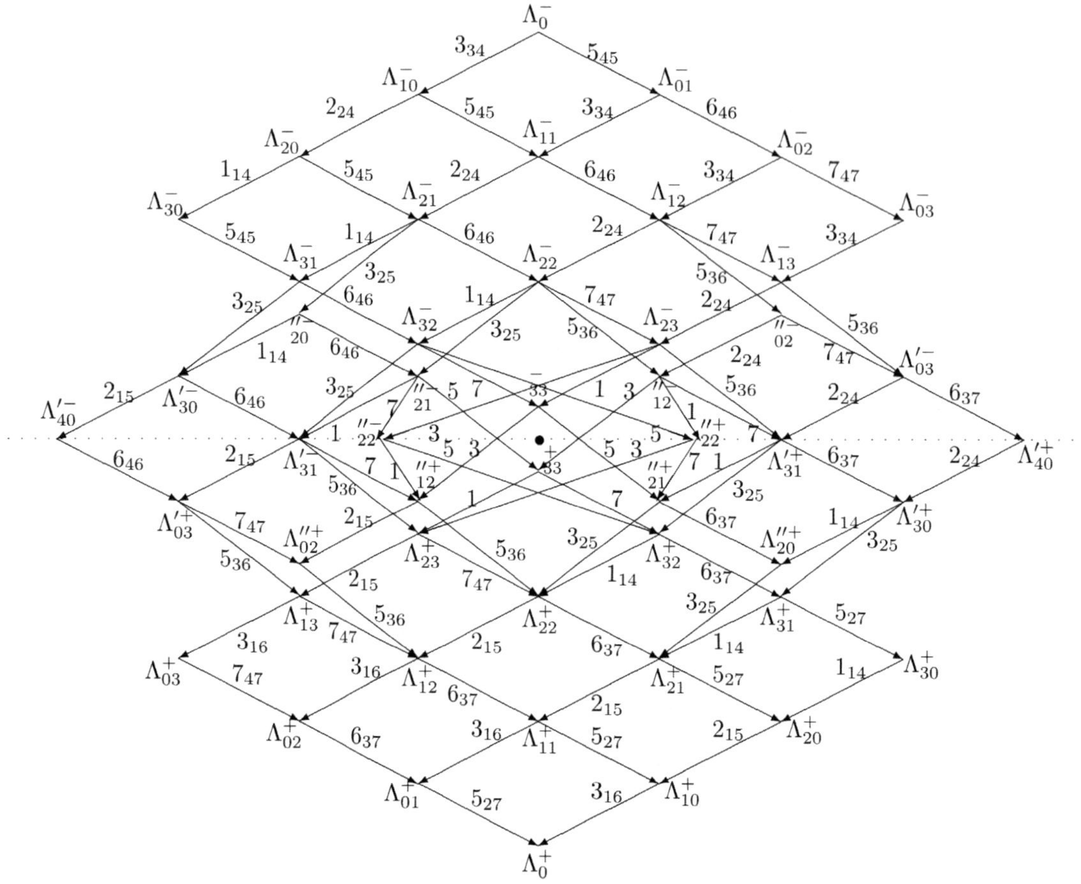

Fig. 12. Reduced multiplets R_4^4 for $su(4,4)$, $su^*(8)$ and $sl(8,\mathbb{R})$

Further reduction of multiplets

There are nine physically relevant and essentially different further reductions of multiplets denoted by R^3_{ab}, $(a,b) = (13),(14),(15),(16),(17),(24),(25),(26),(35)$. They contain 36 ERs/GVMs each and were given in [7]. Here we give only type R^4_{13} :

$$\begin{aligned}
\chi_0^\pm &= \{\,(0,m_2,0,m_5,m_6,m_7)^\pm\,;\,\pm m_\rho)\,\} \qquad (13)\\
\chi_{10}^\pm &= \{\,(0,m_2,m_4,m_{45},m_6,m_7)^\pm\,;\,\pm(m_\rho - m_4)\,\}\\
\chi_{20}^\pm &= \{\,(m_2,0,m_4,m_{2,45},m_6,m_7)^\pm\,;\,\pm(m_\rho - m_{2,4})\,\}\\
\chi_{11}^\pm &= \{\,(0,m_2,m_{45},m_4,m_{56},m_7)^\pm\,;\,\pm(m_\rho - m_{45})\,\}\\
\chi_{21}^\pm &= \{\,(m_2,0,m_{45},m_{2,4},m_{56},m_7)^\pm\,;\,\pm(m_\rho - m_{2,45})\,\}\\
\chi_{12}^\pm &= \{\,(0,m_2,m_{46},m_4,m_5,m_{67})^\pm\,;\,\pm(m_\rho - m_{46})\,\}\\
\chi_{22}^\pm &= \{\,(m_2,0,m_{46},m_{2,4},m_5,m_{67})^\pm\,;\,\pm(m_\rho - m_{2,46})\,\}\\
\chi_{13}^\pm &= \{\,(0,m_2,m_{47},m_4,m_5,m_6)^\pm\,;\,\pm(m_\rho - m_{47})\,\}\\
\chi_{23}^\pm &= \{\,(m_2,0,m_{47},m_{2,4},m_5,m_6)^\pm\,;\,\pm(m_\rho - m_{2,47})\,\}\\
\chi_{00}'^\pm &= \{\,(0,m_{2,4},m_5,0,m_{46},m_7)^\pm\,;\,\pm(m_\rho - m_5 - 2m_4)\,\}\\
\chi_{10}'^\pm &= \{\,(m_2,m_4,m_5,m_2,m_{46},m_7)^\pm\,;\,\pm(m_\rho - m_{2,5} - 2m_4)\,\}\\
\chi_{01}'^\pm &= \{\,(0,m_{2,4},m_{56},0,m_{45},m_{67})^\pm\,;\,\pm(m_\rho - m_{56} - 2m_4)\,\}
\end{aligned}$$

$$\begin{aligned}
\chi'^{\pm}_{11} &= \{\,(m_2, m_4, m_{56}, m_2, m_{45}, m_{67})^{\pm}\,;\ \pm(m_\rho - m_{2,56} - 2m_4)\,\} \\
\chi'^{\pm}_{02} &= \{\,(0, m_{2,4}, m_{57}, 0, m_{45}, m_6)^{\pm}\,;\ \pm(m_\rho - m_{57} - 2m_4)\,\} \\
\chi''^{\pm}_{02} &= \{\,(0, m_{2,45}, m_6, 0, m_4, m_{57})^{\pm}\,;\ \pm(m_\rho - m_6 - 2m_{45})\,\} \\
\chi'^{\pm}_{12} &= \{\,(m_2, m_4, m_{57}, m_2, m_{45}, m_6)^{\pm}\,;\ \pm(m_\rho - m_{2,57} - 2m_4)\,\} \\
\chi'^{\pm}_{03} &= \{\,(0, m_{2,45}, m_{67}, 0, m_4, m_{56})^{\pm}\,;\ \pm(m_\rho - m_{67} - 2m_{45})\,\} \\
\chi'^{\pm}_{40} &= \{\,(0, m_4, m_5, 0, m_{2,46}, m_7)^{\pm}\,;\ \pm(m_\rho - m_5 - 2m_{2,4})\,\}\ ,
\end{aligned}$$

here $m_\rho = \frac{1}{2}(2m_2 + 4m_4 + 3m_5 + 2m_6 + m_7)$. Their diagram is given in Fig. 13.

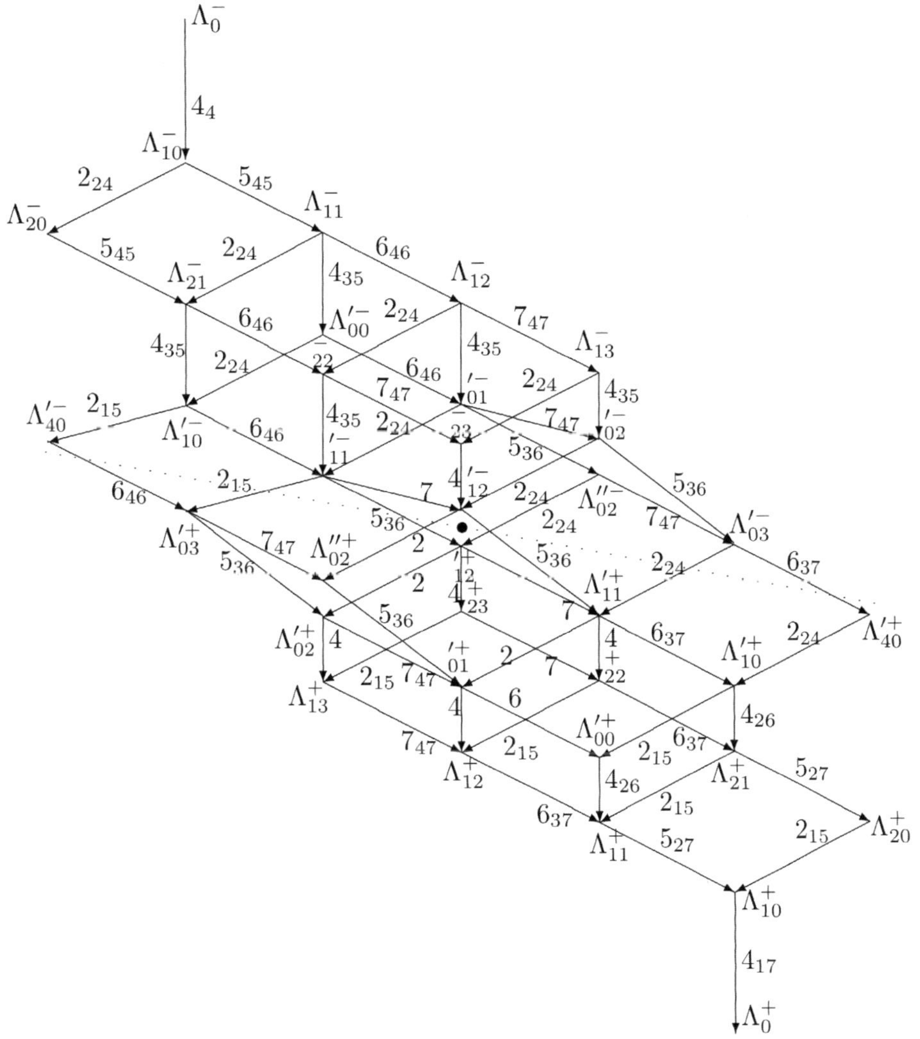

Fig. 13. Reduced multiplets R^4_{13} for $su(4,4)$, $su^*(8)$ and $sl(8,\mathbb{R})$

Yet further reduction of multiplets

There are six physically relevant and essentially different further reductions of multiplets denoted by R^3_{abc}, $(a,b,c) = (1,3,5),(1,3,6),(1,3,7),(1,4,6),(1,4,7),(2,4,6)$. They contain 26 ERs/GVMs each and were given in [7]. Here we give only two types. First we give R^3_{135} :

$$\begin{aligned}
\chi_0^\pm &= \{\, (0,m_2,0,0,m_6,m_7)^\pm\,;\ \pm m_\rho\,\} \qquad (14)\\
\chi_{11}^\pm &= \{\, (0,m_2,m_4,m_4,m_6,m_7)^\pm\,;\ \pm(m_\rho - m_4)\,\}\\
\chi_{21}^\pm &= \{\, (m_2,0,m_4,m_{2,4},m_6,m_7)^\pm\,;\ \pm(m_\rho - m_{2,4})\,\}\\
\chi_{12}^\pm &= \{\, (0,m_2,m_{4,6},m_4,0,m_{67})^\pm\,;\ \pm(m_\rho - m_{4,6})\,\}\\
\chi_{22}^\pm &= \{\, (m_2,0,m_{4,6},m_{2,4},0,m_{67})^\pm\,;\ \pm(m_\rho - m_{2,4,6})\,\}\\
\chi_{13}^\pm &= \{\, (0,m_2,m_{4,67},m_4,0,m_6)^\pm\,;\ \pm(m_\rho - m_{4,67})\,\}\\
\chi_{23}^\pm &= \{\, (m_2,0,m_{4,67},m_{2,4},0,m_6)^\pm\,;\ \pm(m_\rho - m_{2,4,67})\,\}\\
\chi_{00}^{\prime\pm} &= \{\, (0,m_{2,4},0,0,m_{4,6},m_7)^\pm\,;\ \pm(m_\rho - 2m_4)\,\}\\
\chi_{01}^{\prime\pm} &= \{\, (0,m_{2,4},m_6,0,m_4,m_{67})^\pm\,;\ \pm(m_\rho - m_6 - 2m_4)\,\}\\
\chi_{02}^{\prime\pm} &= \{\, (0,m_{2,4},m_{67},0,m_4,m_6)^\pm\,;\ \pm(m_\rho - m_{67} - 2m_4)\,\}\\
\chi_{30}^{\prime\pm} &= \{\, (m_2,m_4,0,m_2,m_{4,6},m_7)^\pm\,;\ \pm(m_\rho - m_2 - 2m_4)\,\}\\
\chi_{40}^{\prime\pm} &= \{\, (0,m_4,0,0,m_{26},m_7)^\pm\,;\ \pm(m_\rho - 2m_{2,4})\,\}\\
\chi_{31}^{\prime\pm} &= \{\, (m_2,m_4,m_6,m_2,m_4,m_{67})^\pm\,;\ \pm(m_\rho - m_{2,6} - 2m_4) = \pm\tfrac{1}{2}m_7\,\}\ ,
\end{aligned}$$

here $m_\rho = m_2 + 2m_4 + m_6 + \frac{1}{2}m_7$. The multiplets are given in Fig. 3-135. Note that the differential operator (of order m_7) from χ_{31}^- to χ_{31}^+ is a degeneration of an integral Knapp-Stein operator.

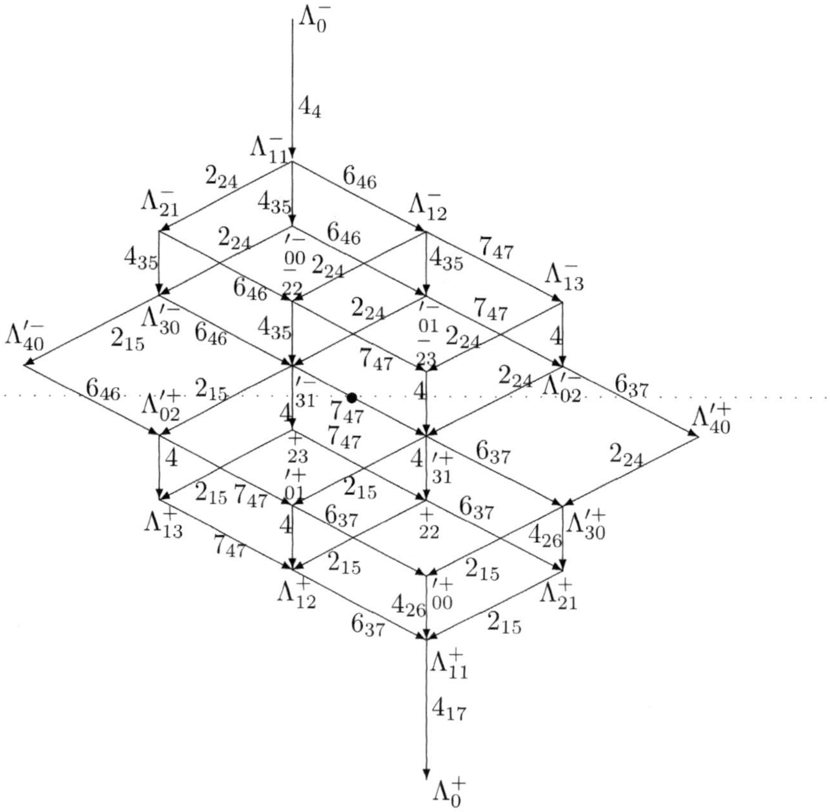

Fig. 14. Reduced multiplets of type R^4_{135} for $su(4,4)$, $su^*(8)$ and $sl(8,\mathbb{R})$

Then we give type R^4_{137} :

$$\begin{aligned}
\chi_0^\pm &= \{\,(0,m_2,0,m_5,m_6,0)^\pm\,;\ \pm m_\rho\,\} \\
\chi_{10}^\pm &= \{\,(0,m_2,m_4,m_{45},m_6,0)^\pm\,;\ \pm(m_\rho - m_4)\,\} \\
\chi_{20}^\pm &= \{\,(m_2,0,m_4,m_{2,45},m_6,0)^\pm\,;\ \pm(m_\rho - m_{2,4})\,\} \\
\chi_{11}^\pm &= \{\,(0,m_2,m_{45},m_4,m_{56},0)^\pm\,;\ \pm(m_\rho - m_{45})\,\} \\
\chi_{21}^\pm &= \{\,(m_2,0,m_{45},m_{2,4},m_{56},0)^\pm\,;\ \pm(m_\rho - m_{2,45})\,\} \\
\chi_{12}^\pm &= \{\,(0,m_2,m_{46},m_4,m_5,m_6)^\pm\,;\ \pm(m_\rho - m_{46})\,\} \\
\chi_{22}^\pm &= \{\,(m_2,0,m_{46},m_{2,4},m_5,m_6)^\pm\,;\ (m_\rho - m_{2,46})\,\} \\
\chi'^\pm_{00} &= \{\,(0,m_{2,4},m_5,0,m_{46},0)^\pm\,;\ \pm(m_\rho - m_5 - 2m_4)\,\} \\
\chi'^\pm_{01} &= \{\,(0,m_{2,4},m_{56},0,m_{45},m_6)^\pm\,;\ \pm(m_\rho - m_{56} - 2m_4)\,\} \\
\chi'^\pm_{30} &= \{\,(m_2,m_4,m_5,m_2,m_{46},0)^\pm\,;\ \pm(m_\rho - m_{2,5} - 2m_4)\,\} \\
\chi'^\pm_{03} &= \{\,(0,m_{2,45},m_6,0,m_4,m_{56})^\pm\,;\ \pm(m_\rho - m_6 - 2m_{45})\,\} \\
\chi'^\pm_{40} &= \{\,(0,m_4,m_5,0,m_{2,46},0)^\pm\,;\ \pm(m_\rho - m_5 - 2m_{2,4})\,\} \\
\chi'^\pm_{31} &= \{\,(m_2,m_4,m_{56},m_2,m_{45},m_6)^\pm\,;\ (m_\rho - m_{2,56} - 2m_4) = \pm\tfrac{1}{2}m_5\,\}\ ,
\end{aligned} \tag{15}$$

here $m_\rho = m_2 + 2m_4 + \frac{3}{2}m_5 + m_6$. The multiplets are given in Fig. 15. Note that the differential operator (of order m_5) from χ^-_{31} to χ^+_{31} is a degeneration of an integral Knapp-Stein operator.

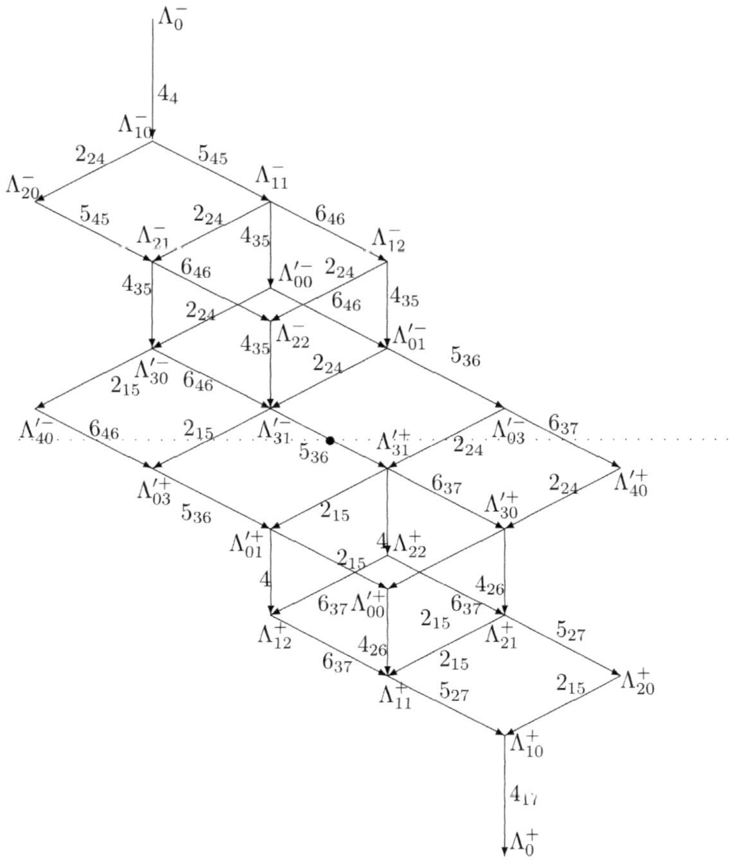

Fig. 15. Reduced multiplets of type R^4_{137} for $su(4,4)$, $su^*(8)$ and $sl(8,\mathbb{R})$

Last reduction of multiplets

There are further reductions of the multiplets - quadruple, etc., but only one quadruple reduction contains representations of physical interest. Namely, this is the multiplet R^4_{1357}, which may be obtained from the main multiplet by setting formally $m_1 = m_3 = m_5 = m_7 = 0$. These multiplets contain 19 ERs/GVMs whose signatures can be given in the following manner:

$$\begin{aligned}
\chi_0^\pm &= \{\, (0, m_2, 0, 0, m_6, 0)^\pm \,;\ \pm m_\rho = \pm(m_2 + 2m_4 + m_6) \,\} \qquad (16)\\
\chi_{11}^\pm &= \{\, (0, m_2, m_4, m_4, m_6, 0)^\pm \,;\ \pm(m_\rho - m_4) = \pm m_{2,4,6} \,\}\\
\chi_{21}^\pm &= \{\, (m_2, 0, m_4, m_{2,4}, m_6, 0)^\pm \,;\ \pm(m_\rho - m_{2,4}) = \pm m_{4,6} \,\}\\
\chi_{12}^\pm &= \{\, (0, m_2, m_{46}, m_4, 0, m_6)^\pm \,;\ \pm(m_\rho - m_{4,6}) = \pm m_{2,4} \,\}\\
\chi_{22}^\pm &= \{\, (m_2, 0, m_{46}, m_{2,4}, 0, m_6)^\pm \,;\ \pm(m_\rho - m_{2,4,6}) = \pm m_4 \,\}\\
\chi'^\pm_{00} &= \{\, (0, m_{2,4}, 0, 0, m_{46}, 0)^\pm \,;\ \pm(m_\rho - 2m_4) = \pm m_{2,6} \,\}\\
\chi'^\pm_{01} &= \{\, (0, m_{2,4}, m_6, 0, m_4, m_6)^\pm \,;\ \pm(m_\rho - m_6 - 2m_4) = \pm m_2 \,\}\\
\chi'^\pm_{30} &= \{\, (m_2, m_4, 0, m_2, m_{46}, 0)^\pm \,;\ \pm(m_\rho - m_2 - 2m_4) = \pm m_6 \,\}\\
\chi'^\pm_{40} &= \{\, (0, m_4, 0, 0, m_{2,46}, 0)^\pm \,;\ \pm(m_\rho - 2m_{2,4}) = \pm(m_6 - m_2) \,\}\\
\chi_{31} &= \{\, (m_2, m_4, m_6, m_2, m_4, m_6) \,;\ 0 \,\}
\end{aligned}$$

The multiplets are given in Fig. 3-1357:

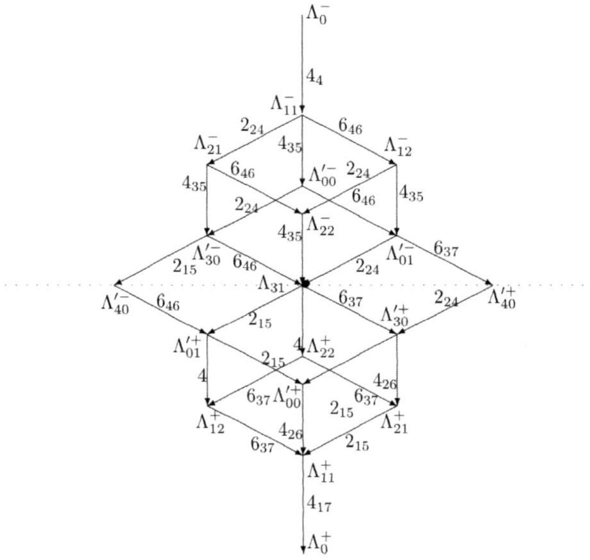

Fig. 16. Reduced multiplets of type R^4_{1357} for $su(4,4)$, $su^*(8)$ and $sl(8,\mathbb{R})$

Note that the ER χ_{31} is not in a pair – it has $c = 0$ and its $\mathcal{M}$ signature is self conjugated. It is placed in the middle of the figure as the bullet. That ER contains the *minimal irreps* characterized by three positive integers which are denoted in this context as m_2, m_4, m_6. Each such irrep is the kernel of the three invariant differential operators $\mathcal{D}^{m_2}_{15}$, $\mathcal{D}^{m_4}_{26}$, $\mathcal{D}^{m_6}_{37}$, which are of order m_2, m_4, m_6, resp., and correspond to the noncompact roots α_{15}, α_{26}, α_{37}, resp.

Acknowledgments

The author thanks the Organizers for the kind invitation to give a plenary talk at the International Conference on Integrable Systems and Quantum Symmetries, Prague, June 2014.

References

[1] V.K. Dobrev, Rev. Math. Phys. **20** (2008) 407-449.
[2] V.K. Dobrev, J. Phys. A: Math. Theor. **42** (2009) 285203.
[3] V.K. Dobrev, Physics of Atomic Nuclei, **76**, No 8, 983-990 (2013).
[4] V.K. Dobrev, Invariant Differential Operators for Non-Compact Lie Groups: the Sp(n,R) Case, in: Springer Proceedings in Mathematics and Statistics, Vol. 36 (ISBN 978-4-431-54269-8), (Springer, Tokyo-Heidelberg, 2013) pp. 311-335.
[5] V.K. Dobrev, J. High Energy Phys. 02 (2013) 015.
[6] V.K. Dobrev, Invariant Differential Operators for Non-Compact Lie Groups: the Reduced SU(3,3) Multiplets, arXiv:1312.5998, Plenary talk at the International Workshop 'Supersymmetries and Quantum Symmetries', Dubna, July 29 - August 3, 2013.
[7] V.K. Dobrev, Invariant Differential Operators for Non-Compact Lie Groups: the Reduced SU(4,4) Multiplets, arXiv:1402.0190, Lectures at "Third International School on Symmetry and Integrable Systems" and at "Humboldt Kolleg on Symmetry and Integrable Systems", Tsakhkadzor, Armenia, July 2013.
[8] V.K. Dobrev, G. Mack, V.B. Petkova, S.G. Petrova and I.T. Todorov, *Harmonic Analysis on the n-Dimensional Lorentz Group and Its Applications to Conformal Quantum Field Theory*, Lecture Notes in Physics, Vol. 63 (Springer, Berlin, 1977).
[9] A.W. Knapp, *Representation Theory of Semisimple Groups (An Overview Based on Examples)*, (Princeton Univ. Press, 1986).
[10] R.P. Langlands, *On the classification of irreducible representations of real algebraic groups*, Math. Surveys and Monographs, Vol. 31 (AMS, 1988), first as IAS Princeton preprint (1973).
[11] A.W. Knapp and G.J. Zuckerman, "Classification theorems for representations of semisimple groups", in: Lecture Notes in Math., Vol. 587 (Springer, Berlin, 1977) pp. 138-159; Ann. Math. **116** (1982) 389-501.
[12] V.K. Dobrev, Rept. Math. Phys. **25**, 159-181 (1988) ; first as ICTP Trieste preprint IC/86/393 (1986).
[13] V.K. Dobrev, Lett. Math. Phys. **9**, 205-211 (1985).
[14] I.N. Bernstein, I.M. Gel'fand and S.I. Gel'fand, Funkts. Anal. Prilozh. **5** (1) (1971) 1-9; English translation: Funct. Anal. Appl. **5** (1971) 1-8.
[15] J. Dixmier, *Enveloping Algebras*, (North Holland, New York, 1977).
[16] Harish-Chandra, Ann. Math. **116** (1966) 1-111.
[17] A.W. Knapp and E.M. Stein, Ann. Math. **93** (1971) 489-578; Inv. Math. **60** (1980) 9-84.
[18] V.K. Dobrev and V.B. Petkova, Rept. Math. Phys. **13**, 233-277 (1978)
[19] I.M. Gelfand and M.A. Naimark, Acad. Sci. USSR. J. Phys. **10** (1946) 93-94.
[20] V. Bargmann, Annals Math. **48**, (1947) 568-640.
[21] V.K. Dobrev, J. Phys. A: Math. Theor. **41** (2008) 425206,
[22] V.K. Dobrev and V.B. Petkova, Lett. Math. Phys. **9** (1985) 287-298.

Journal of Physics: Conference Series **563** (2014) 012009 doi:10.1088/1742-6596/563/1/012009

Symplectic dual pair related to $\mathfrak{so}_{a_1,\ldots,a_{n-1}}(n)$

Alina Dobrogowska and Grzegorz Jakimowicz
Institute of Mathematics, University of Białystok
Akademicka 2, 15-267 Białystok, Poland

E-mail: `alaryzko@alpha.uwb.edu.pl, gjakim@alpha.uwb.edu.pl`

Abstract. We investigate a symplectic dual pair related to a Lie algebra $\mathfrak{so}_{a_1,\ldots,a_{n-1}}(n)$ of deformed skew-symmetric $n \times n$ matrices. The deformations are parametrized by a sequence of real numbers $a_1, \ldots, a_{n-1}$.

1. Introduction

In this paper we construct a symplectic dual pair related to a Lie algebra $\mathfrak{so}_{a_1,\ldots,a_{n-1}}(n)$ of deformed skew-symmetric $n \times n$ matrices, a Lie algebra $\mathfrak{sl}(2,\mathbb{R})$ and a Lie algebra $\mathfrak{p}(1,1)$ of Poincaré group for the (1+1)-dimensional space-time.

We begin by recalling the definitions which will be used in what follows.

Definition 1 *A Poisson manifold* $(P, \{\cdot,\cdot\})$ *is a smooth manifold* P *equipped with a Poisson bracket* $\{\cdot,\cdot\}$.

Definition 2 *Let* $(P_1, \{\cdot,\cdot\}_1)$, $(P_2, \{\cdot,\cdot\}_2)$ *be two Poisson manifolds. A smooth map* $\mathcal{J} : P_1 \longrightarrow P_2$ *is called a Poisson map if*

$$\{\mathcal{J}^* f, \mathcal{J}^* g\}_1 = \mathcal{J}^* \{f, g\}_2, \tag{1}$$

where $f, g \in C^\infty(P_2)$.

Definition 3 *A symplectic manifold* (M, ω) *is a smooth manifold* M *equipped with a skew-symmmetric non-degenerate bilinear form* ω *(symplectic form).*

Definition 4 *For* (M, ω) *a symplectic manifold and* $(P_1, \{\cdot,\cdot\}_1)$, $(P_2, \{\cdot,\cdot\}_2)$ *two Poisson manifolds, a symplectic dual pair is a pair* $\mathcal{J}_1 : M \longrightarrow P_1$, $\mathcal{J}_2 : M \longrightarrow P_2$ *of smooth Poisson maps defined on the same symplectic manifold* (M, ω)*, such that for all* $f \in C^\infty(P_1)$ *and* $g \in C^\infty(P_2)$ *the Poisson bracket on* M *vanishes*

$$\{\mathcal{J}_1^* f, \mathcal{J}_2^* g\}_\omega = 0, \tag{2}$$

see [1, 5, 6, 9]. It is presented on the diagram

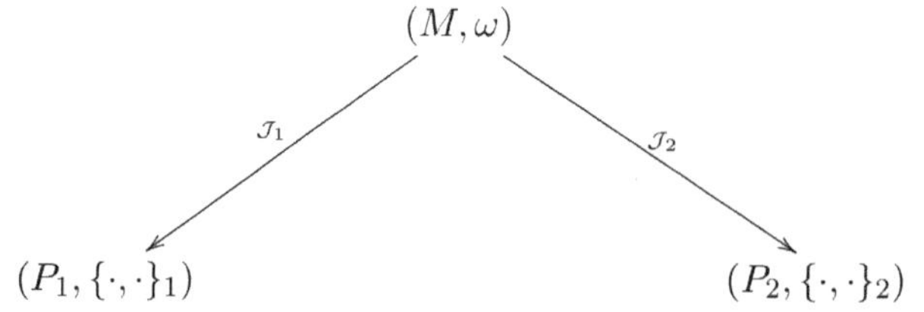

Let $\mathfrak{so}_{a_1,\ldots,a_{n-1}}(n)$ denote the vector space of deformed antisymmetric $n \times n$ matrices of the following form

$$X = \begin{pmatrix} 0 & -a_1x_{12} & -a_1a_2x_{13} & -a_1a_2a_3x_{14} & \ldots & -a_1a_2\cdots a_{n-1}x_{1n} \\ x_{12} & 0 & -a_2x_{23} & -a_2a_3x_{24} & \ldots & -a_2a_3\cdots a_{n-1}x_{2n} \\ x_{13} & x_{23} & 0 & -a_3x_{34} & \ldots & -a_3a_4\cdots a_{n-1}x_{3n} \\ x_{14} & x_{24} & x_{34} & 0 & \ldots & -a_4a_5\cdots a_{n-1}x_{4n} \\ \vdots & \vdots & \vdots & \vdots & \ddots & \vdots \\ x_{1n} & x_{2n} & x_{3n} & x_{4n} & \ldots & 0 \end{pmatrix}, \tag{3}$$

where $x_{ij} \in \mathbb{R}$. The deformations are parametrized by a sequence of real numbers $a_1, \ldots, a_{n-1}$. If all parameters $a_1, \ldots, a_{n-1}$ are non-zero, one has

$$X\eta + \eta X^\top = 0, \tag{4}$$

where η is the following diagonal matrix

$$\eta = \begin{pmatrix} a_1a_2\cdots a_{n-1} & 0 & \ldots & 0 & 0 \\ 0 & a_2\cdots a_{n-1} & \ldots & 0 & 0 \\ \vdots & \vdots & \vdots & \ddots & \vdots \\ 0 & 0 & \ldots & a_{n-1} & 0 \\ 0 & 0 & \ldots & 0 & 1 \end{pmatrix}. \tag{5}$$

The $(\mathfrak{so}_{a_1,\ldots,a_{n-1}}(n), [\cdot,\cdot])$ is a Lie algebra with the standard commutator, see [2, 8]. The considered class of deformed Lie algebras contains several known classical Lie algebras, for example if all $a_i = 1$ then $\mathfrak{so}_{1,\ldots,1}(n) = \mathfrak{so}(n)$, if $a_1 = 0$ and $a_2 = \ldots = a_{n-1} = 1$ then $\mathfrak{so}_{0,1,\ldots,1}(n) = \mathfrak{e}(n-1)$.

The idea of considering these systems comes from [3, 4]. In this paper we present more general case, which reduces to the cases considering in [3, 4] if we put that the parameters $a_3 = a_4 = 1$ and $a_i = 0$ for $i \geq 5$.

2. Poisson structures

Let $\mathcal{L}_+(n)$ be the vector space of strictly upper triangular $n \times n$ matrices ($n \geq 2$)

$$\mathcal{L}_+(n) \ni \rho = \begin{pmatrix} 0 & \rho_{12} & \rho_{13} & \rho_{14} & \ldots & \rho_{1n} \\ 0 & 0 & \rho_{23} & \rho_{24} & \ldots & \rho_{2n} \\ 0 & 0 & 0 & \rho_{34} & \ldots & \rho_{3n} \\ 0 & 0 & 0 & 0 & \ldots & \rho_{4n} \\ \vdots & \vdots & \vdots & \vdots & \ddots & \vdots \\ 0 & 0 & 0 & 0 & \ldots & 0 \end{pmatrix}. \tag{6}$$

Relative to the non-degenerate pairing

$$\langle X, \rho \rangle = \mathrm{Tr}(\rho X), \quad \rho \in \mathcal{L}_+(n), \quad X \in \mathfrak{so}_{a_1,\ldots,a_{n-1}}(n), \tag{7}$$

we will identify $\mathcal{L}_+(n)$ with the dual $(\mathfrak{so}_{a_1,\ldots,a_{n-1}}(n))^*$ of $\mathfrak{so}_{a_1,\ldots,a_{n-1}}(n)$.

The duality $(\mathfrak{so}_{a_1,\ldots,a_{n-1}}(n))^* \cong \mathcal{L}_+(n)$ defines a canonical Lie–Poisson bracket on $\mathcal{L}_+(n)$

$$\{f, g\}_1 = \mathrm{Tr}\left(\rho\left[Df(\rho), Dg(\rho)\right]\right), \tag{8}$$

where $f,g\in C^\infty(\mathcal{L}_+(n))$, see [8]. Moreover if we have

$$Df(\rho):=\begin{pmatrix} 0 & -a_1\frac{\partial f}{\partial\rho_{12}} & -a_1a_2\frac{\partial f}{\partial\rho_{13}} & -a_1a_2a_3\frac{\partial f}{\partial\rho_{14}} & \dots & -a_1a_2\cdots a_{n-1}\frac{\partial f}{\partial\rho_{1n}}\\ \frac{\partial f}{\partial\rho_{12}} & 0 & -a_2\frac{\partial f}{\partial\rho_{23}} & -a_2a_3\frac{\partial f}{\partial\rho_{24}} & \dots & -a_2a_3\cdots a_{n-1}\frac{\partial f}{\partial\rho_{2n}}\\ \frac{\partial f}{\partial\rho_{13}} & \frac{\partial f}{\partial\rho_{23}} & 0 & -a_3\frac{\partial f}{\partial\rho_{34}} & \dots & -a_3a_4\cdots a_{n-1}\frac{\partial f}{\partial\rho_{3n}}\\ \frac{\partial f}{\partial\rho_{14}} & \frac{\partial f}{\partial\rho_{24}} & \frac{\partial f}{\partial\rho_{34}} & 0 & \dots & -a_4a_5\cdots a_{n-1}\frac{\partial f}{\partial\rho_{4n}}\\ \vdots & \vdots & \vdots & \vdots & \ddots & \vdots\\ \frac{\partial f}{\partial\rho_{1n}} & \frac{\partial f}{\partial\rho_{2n}} & \frac{\partial f}{\partial\rho_{3n}} & \frac{\partial f}{\partial\rho_{4n}} & \dots & 0 \end{pmatrix}, \tag{9}$$

then the Lie–Poisson bracket (8) can be rewritten in the form

$$\begin{aligned}\{f,g\}_1=&\sum_{k=1}^{n-1}\sum_{j>k}^{n}\rho_{kj}\left(\sum_{i=1}^{k-1}a_i\dots a_{k-1}\left(\frac{\partial f}{\partial\rho_{ik}}\frac{\partial g}{\partial\rho_{ij}}-\frac{\partial f}{\partial\rho_{ij}}\frac{\partial g}{\partial\rho_{ik}}\right)\right.\\ &\left.+\sum_{i>k}^{j-1}\left(\frac{\partial f}{\partial\rho_{ij}}\frac{\partial g}{\partial\rho_{ki}}-\frac{\partial f}{\partial\rho_{ki}}\frac{\partial g}{\partial\rho_{ij}}\right)+\sum_{i>j}^{n}a_j\dots a_{i-1}\left(\frac{\partial f}{\partial\rho_{ki}}\frac{\partial g}{\partial\rho_{ji}}-\frac{\partial f}{\partial\rho_{ji}}\frac{\partial g}{\partial\rho_{ki}}\right)\right).\end{aligned} \tag{10}$$

Next let us consider the Lie algebra $\mathfrak{sl}(2,\mathbb{R})=\left\{\begin{pmatrix}\frac{1}{2}x & y\\ -z & -\frac{1}{2}x\end{pmatrix}:\ x,y,z\in\mathbb{R}\right\}$. We identify the dual space $\mathfrak{sl}(2,\mathbb{R})^*=\left\{\begin{pmatrix}d_3 & -d_1\\ d_2 & -d_3\end{pmatrix}:\ d_1,d_2,d_3\in\mathbb{R}\right\}$ with $\mathfrak{sl}(2,\mathbb{R})$ via the non-degenerate pairing

$$\langle Y,d\rangle=\mathrm{Tr}(dY),\quad d\in\mathfrak{sl}(2,\mathbb{R})^*,\quad Y\in\mathfrak{sl}(2,\mathbb{R}). \tag{11}$$

Let us recall that the Lie–Poisson bracket on $\mathfrak{sl}(2,\mathbb{R})^*$ is given by the formula

$$\begin{aligned}\{f,g\}_2=2d_3\left(\frac{\partial f}{\partial d_1}\frac{\partial g}{\partial d_2}-\frac{\partial f}{\partial d_2}\frac{\partial g}{\partial d_1}\right)+d_1\left(\frac{\partial f}{\partial d_1}\frac{\partial g}{\partial d_3}-\frac{\partial f}{\partial d_3}\frac{\partial g}{\partial d_1}\right)+\\ +d_2\left(\frac{\partial f}{\partial d_3}\frac{\partial g}{\partial d_2}-\frac{\partial f}{\partial d_2}\frac{\partial g}{\partial d_3}\right),\end{aligned} \tag{12}$$

see [7].

The Lie algebra $\mathfrak{p}(1,1)$ consists of 3×3 matrices of the form

$$\begin{pmatrix}0 & c & h\\ c & 0 & l\\ 0 & 0 & 0\end{pmatrix},\quad c,h,l\in\mathbb{R}. \tag{13}$$

The dual space to the Lie algebra $\mathfrak{p}(1,1)$ can be identified with the space

$$\mathfrak{p}(1,1)^*=\left\{\begin{pmatrix}0 & \frac{1}{2}d_3 & 0\\ \frac{1}{2}d_3 & 0 & 0\\ e_1 & e_2 & 0\end{pmatrix}:\quad e_1,e_2,d_3\in\mathbb{R}\right\} \tag{14}$$

using the non-degenerate pairing

$$\langle Z,e\rangle=\mathrm{Tr}(eZ),\quad e\in\mathfrak{p}(1,1)^*,\quad Z\in\mathfrak{p}(1,1). \tag{15}$$

In the coordinates (e_1,e_2,d_3) the Lie–Poisson bracket for $f,g\in C^\infty(\mathfrak{p}(1,1)^*)$ is given by the formula

$$\{f,g\}_3=e_1\left(\frac{\partial f}{\partial d_3}\frac{\partial g}{\partial e_1}-\frac{\partial f}{\partial e_1}\frac{\partial g}{\partial d_3}\right)+e_2\left(\frac{\partial f}{\partial d_3}\frac{\partial g}{\partial e_2}-\frac{\partial f}{\partial e_2}\frac{\partial g}{\partial d_3}\right). \tag{16}$$

We will consider also the cotangent vector bundle $T^*\mathbb{R}^n$ with canonical symplectic form

$$\omega = \sum_{i=1}^{n} dq_i \wedge dp_i, \tag{17}$$

where $(q_1, \ldots, q_n, p_1, \ldots, p_n) \in T^*\mathbb{R}^n \cong \mathbb{R}^n \times (\mathbb{R}^n)^*$. In this case the Poisson bracket on $T^*\mathbb{R}^n$ takes the standard form

$$\{f, g\}_\omega(q_1, \ldots, q_n, p_1, \ldots, p_n) = \sum_{i=1}^{n} \left(\frac{\partial f}{\partial q_i}\frac{\partial g}{\partial p_i} - \frac{\partial f}{\partial p_i}\frac{\partial g}{\partial q_i} \right), \tag{18}$$

where $f, g \in C^\infty(T^*\mathbb{R}^n)$.

3. Poisson maps

In this section we will introduce the Poisson maps $\mathcal{J}_1 : T^*\mathbb{R}^n \longrightarrow \mathcal{L}_+(n)$, $\mathcal{J}_2 : T^*\mathbb{R}^n \longrightarrow \mathfrak{sl}(2,\mathbb{R})^*$ and $\mathcal{J}_3 : T^*\mathbb{R}^n \longrightarrow \mathfrak{p}(1,1)^*$.

Proposition 1 *The map $\mathcal{J}_1 : T^*\mathbb{R}^n \longrightarrow \mathcal{L}_+(n)$ defined by*

$$\begin{aligned} (\mathcal{J}_1(q_1, \ldots, q_n, p_1, \ldots, p_n))_{ij} &= a_i \ldots a_{j-1} q_i p_j - p_i q_j =: \rho_{ij} \qquad for \qquad i < j, \\ (\mathcal{J}_1(q_1, \ldots, q_n, p_1, \ldots, p_n))_{ij} &= 0 \qquad for \qquad i \geq j, \end{aligned} \tag{19}$$

is a Poisson map.

Proof 1 *For simplicity let us calculate the expression* $\frac{\partial (f \circ \mathcal{J}_1)}{\partial q_i}\frac{\partial (g \circ \mathcal{J}_1)}{\partial p_i} - \frac{\partial (f \circ \mathcal{J}_1)}{\partial p_i}\frac{\partial (g \circ \mathcal{J}_1)}{\partial q_i}$ *for fixed index i. Since the derivatives of ρ_{ij} are*

$$\begin{aligned} \frac{\partial \rho_{ij}}{\partial q_i} &= a_i \ldots a_{j-1} p_j \qquad for \qquad i < j, \\ \frac{\partial \rho_{ji}}{\partial q_i} &= -p_j \qquad for \qquad i > j, \\ \frac{\partial \rho_{ij}}{\partial p_i} &= -q_j \qquad for \qquad i < j, \\ \frac{\partial \rho_{ji}}{\partial q_i} &= a_j \ldots a_{i-1} q_j \qquad for \qquad i > j, \end{aligned}$$

after a direct calculation we obtain

$$\frac{\partial(f\circ\mathcal{J}_1)}{\partial q_i}\frac{\partial(g\circ\mathcal{J}_1)}{\partial p_i}-\frac{\partial(f\circ\mathcal{J}_1)}{\partial p_i}\frac{\partial(g\circ\mathcal{J}_1)}{\partial q_i}=\left(\sum_{j>i}^{n}a_i\dots a_{j-1}p_j\frac{\partial f}{\partial\rho_{ij}}-\sum_{j=1}^{i-1}p_j\frac{\partial f}{\partial\rho_{ji}}\right)$$
$$\left(\sum_{k=1}^{i-1}a_k\dots a_{i-1}q_k\frac{\partial g}{\partial\rho_{ki}}-\sum_{k>i}^{n}q_k\frac{\partial g}{\partial\rho_{ik}}\right)-\left(\sum_{j=1}^{i-1}a_j\dots a_{i-1}q_j\frac{\partial f}{\partial\rho_{ji}}-\sum_{j>i}^{n}q_j\frac{\partial f}{\partial\rho_{ij}}\right)$$
$$\left(\sum_{k>i}^{n}a_i\dots a_{k-1}p_k\frac{\partial g}{\partial\rho_{ik}}-\sum_{k=1}^{i-1}p_k\frac{\partial g}{\partial\rho_{ki}}\right)=-\sum_{j>i}^{n}\sum_{k>i}^{j-1}a_i\dots a_{k-1}\left(a_k\dots a_{j-1}q_kp_j-q_jp_k\right)\frac{\partial f}{\partial\rho_{ij}}\frac{\partial g}{\partial\rho_{ik}}$$
$$+\sum_{k>i}^{n}\sum_{j>i}^{k-1}a_i\dots a_{j-1}\left(a_j\dots a_{k-1}q_jp_k-q_kp_j\right)\frac{\partial f}{\partial\rho_{ij}}\frac{\partial g}{\partial\rho_{ik}}-\sum_{k>i}^{n}\sum_{j=1}^{i-1}\left(a_j\dots a_{k-1}q_jp_k-q_kp_j\right)\frac{\partial f}{\partial\rho_{ji}}\frac{\partial g}{\partial\rho_{ik}}$$
$$+\sum_{j>i}^{n}\sum_{k=1}^{i-1}\left(a_k\dots a_{j-1}q_kp_j-q_jp_k\right)\frac{\partial f}{\partial\rho_{ij}}\frac{\partial g}{\partial\rho_{ki}}-\sum_{j=2}^{i-1}\sum_{k=1}^{j-1}a_j\dots a_{i-1}\left(a_k\dots a_{j-1}q_kp_j-q_jp_k\right)\frac{\partial f}{\partial\rho_{ji}}\frac{\partial g}{\partial\rho_{ki}}$$
$$+\sum_{k=2}^{i-1}\sum_{j=1}^{k-1}a_k\dots a_{i-1}\left(a_j\dots a_{k-1}q_jp_k-q_kp_j\right)\frac{\partial f}{\partial\rho_{ji}}\frac{\partial g}{\partial\rho_{ki}}=\sum_{j>i}^{n}\sum_{k=1}^{i-1}\rho_{kj}\left(\frac{\partial f}{\partial\rho_{ij}}\frac{\partial g}{\partial\rho_{ki}}-\frac{\partial f}{\partial\rho_{ki}}\frac{\partial g}{\partial\rho_{ij}}\right)$$
$$+\sum_{j>i}^{n}\sum_{k>i}^{j-1}a_i\dots a_{k-1}\rho_{kj}\left(\frac{\partial f}{\partial\rho_{ik}}\frac{\partial g}{\partial\rho_{ij}}-\frac{\partial f}{\partial\rho_{ij}}\frac{\partial g}{\partial\rho_{ik}}\right)+\sum_{j=2}^{i-1}\sum_{k=1}^{j-1}a_j\dots a_{i-1}\rho_{kj}\left(\frac{\partial f}{\partial\rho_{ki}}\frac{\partial g}{\partial\rho_{ji}}-\frac{\partial f}{\partial\rho_{ji}}\frac{\partial g}{\partial\rho_{ki}}\right).$$

To obtain the equality (1) for the Poisson map (19) we sum up the above equalities from $i=1$ to $i=n$.

Moreover, if we consider the functions on $T^*\mathbb{R}^n$ given by expression

$$\begin{aligned}D_1&=a_1\dots a_{n-1}q_1^2+a_2\dots a_{n-1}q_2^2+\dots+q_n^2,\\ D_2&=p_1^2+a_1p_2^2+\dots+a_1\dots a_{n-1}p_n^2,\\ D_3&=q_1p_1+q_2p_2+\dots q_np_n,\end{aligned}\tag{20}$$

after simpler calculation we obtain

$$\begin{aligned}\{D_1,D_2\}&=4a_1\dots a_{n-1}D_3,\\ \{D_3,D_1\}&=-2D_1,\\ \{D_3,D_2\}&=2D_2.\end{aligned}\tag{21}$$

We see that if all parameters $a_i\neq 0$ then (D_1,D_2,D_3) form Lie algebra $\mathfrak{sl}(2,\mathbb{R})$, if for some $a_i=0$ then (D_1,D_2,D_3) form Lie algebra $\mathfrak{p}(1,1)$. Summing up we have the following proposition.

Proposition 2 *If all parameters $a_i\neq 0$ then the map $\mathcal{J}_2:T^*\mathbb{R}^n\longrightarrow\mathfrak{sl}(2,\mathbb{R})^*$ defined by*

$$\begin{aligned}(\mathcal{J}_2(q_1,\dots,q_n,p_1,\dots,p_n))_{11}&=-\left(\mathcal{J}_2(q_1,\dots,q_n,p_1,\dots,p_n)\right)_{22}=\frac{1}{2}D_3=:d_3,\\ (\mathcal{J}_2(q_1,\dots,q_n,p_1,\dots,p_n))_{21}&=\frac{1}{2a_1\dots a_{n-1}}D_2=:d_2,\\ (\mathcal{J}_2(q_1,\dots,q_n,p_1,\dots,p_n))_{12}&=-\frac{1}{2}D_1=:-d_1\end{aligned}\tag{22}$$

is a Poisson map.

Journal of Physics: Conference Series **563** (2014) 012009 doi:10.1088/1742-6596/563/1/012009

Proof 2 *After direct calculation we have*

$$
\begin{aligned}
&\{f\circ \mathcal{J}_2, g\circ \mathcal{J}_2\}_\omega(q_1,\dots,q_n,p_1,\dots,p_n)\\
&=\sum_{i=1}^{n}\Bigg(\left(a_i\dots a_{n-1}q_i\frac{\partial f}{\partial d_1}+\frac{1}{2}p_i\frac{\partial f}{\partial d_3}\right)\left(\frac{a_1\dots a_{i-1}}{a_1\dots a_{n-1}}p_i\frac{\partial g}{\partial d_2}+\frac{1}{2}q_i\frac{\partial g}{\partial d_3}\right)\\
&-\left(\frac{a_1\dots a_{i-1}}{a_1\dots a_{n-1}}p_i\frac{\partial f}{\partial d_2}+\frac{1}{2}q_i\frac{\partial f}{\partial d_3}\right)\left(a_i\dots a_{n-1}q_i\frac{\partial f}{\partial d_1}+\frac{1}{2}p_i\frac{\partial f}{\partial d_3}\right)\Bigg)\\
&=\sum_{i=1}^{n}q_ip_i\left(\frac{\partial f}{\partial d_1}\frac{\partial g}{\partial d_2}-\frac{\partial f}{\partial d_2}\frac{\partial g}{\partial d_1}\right)+\frac{1}{2}\sum_{i=1}^{n}a_i\dots a_{n-1}q_i^2\left(\frac{\partial f}{\partial d_1}\frac{\partial g}{\partial d_3}-\frac{\partial f}{\partial d_3}\frac{\partial g}{\partial d_1}\right)\\
&+\frac{1}{2a_1\dots a_{n-1}}\sum_{i=1}^{n}a_1\dots a_{i-1}p_i^2\left(\frac{\partial f}{\partial d_3}\frac{\partial g}{\partial d_2}-\frac{\partial f}{\partial d_2}\frac{\partial g}{\partial d_3}\right)\\
&=\left(\{f,g\}_2\circ\mathcal{J}_2\right)(q_1,\dots,q_n,p_1,\dots,p_n).
\end{aligned}
$$

Proposition 3 *If one or more parameters a_i are equal to zero then the map $\mathcal{J}_3 : T^*\mathbb{R}^n \longrightarrow \mathfrak{p}(1,1)^*$ defined by*

$$
\begin{aligned}
&(\mathcal{J}_3(q_1,\dots,q_n,p_1,\dots,p_n))_{12}=(\mathcal{J}_3(q_1,\dots,q_n,p_1,\dots,p_n))_{21}=\frac{1}{4}D_3=:\frac{1}{2}d_3,\\
&(\mathcal{J}_3(q_1,\dots,q_n,p_1,\dots,p_n))_{31}=\frac{1}{2}(D_2+D_1)=:e_1,\\
&(\mathcal{J}_3(q_1,\dots,q_n,p_1,\dots,p_n))_{32}=\frac{1}{2}(D_2-D_1)=:e_2,\\
&(\mathcal{J}_3(q_1,\dots,q_n,p_1,\dots,p_n))_{ij}=0,\quad for\quad (ij)=(11),(22),(13),(23),(33),
\end{aligned}
\tag{23}
$$

is a Poisson map.

Proof 3 *The proof is completely analogous to the proof above.*

4. Symplectic dual pair

In this section we present the main result of the theorem concerning a symplectic dual pair.

Proposition 4 *If all parameters $a_i \neq 0$ then we have the following symplectic dual pair*

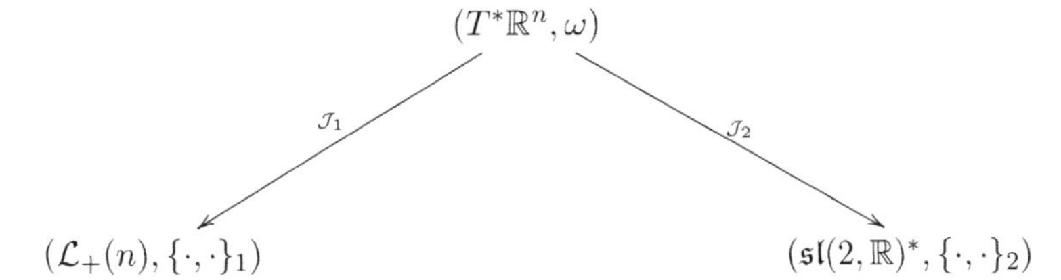

where $\mathcal{J}_1$, $\mathcal{J}_2$ are defined by (19) and (22).

Proof 4 *From (19), we obtain*

$$\rho_{1i}\rho_{2j}-\rho_{1j}\rho_{2i}=a_2\dots a_{i-1}\rho_{12}\rho_{ij}\quad for\quad 2<i<j$$

and we conclude that we have only $2n-3$ functionally independent functions ρ_{1i}, $\rho_{2,j}$, $i=2,\dots,n$, $j=3,\dots,n$ on $T^*\mathbb{R}^n$.

The conditions $\{d_i, f\}_\omega = 0$, $i = 1, 2, 3$ *give the following system of equations*

$$a_1 \dots a_{n-1} q_1 \frac{\partial f}{\partial p_1} + a_2 \dots a_{n-1} q_2 \frac{\partial f}{\partial p_2} + \dots + q_n \frac{\partial f}{\partial p_n} = 0, \tag{24}$$

$$p_1 \frac{\partial f}{\partial q_1} + a_1 p_2 \frac{\partial f}{\partial q_2} + \dots + a_1 \dots a_{n-1} p_n \frac{\partial f}{\partial q_n} = 0, \tag{25}$$

$$q_1 \frac{\partial f}{\partial q_1} + \dots + q_n \frac{\partial f}{\partial q_n} - p_1 \frac{\partial f}{\partial p_1} - \dots - p_n \frac{\partial f}{\partial p_n} = 0. \tag{26}$$

It is easy to see that the $2n-3$ *functions* ρ_{1i}, $\rho_{2,j}$ *are the solutions of the above system of equations. Assume that this system has an additional functionally independent solution* ψ. *From the method of characteristics for the linear partial differential equations, we know that every equation in the system has* $2n-1$ *first integrals which are functionally independent. For example, the functions* ρ_{1i}, $\rho_{2,j}$, q_1, q_2 *form a complete system of first integrals for equation (24). Suppose that* $\psi = f(\rho_{1i}, \rho_{2j}, q_1, q_2)$. *Substituting the function* ψ *to equations (25) and (26), we obtain that* ψ *does not depend explicitly on* q_1 *and* q_2. *This implies that system of equations (24-26) has* $2n-3$ *functionally independent solutions.*

In the opposite direction, from $\{\rho_{ij}, f\}_\omega = 0$ *we obtain*

$$p_i \frac{\partial f}{\partial p_j} - q_j \frac{\partial f}{\partial q_i} + a_i \dots a_{j-1} q_i \frac{\partial f}{\partial q_j} - a_i \dots a_{j-1} p_j \frac{\partial f}{\partial p_i} = 0, \quad i < j. \tag{27}$$

We have $n-1$ *linear independent equations enumerated by* $i = 1$ *and* $j = 2, \dots, n$. *Note that the* j*-th equation in the system has* $2n-1$ *functionally independent solutions* $q_2, \dots, q_{j-1}, q_{j+1}, \dots, q_n, p_2, \dots, p_{j-1}, p_{j+1}, \dots, p_n, d_1, d_2, d_3$. *Using analogous considerations as above we obtain that any solution of the system of equations (27) has the form* $f = g(d_1, d_2, d_2)$, *where* g *is a smooth function.*

Proposition 5 *If one or more parameters* a_i *are equal to zero then we have the following symplectic dual pair*

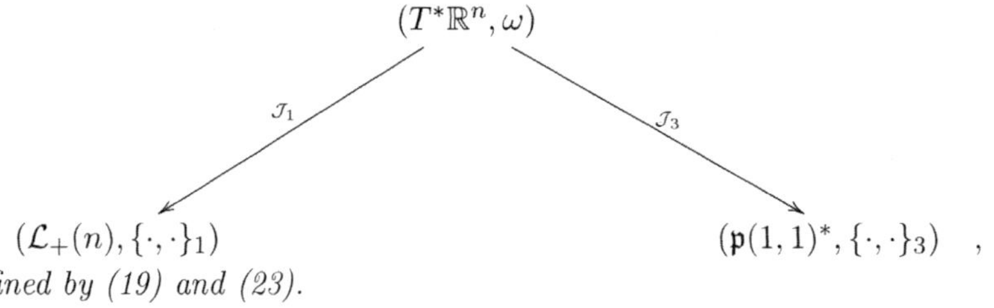

where $\mathcal{J}_1$, $\mathcal{J}_3$ *are defined by (19) and (23).*

Proof 5 *The proof is analogous to the previous proof.*

References

[1] A. Cannas da Silva, A. Weinstein, *Geometric models for noncommutative algebras*, Berkeley Math. Lecture Notes, Amer. Math. Soc. (1999).

[2] A. Dobrogowska, T. Goliński, *Lie bundle on the space of deformed skew-symmetric matrices*, J. Math. Phys. (to appear).

[3] A. Dobrogowska, A. Odzijewicz, *Integrable systems related to deformed so(5)*, SIGMA 10 (2014).

[4] A. Dobrogowska, A. Odzijewicz, *Symplectic Dual Pair Related to Deformed so(5)*, Geometric Methods in Physics, XXXII Workshop 2013, Trends in Mathematics, 121-130, Birkhauser Verlag, Basel, (2014).

[5] M.V. Karasev, *The Maslov quantization conditions in higher cohomology and analogs of notions developed in Lie theory for canonical fibre bundles of symplectic manifolds* I, II Selecta Math. Soviet. 8, 213-234, 235-258 (1989).

[6] N.P. Landsman, *Functorial quantization and the Guillemin-Sternberg conjecture*, Proc. Bialowieza 2002 (arXiv:math-ph/0307059).

[7] J.E. Marsden, T.S. Ratiu, *Introduction to Mechanics and Symmetry: A Basic Exposition of Classical Mechanical Systems*, Springer (1999).
[8] A. Odzijewicz, A. Dobrogowska, *Integrable Hamiltonian systems related to the Hilbert–Schmidt ideal*, J. Geom. Phys., 61, 1426-1445 (2011).
[9] A. Weinstein, *The local structure of Poisson manifolds*, J. Diff. Geom. 18, 523-557 (1983).

Journal of Physics: Conference Series **563** (2014) 012010 doi:10.1088/1742-6596/563/1/012010

Q-representations and unitary representations of the super-Heisenberg group and harmonic superanalysis

Andrzej M.Frydryszak

Institute of Theoretical Physics, University of Wroclaw, Wroclaw, Poland

E-mail: andrzej.frydryszak@ift.uni.wroc.pl

Abstract. We juxtapose two approaches to the representations of the super-Heisenberg group. Physical one, sometimes called concrete approach, based on the super-wave functions depending on the anti-commuting variables, yielding the harmonic superanalysis and recently developed strict theory of unitary representations of the nilpotent super Lie groups covering the unitary representations of the super-Heisenberg group.

1. Introduction

The Heisenberg group plays distinguished role as in theoretical physics as in mathematics, being a fundamental building block of many theories and formalisms [1, 2]. From the physical point of view it comes from the linearization of the group of canonical transformations. Its representations, in fact, arise from the quantization procedure - the canonical quantization. This procedure has its own difficulties, but al the l quantum physics we know and which is supported by experiment, was discovered by the use of this tool, despite the fact that recent decades brought vivid development of new approaches, to mention the deformation quantization only. There are lots of generalizations and applications of the original Heisenberg group and of the harmonic analysis on it, depending on the mathematical or physical context (discrete, supersymmetric, deformed). Here we want to focus only on the generalization coming from the supersymmetry in the context of physics and related to it supermathematics [3, 4]. Firstly we shall discuss supermatrix version of it, coming from the supersymmetric mechanics and in particular from the supersymmetric model of the harmonic oscillator. Such a concrete approach is good enough to produce superized version of the harmonic analysis on the super-Heisenberg group, special superfunctions (super-Hermite polynomials, super-Laguerre etc.). It is related to the so called super-Schroedinger quantization. Within supersymmetric approach it is possible to introduce two types of super-Heisenberg group namely the even super-Heisenberg group and the odd super-Heisenberg group. The even super-Heisenberg group is the one usually discussed in the literature, the odd one, is less frequent. To write (super)representations of the odd super-Heisenberg group one has to introduce a new algebra of "numbers" with the odd unit and and odd imaginary unit in the real and the complex case respectively. Such structure, named the algebra of oddons, has been defined in Ref. [5]. Its particular version is also known under the name super skew field [6] and plays important role in algebraic supergeometry [6].

The plan of the paper is the following. Firstly we shall discuss the super-matrix formulation of the even and odd super Heisenberg group and its Q-representations resulting from the super-

Schrödinger quantization, then we shall briefly comment recent developments in the strict theory of unitary representations of the super Heisenberg group defined as a super Harish-Chandra pair.

2. Super-Heisenberg group

In the so called concrete approach the super-Heisenberg group is introduced as a supermatrix group, related to supersymmetric mechanics formalism in analogy to the way that conventional Heisenberg group arrises in the context of the phase space formulation of classical mechanics [1]. Let us briefly recall this way of defining Heisenberg group and some points of the Schrödinger quantization procedure which yields the harmonic analysis on Heisenberg group. The effect of the canonical quantization procedure is summarized by the Stone-von Neumann theorem.

For the super-Heisenberg group one can follow similar lines, in particular, so called Grassmannian Fourier transform can be defined and superized versions of the Hermite and Laguerre polynomials obtained. The super-Hermite polynomials arise in the super-Srödinger quantization of the fermionic super-oscillator [7]. To fix the notation let us remind some fact concerning the Heisenberg group and phase space, this will form a background to the generalization to the super-Heisenberg group case.

2.1. Heisenberg group

Let us consider conventional phase space $\mathcal{P}_{2n} \ni (p,q)$ with a symplectic form Ω: $\Omega((p,q),(p',q')) = pq' - p'q$ which relates Poisson bracket of classical phase space observables with Hamiltonian vector fields

$$\{f,g\} = \Omega(X_f, X_g) \tag{1}$$

and moreover

$$[X_f, X_g] = X_{\{f,g\}}, \quad \{p^j, q_i\} = \delta_i^j$$

In the extended phase space $\mathcal{P}_{2n+1} \ni (p,q,t)$ one can define Heisenberg algebra by commutation relations

$$[(p,q,t),(p',q',t')] = (0,0,pq'-p'q) = (0,0,\Omega((p,q),(p',q'))), \tag{2}$$

therefore

$$[P_j, Q_k] = \delta_{jk}T. \tag{3}$$

Such commutation relations can be explicitly realized [1] by matrices

$$m(p,q,t) = \begin{pmatrix} 0 & p & t \\ 0 & 0 & q \\ 0 & 0 & 0 \end{pmatrix} = m(p,q,t)m(p',q',t') = m(0,0,pq'), \tag{4}$$

There are two isomorphic forms of the group realization. The first one given by

$$M(p,q,t) = 1 + m(p,q,t) \tag{5}$$

and yields the following form of group multiplication

$$M(p,q,t)M(p',q',t') - M(p+p',q+q',t+t'+pq'). \tag{6}$$

The second one is obtained from the exponentiation

$$e^{m(p,q,t)} = 1 + m(p,q,t) + \frac{1}{2}pq = M(p,q,t+\frac{1}{2}pq) \tag{7}$$

and produces more symmetric form of the multiplication

$$e^{m(p,q,t)}e^{m(p',q',t')} = e^{m(p+p',q+q',t+t'+\frac{1}{2}(pq'-p'q))} \tag{8}$$

The group law can be written in a compact form showing the relation to the symplectic geometry of the phase space, namely

$$(p,q,t)\circ(p',q',t') = \left(p+p',q+q',t+t'+\frac{1}{2}\Omega((p,q),(p',q'))\right) \tag{9}$$

The Schrödinger quantization procedure uses the following construction. We take as the quantum state space of the system the Hilbert space $\mathcal{H} = L^2(R^n, d\mu)$, where the R^n is the configuration space of a classical system, and realize the canonical commutation relations: $[\hat{X}, \hat{P}] = i\hbar$:

$$\hat{X} : f(x) \mapsto xf(x), \qquad \hat{P} : f(x) \mapsto -i\hbar\frac{\partial f}{\partial x}(x) \tag{10}$$

It turns out that his realization is canonical in generic sense. Namely, we have that

$$\hat{X} \rightsquigarrow e^{iq\hat{X}}, \qquad e^{iq\hat{X}} : f(x) \mapsto e^{iqx}f(x) \tag{11}$$

$$\hat{P} \rightsquigarrow e^{ip\hat{P}}, \qquad e^{ip\hat{P}} : f(x) \mapsto f(x+\hbar p), \tag{12}$$

and for elements of the Heisenberg group

$$H_n \ni M(p,q,t) = \begin{pmatrix} 1 & p & t \\ 0 & 1 & q \\ 0 & 0 & 1 \end{pmatrix}$$

one obtains representation

$$\rho : H_n \to U\mathcal{H} \qquad M(0,q,0) \mapsto e^{iqx} \qquad M(0,0,p) \mapsto e^{ip\hat{P}} \tag{13}$$

with

$$\rho_h(p,q,t) = e^{2\pi i h t}e^{2\pi i(hpD+qX)}. \tag{14}$$

The $U\mathcal{H}$ denotes the set of unitary mappings on the $\mathcal{H}$. The $\rho_h(p,q,t)$, $h \in R$, $h \neq 0$ forms one parameter family of the irreducible unitary representations of the H_n. As it was already mentioned this construction is canonical in the sense given by the Stone-von Neumann theorem: *Any irreducible unitary representation of H_n that is nontrivial on the center is equivalent to some ρ_h.*

2.2. The even super-Heisenberg group

To introduce the super-Heisenberg group in the supermatrix form let us follow lines, similar to the ones sketched above. Namely, for supersymmetric system one introduces the graded phase space $\mathcal{P}_{(2n|2m)} \ni (p,q;\Pi,\Theta)$ with $\Pi = (\Pi^j)$, $\Theta = (\Theta_j)$, where $j = 1,\ldots,m$; with the parity $|\Theta_j| = 1 = |\Pi^j|$. For physical models the $\mathcal{P}_{(2n|2m)}$ is Z_2-graded module over Z_2 graded algebra Q [8]. We introduce the supersymplectic form B

$$B(v,v') = -(-1)^{|v||v'|}B(v',v), \text{for homogenous} \quad v,v' \tag{15}$$

The even part of the $\mathcal{P}_{(2n|2m)}$ contains usual phase space, therefore let us here restrict to the new odd sector $(\Pi,\Theta) \in \mathcal{P}_{2m}$. In the odd basis $\{E_j\}_1^{2m}$ one can write

$$v = \Pi^j E_j + \Theta_j E^{m+j} \tag{16}$$

and

$$B(v,\, v') = \Pi^j \Theta'_j + \Theta_j \Pi'^j \tag{17}$$

The generalization of the Heisenberg group related to this sector we shall call the fermionic Heisenberg group. It is a part of the whole super-Heisenberg group.

Let us consider the super-Poison bracket: $f, g, h \in \mathcal{F}(\mathcal{P}_{2m})$, where $f(\Pi, \Theta)$,

$$\{f,\, g\} = -(-1)^{|f||g|}\{g,\, f\} \tag{18}$$

$$(-1)^{|r||h|}\{f\,, \{g,\, h\}\} + (-1)^{|g||f|}\{g\,, \{h,\, f\}\} + (-1)^{|h||g|}\{h\,, \{f,\, g\}\} = 0 \tag{19}$$

for canonical variables it gives

$$\{\Pi^j\,, \Theta_i\} = \delta_i^j. \tag{20}$$

As before we shall pass to the extended phase space $(\Pi,\, \Theta,\, t) \in \mathcal{P}_{(1|2m)}$ and generalized relations

$$[(\Pi,\, \Theta, t),\, (\Pi',\, \Theta', t')]_+ = (0,\, 0,\, \Pi\Theta' + \Theta\Pi') \tag{21}$$

yielding the supercommutator

$$[\hat{\Pi},\, \hat{\Theta}]_+ = \hat{T}. \tag{22}$$

Natural supermatrix realization of fermionic-Heisenberg algebra has the following form

$$\mu(\Pi, \Theta, t) \longrightarrow \left(\begin{array}{cc|c} 0 & t & \Pi \\ 0 & 0 & 0 \\ \hline 0 & \Theta & 0 \end{array}\right); \qquad \mu(\Pi, \Theta, t)^3 = 0 \tag{23}$$

One of the possible forms of the supermatrix realization of the fermionic Heisenberg group is given by

$$M(\Pi, \Theta, t) = 1 + \mu(\Pi, \Theta, t) \tag{24}$$

and it yields the "polarized form" of the multiplication:

$$M(\Pi, \Theta, t) \cdot M(\Pi', \Theta', t') = M(\Pi + \Pi', \Theta + \Theta', t + t' + \Pi\Theta')$$

The supermatrix realization of the whole super-Heisenberg group sH with polarized form of the multiplication reads now

$$(p, q, \Pi, \Theta, t) = \left(\begin{array}{ccc|c} 1 & p & t & \Pi \\ 0 & 1 & q & 0 \\ 0 & 0 & 1 & 0 \\ \hline 0 & 0 & \Theta & 1 \end{array}\right) \tag{25}$$

As before one can take realization coming from the exponentiation, which produces "unpolarized" form of multiplication and relates it to the super-symplectic geometry of the phase super-space

$$(p, q, \Pi, \Theta, t) \diamond (p', q', \Pi', \Theta', t') = (p+p', q+q', \Pi+\Pi', \Theta+\Theta', t+t'+\frac{1}{2}(pq'-qp')+\frac{1}{2}(\Pi\Theta'+\Theta\Pi')). \tag{26}$$

3. Super-Schrödinger representation and super-harmonic analysis

The super-Schrödinger approach to quantization of a supersymmetric system is the generalization of the conventional method of quantization to the situation where one has to deal with anticommuting co-ordinates. This kind of reasoning is widely used in (relativistic) supersymmetric field theory within the superfield formalism. Here, for supermechanics, one considers a generalization of the Hilbert space of the square integrable functions defined on the configuration space of the classical system, what gives the superspace of wave superfunctions integrated in the sense of the Berezin integral. One of possible structures realizing this goal is the Z_2-graded bimodule over Z_2-graded Banach-Grassmann algebra Q[8]. This kind of an approach is a natural one and in fact was the first considered in the literature by Valle [9] and falls into the scheme developed by DeWitt [10]. Here we shall use a slight mutation of the super Hilbert space considered by DeWitt [10], it was introduced in Ref. [8] within the super-GNS construction. This super Hilbert space, as it was mentioned above, is the Z_2-graded Q bimodule, with generalized scalar product with values in Q-algebra [8].

To fix the notation, let

$$q \in Q = Q_0 \oplus Q_1; \qquad q_s q'_r = (-1)^{sr} q'_r q_s, \quad q_s \in Q_s \tag{27}$$

with the following extension of complex conjugation [8]

$$q^* = \bar{q}_0 - i\bar{q}_1; \qquad (qq')^* = q'^* q^* \tag{28}$$

Functions $\mathcal{F}(\zeta)$, $\zeta = (\zeta_1, \ldots, \zeta_m)$, $|\zeta_j| = 1$; $d\zeta = d\zeta_m \ldots d\zeta_1$, where $\zeta_i \in Q_1$

$$< f, g >= \int d\zeta f^*(\zeta) g(\zeta)$$

and $f(\zeta) = \sum_{k,I_k} f_{I_k} \zeta^{I_k}$, where I_k denotes strictly ordered multi-index. Operators conjugated, with respect the above Q-scalar product, to the derivation with respect to ζ and multiplication by ζ are the following

$$\partial_\zeta^+ = i\partial_\zeta, \qquad \hat{\zeta}^+ = -i\zeta$$

Let $D_j = -i\partial_{\zeta_j}$ and $X^j = \zeta^j$, the operator $\Pi^j D_j + \Theta_j X^j$ is formally selfadjoint in $(\mathcal{F}(\zeta), <,>)$ and

$$\exp\left[i(t + \Theta X + \Pi D)\right] f(\zeta) = \exp\left[i(t + \Theta\zeta + \frac{1}{2}\Theta\Pi)\right] f(\zeta + \Pi) \tag{29}$$

or equivalently

$$e^{i(\Theta X + \Pi D)} = e^{\frac{1}{2}\Theta\Pi} e^{i\Theta X} e^{i\Pi D}. \tag{30}$$

The representation π_1 of the fermionic Heisenberg group FH_m defined as

$$\pi_1(\Pi, \Theta, t) f(\zeta) = e^{i(t + \Theta\zeta + \frac{1}{2}\Theta\Pi)} f(\zeta + \Pi)$$

is Q-representation of FH_m. Such Q-representation is defined as a generalization of the notion of representation to the $\star - Q$-algebra case, cf. Ref. [8].

3.1. Grassmannian Fourier-Wigner transform

To illustrate how the generalization of the harmonic analysis on the Heisenberg group can be formulated for the super-Heisenberg group let us restrict to the fermionic Heisenberg group. In this case one can introduce the so called Grassmannian Fourier-Wigner transform [11]. Taking the supermatrix elements of the representation π_1 we get

$$\mathcal{M}(\Pi, \Theta) =< f, \pi_1(\Pi, \Theta)g >, \qquad f, g \in \mathcal{F}(\zeta), \tag{31}$$

Let $V(f, g)$ be a function defined as

$$\begin{aligned} V(f,g)(\Pi, \Theta) &= \mathcal{M}(\Pi, \Theta) =< f, e^{i(\Theta X + \Pi D)} g >= \\ &= \int d\zeta f^*(\zeta - \frac{1}{2}\Pi) e^{i\Theta\zeta} g(\zeta + \frac{1}{2}\Pi) \end{aligned}$$

In analogy to the conventional case [1] above relation gives rise to the definition of the so called Grassmannian Fourier-Wigner transform (GFW) [11] or in other words the super Fourier-Wigner transform, when considered for the whole super-Heisenberg group. Namely,

$$V : \mathcal{F}_m(\zeta) \times \mathcal{F}_m(\zeta) \to \mathcal{F}_{2m}(\zeta) \tag{32}$$

The super Fourier-Wigner transform has analogous properties to the original one, of course, modified by the presence of the Z_2-gradation. For example, if $f_1, f_2, g_1, g_2 \in \mathcal{F}_m(\zeta)$ we have

$$< V(f_1, g_1), V(f_2, g_2) >= (-1)^{|f_1||f_2| + (|f_1| + |f_2|)|g_2|} < g_1, g_2 >< f_1, f_2 >^* . \tag{33}$$

Moreover, let $\alpha, \beta, \gamma, \delta \in Q_1$, then

$$\begin{aligned} & V(\pi_1(\gamma, \delta) f, \pi_1(\alpha, \beta) g)(\Pi, \Theta) - \\ & \exp\left[-\frac{1}{2}(-\Pi\beta - \Theta\alpha + \gamma\Pi + \delta\Theta + \gamma\beta + \delta\alpha)\right] \\ & V(f, g)(\Pi + \alpha - \gamma, \Theta + \beta - \delta). \end{aligned} \tag{34}$$

3.2. Grassmannian Bargmann transform

The Grassmannian Bargmann transform is the specific GFW-transform, defined for spaces with the even dimension m. In such a case one can consider the Grassmannian Gausian $\omega_0 \in \mathcal{F}_m(\zeta)$; $m = 2k$

$$\omega_0 = (PfG)^{-\frac{1}{2}} e^{\frac{1}{2}\zeta G\zeta}, \qquad G_{ij} = -G_{ji}, \tag{35}$$

where PfG denotes the Pfaffian of the antisymmetrix matrix G. The ω_0 is normalized with respect to the Q-scalar product

$$< \omega_0, \omega_0 >= 1.$$

Fixing ω_0 as one of the entries of the super Fourier-Wigner transform we obtain

$$V(\omega_0, f)(\Pi, \Theta) = (PfG)^{-\frac{1}{2}} \int d\zeta e^{(\zeta - \Pi)G(\zeta - \Pi)} e^{i\Theta\zeta - \frac{i}{2}\Theta\Pi} f(\zeta) \tag{36}$$

In terms of the new variable $z_k = G_{kj}\Pi^j + i\Theta_k$ we have

$$V(\omega_0, f)(\Pi, \Theta) = (PfG)^{-\frac{1}{2}} e^{-\frac{1}{4}z^* G^{-1} z} \int d\zeta e^{\frac{1}{2}\zeta G\zeta - \zeta z - \frac{1}{4}z^* G^{-1} z} f(\zeta) \tag{37}$$

This formula gives rise to the definition of the Grassmannian Bargmann (super-Bargmann) transform [11]

$$(\mathbf{B}f)(z) \equiv 2^{-\frac{n}{2}} \int d\zeta e^{\frac{1}{2}\zeta G\zeta - \zeta z - \frac{1}{4}z^* G^{-1} z} f(\zeta) \tag{38}$$

Observing that $Im(-\frac{i}{2}z^*G^{-1}z') = B(v, v')$ and using identification $z = G\Pi + i\Theta \rightarrow (\Pi, \Theta)$ one can rewrite super-Heisenberg group law in the form $(z, t) \cdot (z', t') = (z + z', t + t' + \frac{1}{2}Im(-\frac{i}{2}z^*G^{-1}z'))$. Finally, as in the conventional case [1] the transferred representation β can be defined [11] i.e. $\beta(z, t) \circ \mathbf{B} = \mathbf{B} \circ \pi_1(\Pi, \Theta, t)$. denoting $||z||^2 = \frac{1}{2}z^*G^{-1}z$ we get relation

$$V(\omega_0, f)(\Pi, \Theta) = e^{-\frac{1}{2}||z||^2}(\mathbf{B}f)(z) \tag{39}$$

In view of the above relations it is natural to consider the notion of the Grassmannian Bargmann-Fock space (the super Bargmann-Fock space) $\mathcal{F}(z)$. In the $\mathcal{F}(z)$ we define Q-scalar product

$$< f, g >_{\mathcal{F}} = \int |dz| e^{-\frac{1}{2}||z||^2} f^*(z)g(z), \qquad |dz| = -(\frac{i}{2})^n dz dz^* \tag{40}$$

In particular monomials in z-variables are orthonormal with respect to this Q-scalar product i.e. $< z_{I_k}, z_{J_l} >_{\mathcal{F}} = \delta_{I_k J_l}$. For the Q-representation β, using $w = G\rho + i\sigma$ one can show that

$$(\beta(w)\mathbf{B}f)(z) = e^{-\frac{i}{2}||w||^2} e^{-\frac{i}{2}zG^{-1}w^*}\mathbf{B}f(z) \tag{41}$$

3.3. Grassmannian special functions

The Grassmannian Hermite (the super Hermite) polynomials arise in the super-Schrödinger quantization of the fermionic harmonic oscillator, they can be defined by the Rodrigues-like formula [12]. Namely, the components of the Grassmannian Hermite polynomials are given by

$$h_k^{I_k} = H_k e^{-\frac{1}{2}\zeta G \zeta} \partial^{I_k} e^{\zeta G \zeta}. \tag{42}$$

It is interesting that the super-Bargmann transform, as in the conventional case, relates polynomial base in the $\mathcal{F}(z)$ and the super-Hermite polynomials in the super-Fock Q-module. $(\mathbf{B}h_k)_{I_k}(z) = 2^{\frac{n}{2}} H_k z_{I_k}$. Moreover, using the super Fourier-Wigner transform of the super-Hermite polynomials one can obtain super-Laguerre polynomials [11].

4. Odd super-Heisenberg group

It this section we shall discuss the odd super-Heisenberg group. Let us recall that generic super-phase space can be of two types, depending wheter the supersymmetric system is defined by the even or the odd Hamiltonian. The models described by the even Hamiltonian are more frequent in the supersymmetry. In such models parity of super-coordinates and canonical super-momenta is the same: $|q_i| = |p^i| = 0$ and $|\Theta_j| = |\Pi^j| = 1$. This case is covered by the super-symplectic geometry However, as it was recognized by Leites [13] there exists another option for nontrivial geometry defined by the odd structure, the so called periplectic form. Systems falling into this type are defined by the odd Hamiltonian and parity of the super-coordinates and canonical super-momenta is opposite [14] i.e. $|q_i| = 0 = |\Pi^i| + 1 (\text{mod } 2)$ and $|\Theta_j| = 1 = |p^j| + 1$. Hence, in generall we can consider the following sectors

$$\begin{array}{c|c} (p, q) & (p, \Theta) \\ \hline (\Pi, q) & (\Pi, \Theta) \end{array} \tag{43}$$

The (even) super-Heisenberg group is defined on the "diagonal" of the above diagram, the odd super-Heisenberg group we want to describe now is defined on the "off diagonal" sectors. We wish to fix our attention on the one of them: (Θ_j, p^j). Both odd sectors can be mapped one into the other by the parity shift operator, preserving the type of the periplectic supergeometry they have. Related part of the odd super-Heisenberg group we shall briefly call the odd Heisenberg group $\mathcal{OH}_m$.

To introduce such an odd Heisenberg group we need the odd super-symplectic form $\overset{(1)}{B}(p, \Theta)$, with property that $\overset{(1)}{B}(v_s, w_r) \in Q_{s+r+1}$. Let us define it as a nondegenerate Q-left linear B form, such that

$$\overset{(1)}{B}(v_r, w_s) = -(-1)^{(r+1)(s+1)} \overset{(1)}{B}(w_s, v_r) \tag{44}$$

The $\overset{(1)}{B}$ defined in natural base can be written as

$$\overset{(1)}{B}(v, v') = p^j \Theta'_j - \Theta_j p'^j \tag{45}$$

For the odd supermechanical systems there is naturally defined the odd Poisson bracket, having the properties of the odd graded bracket (called also antibracket) i.e.

$$A = A_0 \oplus A_1; \quad [\cdot, \cdot]_1 : A_s \times A_r \to A_{s+r+1} \tag{46}$$

$$[a, b]_1 = -(-1)^{(a+1)(b+1)}[b, a]_1 \tag{47}$$

$$\sum_{cycl}(-1)^{(a+1)(b+1)}[a, [b, c]_1]_1 = 0 \tag{48}$$

Let us here note that in the graded algebra with the odd associative multiplication, such an odd graded bracket can be realized as graded odd commutator, what we will need to proceed with the canonical super-quantization procedure for the odd systems. To consistently extend the odd super-phase space we have to introduce the odd "time" parameter τ to have the odd Heisenberg group law defined in the connection to the odd geometry $(v, \tau) \bullet (v', \tau') = (v + v', \tau + \tau' + \frac{1}{2} \overset{(1)}{B}(v, v'))$. The odd time can be realized as Q-number i.e. $\tau \in Q_1$, but in that case $\tau^2 = 0$ and we lose the physical interpretation and possibility to use τ as an evolution parameter. On the other hand, as we have already mentioned the odd multiplication seems to be necessary to consistently quantize the odd systems. To solve these difficulties one has to introduce a new structure. In the next section we shall show that the so called algebra of oddons [5] can fulfill our needs.

4.1. Oddons

The oddons were introduced in Ref. [11] in the context of the harmonic superanalysis on the odd Heisenberg group, but it turns out that similar structure appears in the algebraic supergeometry [6]. We shall briefly describe real and complex oddons.

4.1.1. Real Oddons The algebra of real oddons $\mathcal{O}_\mathcal{R}$ is generated by the unit $1 \in Q_R$ and a new generator $\hat{1}$ such that:

$$\hat{1}^2 = 1, \quad |\hat{1}| = 1, \text{i.e.} \quad \hat{1} q_s = (-1)^s q_s \hat{1} \tag{49}$$

$$\mathcal{O}_R = Q_R \oplus \hat{1} Q_R \tag{50}$$

Any oddon can be written as $r = a + \hat{1}b$, $a, b \in Q_R$ or as a pair $(a, b) \in Q_R^{(1|1)}$ and can be decomposed into the homogenous parts.

$$a + \hat{1}b = (a_0 + \hat{1}b_1) + (a_1 + \hat{1}b_0), \qquad a_s, b_s \in Q_s; \quad s - 0, 1 \tag{51}$$

With the use of parity mapping J on the Q algebra, the product of two real oddons can be written in the following form

$$rr' = (a + \hat{1}b)(a' + \hat{1}b') = aa' + J(b)b' + \hat{1}(J(a)b' + ba'), \text{where} \quad J(a) = J(a_0 + a_1) = a_0 - a_1 \tag{52}$$

This multiplication is associative, but because of the presence of $\hat{1}$, $\hat{1}^2 \neq 0$ it is not graded commutative, what can be easily verified. Interesting sub algebras in the oddonic algebra are the following:

- the subalgebra of even oddons $r_0 = a_0 + \hat{1}b_1$; $a_0 \in Q_0$, $b_1 \in Q_1$
- the subalgebra oddons over the body of Q_R i.e. over real numbers $r = a + \hat{1}b$; $a, b \in R$. In this case $r \in R^{(1|1)}$.

One can show, that for homogeneous oddons with nonvanishing body of the even component there exists the inverse. Finally, let us observe that in the oddon algebra we can introduce the new odd product

$$r * r' = r \cdot \hat{1} \cdot r', \tag{53}$$

which we shall use to generalize structures needed for odd super-Heisenberg group and its super-Schrödinger quantization.

The reall Oddons in a way, are graded generalization of so called split numbers or double numbers. One can also consider in this context a counterpart of the graded dual numbers, where additional generator is nilpotent.

4.1.2. Complex Oddons To separate information about new odd unit $\hat{1}$ and the complexification we have firstly described real version of oddons. The definition of the complex oddons requires the introduction of an "odd imaginary unit" $\hat{\imath}$ such that

$$\hat{\imath}^2 = -1, \quad |\hat{\imath}| = 1, \qquad \text{i.e.} \quad \hat{\imath}q_s = (-1)^s q_s \hat{\imath} \tag{54}$$

So, this time we have that $\mathcal{O} = Q \oplus \hat{\imath}Q$. As for real oddons, here also the multiplication

$$zz' = (a + \hat{\imath}b)(a' + \hat{\imath}b') = aa' - J(b)b' + \hat{\imath}(J(a)b' + ba'), \text{where} \quad J(a) = J(a_0 + a_1) = a_0 - a_1 \tag{55}$$

is associative, but is not graded commutative. As in the case of real oddons, for homogeneous elements with nonvanishing body of the even component there exists the inverse. There are also distinguished subalgebras similar to the mentioned above subalgebras of the real oddons.

4.2. Odd Heisenberg group

Let us extend the phase space of the odd system by the odd time parameter τ, but with τ taking oddonic values i.e. $\tau = t\hat{1} \in \mathcal{O}$ and $t \in R$. The group law for the odd Heisenberg group now can be defined by

$$(v, \tau) \bullet (v', \tau') = (v + v', \tau + \tau' + \frac{1}{2} \overset{(1)}{B}(v, v')) \tag{56}$$

The supermatrix realization (with the oddonic entries) of the generator takes similar form to the previous one. We have

$$\mu(p, \Theta, \tau) = \left(\begin{array}{cc|c} 0 & p & \tau \\ 0 & 0 & \Theta \\ \hline 0 & 0 & 0 \end{array}\right) \tag{57}$$

With the use of the odd multiplication we can define odd exponents e_*^q and realize the odd Heisenberg group with multiplication defined by the odd supersymplectic form

$$e_*^{\mu(p,\Theta,\tau)} * e_*^{\mu(p',\Theta',\tau')} = e_*^{\mu(p+p',\Theta+\Theta',\tau+\tau'+\frac{1}{2}(p\Theta'-\Theta p'))} \tag{58}$$

5. Oddonic super-Schrödinger representation

The $\mathcal{O}$-representations of the odd Heisenberg group have similar structure to the super-Schrödinger representations of the even super-Heisenberg group. Here we have to consider relevant objects over oddons instead of the Q algebra. In particular

- $< f, g >= \int d\zeta f^*(\zeta)g(\zeta)$, $\mathcal{O}_{\mathcal{C}}$-valued,
- $D_j = \hat{\imath}\partial_{\zeta_j}$ and $X^j = \zeta^j$, $\zeta \in Q_1$
- $p^i = \hat{1}\Pi^i$, $\Pi^i \in Q_1$

The odd Heisenberg group $\mathcal{O}$-representation $\pi(p, \Theta, t)$ is based on the relation

$$exp_*[\hat{\imath}(t + \Theta X + \Pi D)]f(\zeta) = exp_*[i(t + \Theta\zeta + \frac{1}{2}\Theta * p)]f(\zeta + \hat{1}p) \tag{59}$$

Because $p^i = \hat{1}\Pi^i$ the algebraic form of formulas obtained for the odd Heisenberg group turns out to be preserved. In particular, the Grassmannian Fourier-Wigner transform in the oddonic case takes the following form

$$V(f, g)(p, \Theta) = \int d\zeta f^*(\zeta - \frac{1}{2}\hat{1}p)e^{i\Theta\zeta}g(\zeta + \frac{1}{2}\hat{1}p) \tag{60}$$

The odd exponent allows to define the odd Gaussian $e_*^{\zeta\hat{G}\zeta}$, where $\hat{G} = \hat{1}G$, $G_{ij} = -G_{ji}$ and to introduce the odd Bargmann transform

$$\hat{\mathbf{B}}f(z) = 2^{-\frac{n}{4}} \int d\zeta e_*^{\frac{1}{2}\zeta\hat{G}\zeta - \zeta z - \frac{1}{4}z\hat{G}^{-1}z} f(z), \qquad z_k = G_{kj}p^j + \hat{\imath}\Theta_k. \tag{61}$$

Therefore, one gets similar structures to the ones known for the fermionic Heisenberg group.

6. Unitary representations of super-Heisenberg group

In this section we want to briefly compare, presented above, physical description of super-Heisenberg group with the strict mathematical one. In the physical approach it is, in many cases, enough to define supergroup as a supermatrix group and to work with formalism where "representation" of the group is obtained with the use of the anti-commuting variables, and the definition of the super-Hilbert space includes Z_2-graded algebra valued "scalar product". In the strict approach the Lie supergroup is a group carrying the supermanifold structure and the relevant representation theory is formulated in such a way that notion of unitary representation can be used. The theory of unitary representations of the nilpotent Lie supergroups is surprisingly new, it was formulated in 2006, comparing to the beginings of the supersymmetry, which was invented in the Seventies of the last Century. This somehow gauges the scale of difficulties related with the supergroups. The progress achieved in the last decade is related to the use of the simplification coming from identification of the super Lie group with the super Harish-Chandra pair.

Let us make short comparison of differences in definitions of the main object used in the physical and strict approaches respectively.

6.1. Super Lie group vs. super Harish-Chandra pair

In the strict approach instead of the explicit supermatrix groups one considers the Lie supergroup: object which is a group and supermanifold of dimension $(n|m)$, so locally it can be viewed as $(R^n, C^\infty \otimes \bigwedge(\theta_1, \ldots, \theta_m))$. This object is not easy to handle but can be replaced by the super Harish-Chandra pair: a pair $(G_0, \mathfrak{g})$, where G_0 is a Lie group with the lie algebra $\mathfrak{g}_0$ and $\mathfrak{g} = \mathfrak{g}_0 \oplus \mathfrak{g}_1$ is a Lie superalgebra, moreover there are imposed compatibility conditions:

- G acts on $\mathfrak{g}$ by R-linear automorphisms
- $\sigma(G_0)|_{\mathfrak{g}_0} = Ad, \qquad d\sigma(X)Y = [X, Y]$

To make contact with our previous considerations, let us illustrate the super Harish-Chandra pair notion on the example of the super Heisenberg group [6] $SH_3 = (H_3, \mathfrak{sh}(3|1))$

$$H_3 = \{\left(\begin{array}{ccc|c} 1 & x & t & 0 \\ 0 & 1 & y & 0 \\ 0 & 0 & 1 & 0 \\ \hline 0 & 0 & 0 & 1 \end{array}\right) : x, y, t \in R\}$$

$$\mathfrak{sh}(3|1) = \{\left(\begin{array}{ccc|c} 0 & a & c & \alpha \\ 0 & 0 & b & 0 \\ 0 & 0 & 0 & 0 \\ \hline 0 & 0 & 0 & 0 \end{array}\right) : a, d, c, \alpha \in R\}$$

Details of the unitary representation of such defined super-Heisenberg group can be found in Ref. [6].

6.2. Unitary representations of super Lie groups

There are various definitions of super-Hilbert space used in the literature. The definition of the super-Hilbert space relevant to study unitary representations of the Lie supergroups has been observed in the context of the supercoherent state representations [15], and has been used in mathematical approach to supersymmetry [16, 17, 6]. In such an approach the super-Hilbert space is defined as a Z_2-graded complex Hilbert space $\mathcal{H} = \mathcal{H}_0 \oplus \mathcal{H}_1$ such that

$$< v, w >= \begin{cases} 0, & |v| = |w| + 1 \\ (v, w), & v, w \in \mathcal{H}_0 \\ i(v, w), & v, w \in \mathcal{H}_1 \end{cases}$$

where $(.,.)$ idenotes the scalar product in $\mathcal{H}$ and $< .,. >$ is the even super Hermitian form: $\overline{< v, w >} = (-1)^{|v||w|} < w, v >$.

On such super-Hilbert spaces there are defined unitary representations of Lie supergroups[18]. In particular, as it was shown by Salmasian in 2010 [17], there is valid the generalization of the Stone-von Neumann to the case of the super-Heisenberg group i.e. the super Stone-von Neuman theorem [17]. Namely, let $\chi : R \to C^\times$ be defined by $\chi(t) = e^{i\beta t}$, where $\beta > 0$ then for the supersymplectic form Ω we have implications

- if $\Omega|_{V_1 \times V_1}$ is positive definite then exists a unique unitary representation of SH with character χ.
- if $\Omega|_{V_1 \times V_1}$ is not positive definite then SH does not have any unitary representation with character χ.

7. Final Comments

We have described in brief, how the idea of the representation of the super-Heisenberg group and the generalization of the harmonic analysis on it can be realized. One approach, uses ideas which are present in the formulation of the supersymmetric field theory. Such physical approach yields the supermodule over superalgebra description of the representations. The strict mathematical approach allows to keep the notion of unitary representation. It turned out to be much more difficult to obtain and has been formulated only recently.

References

[1] Folland G B 1989 *Harmonic Analysis in Phase Space* (Princeton University Press)
[2] Howe R 1980 *Bull. Amer. Math. Soc. (N.S.)* **3** 821
[3] Manin Y I 1988 *Gauge Field Theory and Complex Geometry* (NY: Springer)
[4] Leites D A 1980 *Russ. Math. Surv.* **35** 1
[5] Frydryszak A M 1998 *Lett. Math. Phys.* **44** 89
[6] Kwok S in Ferrara S et al. (eds) 2011 *Supersymmetry in Mathematics and Physics* Lecture Notes in Mathematics 2027 (Berlin: Springer)
[7] Frydryszak A M 1989 *Lett. Math. Phys.* **18** 87
[8] Frydryszak A M and Jakobczyk L 1988 *Lett. Math. Phys.* **16** 101
[9] Valle J W F 1979 *Int. J. Theor. Phys.* **18** 923
[10] DeWitt B 1984 *Supermanifolds* (Cambridge Monographs on Mathematical Physics) (Cambridge, MA: CUP)
[11] Frydryszak A M 1992 *Lett. Math. Phys.* **26** 105
[12] Frydryszak A M 1990 *Lett. Math. Phys.* **20** 159
[13] Leites D A 1977 *Dokl. Akad. Nauk. SSSR* **236** 804
[14] Frydryszak A M 1993 *J. Phys. A: Math. Gen.* **26** 7227
[15] El Gradeshi A, Neto M 1996 *Comm. Math. Phys.* **175** 521
[16] Deligne P and Morgan J W 1999 *Notes on Supersymmetry (Following Joseph Bernstein)* Quantum Fields and Strings: A Course for Mathematicians, vols. 1,2 (Princeton, NJ: AMS)
[17] Salmasian H 2010 *Comm. Math. Phys.* **297** 189
[18] Carmeli C, Cassinelli G, Toigo A and Varadarajan V S 2006 *Commun. Math. Phys.* **263** 217

Journal of Physics: Conference Series **563** (2014) 012011 doi:10.1088/1742-6596/563/1/012011

Bethe vectors for XXX-spin chain

Čestmír Burdík[1], **Jan Fuksa**[1,2], **Alexei Isaev**[2]

[1] Department of mathematics, Faculty of nuclear sciences and physical engineering, Czech Technical University in Prague, Trojanova 13, 120 00 Prague, Czech Republic

[2] Bogoliubov Laboratory of Theoretical Physics, JINR, Joliot-Curie 6, 141980 Dubna, Moscow region, Russia

E-mail: burdices@kmlinux.fjfi.cvut.cz, fuksajan@fjfi.cvut.cz, isaevap@theor.jinr.ru

Abstract. The paper deals with algebraic Bethe ansatz for XXX-spin chain. Generators of Yang-Baxter algebra are expressed in basis of free fermions and used to calculate explicit form of Bethe vectors. Their relation to N-component models is used to prove conjecture about their form in general. Some remarks on inhomogeneous XXX-spin chain are included.

1. Introduction

Algebraic Bethe ansatz has turned out as remarkably sufficient tool in the theory of quantum integrable systems. Its origins come up to the 80's and are connected mainly with Leningrad shool. Since that time, it was used successfuly to solve an amount of models.

In this text we are going to discuss some features of well-known model solved by this method, the so called XXX-spin chain.

In section 2, we repeat some properties of XXX-spin chain in terms of algebraic Bethe ansatz. In sections 3, we introduce free fermions and use them in section 4 to express generators of Yang-Baxter algebra. The aim of this text is to calculate an explicit form of Bethe vectors for XXX-spin chain in fermionic basis which is discussed in sections 5. To prove our conjectures about their form, we are forced to use N-component model in section 6. At the end of this text, in section 7, we mention some results for inhomogeneous case.

2. Algebraic Bethe ansatz for XXX-spin chain

Suppose we have a chain of L nodes. A local Hilbert space $h_j = \mathbb{C}^2$ corresponds to the j-th node. The total Hilbert space of the chain is

$$\mathscr{H} = \prod_{j=1}^{L} \otimes h_j = \prod_{j=1}^{L} \otimes \mathbb{C}^2. \tag{1}$$

Let A be an operator acting on $h = \mathbb{C}^2$. We use the following notation for operator A acting nontrivially only in the space $h_j \subset \mathscr{H}$ and trivially in others throughout the text

$$A_j = \mathbb{I}^{\otimes(j-1)} \otimes A \otimes \mathbb{I}^{\otimes(L-j)}. \tag{2}$$

Journal of Physics: Conference Series **563** (2014) 012011 doi:10.1088/1742-6596/563/1/012011

The basic tool of algebraic Bethe ansatz is Lax operator which is an parameter depending object acting on the tensor product $V_a \otimes h_i$

$$L_{a,i}(\lambda): \quad V_a \otimes h_i \to V_a \otimes h_i \tag{3}$$

explicitly defined as

$$L_{a,i}(\lambda) = (\lambda + \frac{1}{2})\mathbb{I}_{a,i} + \sum_{\alpha=1}^{3} \sigma_a^\alpha S_i^\alpha \tag{4}$$

where σ_a^α are usual Pauli matrices acting in V_a, $S_i^\alpha = \frac{1}{2}\sigma_i^\alpha$ are spin operators on the i-th node and $\mathbb{I}_{a,i}$ is identity matrix in $V_a \otimes h_i$. $L_{a,i}(\lambda)$ can be expressed as a matrix in the auxiliary space V_a

$$L_{a,i}(\lambda) = \begin{pmatrix} \lambda + \frac{1}{2} + S_i^z & S_i^- \\ S_i^+ & \lambda + \frac{1}{2} - S_i^z \end{pmatrix}. \tag{5}$$

Its matrix elements form an associative algebra of local operators acting in the quantum space h_i. Introducing permutation operator P

$$P = \frac{1}{2}\left(\mathbb{I} \otimes \mathbb{I} + \sum_{\alpha=1}^{3} \sigma^\alpha \otimes \sigma^\alpha\right) \tag{6}$$

(here $\mathbb{I}$ denotes 2×2 unit matrix), we can rewrite it as

$$L_{a,i}(\lambda) = \lambda \mathbb{I}_{a,i} + P_{a,i}. \tag{7}$$

Assume two Lax operators $L_{a,i}(\lambda)$ resp. $L_{b,i}(\mu)$ in the same quantum space h_i but in different auxiliary spaces V_a resp. V_b. The product of $L_{a,i}(\lambda)$ and $L_{b,i}(\mu)$ makes sense in the tensor product $V_a \otimes V_b \otimes h_i$. It turns out that there is an operator $R_{ab}(\lambda - \mu)$ acting nontrivially in $V_a \otimes V_b$ which intertwines Lax operators in the following way

$$R_{ab}(\lambda - \mu) L_{a,i}(\lambda) L_{b,i}(\mu) = L_{b,i}(\mu) L_{a,i}(\lambda) R_{ab}(\lambda - \mu). \tag{8}$$

Relation (8) describes the so called fundamental commutation relation. The explicit expression for $R_{ab}(\lambda - \mu)$ is

$$R_{ab}(\lambda - \mu) = (\lambda - \mu)\mathbb{I}_{a,b} + P_{a,b} \tag{9}$$

where $\mathbb{I}_{a,b}$ resp. $P_{a,b}$ is identity resp. permutation operator in $V_a \otimes V_b$. In matrix form

$$R_{ab}(\lambda - \mu) = \begin{pmatrix} \lambda - \mu + 1 & 0 & 0 & 0 \\ 0 & \lambda - \mu & 1 & 0 \\ 0 & 1 & \lambda - \mu & 0 \\ 0 & 0 & 0 & \lambda - \mu + 1 \end{pmatrix}. \tag{10}$$

The operator $R_{ab}(\lambda - \mu)$ is called R-matrix and satisfies Yang-Baxter equation

$$R_{ab}(\lambda - \mu) R_{ac}(\lambda) R_{bc}(\mu) = R_{bc}(\mu) R_{ac}(\lambda) R_{ab}(\lambda - \mu) \tag{11}$$

in $V_a \otimes V_b \otimes V_c$. Comparing (7) and (9), we see that Lax operator and R matrix have exactly the same form.

Yang-Baxter algebra of global operators acting on the Hilbert space $\mathscr{H}$ of the full chain is defined via a monodromy matrix

$$T_a(\lambda) = L_{a,1}(\lambda) L_{a,2}(\lambda) \ldots L_{a,L}(\lambda) \tag{12}$$

which is a product of Lax operators along the chain, i.e. over all quantum spaces h_i. Matrix elements of $T_a(\lambda)$ in the auxiliary space V_a

$$T_a(\lambda) = \begin{pmatrix} A(\lambda) & B(\lambda) \\ C(\lambda) & D(\lambda) \end{pmatrix} \tag{13}$$

provide generators of Yang-Baxter algebra.

Equation (8) provides the following relation for monodromy matrix

$$R_{ab}(\lambda-\mu)T_a(\lambda)T_b(\mu) = T_b(\mu)T_a(\lambda)R_{ab}(\lambda-\mu) \tag{14}$$

which describes commutation relations for generators of Yang-Baxter algebra $A(\lambda), B(\lambda), C(\lambda)$ and $D(\lambda)$. We call it global fundamental commutation relation. Equation (14) implies commutativity of transfer matrices

$$\tau(\lambda)\tau(\mu) = \tau(\mu)\tau(\lambda) \tag{15}$$

where transfer matrix $\tau(\lambda)$ is defined as the trace of monodromy matrix over auxiliary space V_a

$$\tau(\lambda) \equiv Tr_a T_a(\lambda) = A(\lambda) + D(\lambda). \tag{16}$$

Obviously, transfer matrix $\tau(\lambda)$ is a polynomial of degree L in λ

$$\tau(\lambda) = 2\Big(\lambda + \frac{1}{2}\Big)^L + \sum_{k=0}^{L-2} \lambda^k Q_k. \tag{17}$$

Due to commutativity (15) of transfer matrices, we see that operators Q_k mutually commute

$$[Q_j, Q_k] = 0. \tag{18}$$

Hamiltonian of the chain is expresed in terms of Pauli matrices resp. permutation operators

$$H = \sum_{k=1}^{L}\sum_{\alpha=1}^{3} S_k^\alpha S_{k+1}^\alpha = \frac{1}{2}\sum_{k=1}^{L} P_{k,k+1} - \frac{L}{4} \tag{19}$$

where we impose cyclic condition $S_{L+1} = S_1$ resp. $P_{L,L+1} = P_{L,1}$. It can be expressed as a function of transfer matrix

$$H = \frac{1}{2}\frac{d}{d\lambda}\ln\tau(\lambda)\Big|_{\lambda=-1/2} - \frac{L}{4}. \tag{20}$$

This is the reason why we can say that transfer matrix $\tau(\lambda)$ is a generating function for commuting conserved charges.

2.1. Global fundamental commutation relations

Let us express global commutation relations (14) for generators of Yang-Baxter algebra. After a simple factorization, R-matrix (10) can be written as

$$R_{ab}(\lambda) = \begin{pmatrix} f(\lambda) & 0 & 0 & 0 \\ 0 & 1 & g(\lambda) & 0 \\ 0 & g(\lambda) & 1 & 0 \\ 0 & 0 & 0 & f(\lambda) \end{pmatrix} \tag{21}$$

where $f(\lambda) = \frac{\lambda+1}{\lambda}$ and $g(\lambda) = \frac{1}{\lambda}$. The matrices $T_a(\lambda)$ resp. $T_b(\mu)$ take the form

$$T_a(\lambda) = \begin{pmatrix} A(\lambda) & & B(\lambda) & \\ & A(\lambda) & & B(\lambda) \\ C(\lambda) & & D(\lambda) & \\ & C(\lambda) & & D(\lambda) \end{pmatrix} \quad \text{resp.} \quad T_b(\mu) = \begin{pmatrix} A(\mu) & B(\mu) & & \\ C(\mu) & D(\mu) & & \\ & & A(\mu) & B(\mu) \\ & & C(\mu) & D(\mu) \end{pmatrix}. \tag{22}$$

Comparing matrix elements of (14) on the positions $(1,1)$, $(1,4)$, $(4,1)$, $(4,4)$ we obtain

$$[A(\lambda), A(\mu)] = [B(\lambda), B(\mu)] = [C(\lambda), C(\mu)] = [D(\lambda), D(\mu)] = 0. \tag{23}$$

Comparing $(2,3)$ resp. $(2,2)$

$$[B(\lambda), C(\mu)] = g(\lambda, \mu)\left(D(\mu)A(\lambda) - D(\lambda)A(\mu)\right), \tag{24}$$
$$[A(\lambda), D(\mu)] = g(\lambda, \mu)\left(C(\mu)B(\lambda) - C(\lambda)B(\mu)\right). \tag{25}$$

From comparison of matrix elements $(1,3)$, $(3,4)$, $(2,1)$ resp. $(4,3)$ we obtain

$$A(\mu)B(\lambda) = f(\lambda, \mu)B(\lambda)A(\mu) + g(\mu, \lambda)B(\mu)A(\lambda), \tag{26}$$
$$D(\mu)B(\lambda) = f(\mu, \lambda)B(\lambda)D(\mu) + g(\lambda, \mu)B(\mu)D(\lambda), \tag{27}$$
$$A(\mu)C(\lambda) = f(\mu, \lambda)C(\lambda)A(\mu) + g(\lambda, \mu)C(\mu)A(\lambda), \tag{28}$$
$$D(\mu)C(\lambda) = f(\lambda, \mu)C(\lambda)D(\mu) + g(\mu, \lambda)C(\mu)D(\lambda), \tag{29}$$

where

$$f(\lambda, \mu) = f(\lambda - \mu) = \frac{\lambda - \mu + 1}{\lambda - \mu}, \qquad g(\lambda, \mu) = g(\lambda - \mu) = \frac{1}{\lambda - \mu}. \tag{30}$$

We see that $g(\mu, \lambda) = -g(\lambda, \mu)$.

2.2. Eigenstates of transfer matrix

Commutation relations (23)-(30) for Yang-Baxter algebra together with an assumption that the Hilbert space $\mathscr{H}$ has structure of a Fock space are sufficient to encover spectrum of the transfer matrix $\tau(\lambda)$.

There is a pseudovacuum vector $|0\rangle \in \mathscr{H}$ such that $C(\lambda)\,|0\rangle = 0$ which is an eigenvector of operators $A(\lambda)$ and $D(\lambda)$

$$A(\lambda)\,|0\rangle = \alpha(\lambda)\,|0\rangle\,, \qquad D(\lambda)\,|0\rangle = \delta(\lambda)\,|0\rangle\,. \tag{31}$$

It is a tensor product of local pseudovacua

$$|0\rangle = |0\rangle_1 \otimes |0\rangle_2 \otimes \cdots \otimes |0\rangle_L \tag{32}$$

where $|0\rangle_k = \left(\begin{smallmatrix}1\\0\end{smallmatrix}\right)$. It can be easily seen that $\alpha(\lambda) = (\lambda+1)^L$, $d(\lambda) = \lambda^L$.

Other eigenstates of transfer matrix (16) are of the form

$$|\lambda_1, \ldots, \lambda_M\rangle \equiv B(\lambda_1)B(\lambda_2)\ldots B(\lambda_M)\,|0\rangle \tag{33}$$

with eigenvalue

$$\Lambda(\lambda, \{\lambda\}) = \alpha(\lambda)\prod_{i=1}^{M} f(\lambda_i, \lambda) + \delta(\lambda)\prod_{i=1}^{M} f(\lambda, \lambda_i). \tag{34}$$

They are called Bethe vectors. For $M \in \mathbb{N}$, we refer to Bethe vector $|\lambda_1, \ldots, \lambda_M\rangle$ as M-magnon state. To get eigenstates of the transfer matrix, parameters $\{\lambda\} = \{\lambda_1, \ldots, \lambda_M\}$ have to satisfy Bethe equations for all $i \in 1, \ldots, M$

$$\alpha(\lambda_i)g(\lambda, \lambda_i)\prod_{\substack{k=1\\k\neq i}}^{M} f(\lambda_k, \lambda_i) + \delta(\lambda_i)g(\lambda_i, \lambda)\prod_{\substack{k=1\\k\neq i}}^{M} f(\lambda_i, \lambda_k) = 0. \tag{35}$$

Explicit form of Bethe equations (35) is

$$\left(\frac{\lambda_i + 1}{\lambda_i}\right)^L = \prod_{\substack{k=1\\k\neq i}}^{M} \frac{\lambda_i - \lambda_k + 1}{\lambda_i - \lambda_k - 1}. \tag{36}$$

3. Free Fermions

Our first aim is to express Bethe vectors in fermionic basis. We start with definition of free fermions. For tensor product of L copies of $\mathbb{C}^2$ we define free fermions as

$$\psi_k \equiv \left(\prod_{j=1}^{k-1} \sigma_j^z\right)\sigma_k^+, \qquad \bar{\psi}_k \equiv \left(\prod_{j=1}^{k-1} \sigma_j^z\right)\sigma_k^-. \tag{37}$$

Commutation relations for the fermions (37) are of the form

$$[\bar{\psi}_i, \psi_j]_+ = \delta_{ij}\mathbb{I}, \quad [\bar{\psi}_i, \bar{\psi}_j]_+ = 0, \quad [\psi_i, \psi_j]_+ = 0. \tag{38}$$

It is a straightforward task to check the following identities

$$\bar{\psi}_{k+1}\psi_k + \bar{\psi}_k\psi_{k+1} + \bar{\psi}_k\bar{\psi}_{k+1} + \psi_{k+1}\psi_k = \sigma_k^x\sigma_{k+1}^x, \tag{39}$$

$$\bar{\psi}_{k+1}\psi_k + \bar{\psi}_k\psi_{k+1} - \bar{\psi}_k\bar{\psi}_{k+1} - \psi_{k+1}\psi_k = \sigma_k^y\sigma_{k+1}^y, \tag{40}$$

$$[\psi_k, \bar{\psi}_k] = \sigma_k^z, \tag{41}$$

$$(1 - 2\bar{\psi}_k\psi_k)(1 - 2\bar{\psi}_{k+1}\psi_{k+1}) = \sigma_k^z\sigma_{k+1}^z. \tag{42}$$

4. Fermionic realization of monodromy matrix

Equation (7) provides us an easy expression for Lax operator. Identity operator $\mathbb{I}$ is a member of algebra of fermions. Therefore, it remains to know a fermionic realization only for permutation operator $P_{a,i}$.

Let us start with permutation operator $P_{k,k+1}$ permuting just the neighboring vector spaces h_k and h_{k+1}. Due to identities (39)-(42) and definition of permutation operator (6), it is straightforward to check that

$$P_{k,k+1} = \mathbb{I} + \bar{\psi}_{k+1}\psi_k + \bar{\psi}_k\psi_{k+1} - \bar{\psi}_k\psi_k - \bar{\psi}_{k+1}\psi_{k+1} + 2\bar{\psi}_k\psi_k\bar{\psi}_{k+1}\psi_{k+1}. \tag{43}$$

Permutation operator $P_{j,k}$ in non-neighboring vector spaces h_j, h_k where $j < k - 1$, becomes a non-local in terms of fermions. Using properties of Pauli matrices, it can be rewritten as

$$P_{j,k} = \frac{1}{2}(\mathbb{I} + \sigma_j^z\sigma_k^z) + (\sigma_j^+\sigma_k^- + \sigma_j^-\sigma_k^+). \tag{44}$$

The first part is local even in the terms of fermions

$$\frac{1}{2}(\mathbb{I} + \sigma_j^z\sigma_k^z) = \mathbb{I} - \bar{\psi}_k\psi_k - \bar{\psi}_j\psi_j + 2\bar{\psi}_k\psi_k\bar{\psi}_j\psi_j, \tag{45}$$

Journal of Physics: Conference Series **563** (2014) 012011 doi:10.1088/1742-6596/563/1/012011

but the second part is nonlocal

$$\sigma_j^+\sigma_k^- + \sigma_j^-\sigma_k^+ = (\psi_j\bar{\psi}_k + \bar{\psi}_j\psi_k)\prod_{l=j}^{k-1}\sigma_l^z = (\psi_j\bar{\psi}_k + \bar{\psi}_j\psi_k)\prod_{l=j}^{k-1}(\mathbb{I} - 2\bar{\psi}_l\psi_l). \tag{46}$$

The nonlocality of $P_{j,k}$ resp. $R_{j,k}(\lambda)$ is a serious problem. There appear difficulties when we attempt to express monodromy matrix (12) in terms of such nonlocal operators. We need to avoid the nonlocality.

Let us remind that $L_{a,i}(\lambda) = R_{a,i}(\lambda)$. For R-matrix $R_{ab}(\lambda)$ satisfying Yang-Baxter equation (11) we can define the matrix $\hat{R}_{ab}(\lambda) = R_{ab}(\lambda)P_{ab}$ which satisfies

$$\hat{R}_{ab}(\lambda)\hat{R}_{bc}(\lambda+\mu)\hat{R}_{ab}(\mu) = \hat{R}_{ab}(\mu)\hat{R}_{bc}(\lambda+\mu)\hat{R}_{ab}(\lambda). \tag{47}$$

Substituting $L_{a,i}(\lambda) = \hat{R}_{a,i}(\lambda)P_{a,i}$ in monodromy matrix (12), we obtain very convenient expression

$$\begin{aligned} T_a(\lambda) = L_{a,1}(\lambda)L_{a,2}(\lambda)\dots L_{a,L}(\lambda) &= \hat{R}_{a,1}(\lambda)P_{a,1}\hat{R}_{a,2}(\lambda)P_{a,2}\dots\hat{R}_{a,L}(\lambda)P_{a,L} = \\ &= \hat{R}_{a,1}(\lambda)\hat{R}_{1,2}(\lambda)\dots\hat{R}_{L-1,L}(\lambda)P_{L-1,L}\dots P_{1,2}P_{a,1}. \end{aligned} \tag{48}$$

It contains operators $\hat{R}_{k,k+1}$ resp. $P_{k,k+1}$ acting only in the neighboring spaces $h_k \otimes h_{k+1}$. Fermionic realization of R-matrix $\hat{R}_{k,k+1}(\lambda)$ is

$$\hat{R}_{k,k+1}(\lambda) = \lambda P_{k,k+1} + \mathbb{I} = (\lambda+1)\mathbb{I} + \lambda(\bar{\psi}_{k+1}\psi_k + \bar{\psi}_k\psi_{k+1} - \bar{\psi}_k\psi_k - \bar{\psi}_{k+1}\psi_{k+1} + 2\bar{\psi}_k\psi_k\bar{\psi}_{k+1}\psi_{k+1}). \tag{49}$$

To get fermionic represenation of Yang-Baxter algebra means to express monodromy matrix (48) as 2×2 matrix in the auxiliary space $V_a = \mathbb{C}^2$. For this purpose, we rewrite (48) as

$$T_a(\lambda) = \hat{R}_{a,1}(\lambda)X(\lambda)P_{a,1} \tag{50}$$

where the operator $X(\lambda)$

$$X(\lambda) = \hat{R}_{1,2}(\lambda)\dots\hat{R}_{L-1,L}(\lambda)P_{L-1,L}\dots P_{1,2} \tag{51}$$

acts nontrivially only in the quantum spaces $\mathscr{H} = h_1 \otimes \cdots \otimes h_L$ and is a scalar in the auxiliary space V_a. Moreover, we know due to equations (43) and (49) how to express $X(\lambda)$ in the terms of fermions.

Permutation matrix (6) can be rewritten as

$$\begin{aligned} P_{a,1} &= \frac{1}{2}\Big(\mathbb{I}\otimes\mathbb{I} + \sigma^x\otimes\sigma^x + \sigma^y\otimes\sigma^y + \sigma^z\otimes\sigma^z\Big) = \\ &= \frac{1}{2}\left[\begin{pmatrix}\mathbb{I} & 0\\ 0 & \mathbb{I}\end{pmatrix} + \begin{pmatrix}0 & \sigma^x\\ \sigma^x & 0\end{pmatrix} + \begin{pmatrix}0 & -i\sigma^y\\ i\sigma^y & 0\end{pmatrix} + \begin{pmatrix}\sigma^z & 0\\ 0 & -\sigma^z\end{pmatrix}\right] = \\ &= \begin{pmatrix}\frac{1}{2}(\mathbb{I}+\sigma^z) & \frac{1}{2}(\sigma^x - i\sigma^y)\\ \frac{1}{2}(\sigma^x + i\sigma^y) & \frac{1}{2}(\mathbb{I}-\sigma^z)\end{pmatrix} = \begin{pmatrix}\psi_1\bar{\psi}_1 & \bar{\psi}_1\\ \psi_1 & \bar{\psi}_1\psi_1\end{pmatrix} = \begin{pmatrix}\mathbb{I} - N_1 & \bar{\psi}_1\\ \psi_1 & N_1\end{pmatrix} \end{aligned} \tag{52}$$

where we have used (37), (41), and $N_1 = \bar{\psi}_1\psi_1$. Hence, we get

$$\hat{R}_{a,1}(\lambda) = \mathbb{I}_{a,i} + \lambda P_{a,1} = \begin{pmatrix}(\lambda+1)\mathbb{I} - \lambda N_1 & \lambda\bar{\psi}_1\\ \lambda\psi_1 & \lambda N_1 + \mathbb{I}\end{pmatrix}. \tag{53}$$

Using (52) and (53), monodromy matrix (50) can be written in the following form

$$T_a(\lambda) = \begin{pmatrix} (\lambda+1)\mathbb{I} - \lambda N_1 & \lambda\bar{\psi}_1 \\ \lambda\psi_1 & \lambda N_1 + \mathbb{I} \end{pmatrix} X(\lambda) \begin{pmatrix} \mathbb{I} - N_1 & \bar{\psi}_1 \\ \psi_1 & N_1 \end{pmatrix} = \begin{pmatrix} A(\lambda) & B(\lambda) \\ C(\lambda) & D(\lambda) \end{pmatrix} \tag{54}$$

where

$$A(\lambda) = (\lambda + 1 - \lambda N_1)X(\lambda)(1 - N_1) + \lambda\bar{\psi}_1 X(\lambda)\psi_1, \tag{55}$$
$$B(\lambda) = (\lambda + 1 - \lambda N_1)X(\lambda)\bar{\psi}_1 + \lambda\bar{\psi}_1 X(\lambda)N_1, \tag{56}$$
$$C(\lambda) = \lambda\psi_1 X(\lambda)(1 - N_1) + (\lambda N_1 + 1)X(\lambda)\psi_1, \tag{57}$$
$$D(\lambda) = \lambda\psi_1 X(\lambda)\bar{\psi}_1 + (\lambda N_1 + 1)X(\lambda)N_1. \tag{58}$$

5. Fermionic realization of Bethe vectors

The goal of our text is to find expression for Bethe vectors (33). For this purpose, fermionic realization (56) of the creation operator $B(\lambda)$ is convenient. The operator $X(\lambda) = \hat{R}_{12}(\lambda)\dots\hat{R}_{L-1,L}(\lambda)P_{L-1,L}\dots P_{12}$ can be written in terms of fermions due to equations (43) and (49). It can be easily seen that

$$\psi_k\left|0\right\rangle = 0 \tag{59}$$

for all $k = 1, \dots, L$.

If we are able to write $B(\lambda)$ in normal form our work would be simple. Unfortunately, it seems as a rather difficult task. Instead, we have to use the "weak approach," i.e. to apply $B(\lambda)$ on the pseudovacuum $|0\rangle$ and try to commute the fermions ψ_k to the right and see what remains.

For our purposes, we need the following set of useful identities, which follow from equations (52) and (53)

$$P_{k,k+1}\,\bar{\psi}_k = \bar{\psi}_{k+1} - \bar{\psi}_{k+1}N_k + \bar{\psi}_k N_{k+1}, \tag{60}$$
$$\hat{R}_{k,k+1}(\lambda)\bar{\psi}_{k+1} = \bar{\psi}_{k+1} + \lambda\bar{\psi}_k + \lambda\bar{\psi}_{k+1}N_k - \bar{\psi}_k N_{k+1} \tag{61}$$

where $N_k = \bar{\psi}_k\psi_k$, again. We can see that

$$P_{k,k+1}\bar{\psi}_k\left|0\right\rangle = \bar{\psi}_{k+1}\left|0\right\rangle, \tag{62}$$
$$\hat{R}_{k,k+1}(\mu)\bar{\psi}_{k+1}\left|0\right\rangle = \bar{\psi}_{k+1}\left|0\right\rangle + \mu\bar{\psi}_k\left|0\right\rangle, \tag{63}$$

and

$$\hat{R}_{k,k+1}(\mu)\left|0\right\rangle = (\mu+1)\left|0\right\rangle, \tag{64}$$
$$P_{k,k+1}\left|0\right\rangle = \left|0\right\rangle. \tag{65}$$

For higher magnons, we need also

$$\hat{R}_{k,k+1}(\lambda)\bar{\psi}_k\bar{\psi}_{k+1} = (\lambda+1)\bar{\psi}_k\bar{\psi}_{k+1} \tag{66}$$

and

$$\hat{R}_{l,l+1}(\lambda)\dots\hat{R}_{k-1,k}(\lambda)\bar{\psi}_k\left|0\right\rangle = \lambda^{k-l}\bar{\psi}_l\left|0\right\rangle + \frac{\lambda^{k-l}}{\lambda+1}\sum_{j=1}^{k-l}\left(\frac{\lambda+1}{\lambda}\right)^j\bar{\psi}_{l+j}\left|0\right\rangle, \tag{67}$$

$$\hat{R}_{l,l+1}(\lambda)\dots\hat{R}_{k,k+1}(\lambda)\bar{\psi}_k\left|0\right\rangle = \hat{R}_{l,l+1}(\lambda)\dots\hat{R}_{k-1,k}(\lambda)\bar{\psi}_k\left|0\right\rangle + \lambda(\lambda+1)^{k-l}\bar{\psi}_{k+1}\left|0\right\rangle. \tag{68}$$

5.1. 1-magnon

It is obvious that the second term in (56) annihilates the pseudovacuum state. Then, using (67), we get for 1-magnon state

$$\begin{aligned}|\mu\rangle \equiv B(\mu)\,|0\rangle &= (\mu+1-\mu N_1)\hat{R}_{12}(\mu)\ldots\hat{R}_{L-1,L}(\mu)P_{L-1,L}\ldots P_{12}\bar{\psi}_1\,|0\rangle = \\ &= (\mu+1-\mu N_1)\hat{R}_{12}(\mu)\ldots\hat{R}_{L-1,L}(\mu)\bar{\psi}_L\,|0\rangle = \\ &= (\mu+1-\mu N_1)\Big[\mu^{L-1}\bar{\psi}_1\,|0\rangle + \frac{\mu^{L-1}}{\mu+1}\sum_{j=1}^{L-1}\Big(\frac{\mu+1}{\mu}\Big)^j\bar{\psi}_{j+1}\Big]\,|0\rangle = \\ &= \frac{\mu^L}{\mu+1}\sum_{k=1}^{L}\left(\frac{\mu+1}{\mu}\right)^k\bar{\psi}_k\,|0\rangle = n(\mu)\sum_{k=1}^{L}[\mu]^k\bar{\psi}_k\,|0\rangle \end{aligned} \tag{69}$$

where we introduce concise notation

$$[\mu] = \frac{\mu+1}{\mu}, \quad \text{and} \quad n(\mu) = \frac{\mu^L}{\mu+1}. \tag{70}$$

5.2. 2-magnon

Using (67) and (68) we obtain

$$\begin{aligned}B(\lambda)\bar{\psi}_k\,|0\rangle &= (\lambda+1)\lambda^{L-2}\sum_{m=0}^{k-2}[\lambda]^m\bar{\psi}_{m+1}\bar{\psi}_k\,|0\rangle + \\ &+\frac{\lambda^{L-2}}{\lambda+1}\sum_{m=0}^{k-2}\sum_{j=1}^{L-k}[\lambda]^{j+m}\bar{\psi}_{m+1}\bar{\psi}_{k+j}\,|0\rangle + \frac{\lambda^L}{\lambda+1}[\lambda]^{k-1}\sum_{j=1}^{L-k}[\lambda]^j\bar{\psi}_k\bar{\psi}_{k+j}\,|0\rangle .\end{aligned} \tag{71}$$

We get for 2-magnon state using (71)

$$\begin{aligned}|\lambda,\mu\rangle &\equiv B(\lambda)B(\mu)\,|0\rangle = n(\mu)\sum_{k=1}^{L}[\mu]^k B(\lambda)\bar{\psi}_k\,|0\rangle = n(\mu)n(\lambda)\Bigg[\sum_{k=2}^{L}\sum_{m=0}^{k-2}[\mu]^k[\lambda]^{m+2}\bar{\psi}_{m+1}\bar{\psi}_k + \\ &+\frac{1}{\lambda(\lambda+1)}\sum_{k=2}^{L-1}\sum_{m=0}^{k-2}\sum_{j=1}^{L-k}[\mu]^k[\lambda]^{j+m+1}\bar{\psi}_{m+1}\bar{\psi}_{k+j} + \sum_{k=1}^{L-1}\sum_{j=1}^{L-k}[\mu]^k[\lambda]^{k+j-1}\bar{\psi}_k\bar{\psi}_{k+j}\Bigg]\,|0\rangle = \\ &= n(\mu)n(\lambda)\Bigg\{\sum_{1\le r<s\le L}\Big[[\mu]^s[\lambda]^{r+1} + [\mu]^r[\lambda]^{s-1}\Big]\bar{\psi}_r\bar{\psi}_s\,|0\rangle + \\ &+\frac{1}{\lambda(\lambda+1)}\sum_{s=3}^{L}\sum_{r=1}^{s-2}\sum_{k=r+1}^{s-1}[\mu]^k[\lambda]^{s+r-k}\bar{\psi}_r\bar{\psi}_s\,|0\rangle\Bigg\} = \\ &= n(\mu)n(\lambda)\sum_{1\le r<s\le L}\Bigg[[\mu]^s[\lambda]^{r+1} + [\mu]^r[\lambda]^{s-1} + \frac{1}{\lambda(\lambda+1)}\sum_{k=r+1}^{s-1}[\mu]^k[\lambda]^{s+r-k}\Bigg]\bar{\psi}_r\bar{\psi}_s\,|0\rangle .\end{aligned} \tag{72}$$

The finite sum in (72) can be calculated explicitly by means of geometric progression

$$\begin{aligned}&\frac{1}{\lambda(\lambda+1)}\sum_{k=r+1}^{s-1}[\mu]^k[\lambda]^{s+r-k} = \\ &= \frac{\mu}{\lambda(\lambda-\mu)}[\mu]^{r+1}[\lambda]^{s-1}\left(\left(\frac{[\mu]}{[\lambda]}\right)^{s-r-1} - 1\right) = \frac{\mu}{\lambda(\lambda-\mu)}\left([\mu]^s[\lambda]^r - [\mu]^{r+1}[\lambda]^{s-1}\right).\end{aligned} \tag{73}$$

Journal of Physics: Conference Series **563** (2014) 012011 doi:10.1088/1742-6596/563/1/012011

Substitution of (73) into (72) gives formula for 2-magnon

$$B(\lambda)B(\mu)|0\rangle = n(\mu)n(\lambda)\sum_{1\le r<s\le L}\Bigg[[\mu]^s[\lambda]^{r+1}+[\mu]^r[\lambda]^{s-1}+ + \frac{\mu}{\lambda(\lambda-\mu)}\left([\mu]^s[\lambda]^r-[\mu]^{r+1}[\lambda]^{s-1}\right)\Bigg]\bar\psi_r\bar\psi_s|0\rangle = = n(\mu)n(\lambda)\sum_{1\le r<s\le L}\left[[\lambda]^r[\mu]^s\frac{\lambda-\mu+1}{\lambda-\mu}+[\mu]^r[\lambda]^s\frac{\mu-\lambda+1}{\mu-\lambda}\right]\bar\psi_r\bar\psi_s|0\rangle. \tag{74}$$

5.3. 3-magnon

For 3-magnon we need at first

$$B(\nu)\bar\psi_r\bar\psi_s|0\rangle = (\nu+1-\nu N_1)X_{12\dots L}(\nu)\bar\psi_1\bar\psi_r\bar\psi_s = = \nu^{L-3}(\nu+1)^2\sum_{m=0}^{r-2}[\nu]^m\bar\psi_{m+1}\bar\psi_r\bar\psi_s|0\rangle+\nu^{L-3}\sum_{l=1}^{s-r-1}\sum_{m=0}^{r-2}[\nu]^{l+m}\bar\psi_{m+1}\bar\psi_{r+l}\bar\psi_s|0\rangle+ +\nu^{L-3}\sum_{j=1}^{L-s}\sum_{m=0}^{r-2}[\nu]^{j+m}\bar\psi_{m+1}\bar\psi_r\bar\psi_{s+j}|0\rangle+ +\nu^{L-3}(\nu+1)^{-2}\sum_{j=1}^{L-s}\sum_{l=1}^{s-r-1}\sum_{m=0}^{r-2}[\nu]^{j+l+m}\bar\psi_{m+1}\bar\psi_{r+l}\bar\psi_{s+j}|0\rangle+ +\nu^{L-s+r-1}(\nu+1)^{s-r-2}\sum_{j=1}^{L-s}\sum_{m=0}^{r-2}[\nu]^{j+m}\bar\psi_{m+1}\bar\psi_s\bar\psi_{s+j}|0\rangle+ +\nu^{L-s+2}(\nu+1)^{s-3}\sum_{j=1}^{L-s}[\nu]^j\bar\psi_r\bar\psi_s\bar\psi_{s+j}|0\rangle+\nu^{L-r}(\nu+1)^{r-1}\sum_{l=1}^{s-r-1}[\nu]^l\bar\psi_r\bar\psi_{r+l}\bar\psi_s|0\rangle+ +\nu^{L-r}(\nu+1)^{r-3}\sum_{j=1}^{L-s}\sum_{l=1}^{s-r-1}[\nu]^{j+l}\bar\psi_r\bar\psi_{r+l}\bar\psi_{s+j}|0\rangle. \tag{75}$$

3-magnon state is obtained from 2-magnon state (74)

$$|\nu,\mu,\lambda\rangle \equiv B(\nu)B(\mu)B(\lambda)|0\rangle = n(\mu)n(\lambda)\sum_{1\le r<s\le L}K_2(s,r)B(\nu)\bar\psi_r\bar\psi_s|0\rangle \tag{76}$$

where we denote for more comfort

$$K_2(s,r) = [\mu]^s[\lambda]^r\frac{\lambda-\mu+1}{\lambda-\mu}+[\mu]^r[\lambda]^s\frac{\mu-\lambda+1}{\mu-\lambda}. \tag{77}$$

Using(75), we get

$$|\nu,\mu,\lambda\rangle = n(\nu)n(\mu)n(\lambda) \sum_{1<q<r<s\leq L} \Bigg[[\nu]^{q+2}K_2(s,r) + \frac{1}{\nu^2}\sum_{l=1}^{r-q-1}[\nu]^{l+q}K_2(s,r-l) +$$
$$+\frac{1}{\nu^2}\sum_{j=1}^{s-r-1}[\nu]^{j+q}K_2(s-j,r) + \frac{1}{\nu^2(\nu+1)^2}\sum_{j=1}^{s-r-1}\sum_{l=1}^{r-q-1}[\nu]^{j+l+q}K_2(s-j,r-l)+$$
$$+\frac{1}{(\nu+1)^2}\sum_{l=q+1}^{r-1}[\nu]^{s-l+q}K_2(r,l) + [\nu]^{s-2}K_2(r,q) + [\nu]^r K_2(s,q)+$$
$$+\frac{1}{(\nu+1)^2}\sum_{l=r+1}^{s-1}[\nu]^{s+r-l}K_2(l,q)\Bigg]\bar{\psi}_q\bar{\psi}_r\bar{\psi}_s|0\rangle = n(\nu)n(\mu)n(\lambda)\times$$
$$\sum_{1\leq q<r<s\leq L}\sum_{\sigma\in S_3}\sigma\Big([\nu]^q[\mu]^r[\lambda]^s\frac{\nu-\mu+1}{\nu-\mu}\cdot\frac{\nu-\lambda+1}{\nu-\lambda}\cdot\frac{\mu-\lambda+1}{\mu-\lambda}\Big)\bar{\psi}_q\bar{\psi}_r\bar{\psi}_s|0\rangle \tag{78}$$

where S_3 denotes symmetric group.

5.4. M-magnon

From results (69), (74) and (78) we can conjecture that general M-magnon state is of the form

$$|\lambda_1,\ldots,\lambda_M\rangle \equiv B(\lambda_1)\cdots B(\lambda_M)|0\rangle =$$
$$= \Big(\prod_{i=1}^{M} n(\lambda_i)\Big)\sum_{1\leq k_1<\ldots<k_M\leq L}\sum_{\sigma_\lambda\in S_M}\sigma_\lambda\Big(\prod_{i<j}^{M}\frac{\lambda_i-\lambda_j+1}{\lambda_i-\lambda_j}\prod_{i=1}^{M}[\lambda_i]^{k_i}\Big)\bar{\psi}_{k_1}\cdots\bar{\psi}_{k_M}|0\rangle \tag{79}$$

where σ_λ is a permutation of the parameters $\{\lambda_1,\ldots,\lambda_M\}$ and $\sum_{\sigma_\lambda\in S_M}$ is the sum over all such permutations. We are going to provide a proof of this conjecture in more general form below.

6. N-component model

In the literature, cf. [2, 5], the so-called two-component model appears. It was introduced to avoid problems with computation of correlation functions for local operators attached to some node x of the chain in the Yang-Baxter algebra generated by (13).

The chain $[1,\ldots,L]$ is divided into two components $[1,\ldots,x]$ and $[x+1,\ldots,L]$. Then, the Hilbert space is splitted into two parts $\mathscr{H} = \mathscr{H}_1\otimes\mathscr{H}_2$ where $\mathscr{H}_1 = h_1\otimes\cdots\otimes h_x$ and $\mathscr{H}_2 = h_{x+1}\otimes\cdots\otimes h_L$. The pseudovacuum $|0\rangle\in\mathscr{H}$ is of the form $|0\rangle = |0\rangle_1\otimes|0\rangle_2$ where $|0\rangle_1\in\mathscr{H}_1$ and $|0\rangle_2\in\mathscr{H}_2$. We define on $V_a\otimes\mathscr{H}_1\otimes\mathscr{H}_2$ a monodromy matrix for each component

$$T_1(\lambda) = L_{a,1}(\lambda)\cdots L_{a,x}(\lambda) = \begin{pmatrix} A_1(\lambda) & B_1(\lambda)\\ C_1(\lambda) & D_1(\lambda)\end{pmatrix}, \tag{80}$$

resp.

$$T_2(\lambda) = L_{a,x+1}(\lambda)\cdots L_{a,L}(\lambda) = \begin{pmatrix} A_2(\lambda) & B_2(\lambda)\\ C_2(\lambda) & D_2(\lambda)\end{pmatrix}. \tag{81}$$

Each of these monodromy matrices satisfies exactly the same commutation relations (14) as original undivided monodromy matrix (12). Moreover, we have

$$A_j(\lambda)|0\rangle_j = \alpha_j(\lambda)|0\rangle_j, \qquad D_j(\lambda)|0\rangle_j = \delta_j(\lambda)|0\rangle_j, \qquad C_j(\lambda)|0\rangle_j = 0, \tag{82}$$

and operators corresponding to different components mutually commute. From construction, we see that

$$\alpha(\lambda) = \alpha_1(\lambda)\alpha_2(\lambda), \qquad \delta(\lambda) = \delta_1(\lambda)\delta_2(\lambda). \tag{83}$$

The full monodromy matrix $T(\lambda)$ for the complete chain $[1, \ldots, L]$ is

$$T(\lambda) = T_1(\lambda)\,T_2(\lambda) = \begin{pmatrix} A_1(\lambda)A_2(\lambda) + B_1(\lambda)C_2(\lambda) & A_1(\lambda)B_2(\lambda) + B_1(\lambda)D_2(\lambda) \\ C_1(\lambda)A_2(\lambda) + D_1(\lambda)C_2(\lambda) & C_1(\lambda)B_2(\lambda) + D_1(\lambda)D_2(\lambda) \end{pmatrix}, \tag{84}$$

and the M-magnon state is represented in the form

$$|\lambda_1, \ldots, \lambda_M\rangle = \prod_{k=1}^{M} B(\lambda_k)\,|0\rangle = \prod_{k=1}^{M}\Big(A_1(\lambda_k)B_2(\lambda_k) + B_1(\lambda_k)D_2(\lambda_k)\Big)\,|0\rangle_1 \otimes |0\rangle_2\,. \tag{85}$$

Izergin and Korepin [5] state that Bethe vectors of the full model can be expressed in terms of Bethe vectors of its components. To obtain this expression we should commute in (85) all operators $A_1(\lambda_k)$ and $D_2(\lambda_k)$ to the right with the help of (26) and (27) and then use (82).

Proposition 1. *An arbitrary Bethe vector corresponding to the full system can be expressed in terms of the Bethe vectors of the first and second component. Let $I = \{\lambda_1, \ldots, \lambda_M\}$ be a finite set of spectral parameters. To concise notation below we will consider the set I as a finite set of indices $I = \{1, \ldots, M\}$, then*

$$\prod_{k\in I} B(\lambda_k)\,|0\rangle =$$
$$\sum_{I_1\cup I_2}\prod_{k_1\in I_1}\Big(\delta_2(\lambda_{k_1})B_1(\lambda_{k_1})\Big)\prod_{k_2\in I_2}\Big(\alpha_1(\lambda_{k_2})B_2(\lambda_{k_2})\Big)\,|0\rangle_1\otimes|0\rangle_2\prod_{k_1\in I_1}\prod_{k_2\in I_2} f(\lambda_{k_1},\lambda_{k_2}) \tag{86}$$

where $f(\lambda_{k_1}, \lambda_{k_2})$ is defined in (30) *and the summation is performed over all divisions of index set I into two disjoint subsets I_1 and I_2 where $I = I_1 \cup I_2$.*

For detailed proof, please, see e.g. [2]. This result can be straightforwardly generalized to arbitrary number of components $N \leq L$.

Proposition 2. *An arbitrary Bethe vector of the full system can be expressed in terms of the Bethe vectors of its components. For $N \leq L$ components the Bethe vector is of the form*

$$\prod_{k\in I} B(\lambda_k)\,|0\rangle = \sum_{I_1\cup I_2\cup\cdots\cup I_N}\prod_{k_1\in I_1}\prod_{k_2\in I_2}\cdots\prod_{k_N\in I_N}\prod_{1\le i<j\le N}\Big(\alpha_i(\lambda_{k_j})\delta_j(\lambda_{k_i})f(\lambda_{k_i},\lambda_{k_j})\Big)$$
$$\times\, B_1(\lambda_{k_1})\,|0\rangle_1 \otimes B_2(\lambda_{k_2})\,|0\rangle_2 \otimes \cdots \otimes B_N(\lambda_{k_N})\,|0\rangle_N \tag{87}$$

where summation is performed over all divisions of the set I into its N mutually disjoint subsets $I_1, I_2, \ldots, I_N$.

The proof is simply performed using (86) by induction on number of components N. More details will be included in [3] where more general inhomogeneous XXZ-chain is concerned.

6.1. Bethe vectors explicitly

By assumption we have a chain of length L. Let us divide it in L components, i.e. into L 1-chains. Using proposition 2 we get for M-magnon (Bethe vector) with $M \leq L$:

$$\prod_{k=1}^{M} B(\lambda_k)\,|0\rangle = \sum_{I_1\cup I_2\cup\cdots\cup I_L}\prod_{k_1\in I_1}\prod_{k_2\in I_2}\cdots\prod_{k_L\in I_L}\prod_{1\le i<j\le L}\Big(\alpha_i(\lambda_{k_j})\delta_j(\lambda_{k_i})f(\lambda_{k_i},\lambda_{k_j})\Big)$$
$$\times\, B_1(\lambda_{k_1})\,|0\rangle_1 \otimes B_2(\lambda_{k_2})\,|0\rangle_2 \otimes \cdots \otimes B_L(\lambda_{k_L})\,|0\rangle_L\,. \tag{88}$$

It holdsd for 1-chain, which is a chain with Hilbert space $h = \mathbb{C}^2$, that

$$B(\lambda)B(\mu)\left|0\right\rangle = 0. \tag{89}$$

Therefore, the sum over all divisions of $\{1,\ldots,M\}$ into L subsets contains just divisions into subsets containing at most one element, i.e. $|I_j| = 0, 1$. Moreover, only M of them is nonempty, let us denote them $I_{n_1}, I_{n_2}, \ldots, I_{n_M}$. We have to sum over all possible combinations of such sets, i.e. over all M-tuples $n_1 < n_2 < \cdots < n_M$; and then, to sum over all distributions of parameters $\lambda_1, \lambda_2, \ldots, \lambda_M$ into the sets $I_{n_1}, \ldots, I_{n_M}$.

After all, we get

$$\begin{aligned}\prod_{k=1}^{M} B(\lambda_k)\left|0\right\rangle = \sum_{1\le n_1<n_2<\cdots<n_M\le L}\ \sum_{\sigma_\lambda\in S_M} \sigma_\lambda\Bigg(\prod_{j=1}^{M}\Big(\prod_{i=1}^{n_j-1}\alpha_i(\lambda_j)\prod_{i=n_j+1}^{L}\delta_i(\lambda_j)\prod_{i=1}^{j-1}f(\lambda_i,\lambda_j)\Big)\\ \times B_{n_1}(\lambda_1)B_{n_2}(\lambda_2)\cdots B_{n_M}(\lambda_M)\Bigg)\left|0\right\rangle_1\otimes\left|0\right\rangle_2\otimes\cdots\otimes\left|0\right\rangle_L.\end{aligned} \tag{90}$$

Moreover, it holds for 1-chain that $B(\lambda) = B$ is parameter independent and eigenvalues $\alpha_i(\lambda) = a(\lambda)$, $\delta_i(\lambda) = d(\lambda)$ are still the same for all components $i = 1,\ldots,L$, where $a(\lambda) = \lambda+1$ and $d(\lambda) = \lambda$. We get

$$\begin{aligned}\prod_{k=1}^{M} B(\lambda_k)\left|0\right\rangle &= \sum_{1\le n_1<n_2<\cdots<n_M\le L}\ \sum_{\sigma\in S_M} \sigma_\lambda\left(\prod_{j=1}^{M} a(\lambda_j)^{n_j-1}d(\lambda_j)^{L-n_j}\prod_{i=1}^{j-1}f(\lambda_i,\lambda_j)\right)\\ &\quad\times B_{n_1}B_{n_2}\cdots B_{n_M}\left|0\right\rangle_1\otimes\left|0\right\rangle_2\otimes\cdots\otimes\left|0\right\rangle_L =\\ &= \prod_{j=1}^{M}\frac{d(\lambda_j)^L}{a(\lambda_j)}\sum_{1\le n_1<n_2<\cdots<n_M\le L}\ \sum_{\sigma\in S_M}\sigma_\lambda\left(\prod_{1\le i<j\le M} f(\lambda_i,\lambda_j)\prod_{j=1}^{M}\left(\frac{a(\lambda_j)}{d(\lambda_j)}\right)^{n_j}\right)\\ &\quad\times B_{n_1}B_{n_2}\cdots B_{n_M}\left|0\right\rangle_1\otimes\left|0\right\rangle_2\otimes\cdots\otimes\left|0\right\rangle_L.\end{aligned} \tag{91}$$

Again, the reader is refered for more details to [3].

Let us return to (79). To prove its validity, it is sufficient to realize that $\psi_{k_1}\psi_{k_2}\ldots\psi_{k_M}\left|0\right\rangle = B_{k_1}B_{k_2}\ldots B_{k_M}\left|0\right\rangle$ for $k_1 < k_2 < \cdots < k_M$.

7. Some remarks on inhomogeneous XXX-chain

We start from inhomogeneous monodromy matrix

$$T_a^{\vec{\xi}}(\lambda) = L_{a,1}(\lambda+\xi_1)L_{a,2}(\lambda+\xi_2)\cdots L_{a,L}(\lambda+\xi_L) \tag{92}$$

where $L_{a,j}(\lambda)$ are Lax operators defined in (4) and $\vec{\xi} = (\xi_1\ldots,\xi_L)$ is an inhomogenity vector. Expressing $T_a^{\vec{\xi}}(\lambda)$ in the auxiliary space V_a, we get

$$T_a^{\vec{\xi}}(\lambda) = \begin{pmatrix} A^{\vec{\xi}}(\lambda) & B^{\vec{\xi}}(\lambda)\\ C^{\vec{\xi}}(\lambda) & D^{\vec{\xi}}(\lambda)\end{pmatrix} \tag{93}$$

where, again, operators $A^{\vec{\xi}}(\lambda)$, $B^{\vec{\xi}}(\lambda)$, $C^{\vec{\xi}}(\lambda)$ and $D^{\vec{\xi}}(\lambda)$ are acting in $\mathscr{H} = h_1\otimes\cdots\otimes h_L$. Acting on pseudovacuum vector $\left|0\right\rangle \in \mathscr{H}$ we get

$$A^{\vec{\xi}}(\lambda)\left|0\right\rangle = \alpha^{\vec{\xi}}(\lambda)\left|0\right\rangle, \tag{94}$$

$$D^{\vec{\xi}}(\lambda)\left|0\right\rangle = \delta^{\vec{\xi}}(\lambda)\left|0\right\rangle, \tag{95}$$

$$C^{\vec{\xi}}(\lambda)\left|0\right\rangle = 0 \tag{96}$$

where

$$\alpha^{\vec{\xi}}(\lambda) = a(\lambda+\xi_1)a(\lambda+\xi_2)\cdots a(\lambda+\xi_L), \tag{97}$$

$$\delta^{\vec{\xi}}(\lambda) = d(\lambda+\xi_1)d(\lambda+\xi_2)\cdots d(\lambda+\xi_L). \tag{98}$$

Here, functions $a(\lambda)=\lambda+1$ and $d(\lambda)=\lambda$.

For inhomogeneous version, we can introduce N-component model as well as for homogeneous Bethe ansatz. For 2-compotent model, for example, we have

$$T_a^{\vec{\xi}}(\lambda) = \underbrace{L_{a,1}(\lambda+\xi_1)\cdots L_{a,x}(\lambda+\xi_x)}_{\text{1st component}}\underbrace{L_{a,x+1}(\lambda+\xi_{x+1})\cdots L_{a,L}(\lambda+\xi_L)}_{\text{2nd component}} = T_a^{\vec{\xi}_1}(\lambda)T_a^{\vec{\xi}_2}(\lambda) \tag{99}$$

where $\vec{\xi}_1=(\xi_1,\ldots,\xi_x)$ resp. $\vec{\xi}_2=(\xi_{x+1},\ldots,\xi_L)$. We have

$$A^{\vec{\xi}}(\lambda)\left|0\right\rangle = \alpha_1^{\vec{\xi}_1}(\lambda)\alpha_2^{\vec{\xi}_2}(\lambda)\left|0\right\rangle, \qquad D^{\vec{\xi}}(\lambda)\left|0\right\rangle = \delta_1^{\vec{\xi}_1}(\lambda)\delta_2^{\vec{\xi}_2}(\lambda)\left|0\right\rangle. \tag{100}$$

The very important property of inhomogeneous chain is, that its operators $A^{\vec{\xi}}(\lambda)$, $B^{\vec{\xi}}(\lambda)$, $C^{\vec{\xi}}(\lambda)$ and $D^{\vec{\xi}}(\lambda)$ satisfy the same fundamental commutation relations as homogeneous chain (23)-(30). Therefore, analogy of propositions 1 and 2 can be easily formulated.

Proposition 3. *Let $N \le L$. An arbitrary Bethe vector of the full system can be expressed in terms of Bethe vectors of its N components*

$$\prod_{k\in I} B^{\vec{\xi}}(\lambda_k)\left|0\right\rangle = \sum_{I_1\cup\cdots\cup I_N}\prod_{k_1\in I_1}\cdots\prod_{k_N\in I_N}\prod_{1\le i<j\le N}\left(\alpha_i^{\vec{\xi}_i}(\lambda_{k_j})\delta_j^{\vec{\xi}_j}(\lambda_{k_i})f(\lambda_{k_i},\lambda_{k_j})\right)$$
$$\times B_1^{\vec{\xi}_1}(\lambda_{k_1})B_2^{\vec{\xi}_2}(\lambda_{k_2})\cdots B_N^{\vec{\xi}_N}(\lambda_{k_N})\left|0\right\rangle. \tag{101}$$

To get explicit formula for Bethe vectors we have to divide the chain into L components of length 1 as we did in the last section. We get for M-magnon

$$\prod_{k=1}^{M} B^{\vec{\xi}}(\lambda_k)\left|0\right\rangle =$$
$$= \sum_{1\le n_1<\cdots<n_M\le L}\ \sum_{\sigma_\lambda\in S_M}\sigma_\lambda\Bigg(\prod_{j=1}^{M}\Big(\prod_{i=1}^{n_j-1}\alpha_i^{\xi_i}(\lambda_j)\prod_{i=n_j+1}^{L}\delta_i^{\xi_i}(\lambda_j)\prod_{i=1}^{j-1}f(\lambda_i,\lambda_j)\Big)$$
$$\times B_{n_1}^{\xi_{n_1}}(\lambda_1)\cdots B_{n_M}^{\xi_{n_M}}(\lambda_M)\Bigg)\left|0\right\rangle =$$
$$= \sum_{1\le n_1<\cdots<n_M\le L}\ \sum_{\sigma_\lambda\in S_M}\sigma_\lambda\Bigg(\prod_{j=1}^{M}\Big(\prod_{i=1}^{n_j-1}\alpha_i^{\xi_i}(\lambda_j)\prod_{i=n_j+1}^{L}\delta_i^{\xi_i}(\lambda_j)\prod_{i=1}^{j-1}f(\lambda_i,\lambda_j)\Big)\Bigg)B_{n_1}\cdots B_{n_M}\left|0\right\rangle =$$
$$= \prod_{j=1}^{M}\prod_{i=1}^{L}d(\lambda_j+\xi_i)\sum_{1\le n_1<\cdots<n_M\le L}\ \sum_{\sigma_\lambda\in S_M}\sigma_\lambda\left(\prod_{j=1}^{M}\frac{1}{a(\lambda_j+\xi_{n_j})}\prod_{i=1}^{n_j}\frac{a(\lambda_j+\xi_i)}{d(\lambda_j+\xi_i)}\prod_{i=1}^{j-1}f(\lambda_i,\lambda_j)\right)$$
$$\times B_{n_1}\cdots B_{n_M}\left|0\right\rangle. \tag{102}$$

where, again, B-operators $B_{n_j}^{\xi_{n_j}}(\lambda)=B_{n_j}$ are parameter independent for 1-chains. For more details, see [3].

8. Final remarks

We showed in this text explicit expressions for Bethe vectors of XXX-spin chain, both in fermionic and usual representation. We discussed also inhomogeneous version. We refer the reader for more details to [3] where we plan to discuss Bethe vectors for XXZ-chain in more detailed form.

Acknowledgments

JF thanks CTU for support via Project SGS12/198/OHK4/3T/14. The work of one of the authors (API) was supported by the grant RFBR 14-01-00474.

[1] Faddeev L D 1996 How Algebraic Bethe Ansatz works for integrable model `arXiv:hep-th/9605187`
[2] Slavnov N A 2007 Algebraic Bethe ansatz and quantum integrable models *Uspekhi Fiz. Nauk* **62** 727
[3] Fuksa J, Isaev A P and Slavnov N A *in preparation*
[4] Jordan P and Wigner E 1928 *Z. Phys.* **47** 631
[5] Izergin A G and Korepin V E 1984 The quantum inverse scattering method approach to correlation functions *Comm. Math. Phys.* **94** No.1 67-92
[6] Göhmann F and Korepin V E 2000 Solution of the quantum inverse problem *J. Phys. A: Math. Gen.* **33** 1199-1220

Journal of Physics: Conference Series **563** (2014) 012012 doi:10.1088/1742-6596/563/1/012012

On the derivation of Darboux form for the action-angle dual of trigonometric BC_n Sutherland system

Tamás F Görbe

Department of Theoretical Physics, University of Szeged, Tisza Lajos krt 84-86, H-6720 Szeged, Hungary

E-mail: tfgorbe@physx.u-szeged.hu

Abstract. Recently Fehér and the author have constructed the action-angle dual of the trigonometric BC_n Sutherland system via Hamiltonian reduction. In this paper a reduction-based calculation is carried out to verify the canonical Poisson bracket relations on the phase space of this dual model. Hence the material serves complementary purposes whilst it can also be regarded as a suitable modification of the hyperbolic case previously sorted out by Pusztai.

1. Introduction

The integrable one-dimensional many-body systems of Calogero, Moser, and Sutherland and generalized versions of them have proven to be a fruitful source of both diverse physical applications and connections between seemingly distant areas of mathematics. For details, see e.g. [1, 2, 3]. Among the numerous aspects of these models their duality relations are rather interesting. Two Liouville integrable many-body Hamiltonian systems (M, ω, H) and $(\tilde{M}, \tilde{\omega}, \tilde{H})$ with Darboux coordinates q, p and λ, ϑ, respectively, are said to be duals of each other if there is a global symplectomorphism $\mathcal{R}\colon M \to \tilde{M}$ of the phase spaces, which exchanges the canonical coordinates with the action-angle variables for the Hamiltonians. Practically, this means that $H \circ \mathcal{R}^{-1}$ depends only on λ, while $\tilde{H} \circ \mathcal{R}$ only on q. In more detail, q are the particle positions for H and action variables for $\tilde{H}$, and similarly, λ are the positions of particles modelled by the Hamiltonian $\tilde{H}$ and action variables for H.

A notable work has been done by Ruijsenaars [4, 5] in constructing action-angle duality maps for models with rational, hyperbolic, and trigonometric potentials associated with the A_n root system. Many of these dualities have been interpreted in terms of Hamiltonian reduction [6, 7].

The suspected existence of action-angle duality between models related other root systems has been confirmed by Pusztai [8] proving the hyperbolic BC_n Sutherland [9] and the rational BC_n Ruijsenaars – Schneider – van Diejen (RSvD) [10] systems to be in duality.

In a recent paper by Fehér and the author [11] earlier results [8, 12] have been generalized to obtain a new dual pair involving the trigonometric BC_n Sutherland system. This was achieved by applying Hamiltonian reduction to the cotangent bundle $T^*\mathrm{U}(2n)$ with respect to the symmetry group $G_+ \times G_+$ with $G_+ \simeq \mathrm{U}(n) \times \mathrm{U}(n)$. The systems in duality arose as two cross sections of the orbits of the symmetry group in the level surface of the momentum map since these cross sections were identified with the phase spaces of the trigonometric BC_n Sutherland and a rational BC_n RSvD-type systems. The aim of this paper is to provide detailed calculations proving that under this identification the coordinates λ, ϑ – introduced on a dense submanifold of the phase space of the dual model – are canonical (Darboux) coordinates as stated in [11].

Section 2 is a selective review of [11] devoted to establishing context and introducing necessary notations for succeeding calculations. The core of the paper is Section 3 which contains a series of lemmas culminating in the main result. Concluding the paper, Section 4 gives a brief discussion of the outcome and its relation to other cases considered formerly.

2. Context and notations

Choose an arbitrary positive integer, n. Let G and $\mathcal{G}$ denote the unitary group $\mathrm{U}(2n)$ and its Lie algebra, respectively. The Lie algebra $\mathcal{G}$ can be equipped with the Ad-invariant bilinear form

$$\langle \cdot, \cdot \rangle \colon \mathcal{G} \times \mathcal{G} \to \mathbb{R}, \quad (Y_1, Y_2) \mapsto \langle Y_1, Y_2 \rangle = \mathrm{tr}(Y_1 Y_2), \tag{1}$$

which allows one to identify $\mathcal{G}$ with its dual space $\mathcal{G}^*$ in the usual manner. The cotangent bundle T^*G can be trivialized using left-translations

$$T^*G \cong G \times \mathcal{G}^* \cong G \times \mathcal{G} = \{(y, Y) \mid y \in G, \ Y \in \mathcal{G}\}. \tag{2}$$

Then the canonical symplectic form of T^*G can be written as $\Omega^{T^*G} = -\mathrm{d}\langle y^{-1}\mathrm{d}y, Y \rangle$, and it can be evaluated locally according to the formula

$$\Omega^{T^*G}_{(y,Y)}(\Delta y \oplus \Delta Y, \Delta' y \oplus \Delta' Y) = \langle y^{-1}\Delta y, \Delta' Y \rangle - \langle y^{-1}\Delta' y, \Delta Y \rangle + \langle [y^{-1}\Delta y, y^{-1}\Delta' y], Y \rangle, \tag{3}$$

where $\Delta y \oplus \Delta Y, \Delta' y \oplus \Delta' Y \in T_{(y,Y)}T^*G$ are arbitrary tangent vectors at a point $(y, Y) \in T^*G$. By introducing the $2n \times 2n$ Hermitian, unitary matrix

$$C = \begin{bmatrix} \mathbf{0}_n & \mathbf{1}_n \\ \mathbf{1}_n & \mathbf{0}_n \end{bmatrix} \in G, \tag{4}$$

where $\mathbf{1}_n$ and $\mathbf{0}_n$ denote the identity and null matrices of size n, respectively, an involutive automorphism of G can be defined as conjugation with C

$$\Gamma \colon G \to G, \quad y \mapsto \Gamma(y) = CyC^{-1}. \tag{5}$$

The fix-point subgroup of Γ in G is

$$G_+ = \{y \in G \mid \Gamma(y) = y\} \cong \mathrm{U}(n) \times \mathrm{U}(n). \tag{6}$$

Let Γ stand for the induced involution of the Lie algebra $\mathcal{G}$, too. Hence $\mathcal{G}$ can be decomposed as

$$\mathcal{G} = \mathcal{G}_+ \oplus \mathcal{G}_-, \quad Y = Y_+ + Y_-, \tag{7}$$

where $\mathcal{G}_\pm$ are the eigenspaces of Γ corresponding to the eigenvalues ± 1, respectively.

In [11] a reduction of T^*G based on the symmetry group $G_+ \times G_+$ was performed by using the shifting trick of symplectic reduction [13]. For that a coadjoint orbit of the symmetry group must be prepared. To any vector $V \in \mathbb{C}^{2n}$ that satisfies $CV + V = 0$ associate an element $\upsilon^\ell_{\mu,\nu}(V)$ of $\mathcal{G}_+$ by the definition

$$\upsilon^\ell_{\mu,\nu}(V) = \mathrm{i}\mu\big(VV^\dagger - \mathbf{1}_{2n}\big) + \mathrm{i}(\mu - \nu)C, \tag{8}$$

where $\mu, \nu \in \mathbb{R}$ are real parameters. The set

$$\mathcal{O}^\ell = \big\{\upsilon^\ell \in \mathcal{G}_+ \mid \exists\, V \in \mathbb{C}^{2n}, \ V^\dagger V = 2n, \ CV + V = 0, \ \upsilon^\ell = \upsilon^\ell_{\mu,\nu}(V)\big\} \tag{9}$$

represents a coadjoint orbit of G_+ of dimension $2(n-1)$. Let $\mathcal{O}^r := \{\upsilon^r\}$ denote the one-point coadjoint orbit of G_+ containing the element $\upsilon^r = -\mathrm{i}\kappa C$ with some constant $\kappa \in \mathbb{R}$ and consider

$$\mathcal{O} = \mathcal{O}^\ell \oplus \mathcal{O}^r \subset \mathcal{G}_+ \oplus \mathcal{G}_+ \cong (\mathcal{G}_+ \oplus \mathcal{G}_+)^*, \tag{10}$$

which is a coadjoint orbit of $G_+ \times G_+$. The initial phase space for symplectic reduction is

$$P = T^*G \times \mathcal{O} \quad \text{with the symplectic form} \quad \Omega = \Omega^{T^*G} + \Omega^{\mathcal{O}}, \tag{11}$$

where $\Omega^{\mathcal{O}}$ is the Kirillov – Kostant – Souriau symplectic form on the coadjoint orbit $\mathcal{O}$.
For any point $x = (y, Y, \upsilon^\ell, \upsilon^r) \in P$ and smooth functions $f, f' \in C^\infty(P)$

$$\Omega_x((\boldsymbol{X}_f)_x, (\boldsymbol{X}_{f'})_x) = \Omega^{T^*G}_{(y,Y)}(\Delta y \oplus \Delta Y, \Delta' y \oplus \Delta' Y) + \langle [D_{\upsilon^\ell}, D'_{\upsilon^\ell}], \upsilon^\ell \rangle, \tag{12}$$

where $(\boldsymbol{X}_f)_x = \Delta y \oplus \Delta Y \oplus \Delta\upsilon^\ell \oplus 0$, $(\boldsymbol{X}_{f'})_x = \Delta' y \oplus \Delta' Y \oplus \Delta'\upsilon^\ell \oplus 0 \in T_x P$ and $\Delta\upsilon^\ell = [D_{\upsilon^\ell}, \upsilon^\ell]$, $\Delta'\upsilon^\ell = [D'_{\upsilon^\ell}, \upsilon^\ell]$ with some $\mathcal{G}_+$-valued $D_{\upsilon^\ell}, D'_{\upsilon^\ell}$. The natural symplectic action of $G_+ \times G_+$ on P is defined by

$$\Phi_{(g_L, g_R)}(y, Y, \upsilon^\ell, \upsilon^r) = \big(g_L y g_R^{-1}, g_R Y g_R^{-1}, g_L \upsilon^\ell g_L^{-1}, \upsilon^r\big). \tag{13}$$

The corresponding momentum map $J\colon P \to \mathcal{G}_+ \oplus \mathcal{G}_+$ is given by the formula

$$J(y, Y, \upsilon^\ell, \upsilon^r) = \big((yYy^{-1})_+ + \upsilon^\ell\big) \oplus \big(-Y_+ + \upsilon^r\big). \tag{14}$$

The reduced phase space is

$$P_{\text{red}} = J^{-1}(0)/(G_+ \times G_+), \tag{15}$$

which is a smooth symplectic manifold.

One of the main results in [11] was the construction of a semi-global cross section of symmetry group orbits in the momentum constraint surface $J^{-1}(0)$, that is a model of the reduced phase space (15). This was done by solving the momentum equation $J(y, Y, \upsilon^\ell, \upsilon^r) = \mathbf{0}_{2n} \oplus \mathbf{0}_{2n}$ through the diagonalization of the $(\mathcal{G}_-)$-part of the Lie algebra component. In particular, the following matrix similarity was demonstrated

$$Y \sim \mathrm{i}h(\lambda)\Lambda(\lambda)h(\lambda)^{-1}, \tag{16}$$

where $\Lambda(\lambda) = \operatorname{diag}(\lambda, -\lambda)$ with $\lambda = (\lambda_1, \dots, \lambda_n) \in \mathbb{R}^n$ subject to $\lambda_1 > \dots > \lambda_n > |\kappa|$ and $h(\lambda)$ is the unitary matrix

$$h(\lambda) = \begin{bmatrix} \alpha(\operatorname{diag}(\lambda)) & \beta(\operatorname{diag}(\lambda)) \\ -\beta(\operatorname{diag}(\lambda)) & \alpha(\operatorname{diag}(\lambda)) \end{bmatrix}, \tag{17}$$

with the real functions $\alpha(x), \beta(x)$ defined on the interval $[|\kappa|, \infty) \subset \mathbb{R}$ by the formulae

$$\alpha(x) = \frac{\sqrt{x + \sqrt{x^2 - \kappa^2}}}{\sqrt{2x}}, \quad \beta(x) = \kappa \frac{1}{\sqrt{2x}} \frac{1}{\sqrt{x + \sqrt{x^2 - \kappa^2}}}, \tag{18}$$

if $\kappa \neq 0$. For $\kappa = 0$, set $h(\lambda) = \mathbf{1}_{2n}$. This approach enables one to define the smooth map

$$\mathfrak{L}\colon P_0 \to \mathbb{R}^n, \quad (y, Y, \upsilon^\ell, \upsilon^r) \mapsto \lambda, \tag{19}$$

which descends to a smooth map $\mathfrak{L}_{\text{red}}\colon P_{\text{red}} \to \mathbb{R}^n$. The image of the constraint surface $J^{-1}(0)$ under the map $\mathfrak{L}$ (19) turned out to be the closure of the domain

$$C_2 = \left\{ \lambda \in \mathbb{R}^n \,\middle|\, \begin{matrix} \lambda_a - \lambda_{a+1} > 2\mu, \\ (a = 1, \dots, n-1) \end{matrix} \quad \text{and} \quad \lambda_n > \nu \right\}. \tag{20}$$

Introduce the vector $F \in \mathbb{C}^{2n}$ by the formulae

$$\begin{aligned} F_a &= \left[1 - \frac{\nu}{\lambda_a}\right]^{\frac{1}{2}} \prod_{\substack{b=1 \\ (b\neq a)}}^{n} \left[1 - \frac{2\mu}{\lambda_a - \lambda_b}\right]^{\frac{1}{2}} \left[1 - \frac{2\mu}{\lambda_a + \lambda_b}\right]^{\frac{1}{2}}, \quad a \in \{1, \ldots, n\}, \\ F_{n+a} &= e^{\mathrm{i}\vartheta_a} \left[1 + \frac{\nu}{\lambda_a}\right]^{\frac{1}{2}} \prod_{\substack{b=1 \\ (b\neq a)}}^{n} \left[1 + \frac{2\mu}{\lambda_a - \lambda_b}\right]^{\frac{1}{2}} \left[1 + \frac{2\mu}{\lambda_a + \lambda_b}\right]^{\frac{1}{2}}. \end{aligned} \tag{21}$$

and the $2n \times 2n$ matrices $A(\lambda, \vartheta)$ and $B(\lambda, \vartheta)$ by

$$A_{j,k}(\lambda, \vartheta) = \frac{2\mu F_j \overline{(CF)}_k - 2(\mu - \nu) C_{j,k}}{2\mu - \Lambda_j + \Lambda_k}, \quad j, k \in \{1, \ldots, 2n\}, \tag{22}$$

and

$$B(\lambda, \vartheta) = -\big(h(\lambda) A(\lambda, \vartheta) h(\lambda)\big)^{\dagger}. \tag{23}$$

These are unitary matrices satisfying $\Gamma(A) = A^{-1}$, $\Gamma(B) = B^{-1}$. The matrix B can be diagonalized using some $\eta \in G_+$

$$B = \eta \operatorname{diag}(\exp(2\mathrm{i}q), \exp(-2\mathrm{i}q)) \eta^{-1}, \tag{24}$$

where $q = q(\lambda, \vartheta) \in \mathbb{R}^n$ is unique and subject to $\pi/2 > q_1 > \cdots > q_n > 0$. Relying on (24) set

$$y(\lambda, \vartheta) = \eta \operatorname{diag}(\exp(\mathrm{i}q), \exp(-\mathrm{i}q)) \eta^{-1}, \tag{25}$$

and introduce the vector $V(\lambda, \vartheta) \in \mathbb{C}^{2n}$ by

$$V(\lambda, \vartheta) = y(\lambda, \vartheta) h(\lambda) F(\lambda, \vartheta). \tag{26}$$

It was also shown in [11] that $V + CV = 0$ and $|V|^2 = 2n$ ensuring that $v^{\ell}_{\mu,\nu}(V) \in \mathcal{O}^{\ell}$ (9).
Theorem 4.1 of [11] claims that the set

$$\tilde{S}^0 := \{(y(\lambda, \vartheta), ih(\lambda)\Lambda(\lambda)h(\lambda)^{-1}, v^{\ell}_{\mu,\nu}(V(\lambda, \vartheta)), v^r) \mid (\lambda, e^{\mathrm{i}\vartheta}) \in C_2 \times \mathbb{T}^n\}. \tag{27}$$

is contained in the constraint surface $J^{-1}(0)$ and provides a cross-section for the $G_+ \times G_+$-action restricted to $\mathfrak{L}^{-1}(C_2) \subset J^{-1}(0)$. In particular, $C_2 \subset \mathfrak{L}(J^{-1}(0))$ and $\tilde{S}^0$ intersects every gauge orbit in $\mathfrak{L}^{-1}(C_2)$ precisely in one point. Since the elements of $\tilde{S}^0$ are parametrized by $C_2 \times \mathbb{T}^n$ in a smooth and bijective manner, the following identifications were gained

$$\mathfrak{L}^{-1}_{\text{red}}(C_2) \simeq \tilde{S}^0 \simeq C_2 \times \mathbb{T}^n. \tag{28}$$

Let $\tilde{\sigma}_0$ denote the tautological injection

$$\tilde{\sigma}_0 \colon \tilde{S}^0 \to P. \tag{29}$$

This way $C_2 \times \mathbb{T}^n$ yields a model of an open submanifold $\mathfrak{L}^{-1}(C_2)$ of P_{red} corresponding to the open submanifold $\mathfrak{L}^{-1}(C_2) \subset J^{-1}(0)$ was obtained. The purpose of this paper is to show that the pull-back $\tilde{\sigma}_0^*(\Omega)$ of the symplectic form Ω (11) is

$$\tilde{\sigma}_0^*(\Omega) = \sum_{a=1}^{n} \mathrm{d}\lambda_a \wedge \mathrm{d}\vartheta_a \tag{30}$$

by computing the Poisson brackets

$$\{\lambda_a, \lambda_b\}, \quad \{\lambda_a, \vartheta_b\}, \quad \{\vartheta_a, \vartheta_b\}, \qquad a, b \in \{1, \dots, n\}. \tag{31}$$

Now, consider the reduced functions $f_j^{\rm red} = \tilde{\sigma}_0^*(f_j)$ for some $f_j \in C^\infty(P)^{G_+ \times G_+}$ $(j = 1, 2)$. Then the definition of symplectic reduction implies

$$\tilde{\sigma}_0^*(\{f_1, f_2\}) = \{f_1^{\rm red}, f_2^{\rm red}\}, \tag{32}$$

where the Poisson bracket on the left-hand-side is computed on (P, Ω) (11). The idea is to extract the required Poisson brackets in (31) from equality (32) applied to various choices of f_1, f_2. Note that $\{f_1, f_2\} = \Omega(\boldsymbol{X}_{f_2}, \boldsymbol{X}_{f_1})$ with the corresponding Hamiltonian vector fields.

3. Calculation of Poisson brackets

The following verification is an appropriate adaptation of an argument presented by Pusztai in [14] which since has been applied in the simpler case of A_n root system in [15]. Differences between these earlier results and the calculations below are highlighted in the Discussion.

Consider the following families of real-valued smooth functions on the phase space P (11)

$$\varphi_m(y, Y, \upsilon^\ell, \upsilon^r) := \frac{1}{m}\mathrm{Re}\big(\mathrm{tr}(Y^m)\big), \quad m \in \mathbb{N}, \tag{33}$$

$$\chi_k(y, Y, \upsilon^\ell, \upsilon^r) := \mathrm{Re}\big(\mathrm{tr}(Y^k y^{-1} Z(\upsilon^\ell) y C)\big), \quad k \in \mathbb{N}_0, \tag{34}$$

where $Z(\upsilon^\ell) = (\mathrm{i}\mu)^{-1}\upsilon^\ell_{\mu,\nu}(V) + \mathbf{1}_N - (1 - \nu/\mu)C = VV^\dagger$. The corresponding reduced functions on $\tilde{S}^0$ are

$$\varphi_m^{\rm red}(\lambda, \vartheta) = \begin{cases} 0, & \text{if } m \text{ is odd}, \\ (-1)^{\frac{m}{2}}\dfrac{2}{m}\displaystyle\sum_{j=1}^n \lambda_j^m, & \text{if } m \text{ is even}, \end{cases} \tag{35}$$

and

$$\chi_k^{\rm red}(\lambda, \vartheta) = \begin{cases} (-1)^{\frac{k+1}{2}} 2 \displaystyle\sum_{a=1}^n \lambda_a^k \left[1 - \frac{\kappa^2}{\lambda_a^2}\right]^{\frac{1}{2}} |X_a| \sin(\vartheta_a), & \text{if } k \text{ is odd}, \\ (-1)^{\frac{k}{2}} 2 \displaystyle\sum_{a=1}^n \lambda_a^k \left[1 - \frac{\kappa^2}{\lambda_a^2}\right]^{\frac{1}{2}} |X_a| \cos(\vartheta_a) - \kappa \lambda_a^{k-1}\big(|F_a|^2 - |F_{n+a}|^2\big), & \text{if } k \text{ is even}, \end{cases} \tag{36}$$

where

$$X_a = F_a \overline{F}_{n+a} = e^{-\mathrm{i}\vartheta_a}\left[1 - \frac{\nu^2}{\lambda_a^2}\right]^{\frac{1}{2}} \prod_{\substack{b=1 \\ (b \neq a)}}^n \left[1 - \frac{4\mu^2}{(\lambda_a - \lambda_b)^2}\right]^{\frac{1}{2}} \left[1 - \frac{4\mu^2}{(\lambda_a + \lambda_b)^2}\right]^{\frac{1}{2}}. \tag{37}$$

Now let us take an arbitrary point $x = (y, Y, \upsilon^\ell, \upsilon^r) \in P$ and an arbitrary tangent vector $\delta x = \delta y \oplus \delta Y \oplus \delta \upsilon^\ell \oplus 0 \in T_x P$. The derivative of φ_m can be easily obtained and has the form

$$(\mathrm{d}\varphi_m)_x(\delta x) = \begin{cases} 0, & \text{if } m \text{ is odd}, \\ \langle Y^{m-1}, \delta Y \rangle, & \text{if } m \text{ is even}. \end{cases} \tag{38}$$

The derivative of χ_k can be written as

$$\begin{aligned}(\mathrm{d}\chi_k)_x(\delta x) =& \left\langle \frac{[[Y^k, C]_\pm, y^{-1}Z(v^\ell)y]}{2}, y^{-1}\delta y \right\rangle \\ &+ \left\langle \sum_{j=0}^{k-1} \frac{Y^{k-j-1}[y^{-1}Z(v^\ell)y, C]_\pm Y^j}{2}, \delta Y \right\rangle \\ &+ \left\langle \frac{y[C, Y^k]_\pm y^{-1} + Cy[C, Y^k]_\pm y^{-1}C}{4\mathrm{i}\mu}, \delta v^\ell \right\rangle,\end{aligned} \tag{39}$$

where $[A, B]_\pm := AB \pm BA$ with the sign of $(-1)^k$. The Hamiltonian vector field of φ_m is

$$(\boldsymbol{X}_{\varphi_m})_x = \Delta y \oplus \Delta Y \oplus \Delta v^\ell \oplus 0 = yY^{m-1} \oplus 0 \oplus 0 \oplus 0, \tag{40}$$

while the Hamiltonian vector field corresponding to χ_k is

$$(\boldsymbol{X}_{\chi_k})_x = \Delta' y \oplus \Delta' Y \oplus \Delta' v^\ell \oplus 0, \tag{41}$$

where

$$\Delta' y = \frac{y}{2} \sum_{j=0}^{k-1} Y^{k-j-1}[y^{-1}Z(v^\ell)y, C]_\pm Y^j, \tag{42}$$

$$\Delta' Y = \frac{1}{2}[[Y^k, y^{-1}Z(v^\ell)y]_\pm, C], \tag{43}$$

$$\Delta' v^\ell = \frac{1}{4\mathrm{i}\mu}[(y[C, Y^k]_\pm y^{-1} + Cy[C, Y^k]_\pm y^{-1}C), v^\ell]. \tag{44}$$

Lemma 1. $\{\lambda_a, \lambda_b\} = 0$ *for any* $a, b \in \{1, \dots, n\}$.

Proof. Using (38) one has $\{\varphi_m, \varphi_l\} \equiv 0$ for any $m, l \in \mathbb{N}$ which implies that $\{\varphi_m^{\mathrm{red}}, \varphi_l^{\mathrm{red}}\} \equiv 0$. Let $m, l \in \mathbb{N}$ be arbitrary even numbers. Direct calculation of the Poisson bracket $\{\varphi_m^{\mathrm{red}}, \varphi_l^{\mathrm{red}}\}$ using (35) and the Leibniz rule results in the formula

$$\{\varphi_m^{\mathrm{red}}, \varphi_l^{\mathrm{red}}\} = (-1)^{\frac{m+l}{2}} 4 \sum_{a,b=1}^{n} \lambda_a^{m-1}\{\lambda_a, \lambda_b\}\lambda_b^{l-1}. \tag{45}$$

By introducing the $n \times n$ matrices

$$\boldsymbol{P}_{a,b} := \{\lambda_a, \lambda_b\} \quad \text{and} \quad \boldsymbol{U}_{a,b} := \lambda_a^{2b-1}, \qquad a, b \in \{1, \dots, n\} \tag{46}$$

and choosing m and l from the set $\{1, \dots, 2n\}$, the equation $\{\varphi_m^{\mathrm{red}}, \varphi_l^{\mathrm{red}}\} \equiv 0$ can be cast into the matrix equation

$$(-1)^{\frac{m+l}{2}} \boldsymbol{U}^\dagger \boldsymbol{P} \boldsymbol{U} = \boldsymbol{0}_n. \tag{47}$$

Since $\boldsymbol{U}$ is an invertible Vandermonde-type matrix it follows from (47) that $\boldsymbol{P} = \boldsymbol{0}_n$ which reads as $\{\lambda_a, \lambda_b\} = 0$ for all $a, b \in \{1, \dots, n\}$. □

Lemma 2. $\{\lambda_a, \vartheta_b\} = \delta_{a,b}$ *for any* $a, b \in \{1, \dots, n\}$.

Proof. By choosing two even numbers, k and m, and calculating the Poisson bracket $\{\chi_k, \varphi_m\}$ at an arbitrary point $x = (y, Y, v^\ell, v^r) \in P$ the results (40)-(44) imply that

$$\{\chi_k, \varphi_m\}(x) = \chi_{k+m-1}(x) + \frac{1}{2}\mathrm{tr}\big((Y^k C Y^{m-1} - Y^{m-1} C Y^k) y^{-1} Z(v^\ell) y\big). \tag{48}$$

The computation of the reduced form of (48) shows that

$$\{\chi_k^{\mathrm{red}}, \varphi_m^{\mathrm{red}}\} = 2\chi_{k+m-1}^{\mathrm{red}}. \tag{49}$$

By utilizing (35), (36) and the result of the previous lemma one can write the l.h.s. of (49) as

$$\{\chi_k^{\mathrm{red}}, \varphi_m^{\mathrm{red}}\} = (-1)^{\frac{k+m}{2}} 4 \sum_{b=1}^{n} \lambda_b^k \left[1 - \frac{\kappa^2}{\lambda_b^2}\right]^{\frac{1}{2}} |X_b(\lambda)| \sin(\vartheta_b) \sum_{a=1}^{n} \{\lambda_a, \vartheta_b\} \lambda_a^{m-1}. \tag{50}$$

Now, returning to equation (49) together with (50) one can obtain the following equivalent form

$$\sum_{b=1}^{n} \lambda_b^k \left[1 - \frac{\kappa^2}{\lambda_b^2}\right]^{\frac{1}{2}} |X_b(\lambda)| \sin(\vartheta_b) \left(\sum_{a=1}^{n} \{\lambda_a, \vartheta_b\} \lambda_a^{m-1} - \lambda_b^{m-1} \right) = 0. \tag{51}$$

By introducing the $n \times n$ matrices

$$\boldsymbol{V}_{b,d} := \left[1 - \frac{\kappa^2}{\lambda_b^2}\right]^{\frac{1}{2}} |X_b(\lambda)| \sin(\vartheta_b) \left(\sum_{a=1}^{n} \{\lambda_a, \vartheta_b\} \lambda_a^{2d-1} - \lambda_b^{2d-1} \right), \quad b, d \in \{1, \dots, n\} \tag{52}$$

and using the Vandermonde-type matrix $\boldsymbol{U}$ defined in (46) one is able to write (51) into the matrix equation $\boldsymbol{U}^\dagger \boldsymbol{V} = \boldsymbol{0}_n$. Since $\boldsymbol{U}$ is invertible $\boldsymbol{V} = \boldsymbol{0}_n$ and therefore in the dense subset of $C_2 \times \mathbb{T}^n$ where $\sin(\vartheta_b) \neq 0$ the following holds

$$\sum_{a=1}^{n} \{\lambda_a, \vartheta_b\} \lambda_a^{m-1} - \lambda_b^{m-1} = 0, \quad \forall b \in \{1, \dots, n\}. \tag{53}$$

With the matrices $\boldsymbol{U}$ and

$$\boldsymbol{Q}_{b,a} := \{\lambda_a, \vartheta_b\}, \quad a, b \in \{1, \dots, n\} \tag{54}$$

equation (53) can be written equivalently as $\boldsymbol{QU} - \boldsymbol{U} = \boldsymbol{0}_n$, which immediately implies that $\boldsymbol{Q} = \boldsymbol{1}_n$. Due to the continuity of Poisson bracket $\boldsymbol{Q} = \boldsymbol{1}_n$ must hold for every point in $C_2 \times \mathbb{T}^n$, therefore one has $\{\lambda_a, \vartheta_b\} = \delta_{a,b}$ for all $a, b \in \{1, \dots, n\}$. □

Lemma 3. *$\{\vartheta_a, \vartheta_b\} = 0$ for any $a, b \in \{1, \dots, n\}$.*

Proof. Let k and l be two arbitrarily chosen odd integers, and set $f = \chi_l$ and $f' = \chi_k$ in (12). First, one can calculate the Poisson bracket $\{\chi_k^{\mathrm{red}}, \chi_l^{\mathrm{red}}\}$ indirectly, that is, work out the Poisson bracket $\{\chi_k, \chi_l\} = \Omega(\boldsymbol{X}_{\chi_l}, \boldsymbol{X}_{\chi_k})$ explicitly and restrict it to the gauge (27). The first term on the right-hand side of equation (12), namely $\langle y^{-1}\Delta y, \Delta' Y\rangle$ can be written as

$$\begin{aligned}
\langle y^{-1}\Delta y, \Delta' Y\rangle =& (-1)^{\frac{k+l+2}{2}} 2l \sum_{a=1}^{n} \lambda_a^{k+l-1} \left[1 - \frac{\kappa}{\lambda_a^2}\right] |X_a(\lambda)|^2 \sin(2\vartheta_a) \\
& (-1)^{\frac{k+l+2}{2}} 2 \sum_{\substack{a,b=1 \\ (a \neq b)}}^{n} \lambda_a^k \lambda_b^l \left[1 - \frac{\kappa^2}{\lambda_a^2}\right]^{\frac{1}{2}} \left[1 - \frac{\kappa^2}{\lambda_b^2}\right]^{\frac{1}{2}} |X_a||X_b| \frac{\sin(\vartheta_a - \vartheta_b)}{\lambda_a + \lambda_b} \\
& (-1)^{\frac{k-l+2}{2}} 2 \sum_{\substack{a,b=1 \\ (a \neq b)}}^{n} \lambda_a^k \lambda_b^l \left[1 - \frac{\kappa^2}{\lambda_a^2}\right]^{\frac{1}{2}} \left[1 - \frac{\kappa^2}{\lambda_b^2}\right]^{\frac{1}{2}} |X_a||X_b| \frac{\sin(\vartheta_a + \vartheta_b)}{\lambda_a - \lambda_b}.
\end{aligned} \tag{55}$$

Due to antisymmetry in the indices the second term can be gained by interchanging k and l

$$\begin{aligned}\langle y^{-1}\Delta' y, \Delta Y\rangle =&(-1)^{\frac{k+l+2}{2}} 2\,k \sum_{a=1}^{n} \lambda_a^{k+l-1}\left[1-\frac{\kappa}{\lambda_a^2}\right] |X_a(\lambda)|^2 \sin(2\vartheta_a)\\ &(-1)^{\frac{k-l+2}{2}} 2 \sum_{\substack{a,b=1\\(a\neq b)}}^{n} \lambda_a^k\lambda_b^l\left[1-\frac{\kappa^2}{\lambda_a^2}\right]^{\frac{1}{2}}\left[1-\frac{\kappa^2}{\lambda_b^2}\right]^{\frac{1}{2}} |X_a||X_b| \frac{\sin(\vartheta_a-\vartheta_b)}{\lambda_a+\lambda_b}\\ &(-1)^{\frac{k+l+2}{2}} 2 \sum_{\substack{a,b=1\\(a\neq b)}}^{n} \lambda_a^k\lambda_b^l\left[1-\frac{\kappa^2}{\lambda_a^2}\right]^{\frac{1}{2}}\left[1-\frac{\kappa^2}{\lambda_b^2}\right]^{\frac{1}{2}} |X_a||X_b| \frac{\sin(\vartheta_a+\vartheta_b)}{\lambda_a-\lambda_b}.\end{aligned} \tag{56}$$

One can easily check that the third term in (12) vanishes. The last term of (12) takes the form

$$\begin{aligned}\langle [D_v, D_v'], v\rangle =&(-1)^{\frac{k+l+2}{2}} 4 \sum_{\substack{a,b=1\\(a\neq b)}}^{n} \lambda_a^k\lambda_b^l\left[1-\frac{\kappa^2}{\lambda_a^2}\right]^{\frac{1}{2}}\left[1-\frac{\kappa^2}{\lambda_b^2}\right]^{\frac{1}{2}} |X_a||X_b| \frac{\sin(\vartheta_a-\vartheta_b)}{\big(4\mu^2-(\lambda_a+\lambda_b)^2\big)(\lambda_a+\lambda_b)}\\ &(-1)^{\frac{k-l+2}{2}} 4 \sum_{\substack{a,b=1\\(a\neq b)}}^{n} \lambda_a^k\lambda_b^l\left[1-\frac{\kappa^2}{\lambda_a^2}\right]^{\frac{1}{2}}\left[1-\frac{\kappa^2}{\lambda_b^2}\right]^{\frac{1}{2}} |X_a||X_b| \frac{\sin(\vartheta_a+\vartheta_b)}{\big(4\mu^2-(\lambda_a-\lambda_b)^2\big)(\lambda_a-\lambda_b)}.\end{aligned} \tag{57}$$

As a result of this indirect calculation one obtains the following expression for $\{\chi_k^{\rm red}, \chi_l^{\rm red}\}$

$$\begin{aligned}\{\chi_k^{\rm red}, \chi_l^{\rm red}\} =& (-1)^{\frac{k-l+2}{2}} 2(k-l) \sum_{a=1}^{n} \lambda_a^{k+l-1}\left[1-\frac{\kappa^2}{\lambda_a^2}\right] |X_a|^2 \sin(2\vartheta_a)\\ &(-1)^{\frac{k+l+2}{2}} 16\mu^2 \sum_{\substack{a,b=1\\(a\neq b)}}^{n} \lambda_a^k\lambda_b^l\left[1-\frac{\kappa^2}{\lambda_a^2}\right]^{\frac{1}{2}}\left[1-\frac{\kappa^2}{\lambda_b^2}\right]^{\frac{1}{2}} |X_a||X_b| \frac{\sin(\vartheta_a-\vartheta_b)}{\big(4\mu^2-(\lambda_a+\lambda_b)^2\big)(\lambda_a+\lambda_b)}\\ &(-1)^{\frac{k-l+2}{2}} 16\mu^2 \sum_{\substack{a,b=1\\(a\neq b)}}^{n} \lambda_a^k\lambda_b^l\left[1-\frac{\kappa^2}{\lambda_a^2}\right]^{\frac{1}{2}}\left[1-\frac{\kappa^2}{\lambda_b^2}\right]^{\frac{1}{2}} |X_a||X_b| \frac{\sin(\vartheta_a+\vartheta_b)}{\big(4\mu^2-(\lambda_a-\lambda_b)^2\big)(\lambda_a-\lambda_b)}.\end{aligned} \tag{58}$$

One can also carry out a direct computation of $\{\chi_k^{\rm red}, \chi_l^{\rm red}\}$ by using basic properties of the

Poisson bracket and the previous two lemmas

$$\{\chi_k^{\rm red}, \chi_l^{\rm red}\} = (-1)^{\frac{k-l+2}{2}} 2(k-l) \sum_{a=1}^{n} \lambda_a^{k+l-1}\left[1-\frac{\kappa^2}{\lambda_a^2}\right]|X_a|^2 \sin(2\vartheta_a)$$
$$(-1)^{\frac{k+l+2}{2}} 16\mu^2 \sum_{\substack{a,b=1\\(a\neq b)}}^{n} \lambda_a^k\lambda_b^l \left[1-\frac{\kappa^2}{\lambda_a^2}\right]^{\frac{1}{2}}\left[1-\frac{\kappa^2}{\lambda_b^2}\right]^{\frac{1}{2}} |X_a||X_b| \frac{\sin(\vartheta_a-\vartheta_b)}{(4\mu^2-(\lambda_a+\lambda_b)^2)(\lambda_a+\lambda_b)}$$
$$(-1)^{\frac{k-l+2}{2}} 16\mu^2 \sum_{\substack{a,b=1\\(a\neq b)}}^{n} \lambda_a^k\lambda_b^l \left[1-\frac{\kappa^2}{\lambda_a^2}\right]^{\frac{1}{2}}\left[1-\frac{\kappa^2}{\lambda_b^2}\right]^{\frac{1}{2}} |X_a||X_b| \frac{\sin(\vartheta_a+\vartheta_b)}{(4\mu^2-(\lambda_a-\lambda_b)^2)(\lambda_a-\lambda_b)}$$
$$(-1)^{\frac{k-l}{2}} 4 \sum_{a,b=1}^{n} \lambda_a^k\lambda_b^l \left[1-\frac{\kappa^2}{\lambda_a^2}\right]^{\frac{1}{2}}\left[1-\frac{\kappa^2}{\lambda_b^2}\right]^{\frac{1}{2}} |X_a||X_b| \cos(\vartheta_a)\cos(\vartheta_b)\{\vartheta_a,\vartheta_b\}. \tag{59}$$

Now it is obvious that (58) and (59) must be equal therefore the extra term must vanish

$$\sum_{a,b=1}^{n} \lambda_a^k\lambda_b^l \left[1-\frac{\kappa^2}{\lambda_a^2}\right]^{\frac{1}{2}}\left[1-\frac{\kappa^2}{\lambda_b^2}\right]^{\frac{1}{2}} |X_a||X_b| \cos(\vartheta_a)\cos(\vartheta_b)\{\vartheta_a,\vartheta_b\} = 0. \tag{60}$$

By utilizing the $n \times n$ matrices

$$\boldsymbol{W}_{a,b} = \lambda_a^b \left[1-\frac{\kappa^2}{\lambda_a^2}\right]^{\frac{1}{2}} |X_a(\lambda)|\cos(\vartheta_a), \quad \boldsymbol{R}_{a,b} = \{\vartheta_a,\vartheta_b\}, \quad a,b \in \{1,\dots,n\} \tag{61}$$

one can reformulate (60) as the matrix equation

$$\boldsymbol{W}^\dagger \boldsymbol{R} \boldsymbol{W} = \boldsymbol{0}_n. \tag{62}$$

Since $\boldsymbol{W}$ is easily seen to be invertible in a dense subset of the phase space $C_2 \times \mathbb{T}^n$, eq. (62) and the continuity of Poisson bracket imply $\boldsymbol{R} = \boldsymbol{0}_n$ for the full phase space, i.e., $\{\vartheta_a, \vartheta_b\} = 0$ for all $a, b \in \{1, \dots, n\}$. □

Lemmas 1, 2, and 3 together imply the following result of [11], whose proof was omitted in that paper to save space.

Theorem 4. *The reduced symplectic structure on* $\tilde{S}^0$ (27)*, given by the pull-back of* Ω (11) *by the map* $\tilde{\sigma}_0$ (29)*, has the canonical form* $\tilde{\sigma}_0^*(\Omega) = \sum_{a=1}^n \mathrm{d}\lambda_a \wedge \mathrm{d}\vartheta_a$.

4. Discussion

In this paper an explicit derivation of the Darboux form (30) was given. The Poisson bracket relations

$$\{\lambda_a, \lambda_b\} = 0, \quad \{\lambda_a, \vartheta_b\} = \delta_{a,b}, \quad \{\vartheta_a, \vartheta_b\} = 0, \qquad a,b \in \{1,\dots,n\} \tag{63}$$

were proved in Lemmas 1, 2, and 3, respectively. As a consequence Theorem 4 was proved.

As mentioned before the method used in this paper has been previously applied to the analogous hyperbolic models associated with the C_n [14] and A_n [7, 15] root systems. In [8] the hyperbolic BC_n case has been settled by "an almost verbatim computation as in the C_n case". In fact, a careful comparison of corresponding equations shows subtle differences as a result of the dissimilar characteristics of the underlying systems. For example, most of the expressions in Section 3 contain factors with the parameter κ which reflects the BC_n feature. As one would expect taking the limit $\kappa \to 0$ turns these formulae into the ones seen in the C_n case. The trigonometric nature of the considered systems can be accounted for another difference when minor complications occur in Lemmas 2 and 3 due to the appearance of trigonometric functions. These issues have been resolved by using density and continuity arguments.

Acknowledgments

The author is grateful to László Fehér for his valuable suggestions. This work was supported in part by the EU and the State of Hungary, co-financed by the European Social Fund in the framework of TÁMOP-4.2.4.A/2-11/1-2012-0001 'National Excellence Program'.

References

[1] Ruijsenaars S N M 1999 *Particles and Fields* ed Semenoff G and Vinet L (New York: Springer) pp 251–352
[2] Etingof P I 2007 *Calogero-Moser systems and representation theory* (Zürich: European Mathematical Society)
[3] Polychronakos A P 2006 *J. Phys.* A **39** 12793
[4] Ruijsenaars S N M 1988 *Commun. Math. Phys.* **115** 127–165
[5] Ruijsenaars S N M 1995 *Publ. RIMS Kyoto Univ.* **31** 247–353
[6] Fock V, Gorsky A, Nekrasov N and Rubtsov V 2000 *JHEP* **07** 028
[7] Fehér L and Klimčík C 2009 *J. Phys.* A **42** 185202
[8] Pusztai B G 2012 *Nucl. Phys.* B **856** 528–551
[9] Olshanetsky M A and Perelomov A M 1976 *Invent. Math.* **37** 93–108
[10] van Diejen J F 1994 *Theor. Math. Phys.* **99** 549–554
[11] Fehér L and Görbe T F 2014 (*Preprint* `arXiv:1407.2057 [math-ph]`)
[12] Fehér L and Ayadi V 2010 *J.Math.Phys.* **51** 103511
[13] Ortega J P and Ratiu T S 2004 *Momentum maps and Hamiltonian reduction* (Basel: Birkhäuser)
[14] Pusztai B G 2011 *Nucl. Phys.* B **853** 139–173
[15] Ayadi V, Fehér L and Görbe T F 2012 *J. Geom. Symmetry Phys.* **27** 27–44

Journal of Physics: Conference Series **563** (2014) 012013 doi:10.1088/1742-6596/563/1/012013

Bosonization and Lie Group Structure

Yuan K. Ha

Department of Physics, Temple University, Philadelphia, Pennsylvania 19122 U.S.A.

Email: yuanha@temple.edu

Abstract. We introduce a concise quantum operator formula for bosonization in which the Lie group structure appears in a natural way. The connection between fermions and bosons is found to be exactly the connection between Lie group elements and the group parameters. Bosonization is an extraordinary way of expressing the equation of motion of a complex fermion field in terms of a real scalar boson in two dimensions. All the properties of the fermion field theory are known to be preserved under this remarkable transformation with substantial simplification and elucidation of the original theory, much like Lie groups can be studied by their Lie algebras.

1. Introduction

Two-dimensional quantum field theories are important because the physics they describe is important. They contain exact symmetries and are often completely integrable. Some have exact solutions obtained without the need of perturbation theory and virtual particles. In contrast, no four-dimensional quantum field theories are known to be exactly solvable and therefore they can only be investigated as perturbative field theories.

Bosonization is a mathematical procedure in which a fermion field theory is expressed in terms of a boson theory with substantial simplification [1]. It is a unique phenomenon in two-dimensions. Bosonization is a search of the self-similarity of a system even though the equations appear to be different. It is therefore a novel symmetry operation in quantum field theory. In bosonization, any symmetry discovered which is not explicitly present in the fermion Lagrangian is a quantum symmetry. In particular, we find that the connection between fermions and bosons is exactly the connection between Lie group elements and the group parameters of the diagonal generators in the theory.

2. Quantum Fields

Fermions and bosons are fundamental fields in quantum field theory. Fermions are fields which obey anti-commutation rules

$$\begin{aligned} \{\psi_\rho(x), \psi_\sigma{}^\dagger(y)\} &= \delta(x-y)\delta_{\rho\sigma} \\ \{\psi_\rho(x), \psi_\sigma(y)\} &= 0. \end{aligned} \tag{1}$$

where $\rho, \sigma = +,-$ are the two components of the fermion $\psi(x)$ in two dimensions. Bosons are fields which obey the commutation rule

$$[\phi(x), \phi(y)] = 0 . \tag{2}$$

The Lagrangian for the free Dirac fermion with mass m is

$$L = i\overline{\psi}(x)\gamma^{\mu}\partial_{\mu}\psi(x) - m\overline{\psi}(x)\psi(x) , \tag{3}$$

with gamma matrices defined below:

$$\gamma^0 = \begin{bmatrix} 0 & -i \\ i & 0 \end{bmatrix}, \qquad \gamma^1 = \begin{bmatrix} 0 & i \\ i & 0 \end{bmatrix}, \qquad \gamma^5 = \begin{bmatrix} 1 & 0 \\ 0 & -1 \end{bmatrix} = -i\gamma^0\gamma^1 . \tag{4}$$

In the Weyl basis, the fermion components are given in the form

$$\psi = \begin{pmatrix} \psi_+ \\ \psi_- \end{pmatrix} = \begin{pmatrix} \psi_R \\ \psi_L \end{pmatrix}, \tag{5}$$

in which the left- and right-handed components are respectively,

$$\psi_L = \frac{1-\gamma^5}{2}\psi , \qquad \psi_R = \frac{1+\gamma^5}{2}\psi . \tag{6}$$

We also introduce the two-dimensional spacetime metric $g^{\mu\nu}$ with components $g^{00} = -g^{11} = 1$. The Lagrangian for the free boson with mass μ is

$$L = \frac{1}{2}\partial_{\mu}\phi(x)\partial^{\mu}\phi(x) - \mu^2\phi^2(x) . \tag{7}$$

3. Fermion Current As Dynamical Variable

The key to bosonization lies in the recognition of the central role of the energy and momentum tensor. In the current algebra approach to quantum field theory, it is known that fermion current can be successfully used as dynamical variable. For the massless free fermion, we define the conventional fermion current operator as

$$j^{\mu}(x) = \overline{\psi}(x)\gamma^{\mu}\psi(x) , \tag{8}$$

in terms of which the energy-momentum tensor takes the elegant form [2]

$$T^{\mu\nu}(x) = \frac{\pi}{2}\left(j^{\mu}(x)j^{\nu}(x) + j^{\nu}(x)j^{\mu}(x) - g^{\mu\nu}j_{\lambda}(x)j^{\lambda}(x)\right) . \tag{9}$$

For the massless free boson, the canonical energy-momentum tensor is

$$T^{\mu\nu}(x) = \frac{1}{2}\partial^{\mu}\phi(x)\partial^{\nu}\phi(x) + \frac{1}{2}\partial^{\nu}\phi(x)\partial^{\mu}\phi(x) - g^{\mu\nu}\frac{1}{2}\partial_{\lambda}\phi(x)\partial^{\lambda}\phi(x). \tag{10}$$

It is obvious that the two energy-momentum tensors have a parallel structure. In addition, the commutator structure of the fermion currents at equal times, given by

$$[j^{\mu}(x), j^{\nu}(y)] = -\frac{i}{\pi}\varepsilon^{\mu\nu}\frac{\partial}{\partial x}\delta(x-y), \tag{11}$$

with $\varepsilon^{01} = -\varepsilon^{10} = 1$, is also very similar to the canonical commutation rule for the boson

$$[\phi(x), \pi(y)] = i\delta(x-y), \tag{12}$$

with $\pi(x) = \partial_0\phi(x)$ being the canonical conjugate of $\phi(x)$. The two energy-momentum tensors in equation (9) and equation (10), and the two commutators in equation (11) and equation (12) are identical, respectively, if the fermion current can be expressed as

$$j^{\mu}(x) = \frac{1}{\sqrt{\pi}}\varepsilon^{\mu\nu}\partial_{\nu}\phi(x). \tag{13}$$

This expression guarantees the fermion current conservation

$$\partial_{\mu}j^{\mu}(x) = 0 \tag{14}$$

due to the antisymmetry property of the tensor $\varepsilon^{\mu\nu}$. In this case, a single massless fermion has the same Hamiltonian as that for a single massless boson. We shall find that equation (13) is a crucial step in the construction of all bosonization.

4. Heisenberg Equation

The dynamics of the fermion field is governed by the Heisenberg equation of motion in quantum theory,

$$[P_{\mu}, \psi(x)] = -i\partial_{\mu}\psi(x), \tag{15}$$

in which the momentum operator P_{μ} is defined through the energy-momentum tensor from equation (9),

$$P^{\mu} = \int T^{0\mu}(x')dx', \qquad T^{0\mu} = \frac{\pi}{2}(j^0 j^{\mu} + j^{\mu} j^0 - g^{0\mu} j_{\lambda} j^{\lambda}). \tag{16}$$

Applying the commutation relation between the current operator and the fermion at equal times,

$$[j^{\mu}(x), \psi(y)] = -(g^{0\mu} + \varepsilon^{0\mu}\gamma^5)\psi(x)\delta(x-y), \tag{17}$$

we obtain the field equations of motion

$$\partial_0 \psi(x) = -i\pi[j^0(x) + \gamma^5 j^1(x)]\psi(x), \tag{18}$$

$$\partial_1 \psi(x) = +i\pi[j^1(x) + \gamma^5 j^0(x)]\psi(x). \tag{19}$$

Together they give the massless Dirac equation as required,

$$i\gamma^\mu \partial_\mu \psi(x) = 0. \tag{20}$$

We may formally integrate equation (19) by noting that the derivative of $\psi(x)$ is always proportionally to the field itself. This leads to an exponential expression for the fermion field of the form

$$\psi(x) = \exp\left[i\pi \int [j^1(x') + \gamma^5 j^0(x')]dx'\right]\psi_0, \tag{21}$$

where ψ_0 is a constant in the form of a spinor. According to equation (13), we may replace the current operators by the boson such that

$$j^0(x) = \frac{1}{\sqrt{\pi}}\partial_1\phi(x), \qquad j^1(x) = -\frac{1}{\sqrt{\pi}}\partial_0\phi(x). \tag{22}$$

The complete fermion expression in components then appears as

$$\psi(x) = \begin{bmatrix} \psi_+(x) \\ \psi_-(x) \end{bmatrix} = \begin{bmatrix} C_+ \exp\left[+i\sqrt{\pi}[\phi(x) - \int_{-\infty}^{x} \partial_0\phi(x')dx']\right] \\ C_- \exp\left[-i\sqrt{\pi}[\phi(x) + \int_{-\infty}^{x} \partial_0\phi(x')dx']\right] \end{bmatrix}, \tag{23}$$

with constants C_+, C_- to be determined by the canonical anticommutation rule in equation (1). This is the explicit derivation of the formal result of bosonization in quantum field theory [3]. A complex fermion is thereby successfully constructed with a real scalar boson.

A remarkable simplification of equation (23) with profound significance can be achieved by introducing the chiral components of the boson

$$\phi_\pm(x) = \pm\frac{1}{\sqrt{2\pi}}\int_0^{\pm\infty} \frac{dk}{\sqrt{2k^0}}\left[c^{(-)}(k)e^{-ik\cdot x} + c^{(+)}(k)e^{ik\cdot x}\right]e^{-\alpha|k|/2}. \tag{24}$$

Here the scalar field $\phi(x)$ is decomposed into positive- and negative-energy parts: $\phi_+(x), \phi_-(x)$; k^0 is the energy frequency and $c^{(+)}(k), c^{(-)}(k)$ are the creation and annihilation operators of the boson with momentum k. The quantity $1/\alpha$ is an ultraviolet cutoff imposed to guarantee the boson to remain finite at high frequencies; α will be taken to zero at the end of any calculation involving the fermion $\psi(x)$. If the fermion components are to depend on the boson, each chiral component of the fermion should match perfectly to the chiral component of the boson with all energy frequencies in each component included. This is only possible if equation (23) takes the form

$$\begin{bmatrix} \psi_+(x) \\ \psi_-(x) \end{bmatrix} = \begin{bmatrix} C_+ \exp[+i\sqrt{4\pi}\phi_+(x)] \\ C_- \exp[-i\sqrt{4\pi}\phi_-(x)] \end{bmatrix}. \tag{25}$$

Accordingly, we find that

$$\phi_\pm(x) = \frac{1}{2}\left[\phi(x) \mp \int_{-\infty}^{x} \pi(x')dx'\right]. \tag{26}$$

The constants C_+, C_- are evaluated and found to be equal to $1/\sqrt{2\pi\alpha}$. The fermion expression therefore takes the ultimate concise form

$$\psi(x) = \begin{bmatrix} \psi_+(x) \\ \psi_-(x) \end{bmatrix} = \frac{1}{\sqrt{2\pi\alpha}} \begin{bmatrix} \exp[+i\sqrt{4\pi}\phi_+(x)] \\ \exp[-i\sqrt{4\pi}\phi_-(x)] \end{bmatrix}. \tag{27}$$

This is the quintessential expression in the existing treatment of bosonization.

With the boson representation defined in equation (27), we find that bilinear terms in the fermion Lagrangian have their corresponding expressions in the following:

$$\begin{aligned}
& i\bar{\psi}\gamma^\mu\partial_\mu\psi \leftrightarrow \frac{1}{2}\partial_\mu\phi\partial^\mu\phi \\
& \bar{\psi}\psi \leftrightarrow -\frac{1}{\pi\alpha}\cos\sqrt{4\pi}\phi \\
& i\bar{\psi}\gamma^5\psi \leftrightarrow \frac{1}{\pi\alpha}\sin\sqrt{4\pi}\phi \\
& \psi\gamma^\mu\psi \leftrightarrow \frac{1}{\sqrt{\pi}}\varepsilon^{\mu\nu}\partial_\nu\phi \\
& \bar{\psi}\gamma^5\gamma^\mu\psi \leftrightarrow \frac{1}{\sqrt{\pi}}\partial^\mu\phi.
\end{aligned} \tag{28}$$

These equivalences allow the bilinear terms appearing in the Lagrangian of a fermion theory to be transformed directly into the corresponding terms in the Lagrangian of a boson theory, while preserving all the key properties of the original fermion theory. Many symmetries in the fermion theory become immediately obvious in the bosonized form [4]. In addition, any hidden symmetry discovered which is not explicitly present in the fermion Lagrangian is a quantum symmetry. Quantum symmetries are symmetries acting on the field operators and they are typically discrete symmetries such as self-duality and self-triality [5].

5. Bosonization Theorem

Bosonization is an extraordinary way of expressing the equation of motion of a complex fermion field in term of a real scalar field with substantial simplification and elucidation of the original theory. Surprisingly, the bosonization formula given in equation (27) contains a natural Lie group structure. The exponential map from boson to fermion is strikingly similar to the exponential map from Lie algebra to

Lie group. Aside from the constant in equation (27), the fermion structure in terms of the boson is basically

$$\begin{aligned} \psi_+(x) &= \exp + i\sqrt{4\pi}\phi_+(x) \\ \psi_-(x) &= \exp - i\sqrt{4\pi}\phi_-(x). \end{aligned} \tag{29}$$

For the $U(1)$ Lie group, the group elements of the unit circle are

$$\begin{aligned} z &= \exp(+i\theta\cdot 1) = \cos\theta + i\sin\theta \\ z &= \exp(-i\theta\cdot 1) = \cos\theta - i\sin\theta. \end{aligned} \tag{30}$$

The quantity θ is a real number and it is the group parameter; z is a complex number which is the group element. The identity 1 is the group generator. For quantum fields, we note that the boson $\phi(x)$ is a real-valued function, whereas the fermion $\psi(x)$ is a complex-valued function. Thus in bosonization, if the boson is identified with the group parameter and the fermion with the group element, then the link between fermion and boson is exactly that given by the property of a Lie group. A subtlety, however, is to be noted. It is the chiral components which are independently bosonized. Each chiral component of the fermion is replaced by the corresponding chiral component of the boson and the complete fermion is reconstructed from these chiral components. This is necessary because the fermion is always a two-component spinor. We therefore arrive at a remarkable theorem in bosonization:

> **Theorem:** For any Lagrangian invariant under $U(1)$ symmetry transformation in two dimensions, if the chiral fermion is the group element, then the chiral boson is the group parameter and vice versa.

6. Interacting Fields

The power of bosonization is recognized by investigating a self-interacting theory, the massless Thirring model [6]. This is an exactly solvable quantum field theory with explicit solution to multi-point correlation functions. The Lagrangian is

$$L = i\overline{\psi}(x)\gamma^\mu\partial_\mu\psi(x) - \frac{g}{2}j^\mu(x)j_\mu(x). \tag{31}$$

For consistency with relativistic requirement, it is necessary that the fermion current in this theory be defined according to [7]

$$j^\mu(x) = \left(g^{0\mu} + \frac{\varepsilon^{0\mu}}{1+g/\pi}\right)\left(\overline{\psi}(x)\gamma^\mu\psi(x)\right), \tag{32}$$

with no summation over the index μ. This is the current which is found to be conserved, i.e.

$$\partial_\mu j^\mu(x) = 0, \tag{33}$$

and therefore it can be expressed in terms of another boson $\tilde{\phi}(x)$ as

$$j^{\mu}(x) = \frac{1}{\sqrt{\pi}} \varepsilon^{\mu\nu} \partial_{\nu} \tilde{\phi}(x). \tag{34}$$

The bosonization formula in this interacting case is now given by

$$\psi(x) = \begin{bmatrix} \psi_{+}(x) \\ \psi_{-}(x) \end{bmatrix} = \frac{1}{\sqrt{2\pi\alpha}} \begin{bmatrix} \exp[+i\sqrt{4\pi}\tilde{\phi}_{+}(x)] \\ \exp[-i\sqrt{4\pi}\tilde{\phi}_{-}(x)] \end{bmatrix}, \tag{35}$$

with the chiral components of the boson being

$$\begin{aligned} \tilde{\phi}_{\pm}(x) &= \frac{1}{2}\left(\tilde{\phi}(x) \mp \int_{-\infty}^{x} \tilde{\pi}(x')dx' \right) \\ \tilde{\phi}_{+}(x) + \tilde{\phi}_{-}(x) &= \tilde{\phi}(x). \end{aligned} \tag{36}$$

The new boson $\tilde{\phi}(x)$ and its canonical conjugate $\tilde{\pi}(x)$ reduce to free fields when the fermion coupling $g = 0$ as follow:

$$\tilde{\phi}(x) = \frac{\phi(x)}{\sqrt{1+g/\pi}}, \qquad \tilde{\pi}(x) = \sqrt{1+g/\pi} \cdot \partial_0 \phi(x). \tag{37}$$

The bosonized Lagrangian from equation (31) is therefore

$$\begin{aligned} L &= \frac{1}{2}\partial_{\mu}\tilde{\phi}(x)\partial^{\mu}\tilde{\phi}(x) + \frac{g}{2}\left(\frac{1}{\pi}\partial_{\alpha}\tilde{\phi}(x)\partial^{\alpha}\tilde{\phi}(x)\right) \\ &= \left(1+\frac{g}{\pi}\right) \times \frac{1}{2}\frac{\partial_{\mu}\phi(x)}{\sqrt{1+g/\pi}}\frac{\partial^{\mu}\phi(x)}{\sqrt{1+g/\pi}} \\ &= \frac{1}{2}\partial_{\mu}\phi(x)\partial^{\mu}\phi(x). \end{aligned} \tag{38}$$

The massless Thirring model is immediately found to be equivalent to a theory of free massless boson as confirmed by examining the correlation functions and other key field theory properties. The chiral fermions and the chiral bosons in the interacting case also conform to a Lie group structure.

For the massive Thirring model, the result of bosonization gives the quantum sine-Gordon theory [8]. The result is consistent with the known solution of the mass spectrum and the scattering matrix in the massive Thirring model. The fundamental fermion of the Thirring model corresponds to the soliton of the sine-Gordon theory. The Lagrangians are, respectively,

$$\begin{aligned} L &= i\overline{\psi}(x)\gamma^{\mu}\partial_{\mu}\psi(x) - \frac{g}{2}j^{\mu}(x)j_{\mu}(x) - m\overline{\psi}(x)\psi(x), \\ L &= \frac{1}{2}\partial_{\mu}\phi(x)\partial^{\mu}\phi(x) + \frac{m}{\pi\alpha}\cos\sqrt{\frac{4\pi}{1+g/\pi}}\phi(x). \end{aligned} \tag{39}$$

7. U(2) Thirring Model

We consider an interacting theory involving two massless fermion species $\psi^1(x), \psi^2(x)$, each of which is a two-component spinor. They form an isovector under global $U(2)$ symmetry transformation. The model is important in the study of many two-dimensional systems and the Kondo problem [9]. There are four chiral components interacting according to the Thirring Lagrangian

$$L = i\bar{\psi}^1(x)\gamma^\mu\partial_\mu\psi^1(x) + i\bar{\psi}^2(x)\gamma^\mu\partial_\mu\psi^2(x) - gJ^{A,\mu}(x)J^A{}_\mu(x). \tag{40}$$

The isospin currents are defined as

$$J^{A,\mu}(x) = \left(g^{0\mu} + \frac{\varepsilon^{0\mu}}{1+g/\pi}\right)\bar{\psi}^a(x)\gamma^\mu\frac{\lambda^A}{2}\psi^a(x), \tag{41}$$

with the indices $a = 1,2$ and $A = 0,1,2,3$. The generators for the $U(2)$ Lie group are the following:

$$\lambda^0 = \begin{bmatrix} 1 & 0 \\ 0 & 1 \end{bmatrix}, \quad \lambda^1 = \begin{bmatrix} 0 & 1 \\ 1 & 0 \end{bmatrix}, \quad \lambda^2 = \begin{bmatrix} 0 & -i \\ i & 0 \end{bmatrix}, \quad \lambda^3 = \begin{bmatrix} 1 & 0 \\ 0 & -1 \end{bmatrix}. \tag{42}$$

They satisfy the $SU(2)$ commutation relations

$$\left[\frac{\lambda^i}{2}, \frac{\lambda^j}{2}\right] = i\varepsilon_{ijk}\frac{\lambda^k}{2}, \qquad i, j, k = 1,2,3. \tag{43}$$

The structure constants are the totally antisymmetric tensor ε_{ijk} with $\varepsilon_{123} = 1$. The dynamics of the $U(2)$ Thirring model shows that only the currents corresponding to the diagonal generators are conserved and they can be expressed in terms of two real scalar fields $\tilde{\phi}^1(x), \tilde{\phi}^2(x)$,

$$\begin{aligned} \partial_\mu J^{A=0,\mu}(x) = 0 \longrightarrow J^{A=0,\mu} &= \frac{1}{\sqrt{\pi}}\varepsilon^{\mu\nu}\partial_\nu\tilde{\phi}^1(x) \\ \partial_\mu J^{A=3,\mu}(x) = 0 \longrightarrow J^{A=3,\mu} &= \frac{1}{\sqrt{\pi}}\varepsilon^{\mu\nu}\partial_\nu\tilde{\phi}^2(x). \end{aligned} \tag{44}$$

Both scalar fields reduce to free fields in the non-interacting case analogous to equation (37).

The bosonization formula in the $U(2)$ Thirring model therefore becomes

$$\psi^a(x) = \begin{bmatrix} \psi^a{}_+(x) \\ \psi^a{}_-(x) \end{bmatrix} - \frac{1}{\sqrt{2\pi\alpha}}\begin{bmatrix} \exp[+i\sqrt{4\pi}\tilde{\phi}^a{}_+(x)] \\ \exp[-i\sqrt{4\pi}\tilde{\phi}^a{}_-(x)] \end{bmatrix}, \tag{45}$$

with each fermion $\psi^a(x)$ corresponding to its own boson $\phi^a(x)$, $a = 1,2$. Again, we find that in this interacting case there is an underlying Lie group structure between the chiral components of the fermions and the chiral components of the bosons as a consequence of the equation of motion.

The energy-momentum tensor for the $U(2)$ Thirring model is appropriately defined as

$$T^{\mu\nu}(x) = \left(1+\frac{g}{\pi}\right)\frac{\pi}{2}\sum_{A=0}^{3} J^{A,\mu}(x)J^{A,\nu}(x) + J^{A,\nu}(x)J^{A,\mu}(x) - g^{\mu\nu}J^{A}{}_{\lambda}(x)J^{A,\lambda}(x), \quad (46)$$

from which the Hamiltonian $T^{00}(x)$ is calculated. In terms of ordinary scalar field notation, we find

$$T^{00}(x) = \frac{1}{2}\left(\partial_0\phi^1\right)^2 + \frac{1}{2}\left(\partial_1\phi^1\right)^2 + \frac{1}{2}\left(\partial_0\phi^2\right)^2 + \frac{1}{2}\left(\partial_1\phi^2\right)^2 - \frac{g}{4\pi^2\alpha^2}\cos\sqrt{\frac{4\pi}{1+g/\pi}}\left(\phi^1 - \phi^2\right). \quad (47)$$

The Lagrangian which yields the above Hamiltonian is therefore

$$L = \frac{1}{2}\partial_\mu\phi^1\partial^\mu\phi^1 + \frac{1}{2}\partial_\mu\phi^2\partial^\mu\phi^2 + \frac{g}{4\pi^2\alpha^2}\cos\sqrt{\frac{4\pi}{1+g/\pi}}\left(\phi^1 - \phi^2\right). \quad (48)$$

By defining the following linear combinations:

$$A = \left(\phi^1 + \phi^2\right)/\sqrt{2}, \qquad B = \left(\phi^1 - \phi^2\right)/\sqrt{2}, \quad (49)$$

we may decouple a free massless boson $A(x)$ and obtain a sine-Gordon theory with boson $B(x)$ through the Lagrangian

$$L = \frac{1}{2}\partial_\mu A\partial^\mu A + \left(\frac{1}{2}\partial_\mu B\partial^\mu B + \frac{g}{4\pi^2\alpha^2}\cos\sqrt{\frac{8\pi}{1+g/\pi}}B\right). \quad (50)$$

This is the decomposition of the Lie algebra $u(2) \approx u(1)\otimes su(2)$ in terms of the bosons and the realization of the isomorphism of the group $U(2) \approx U(1)\otimes SU(2)$ in terms of the fermions. The sine-Gordon Lagrangian in equation (50) is further equivalent to another massive Thirring model represented by a fermion $\chi(x)$ of mass m' with Lagrangian

$$L = i\bar{\chi}(x)\gamma^\mu\partial_\mu\chi(x) - \frac{g'}{2}k^\mu(x)k_\mu(x) - m'\bar{\chi}(x)\chi(x). \quad (51)$$

The fermion current here is given by

$$k^\mu(x) = \left(g'^{0\mu} + \frac{\varepsilon^{0\mu}}{1+g'/\pi}\right)\left(\bar{\chi}(x)\gamma^\mu\chi(x)\right). \quad (52)$$

The conservation of this fermion current then allows it to be expressed in terms of another boson $\tilde{B}(x)$, i.e.

$$\partial_\mu k^\mu(x) = 0 \longrightarrow k^\mu(x) = \frac{1}{\sqrt{\pi}}\varepsilon^{\mu\nu}\partial_\nu\tilde{B}(x). \quad (53)$$

The bosonization formula involved is,

$$\chi(x)=\begin{bmatrix}\chi_+(x)\\ \chi_-(x)\end{bmatrix}=\frac{1}{\sqrt{2\pi\alpha}}\begin{bmatrix}\exp[+i\sqrt{4\pi}\widetilde{B}_+(x)]\\ \exp[-i\sqrt{4\pi}\widetilde{B}_-(x)]\end{bmatrix}, \quad (54)$$

where the fields $\widetilde{B}(x)$ and $B(x)$ are related by

$$\widetilde{B}(x)=\frac{B(x)}{\sqrt{1+g'/\pi}}, \quad (55)$$

with the identification of the parameters $m'=g/4\pi\alpha$ and $g'=(g-\pi)/2$.

8. Dynamical Mass Generation

The $U(2)$ Thirring model with two massless fermions is thus seen to be equivalent to the single massive fermion $U(1)$ Thirring model via bosonization. The coupling constant g in the $U(2)$ fermion theory becomes the mass parameter m' in the $U(1)$ Thirring model. This is dynamical mass generation. The single fermion $\chi(x)$ in the massive Thirring model is a non-linear composite of the original Fermi fields $\psi^1(x),\psi^2(x)$ in the $U(2)$ massless case. It is feasible to generalize to the $U(N)$ Thirring model with N massless fermions [10]. The fermion currents corresponding to the diagonal generators of the $U(N)$ group will be conserved, resulting in N scalar fields to form a N coupled sine-Gordon Lagrangian. Bosonization maintains all the symmetries of the original fermion theory and, in particular, reveals a natural connection between fermion and boson in terms of Lie group structure. It becomes a novel and important symmetry operation in quantum field theory, very much like Lie groups can be studied effectively by their Lie algebras.

References

[1] Ha Y K 1984 *Phys. Rev.* D **29** 1744.
[2] Sugawara H 1968 *Phys. Rev.* **170** 1659.
Sommerfield C M 1968 *Phys. Rev.* **176** 2019.
[3] Mandelstam S 1975 *Phys. Rev.* D **11** 3026.
[4] Ha Y K 2013 *J. Phys.: Conf. Ser.* **474** 012034.
[5] Shankar R 1980 *Phys. Lett.* **92B** 333.
[6] Thirring W 1958 *Ann. Phys. (N.Y.)* **3** 91.
[7] Sommerfield C M 1963 *Ann. Phys. (N.Y.)* **26** 1.
[8] Coleman S 1975 *Phys. Rev.* D **11** 2088.
[9] Andrei N and Lowenstein J H 1979 *Phys. Rev. Lett.* **43** 1698.
[10] Dashen R and Frishman Y 1975 *Phys. Rev.* D **11** 2781.

Journal of Physics: Conference Series **563** (2014) 012014 doi:10.1088/1742-6596/563/1/012014

Vortex solutions in a Witten-type model

Satoru Itaya[1], **Nobuyuki Sawado**[1], **Michitaka Suzuki**[1 2],

[1] Department of Physics, Tokyo University of Science, Noda, Chiba 278-8510, Japan
[2] Application Product Co.,Ltd., Shinagawa, Tokyo 141-0022, Japan

E-mail: str.itaya@gmail.com, sawado@ph.noda.tus.ac.jp

Abstract. Straight line vortex solutions in a Witten's superconducting string model are studied. The model has many parameters and this is the main reason of the complexity. We argue the precise conditions of the parameters for finding the solutions of the model. We obtain the rotationally symmetric solutions for the winding numbers $m = 1-4$ with/without the gauge field. For the higher winding numbers, an energy minimization algorithm is used to investigate non-rotational solutions.

1. Introduction

Superconducting strings in the local $U(1) \otimes U(1)$ symmetry was introduced by Witten in 80s' [1]. There are many studies based on the model for the straight line vortex solutions, and also the closed type vortex called vortons. As has been claimed that the solutions may have rich variety of physical applications for the condensed matter physics and especially for the cosmology [2, 3, 4].

This model contains two Abelian gauge fields $R_\mu^{(a)} (a = 1, 2)$ interacting with two complex scalar fields ϕ and σ with the Lagrangian density

$$\mathcal{L} = -\frac{1}{4} \sum_{a=1,2} F_{\mu\nu}^{(a)} F^{(a)\mu\nu} + D_\mu \phi^* D^\mu \phi + D_\mu \sigma^* D^\mu \sigma - U. \tag{1}$$

Here $F_{\mu\nu}^{(a)} = \partial_\mu R_\nu^{(a)} - \partial_\nu R_\mu^{(a)}$ are the Abelian field strengths, the gauge covariant derivatives of scalars are $D_\mu = \partial_\mu - i g_a R_\mu^{(a)}$ where g_a are the gauge coupling constants. The scalar field potential is defined as

$$U = \frac{\lambda_\phi}{4}(|\phi|^2 - \eta_\phi^2)^2 + \frac{\lambda_\sigma}{4}|\sigma|^2(|\sigma|^2 - 2\eta_\sigma^2) + \beta|\phi|^2|\sigma|^2, \tag{2}$$

where $\lambda_\phi, \lambda_\sigma, \eta_\phi, \eta_\sigma$ and β are positive constants. This theory is invariant under the local $U(1) \otimes U(1)$ transformation.

The theory admits the stationary solutions of straight line vortex. If the parameters in the potential are properly chosen, the field ϕ vanishes in the vortex core and approaches the finite value at the infinity, while σ develops a non-zero condensates value in the core and vanishes at the infinity. The fields ϕ, σ are sometimes referred to as a vortex field and a condensate field, respectively.

Journal of Physics: Conference Series **563** (2014) 012014 doi:10.1088/1742-6596/563/1/012014

Several attempts have been done for finding rotationally symmetric solutions of the model [5, 6, 7]. All of them were only of the winding number $m = 1$ solutions of the global limit, i.e., the gauge coupling was ignored. They were quite confusing because they each suggested different conditions for the model parameters. We have tried to confirm the existence of such solutions by employing the conditions, but often failed to find the solutions. We have to say that existence of such vortex solutions in the model is thus still an open problem.

In this paper, first we propose the correct conditions for the model parameters. Within the full gauged model, finding the solutions is still absent thus we try to find the solutions with the gauge field. In [5, 6, 7], only the vortex with unit winding number $m = 1$ was discussed, thus we try to get solutions of higher winding number solutions. For such solutions, breaking of the rotational symmetry might happen. The numerical simulations are done by an energy minimization algorithm based on a simulated annealing [8]. We shall see the winding number m solutions comprise m constituents.

2. The model parameters

We employ a rotationally symmetric ansatz

$$\phi = \phi_0(r)e^{im\theta}, \sigma = \sigma_0(r)e^{i(\omega t+kz)}, \tag{3}$$

where $m \in \mathbb{Z}$ is the winding number and $\omega, k \in \mathbb{R}$ describe the traveling wave along z axis. The boundary conditions are given as

$$\phi_0(0) = 0, \quad \phi_0(\infty) = \eta_\phi, \quad \sigma_0'(0) = 0, \quad \sigma_0(\infty) = 0. \tag{4}$$

For the moment we consider the global limit of this model thus $g_a = 0$. The gauge fields decouple and the Lagrangian reduces to

$$\mathcal{L} = \partial_\mu\phi^*\partial^\mu\phi + \partial_\mu\sigma^*\partial^\mu\sigma - U \tag{5}$$

so that the internal $U(1) \otimes U(1)$ symmetry becomes global. The ungauged model was already studied in [6, 7]. In [6], following three conditions were proposed for the existence of the solutions

$$\lambda_\phi\eta_\phi^4 > \lambda_\sigma\left(\eta_\sigma^2 + \frac{2(\omega^2-k^2)}{\lambda_\sigma}\right)^2, \quad \beta\eta_\phi^2 - \frac{1}{2}\lambda_\sigma\eta_\sigma^2 - \omega^2 + k^2 > 0,$$
$$k^2 - \omega^2 < \frac{1}{2}\lambda_\sigma\eta_\sigma^2 - \sqrt{\beta\lambda_\phi}\eta_\phi^2 \tag{6}$$

while in [7], a simple estimate was given

$$\beta\eta_\phi^2 - \frac{1}{2}\lambda_\sigma\eta_\sigma^2 < \frac{1}{2}\lambda_\phi\eta_\phi^2. \tag{7}$$

Both conditions are, however, not precise one. In fact, we could not find solutions with the parameters based on these conditions (6) or (7). Thus we start with a study of the proper parameter conditions of the model.

The Euler-Lagrange equations in the theory (5) with the ansatz (3) read

$$r\phi_0'' + \phi_0' - \frac{m^2}{r}\phi_0 - \frac{r}{2}\lambda_\phi(\phi_0^2 - \eta_\phi^2)\phi_0 - r\beta\phi_0\sigma_0^2 = 0, \tag{8}$$
$$r\sigma_0'' + \sigma_0' + r(\omega^2 - k^2)\sigma_0 - \frac{r}{2}\lambda_\sigma(\sigma_0^2 - \eta_\sigma^2)\sigma_0 - r\beta\phi_0^2\sigma_0 = 0. \tag{9}$$

We examine their analytical properties by expanding the fields at the origin and also infinity. The critical one is the expansion of the σ field at the infinity and then we describe it in detail. We introduce a new coordinate $x \equiv \frac{r}{1+r}$ which runs from 0 to 1. At the vicinity of infinity, σ_0 can be expanded as $\sigma_0 = \sum_{i=1} \sigma^{(i)}(1-x)^i$ with the coefficients $\sigma^{(i)}, i = 1, 2, 3, \cdots$. Substituting this expansion for (9), we get a following equation

$$\left(k^2 - \omega^2 + \beta\eta_\phi^2 - \frac{1}{2}\eta_\sigma^2\lambda_\sigma\right)\sigma^{(1)} + \left[\left(k^2 - \omega^2 + \beta\eta_\phi^2 - \frac{1}{2}\eta_\sigma^2\lambda_\sigma\right)(\sigma^{(1)} + \sigma^{(2)})\right](1-x)$$
$$+ \left[\left({\sigma^{(1)}}^2\lambda_\sigma - \frac{2\beta({\sigma^{(1)}}^2\beta + m^2)}{\lambda_\phi} - 1\right)\sigma^{(1)} + \left(k^2 - \omega^2 + \beta\eta_\phi^2 - \frac{1}{2}\eta_\sigma^2\lambda_\sigma\right)(\sigma^{(2)} + \sigma^{(3)})\right](1-x)^2$$
$$+ O((1-x)^3) = 0. \tag{10}$$

Each coefficient of $(1-x)^i$ should be zero, then for the zeroth order we have

$$\lambda_\sigma = \frac{2(k^2 - \omega^2 + \beta\eta_\phi^2)}{\eta_\sigma^2}. \tag{11}$$

From this the first order becomes zero, too. Also for the second order, the $\sigma^{(1)}$ should satisfy

$$\sigma^{(1)} = -\sqrt{\frac{2m^2\beta + \lambda_\phi}{\frac{k^2-\omega^2+\beta\eta_\phi^2}{\eta_\sigma^2}\lambda_\phi - 2\beta^2}}.$$

Since the σ_0 is defined as a real, the $\sigma^{(1)}$ is also real. From this, we obtain the additional condition,

$$\frac{k^2 - \omega^2 + \beta\eta_\phi^2}{\eta_\sigma^2}\lambda_\phi > 2\beta^2. \tag{12}$$

All coefficients of the higher orders are automatically guaranteed once the conditions (11) and (12) are employed. Here we get correct form of the conditions (11) and (12) for the parameters. In the following we perform the numerical analysis with the parameters based on these conditions.

3. The rotationally symmetric solutions

3.1. The ungauged model

First we present the solutions with the rotational symmetry. In [6, 7] the authors presented only solutions with unit winding number, then we investigate the higher winding numbers as well. Unlike the standard Abelian-Higgs model, the model possesses the solutions only with the scalar fields. We solve the Euler-Lagrange equations (8) and (9) in terms of a relaxation technique. In Figure 1 we plot typical profiles of the scalar fields and also each value of energy. One can easily see that for the higher winding numbers the behavior at the origin becomes shallower and also the speed of approaching the vacuum gets slower.

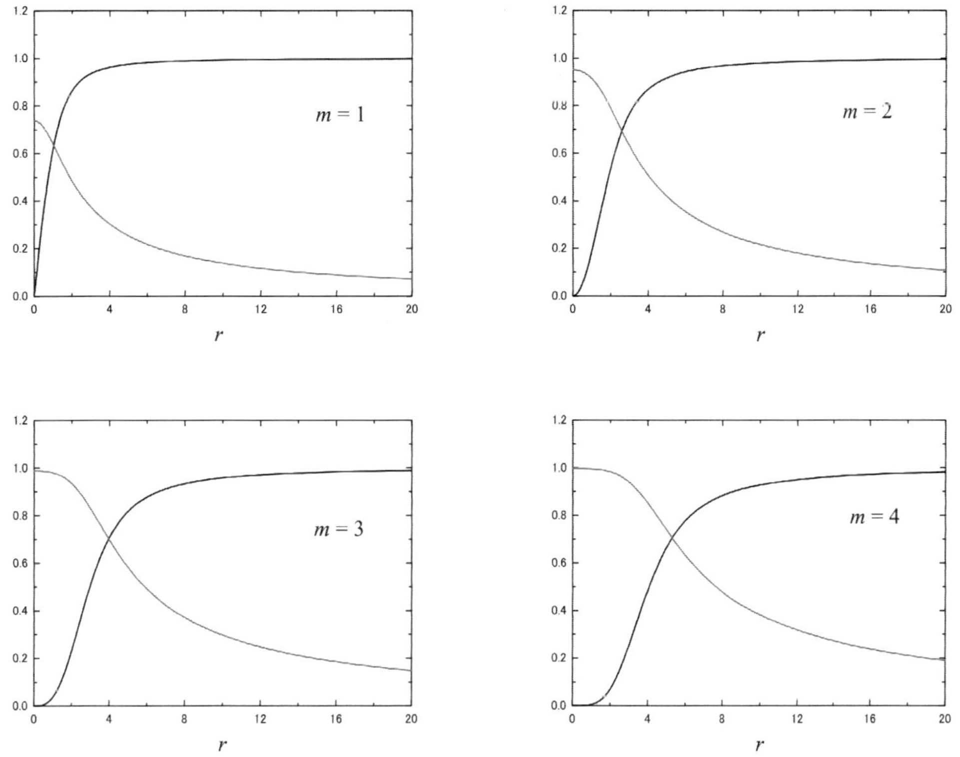

Figure 1. The profiles of $\phi_0(r)$(black line) and $\sigma_0(r)$(red line) of the ungauged model for each winding number for the parameters $\lambda_\phi = 5.5, \lambda_\sigma = 3.0, \eta_\phi = 1.0, \eta_\sigma = 1.0, \beta = 1.5$. The energies are $E_{m=1} = 39.84, E_{m=2} = 132.63, E_{m=3} = 269.83, E_{m=4} = 446.61$, respectively.

3.2. The gauged model

The model (1) originally was proposed as the gauged model, thus we have to introduce the gauge field in our numerical analysis. Physically it is important because the gauge field describes the magnetic field penetrating a superconductor. The gauged model was already studied in [5] but only for the unit winding number.

We assume the rotational symmetry of the fields, and perform the asymptotic expansion of both the scalar fields and the gauge fields. The main difficulty of the problem is that the gauge field coupled with σ has no asymptotic solution at the infinity, while the gauge field coupled with ϕ can exist. This indicates that there are no solutions of the full gauged model (1) at least within the rotational symmetry.

Here we consider the following Lagrangian density

$$\mathcal{L} = -\frac{1}{4}F_{\mu\nu}F^{\mu\nu} + (D_\mu\phi)^*(D^\mu\phi) + (\partial_\mu\sigma)^*(\partial^\mu\sigma) - U. \tag{13}$$

Here $F_{\mu\nu} = \partial_\mu R_\nu - \partial_\nu R_\mu$ is the Abelian field strength, the gauge covariant derivative of scalar field ϕ is $D_\mu\phi = (\partial_\mu - igR_\mu)\phi$, where g is the gauge coupling constant.

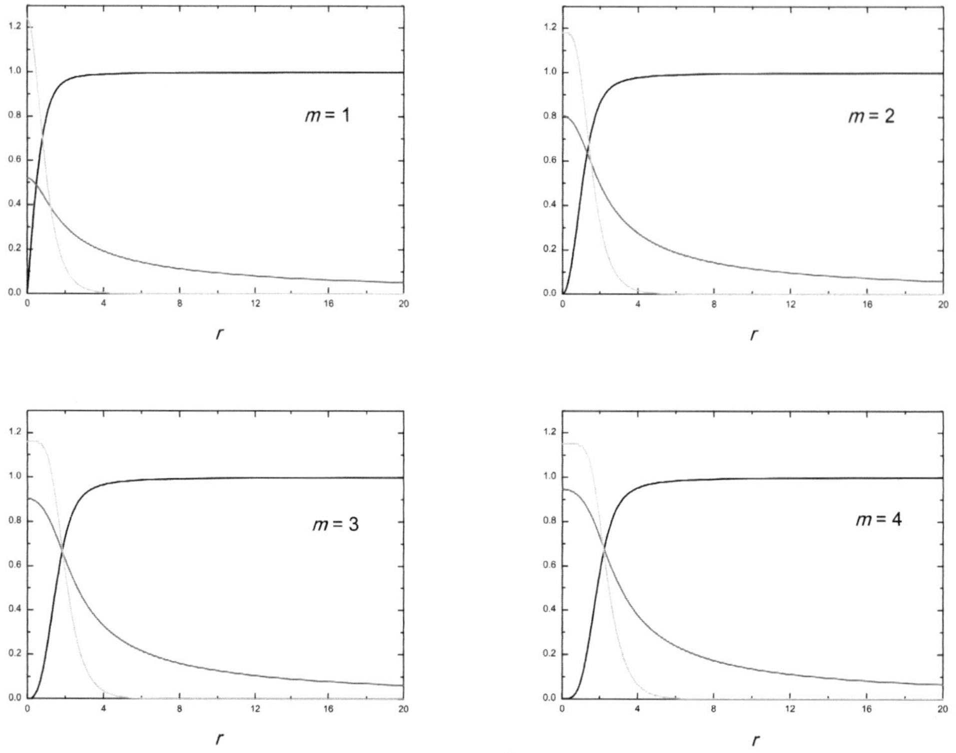

Figure 2. The profiles of $\phi_0(r)$(black line) , $\sigma_0(r)$(red line) and the magnetic field calculated from the gauge field coupled with ϕ(green line) for each winding number for the parameters $\lambda_\phi = 5.5, \lambda_\sigma = 3.0, \eta_\phi = 1.0, \eta_\sigma = 1.0, \beta = 1.5, g = 1.0$. The energies are $E_{m=1} = 7.57, E_{m=2} = 15.14, E_{m=3} = 22.47, E_{m=4} = 29.71$, respectively.

In [5], The conditions for the parameters were suggested

$$\lambda_\phi \eta_\phi^4 > \lambda_\sigma \eta_\sigma^4, \quad \beta\eta_\phi^2 - \frac{1}{2}\lambda_\sigma \eta_\sigma^2 > 0, \quad \sqrt{\lambda_\phi}\eta_\phi < \sqrt{\lambda_\sigma}\eta_\sigma. \tag{14}$$

These conditions are again not precise. In fact, the conditions which ensure the existence of the solutions hold the same as (11),(12) because the gauge coupling has no effect for analytical problems.

For the gauge field, we consider the following rotationally symmetric ansatz,

$$R_\theta = R(r), \tag{15}$$

and the boundary conditions

$$R(0) = R(\infty) = 0. \tag{16}$$

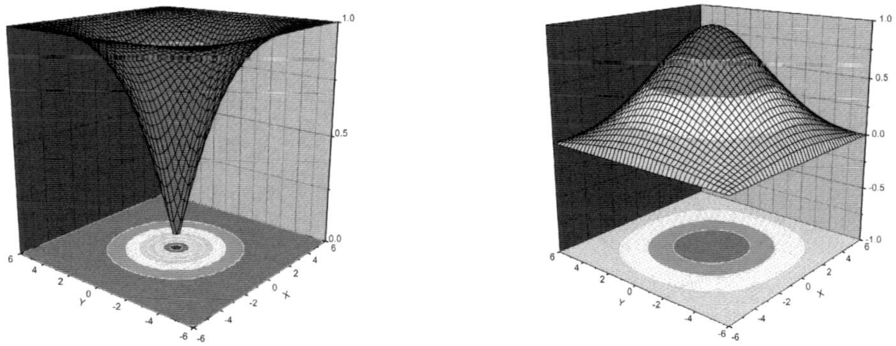

Figure 3. The profiles of ϕ(left) , σ(right) in the case of $m = 1$ for the parameters $\lambda_\phi = 10.0, \lambda_\sigma = 9.0, \eta_\phi = 1.0, \eta_\sigma = 1.0, \beta = 4.5$.

The equations of these fields are given as following forms,

$$r\phi_0'' + \phi_0' - r\left(\frac{m}{r} - gR\right)^2 \phi_0 - \frac{r}{2}\lambda_\phi(\phi_0^2 - \eta_\phi^2)\phi_0 - r\beta\phi_0\sigma_0^2 = 0, \tag{17}$$

$$r\sigma_0'' + \sigma_0' + r(\omega^2 - k^2)\sigma_0 - \frac{r}{2}\lambda_\sigma(\sigma_0^2 - \eta_\sigma^2)\sigma_0 - r\beta\phi_0^2\sigma_0 = 0. \tag{18}$$

$$rR'' + R' - \frac{R}{r} + 2rg\left(\frac{m}{r} - gR\right)\phi_0^2 = 0, \tag{19}$$

Results of the gauged model are shown in Figure 2. The behavior of scalar fields is similar to that of the ungauged model, and the width of the magnetic field gets broader for the higher winding numbers.

4. The solutions without the rotational symmetry

So far, we have discussed the rotationally symmetric solutions by solving the Euler-Lagrange equations. It seems reasonable at least for the unit winding number. For the higher winding numbers, however, the premise may be wrong in the case of the ungauged model. In fact, the solutions of the ungauged model always satisfy

$$E_m > mE_{m=1} \tag{20}$$

which indicates that the winding number m ($m > 1$) solutions have to break into the number m of unit winding number ones. One possibility to avoid such collapse is to relax the assumption of the rotational symmetry. It is known that non-rotationally symmetric solutions with higher winding numbers (or with the higher topological charges) often appear in several field theory models [9, 10]. Here we compute the solutions without the rotational symmetry in terms of an energy minimization scheme. We employ the method called the simulated annealing algorithm [8] in the two dimensional Cartesian grid. As in the case of the symmetric solutions, we suppose that ϕ carries the winding number and σ exhibits the traveling wave along the z axis. The

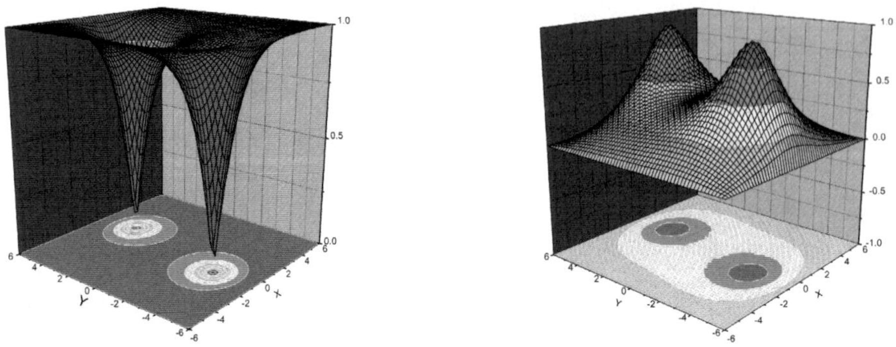

Figure 4. The profiles of ϕ(left) , σ(right) in the case of $m = 2$ for the parameters $\lambda_\phi = 10.0, \lambda_\sigma = 9.0, \eta_\phi = 1.0, \eta_\sigma = 1.0, \beta = 4.5$.

Hamiltonian is thus

$$\mathcal{H} = (\partial_x\phi)^2 + (\partial_y\phi)^2 + (\partial_x\sigma)^2 + (\partial_y\sigma)^2 + (\omega^2 + k^2)|\sigma|^2 + U. \tag{21}$$

Since the fields ϕ and σ are complex scalar fields, in the actual simulation we separate them into the real part and the imaginary part, and the Hamiltonian is minimized about the components ϕ_R, ϕ_I, σ_R and σ_I (where the subscript R, I mean the real and imaginary part, respectively). Here, we consider the bare case (ω=k=0). The profiles of the scalar fields are shown with the contour plot in Figure 3 of $m = 1$ and Figure 4 of $m = 2$. In both result, x and y are from -6 to 6. As was expected that the solutions with $m = 1$ has rotational symmetry. For the higher winding numbers, the solutions split into constituents. Thus we conclude the solutions do not have the rotational symmetry but are composed of two centers. The solutions, however, might be affected by the edge of the area. Of course if we consider the system with finite area such as a superconductor of a small material, the solutions can be stable. More thorough, large-scale numerical analysis is needed to reach the final conclusion. In addition, the choice of the parameters in the non-rotational system is unknown again and should be examined.

In the case of the gauged model, the solutions with the higher winding number may be stable because the energies exhibit $E_m \leqq mE_{m=1}$ (see caption of Figure 2). It means the gauge field might help the stability of higher winding number solutions. Since they are quite subtle, we may have ground states with breaking the rotational symmetry. The analysis is currently studied and the result will be reported elsewhere.

5. Conclusion

In this paper, we obtained several straight line vortex solutions in a Witten-type model. First we found the precise conditions of the model parameters for the existence of the solutions. For the gauged model, the solutions exist when only a gauge field couple with ϕ within the rotational symmetry. We thus obtained the solutions of such semi-local model. In the previous attempts, only the solutions with unit winding number was discussed, and then we investigated several higher winding number solutions as well. For the higher winding numbers, we suggested the possibility that the rotational symmetry should be broken and as a result, the winding number m solutions always split into m constituents.

For the rotational symmetry, only one gauge field is possible to implement. If the symmetry is broken, however, it might not be true. It is certainly interesting that analysis of energy minimization for the gauged model. The result will be reported in near future.

Acknowledgments The authors express their gratitude to the conference organizers of ISQS-22 for kind accommodation and hospitality. We also would like to thank Yasuhide Kobayashi for many useful advices of the numerical analysis. S.I. also acknowledges the financial support of Tokyo University of Science.

References

[1] E. Witten, Nucl. Phys. B **249**, 557 (1985).
[2] R. H. Brandenberger, B. Carter, A. C. Davis and M. Trodden, Phys. Rev. D **54**, 6059 (1996) [hep-ph/9605382].
[3] E. Radu and M. S. Volkov, Phys. Rept. **468**, 101 (2008) [arXiv:0804.1357 [hep-th]].
[4] J. Garaud, E. Radu and M. S. Volkov, Phys. Rev. Lett. **111**, 171602 (2013) [arXiv:1303.3044 [hep-th]].
[5] R. L. Davis and E. P. S. Shellard, Phys. Lett. B **207**, 404 (1988).
[6] Y. Lemperiere and E. P. S. Shellard, Nucl. Phys. B **649**, 511 (2003) [hep-ph/0207199].
[7] Y. Lemperiere and E. P. S. Shellard, Phys. Rev. Lett. **91**, 141601 (2003) [hep-ph/0305156].
[8] M. Hale, O. Schwindt and T. Weidig, Phys. Rev. E **62**, 4333 (2000) [hep-th/0002058].
[9] H. J. de Vega and F. A. Schaposnik, Phys. Rev. D **14**, 1100 (1976).
[10] M. Karliner and I. Hen, arXiv:0901.1489 [hep-th].

Journal of Physics: Conference Series **563** (2014) 012015 doi:10.1088/1742-6596/563/1/012015

Yangian symmetric correlators, R operators and amplitudes

R Kirschner[1]

[1] Institut für Theoretische Physik, Universität Leipzig, PF 100 920, D-04009 Leipzig, Germany.

E-mail: `Roland.Kirschner@itp.uni-leipzig.de`

Abstract. Yangian symmetric correlators can be constructed by the action of Yang-Baxter R operators on trivial basic correlators. The example of a four-point correlator is given in two representations and the construction of the completely connected N point correlator is described. The helicity representation is dicussed and the relation of the four-point correlator to tree-level scattering amplitudes is shown.

1. Spin chains with $g\ell_n$ symmetry

Let us first formulate the representations used as the quantum space at a site of the chain (compare [15]). Jordan-Schwinger type (JS) representations are based on the Heisenberg algebra, i.e. on n canonically conjugated pairs,

$$[p_a, x_b] = \delta_{ab}, \ \ a, b, = 1, .., n. \tag{1}$$

Generators of $g\ell_n$ representations can be built, e.g. as

$$\mathrm{L}^+_{ab} = p_a x_b, \ \ \text{or} \ \ \mathrm{L}^-_{ab} = -x_a p_b \tag{2}$$

and functions of x_a are the elements of the representation space. The two representations $L^{\pm}$ are dual to each other. If we distinguish the variables in one of the representations by the notation y_a, then the duality can be formulated as

$$[\mathrm{L}^+_{x,ab} + \mathrm{L}^-_{y,ab}, (\mathbf{xy})] = 0, \quad (\mathbf{xy}) = \sum_1^n x_a y_a. \tag{3}$$

The representation in terms of functions of x_a decomposes into the ones of definite degree of homogeneity,

$$(\mathbf{xp}) \cdot \psi(\mathbf{x}) = 2\ell\psi(\mathbf{x}). \tag{4}$$

Such a representation can also be described by functions of $n-1$ coordinate ratios, e.g. $x'_a = \frac{x_a}{x_n}, a = 1, ...n-1$.

$$\psi(\mathbf{x}) = x_n^{2\ell}\phi(\mathbf{x}'), \ \mathrm{L}_x \cdot \psi(\mathbf{x}) = x_n^{2\ell}\mathrm{L}_{(x',2\ell)} \ \phi(\mathbf{x}').$$

The matrix elements of $\mathrm{L}_{(x',2\ell)}$ generate a lowest weight representation with the constant representing the lowest weight vector. For generic values of 2ℓ all polynomials in x'_a form an irreducible representation. For the special values of non-negative integer 2ℓ the lowest weight modul is finite-dimensional.

We use the basic ingredients of the Quantum Inverse Scattering Method (QISM) [1, 2, 3, 4, 5]. The generators of JS representations enter the L operators depending on a spectral parameter.

$$\mathrm{L}^+(u) = \mathrm{I}(u) + \mathrm{L}^+ = \mathrm{I}(u) + \mathbf{p} \otimes \mathbf{x}, \quad \mathrm{L}^-(u) = \mathrm{I}u + \mathrm{L}^- = \mathrm{I}(u) - \mathbf{x} \otimes \mathbf{p}. \tag{5}$$

In general, if L_{ab} generate a representation of $g\ell_n$ on V then the L matrix operator, $\mathrm{L}(u) = \mathrm{I}u + \mathrm{L}_{ab}$, acting on the tensor product of this space with the one of the fundamental representation,$V \otimes V_f$, obeys the fundamental Yang-Baxter (YB) relation,

$$\mathcal{R}_{12}(u-v)\mathrm{L}(u) \otimes \mathrm{L}(v) = \mathrm{L}(v) \otimes \mathrm{L}(u)\mathcal{R}_{12}(u-v),$$

$$\mathcal{R}_{ab,ef}(u-v)\,\mathrm{L}_{ec}(u)\,\mathrm{L}_{fd}(v) = \mathrm{L}_{bf}(v)\,\mathrm{L}_{ae}(u)\,\mathcal{R}_{ef,cd}(u-v), \tag{6}$$

with the $n^2 \times n^2$ matrix $\mathcal{R}_{12}(u) = \mathrm{I} + \mathrm{P}_{12}$ involving the permutation in $V_f \otimes V_f$. The simple form of the $\mathrm{L}^{\pm}$ operators of JS type (5) allows to derive easily the following relations:

- elementary canonical transform defined as $\mathcal{C}^{-1}(\mathbf{x}, \mathbf{p})\mathcal{C} = (\mathbf{p}, -\mathbf{x})$:

$$\mathcal{C}^{-1}\mathrm{L}^+(u)\mathcal{C} = \mathrm{L}^-(u) \tag{7}$$

- matrix transposition:

$$\left(\mathrm{L}^+(u)\right)^t = -\mathrm{L}^-(-u-1) \tag{8}$$

- operator transposition:

$$\mathrm{L}^{+T}(u) = -\mathrm{L}^+(-1-u) \tag{9}$$

- inversion:

$$\left(\frac{\mathrm{L}^+(u)}{u}\right)^{-1} = \frac{\mathrm{L}^+(-u-1-(\mathbf{xp}))}{-u-1-(\mathbf{xp})} \tag{10}$$

- basic actions:

$$\mathrm{L}^+(u) \cdot 1 = (u+1)\mathrm{I} \cdot 1, \quad \mathrm{L}^-(u) \cdot 1 = u\mathrm{I} \cdot 1$$
$$\mathrm{L}^+(u) \cdot \delta(\mathbf{x}) = u\mathrm{I} \cdot \delta(\mathbf{x}), \quad \mathrm{L}^-(u) \cdot \delta(\mathbf{x}) = (u+1)\mathrm{I} \cdot \delta(\mathbf{x}). \tag{11}$$

Now we consider the representations and operators referring to the entire chain of N sites. The fundamental Yang-Baxter relation (6) is not only equivalent to the Lie algebra relation but can serve also as the starting point to formulate the related co-algebra, i.e. the underlying structures of the tensor product representations.

The monodromy matrix is defined as acting on $V_1 \otimes V_2 ... \otimes V_N \otimes V_f$ and constructed from the matrix product of the corresponding L operators.

$$\mathrm{T}^{\sigma}_{1,\ldots,N}(\mathbf{u}) = \mathrm{L}^{\sigma_1}_1(u_1)...\mathrm{L}^{\sigma_1}_N(u_N), \qquad \sigma = (\sigma_1, ...\sigma_N), \ \mathbf{u} = (u_1, ...u_N). \tag{12}$$

It is well known that the monodromy matrices obey the fundamental YB relation, i.e. the relation (6) with $L(u)$ substituted by $\mathrm{T}(u)$.

We need still another kind of YB operators, called general because they act on the tensor product of generic representations, in our case with JS type generators acting on functions of the coordinate components. These general YB operators appear as product of two factors. The basic factor can be derived from the L operators by considering a YB relation in the following form as the defining condition,

$$\mathrm{R}^{\sigma_1,\sigma_2}_{12}(u_1-u_2)\mathrm{L}^{\sigma_1}_1(u_1)\mathrm{L}^{\sigma_2}_2(u_2) = \mathrm{L}^{\sigma_1}_1(u_2)\mathrm{L}^{\sigma_2}_2(u_1)\mathrm{R}^{\sigma_1,\sigma_2}_{12}(u_1-u_2). \tag{13}$$

Here L_1 and L_2 act on variables $\mathbf{x}_1$ and $\mathbf{x}_2$, respectively, i.e. on the tensor product of generic representations. They enter the relation in matrix product (unlike the fundamental YB relation where their tensor product enters). We find the explicit solutions

$$\mathrm{R}_{12}^{+-}(u) = (\mathbf{x}_1\mathbf{x}_2)^u, \tag{14}$$

$$\mathrm{R}_{12}^{++}(u) = \int \frac{dc}{c^{1+u}} e^{-c(\mathbf{x}_1\mathbf{p}_2)}. \tag{15}$$

If the action is restricted to homogeneous functions of degree $2\ell_1$ and $2\ell_2$, repectively,

$$\psi(\lambda_1\mathbf{x}_1, \lambda_2\mathbf{x}_2) = \lambda_1^{2\ell_1}\lambda_2^{2\ell_2}\ \psi(\mathbf{x}_1, \mathbf{x}_2), \tag{16}$$

then additional YB relations can be written,

$$\mathrm{R}_{21}^{\sigma_2,\sigma_1}(u_1^{\sigma_1} - u_2^{\sigma_2})\mathrm{L}_1^{\sigma_1}(u_1)\mathrm{L}_2^{\sigma_2}(u_2) = \mathrm{L}_1^{\sigma_1}(u_1)\mathrm{L}_2^{\sigma_2}(u_2)\mathrm{R}_{21}^{\sigma_2,\sigma_1}(u_1^{\sigma_1} - u_2^{\sigma_2}). \tag{17}$$

These are obtained form the above ones by using the inversion relation (10) obeyed by the L matrix operators and replacing the operators of dilatations $(\mathbf{x}_i, \mathbf{p}_i)$ by their eigenvalues $2\ell_i$. Here and in the following we use the notations $u_i^+ = u_i + 2\ell_i,\ u_i^- = u_i - 2\ell_i - n$.

Now the general Yang-Baxter operator is the product of both of the above two types,

$$\mathrm{R}_{12}(u_1, u_1^{\sigma_1}, u_2, u_2^{\sigma_2}) = \mathrm{P}_{12}\mathrm{R}_{12}^{\sigma_1,\sigma_2}(u_1 - u_2)\mathrm{R}_{21}^{\sigma_2,\sigma_1}(u_1^{\sigma_1} - u_2^{\sigma_2}).$$

P_{12} denotes the operator of permutation of the tensor factors in $V_1 \otimes V_2$. The corresponding YB relation reads

$$\mathrm{R}_{12}(u_1, u_1^{\sigma_1}, u_2, u_2^{\sigma_2})\mathrm{L}_1^{\sigma_1}(u_1)\mathrm{L}_2^{\sigma_2}(u_2) = \mathrm{L}_2^{\sigma_2}(u_2)\mathrm{L}_1^{\sigma_1}(u_1)\mathrm{R}_{12}(u_1, u_1^{\sigma_1}, u_2, u_2^{\sigma_2}). \tag{18}$$

The restriction to particular degrees of homogeneity (16) is assumed.

2. Yangian symmetric correlators

We study the matrix-operator eigenvalue condition with the monodromy operators (12) for functions of N points in a space of n dimensions,

$$\mathrm{T}_{1,\dots,N}^{\sigma}(\mathbf{u})\Phi = E(\mathbf{u})\mathrm{I}\ \Phi, \tag{19}$$

and *define* [16] : N point Yangian symmetric correlators $\Phi = \Phi(\mathbf{x}_1, ..., \mathbf{x}_N)$ are homogeneous eigenfunctions of the matrix operator $\mathrm{T}(\mathbf{u})$.

Each n-dimensional point $\mathbf{x}_i, i = 1, ..., N$, is marked with a spectral parameter u_i, a dilatation weight $2\ell_i$ and a signature σ_i.

The notion of Yangian algebras has been introduced in [6]. Its relation to the QISM has been explained in [7]. The Yangian symmetry of scattering amplitudes was understood on the basis of [8, 9].

Symmetric correlators can be generated by R operators [17]. If $\Phi(1, .., N; \sigma, \mathbf{u})$ is a solution of the monodromy eigenvalue relation (19) then also

$$\mathrm{R}_{i,i+1}^{\sigma_i,\sigma_{i+1}}(u_i - u_{i+1}) \cdot \Phi(1, 2, .., N; \sigma, \mathbf{u}) \quad \text{and} \quad \mathrm{R}_{i+1,i}^{\sigma_{i+1},\sigma_i}(u_i^{\sigma_i} - u_{i+1}^{\sigma_{i+1}}) \cdot \Phi(1, 2, .., N; \sigma, \mathbf{u}) \tag{20}$$

obey (19) with parameter permutations.

The Yang-Baxter RLL relations (13, 17) allow to permute the R operators acting on cyclically adjacent points and having particular arguments with the monodromy operator. The

Journal of Physics: Conference Series **563** (2014) 012015 doi:10.1088/1742-6596/563/1/012015

modification of the monodromy operator and the appropriate argument of R can be read off from the *parameter permutation rules* [14] :

$$\mathrm{R}_{12}^{\sigma_1,\sigma_2}(u_1-u_2): \quad (u_1,u_1^\sigma,u_2,u_2^\sigma,...,u_N,u_N^\sigma) \to (u_2,u_1^\sigma,u_1,u_2^\sigma...,u_N,u_N^\sigma),$$

$$\mathrm{R}_{21}^{\sigma_2,\sigma_1}(u_1^\sigma-u_2^\sigma): \quad (u_1,u_1^\sigma,u_2,u_2^\sigma,...,u_N,u_N^\sigma) \to (u_1,u_2^\sigma,u_2,u_1^\sigma...,u_N,u_N^\sigma). \tag{21}$$

The generation of non-trivial solutions can be started from (trivial) basic correlators.

$$\Omega(\sigma) = \prod_{j:\sigma_j=-} \delta(\mathbf{x}_j), \quad \mathrm{T}_{1,...,N}^{\sigma_\mathrm{T}}(\mathbf{u})\ \Omega(\sigma_\mathbf{o}) = \prod_1^N (u_i + \frac{1}{2}(1+\sigma_{T,i}\sigma_{o,i}))\mathrm{I}\ \Omega(\sigma_\mathbf{o})$$

We resort to two standard signature configurations

- regular case : $\sigma_0 = (+,+,...,+), \quad \Omega = 1,$
- uniform case : $\sigma_T = (+,+,...,+), \quad \Omega_{I,J} = \prod_{i\in I}\delta(\mathbf{x}_i).$

Signature cases can be related by elementary canonical transformation $\mathcal{C}$ acting uniformly on the components of the site variables $x_{s,a}, p_{s,a}, s = 1,...,n, a = 1,...N,\ \ \mathcal{C}^{-1}(\mathbf{x}_{s,a},\mathbf{p}_{s,a})\mathcal{C} = (\mathbf{p}_{s,a}, -\mathbf{x}_{s,a})$.

Let us construct an example of a 4-point correlator in both representations: In the uniform case we have

$$\mathrm{R}_{32}^{++}(u_2^+-u_3^+)\mathrm{R}_{12}^{++}(u_4-u_2)\mathrm{R}_{34}^{++}(u_3-u_1)\mathrm{R}_{14}^{++}(u_1-u_4)\delta(\mathbf{x}_2)\delta(\mathbf{x}_4)$$

$$= \Phi^{+-+-}(u_2,u_1^+,u_4,u_3^+,u_1,u_2^+,u_3,u_4^+) =$$

$$\int \frac{dc_{23}dc_{21}dc_{43}dc_{41}}{c_{32}^{1+u_2-u_3-n}c_{21}^{1+u_4-u_2}c_{43}^{1+u_3-u_1}c_{41}^{1+u_1-u_4}}\delta(\mathbf{x}_2-c_{23}\mathbf{x}_3-c_{21}\mathbf{x}_1)\delta(\mathbf{x}_4-c_{41}\mathbf{x}_1-c_{43}\mathbf{x}_3). \tag{22}$$

In the regular case we have

$$\mathrm{R}_{32}^{+-}(u_2^+-u_3^+)\mathrm{R}_{12}^{+-}(u_4-u_2)\mathrm{R}_{34}^{+-}(u_3-u_1)\mathrm{R}_{14}^{+-}(u_1-u_4)\cdot 1 =$$

$$(\mathbf{x}_3\mathbf{x}_2)^{u_2-u_3-n}(\mathbf{x}_1\mathbf{x}_2)^{u_4-u_2}(\mathbf{x}_3\mathbf{x}_4)^{u_3-u_1}(\mathbf{x}_1\mathbf{x}_4)^{u_1-u_4}.$$

The expression is an eigenfunction of the monodromy with the spectral parameters permuted to (u_2, u_4, u_1, u_3).

3. The generic N point correlator

We consider the particular signature configuration $-...-+...+$ and use the uniform representation

$$\Omega_{N,K} = \prod_1^K \delta(\mathbf{x}_i) = \Phi_0(u_1, u_1-n; ...u_K, u_K-n; u_{K+1}, u_{K+1}; ...; u_N, u_N). \tag{23}$$

The last notation shows the spectral parameters u_s as in the monodromy operator and also their combination with the weights u_s^+. As shown above the action of R operators is accompanied by parameter permutations and writing the resulting parameter configuration allows easily to fix the arguments of the R operators such that the result is a symmetric correlator.

Each R_{ji}^{++} operation brings in an integration over c_{ij}. The completely connected correlator of interest involves $K\times(N-K)$ integrations, there is an integral with the variable c_{ij} for each pair of sites i,j of different signature. The integration variables can be regarded as the elements of a

$K \times (N-K)$ matrix $\hat{c}$ and are related to the points of the Grassmannian $G_{N,K}$. The integration contour lies in the maximal Schubert cell.

$$M_{N,K} = \int d^{K(N-K)}c \;\; \phi(c)\; \delta(\hat{C}\mathbf{x}), \tag{24}$$

$$d^{K(N-K)}c = d^{N-K}c_1 ... d^{N-K}c_N, \; d^{N-K}c_i = dc_{i,K+1}...dc_{i,N},$$

$$\hat{C} = (I_{K\times K}, -\hat{c}), \;\; \delta(\hat{C}\mathbf{x}) = \prod_1^K \delta^{(n)}(\mathbf{x}_i - \sum_{j=K+1}^{N} c_{ij}\mathbf{x}_j).$$

We shall see how the symmetry condition (19) fixes ϕ.

This symmetric correlator can be obtained by a $K(N-K)$-fold R operator action on the basic one (23) in several ways. We indicate as a standard way for $N \geq 2K$ the following procedure.

$$M_{N,K} = \Phi_K(u_1, u_{K+1}; ...; u_K, u_{2K}; ...; u_{N-K+1}, u_1^+; ...; u_N, u_K^+), \tag{25}$$

where Φ_K is the result of R operators action in the Kth step starting from the basic correlator (23). The first step is the action by $N-K$ R operators

$$\mathrm{R}_{N,N-1}(u_1^+ - u_N)...\mathrm{R}_{K+2,K+1}(u_1^+ - u_{K+2})\mathrm{R}_{K+1,1}(u_1^+ - u_{K+1})\Phi_0 =$$

$$\Phi_1(u_1, u_{K+1}; ...; u_K, u_K^+; u_{K+1}, u_{K+2}; ...; u_N, u_1^+) =$$

$$\int \frac{dc^1_{1,K+1}dc^1_{K+1,K+2}...dc^1_{N-1,N}}{(c^1_{1,K+1})^{1+u_1^+-u_{K+1}}(c^1_{K+1,K+2})^{1+u_1^+-u_{K+2}}...(c^1_{N-1,N})^{1+u_1^+-u_N}}$$

$$\delta\left(\mathbf{x}_1 - c^1_{1,K+1}\mathbf{x}_{K+1} + \sum_{K+2}^{N}(-1)^{K+j}c^1_{K+1,K+2}...c^1_{j-1,j}\;\mathbf{x}_j\right)\; \prod_2^K \delta(\mathbf{x}_i).$$

We observe that the u^+ parameters of the sites $1, K+1, K+1, ...N$ are shifted cyclically leaving the ones at $2, ...K$ untouched. At the Lth step the action is by

$$\mathrm{R}_{N,N-1}(u_L^+ - v_N)...\mathrm{R}_{K+2,K+1}(u_L^+ - v_{K+2})\mathrm{R}_{K+1,L}(u_L^+ - v_{K+1}),$$

where $v_{K+1}, ..., v_N$ is the sequence of u^+ parameters appearing at the step $L-1$ at the sites $K+1, ..., N$. The action produces the cyclic shift of the sequence of u^+ parameters at the sites $L, K+1, ...N$ leaving the sites $1, ..., L-1, L+1, ...K$ untouched.

By these R actions a completely connected correlator is generated, i.e. there are links between all pairs of sites with different signature. We consider the details for the case $K=2$ where

$$M_{N,2} = \Phi_2(u_1, u_3; u_2, u_4; u_3, u_5; ...; u_{N-1}, u_1 - n; u_N, u_2 - n) =$$

$$\int \frac{dc^1_{1,3}dc^1_{3,4}...dc^1_{N-1,N}}{\mathcal{N}_1}\frac{dc^2_{2,3}dc^2_{3,4}...dc^2_{N-1,N}}{\mathcal{N}_2} \times$$

$$\delta(\mathbf{x}_1 - c^1_{1,3}\mathbf{x}_3 + \sum_4^N(-1)^j S^{(1,2)}_{3,j}\;\mathbf{x}_j)\delta(\mathbf{x}_2 - c^2_{2,3}\mathbf{x}_3 + \sum_4^N(-1)^j c^2_{3,4}...c^2_{j-1,j}\;\mathbf{x}_j),$$

$$\mathcal{N}_1 = (c^1_{1,3})^{1+u_1^+-u_3}(c^1_{3,4})^{1+u_1^+-u_4}...(c^1_{N-1,N})^{1+u_1^+-u_N},$$

$$\mathcal{N}_2 = (c^2_{2,3})^{1+u_2^+-u_4}(c^1_{3,4})^{1+u_2^+-u_5}...(c^1_{N-1,N})^{1+u_2^+-u_1^+}. \tag{26}$$

In the general case we end up with

$$M_{N,K} = \int \frac{d^{N-K}c^1}{\mathcal{N}_1} ... \int \frac{d^{N-K}c^K}{\mathcal{N}_K} \times$$

$$\delta(\mathbf{x}_1 - \sum_1^{N-K} (-1)^j S^{1,K}_{K+1,K+j} \mathbf{x}_{K+j}) ... \delta(\mathbf{x}_K - \sum_1^{N-K} (-1)^j S^{K,K}_{K+1,K+j} \mathbf{x}_{K+j}), \tag{27}$$

where

$$d^{N-K}c^L = dc^L_{L,K+1} dc^L_{K+1,K+2} ... dc^L_{N-1,N},$$

$$\mathcal{N}_L = (c^L_{L,K+1})^{1+u^+_L - u_{K+L}} (c^L_{K+1,K+2})^{1+u^+_L - u_{K+L+1}} ...$$

$$...(c^L_{K-L,K-L+1})^{1+u^+_L - u_N} (c^L_{K-L+1,K-L+2})^{1+u^+_L - u^+_1} ... (c^L_{N-1,N})^{1+u^+_L - u^+_{L-1}}.$$

It is convenient to define $S^{(i_0,i)}_{j_0,j}$ for $1 \leq i_0 < i \leq K < j_0 < j \leq N$ as a sum of products of the integration variables $c^L_{j_1-1,j_1}$ entering by the action of R_{j_1,j_1-1} at the Lth step, $i_0 \leq L \leq i$ by the following iteration

$$S^{(i_0,i)}_{j_0,j} = S^{(i_0,i)}_{j_0,j-1} c^i_{j-1,j} + S^{(i_0,i-1)}_{j_0,j} \tag{28}$$

$$S^{(i,i)}_{j_0,j} = c^i_{i,j_0} c^i_{j_0,j_0+1} ... c^i_{j-1,j}, \; S^{(i_0,i)}_{j_0,j_0} = c^i_{ij_0}.$$

The second relation results in

$$c^i_{j-1,j} = \frac{S^{(i,i)}_{j_0,j}}{S^{(i,i)}_{j_0,j-1}},$$

and then the frist relation can be solved as

$$S^{(i_0,i-1)}_{j_0,j} = \begin{vmatrix} S^{i,i}_{j_0,j-1} & S^{i,i}_{j_0,j} \\ S^{i_0,i}_{j_0,j-1} & S^{i_0,i}_{j_0,j} \end{vmatrix} \frac{1}{S^{i,i}_{j_0,j-1}}.$$

We shall transform the integral (27) with respect to $c^i_{j-1,j}$ into the link form (24) where the integration variables are $c_{i,j} = (-1)^j S^{i,K}_{K+1,K+j}$.

In the case $K = 2$ we have to consider

$$S^{(2,2)}_{3,j} = c^2_{3,4} ... c^2_{j-1,j}, S^{(1,1)}_{3,j} = c^1_{3,4} ... c^1_{j-1,j}, S^{(1,2)}_{3,4} = c^1_{34} + c^2_{34},$$

$$S^{(1,2)}_{3,j} = S^{(1,2)}_{3,j-1} c^2_{j-1,j} + c^1_{3,4} ... c^1_{j-1,j}.$$

Comparing the R operator expression with the expected maximal cell link integral expression (24) we see that the link variables $c_{2j}, c_{1j}, j > 3$ have to be proportional correspondingly to $S^{(2,2)}_{3,j}$ and $S^{(1,2)}_{3,j}$. Therefore we find the expression of the R operation variables in terms of these sums.

$$c^2_{j-1,j} = \frac{S^{(2,2)}_{3,j}}{S^{(2,2)}_{3,j-1}}, \; c^1_{j-1,j} = \frac{S^{(1,1)}_{3,j}}{S^{(1,1)}_{3,j-1}}, \; S^{(1,2)}_{3,j} = S^{(1,2)}_{3,j-1} c^2_{j-1,j} + S^{(1,1)}_{3,j},$$

$$S^{(1,1)}_{3,j} = \frac{1}{S^{(2,2)}_{3,j-1}} \begin{vmatrix} S^{(2,2)}_{3,j-1} & S^{(1,2)}_{3,j-1} \\ S^{(2,2)}_{3,j} & S^{(1,2)}_{3,j} \end{vmatrix}.$$

We identify the link variables $c_{ij}, i = 1, 2; j = 3, ..., N$.

$$c_{13} = c^1_{1,3}, \; c_{14} = -c^1_{1,3} S^{(12)}_{34} = -c^1_{13}(c^1_{34} + c^2_{34}), c_{1j} = (-1)^{j+1} c^1_{13} S^{(1,2)}_{3j}$$

$$c_{23} = c^2_{2,3},\ c_{24} = -c^2_{1,3}S^{(2,2)}_{34} = -c^2_{13}c^2_{34}, c_{2j} = (-1)^{j+1}c^2_{23}S^{(2,2)}_{3j}.$$

Now we can relate the expressions $S^{i,2}_{3j}$ with the link variables.

$$S^{(1,2)}_{3j} = (-1)^{j+1}\frac{c_{1j}}{c_{13}},\quad S^{(2,2)}_{3j} = (-1)^{j+1}\frac{c_{2j}}{c_{23}}.$$

Finally we express the original R integration variables in terms of the link variables.

$$c^2_{23} = c_{23},\ c^2_{j-1,j} = -\frac{c_{2,j}}{c_{2,j-1}}, j \geq 4. \tag{29}$$

$$c^1_{13} = c_{13},\ c^1_{j-1,j} = \frac{S^{(1,1)}_{3,j}}{S^{(1,1)}_{3,j-1}}. \tag{30}$$

We substitute the result for $S^{(11)}_{3j}$ in terms of $S^{(i2)}_{3j}$ and obtain

$$c^1_{j-1,j} = -\frac{c_{2,j-2}\begin{vmatrix} c_{2,j-1} & c_{1,j-1} \\ c_{2,j} & c_{1,j}\end{vmatrix}}{c_{2,j-1}\begin{vmatrix} c_{2,j-2} & c_{1,j-2} \\ c_{2,j-1} & c_{1,j-1}\end{vmatrix}}.$$

The last expression works well for $j \geq 5$. For $j = 4$ one finds the correct result with the substitution $c_{22} = 1, c_{12} = 0$.

We calculate the denominator of the integrand in $M_{N,2}$.

$$\mathcal{N}_2 = c_{23}^{1+u_2^+-u_4}\left(\frac{c_{24}}{c_{23}}\right)^{1+u_2^+-u_5}\cdots\left(\frac{c_{2,N-1}}{c_{2,N-2}}\right)^{1+u_2^+-u_N}\left(\frac{c_{2N}}{c_{2,N-1}}\right)^{1+u_2^+-u_1^+} =$$

$$c_{23}^{u_5-u_4}c_{24}^{u_6-u_5}...c_{2,N-1}^{u_1^+-u_N}c_{2N}^{1+u_2^+-u_1^+},$$

$$\mathcal{N}_1 = c_{13}^{1+u_1^+-u_3}\left(\frac{m_4}{c_{23}c_{13}}\right)^{1+u_1^+-u_4}\left(\frac{c_{23}m_5}{c_{24}m_4}\right)^{1+u_1^+-u_5}\cdots$$

$$\cdots\left(\frac{c_{2,n_2}m_{N-1}}{c_{2,N-2}m_{N_2}}\right)^{1+u_1^+-u_{N-1}}\left(\frac{c_{2,N-2}m_N}{c_{2,N-1}m_{N-1}}\right)^{1+u_1^+-u_N} =$$

$$c_{13}^{u_4-u_3}m_4^{u_5-u_4}...m_{N-1}^{u_N-u_{N-1}}m_N^{1+u_1^+-u_N}\ c_{23}^{u_4-u_5}c_{24}^{u_5-u_6}...c_{2,N-2}^{u_{N-1}-u_N}c_{2,N_1}^{-1+u_N-u_1^+}.$$

Here we have abbreviated the determinants

$$m_j = \begin{vmatrix} c_{2,j-1} & c_{1,j-1} \\ c_{2,j} & c_{1,j}\end{vmatrix},\quad j \geq 4.$$

The result for the integrands denominator is

$$\mathcal{N}_1\mathcal{N}_2 = c_{13}^{u_4-u_3}m_4^{u_5-u_4}...m_{N-1}^{u_N-u_{N-1}}m_N^{1+u_1^+-u_N}\ c_{2,N-1}^{-1}c_{2N}^{1+u_2^+-u_1^+}.$$

Still the transformation is incomplete, the Jacobian must be taken into account. Consider first the factor in the Jacobian

$$\frac{\partial(c^2_{23}c^2_{34}...)}{\partial(c_{23}c_{24}...)}.$$

The matrix comes out triangualar by using (29) and the product of the diagonal elements is

$$\prod_3^N \frac{\partial c_{2j}}{\partial c^2_{j-1,j}} = \frac{1}{c_{23}c_{24}...c_{2N}}.$$

The triangularity simplifies the calculation also in

$$\frac{\partial(c^1_{13}c^1_{34}...)}{\partial(c_{13}c_{14}...)}.$$

Using (30) we see that it equals to

$$\prod \frac{\partial c^1_{j-1,j}}{\partial c_{1j}} = \frac{1}{c_{13}}\frac{c_{23}}{m_4}...\frac{c_{2,N-2}}{m_{N-1}}.$$

The Jacobian of the transformation to the link integration is the product of the above partial Jacobians due to trangularity.

$$\frac{\partial(c^1_{13}c^1_{34}...c^1_{N-1,N}c^2_{23}c^2_{34}...c^2_{N-1,N})}{\partial(c_{13}c_{14}...c_{1N}c_{23}c_{24}...c_{2,N}} = \frac{1}{c_{2,N-1}c_{14}}\frac{1}{m_4...m_{N-1}}.$$

We obtain the completely connected symmetric correlator for $K = 2$ in form of the maximal cell link integral.

$$M_{N,2} = \int \frac{dc^1_{1,3}dc^1_{3,4}...dc^1_{N-1,N}dc^2_{2,3}dc^2_{3,4}...dc^2_{N-1,N}}{(c_{1,3})^{1+u_4-u_3}(34)^{1+u_5-u_4}...(N-1,N)^{1+u_1^+-u_N}c_{2N}^{1+u_2-u_1}} \times$$

$$\delta(\mathbf{x}_1 - \sum_3^N c_{1j}\,\mathbf{x}_j)\delta(\mathbf{x}_2 - \sum_3^N c_{2j}\,\mathbf{x}_j). \quad (31)$$

Here $(j-1,j) = m_j$ denotes the minor of the matrix $\hat{C}$ of the columns $j-1, j$,

$$\hat{C} = \begin{pmatrix} 1 & 0 & -c_{13} & ... & -c_{1N} \\ 0 & 1 & -c_{23} & ... & -c_{2N} \end{pmatrix}.$$

The transformation can be done for general K relying on the iterative relation (28). The result can be written as (24) with

$$\phi(c) = (2,...,K+1)^{-1-u_{K+2}+u_{K+1}}(3,...,K+2)^{-1-u_{K+3}+u_{K+2}}...(N-K,...,N-1)^{-1-u_N+u_{N-1}}$$

$$(N-K+1,...,N)^{-1-u_1^++u_N}(N-K+2,...,N,1)^{-1-u_2^++u_1^+}...(N,1...,K-1)^{-1-u_K^++u_{K-1}^+}. \quad (32)$$

The factors are powers of the subsequent $K \times K$ minors of the rectangular matrix $\hat{C}$ (24).

In application to scattering amplitudes the special case of this form without parameters was proposed in [10] and used extensively in the related literature. The parameter dependent expression has been discussed recently in [13].

The result is by construction an eigenfunction of the monodromy $T(u_1, u_2, ..., u_N)$. The action of the R operators has resulted in permutations of the weight dependent parameters u_i^+; the resulting configuration can be characterized by the permutation

$$\begin{pmatrix} u_1 & u_2 & ... & u_K & u_{K+1} & ... & u_{N-K} & u_{N-K+1} & ... & u_N \\ u_{K+1} & u_{K+2} & ... & u_{2K} & u_{2K+1} & ... & u_N & u_1 - n & ... & u_K - n \end{pmatrix}.$$

The values of v_i^+ are on the second row and the weights attributed to the points are calculated as $2\ell_i = v_i^+ - u_i$.

In the case $K = 2$ we have the list of weights and exponents as

$$2\ell_1 = u_3 - u_1, 2\ell_2 = u_4 - u_2, ..., 2\ell_i = u_{i+2} - u_i, ...2\ell_{N-2} = u_N - u_{N-2},$$

$$2\ell_{N-1} = u_1 - n - u_{N-1}, 2\ell_N = u_2 - n - u_N,$$

$$\alpha_3 = u_4 - u_3, \alpha_4 = u_5 - u_4, ..., \alpha_i = u_{i+1} - u_i, ..., \alpha_{N_1} = u_N - u_{N-1}, \alpha_1 = u_1 - n - u_1, \alpha_2 = u_2 - u_1.$$

The weights are not independent, in any case

$$\sum_{s=1}^{N} 2\ell_s = -2n.$$

The weights do not fix all the N spectral parameters, e.g. u_N is independent. In the case of even $N = 2M$ there are actually two relations,

$$\sum_{k=1}^{M} 2\ell_{2k} = -n, \sum_{k=1}^{M} 2\ell_{2k-1} = -n.$$

Therefore the weights fix only $N - 2$ of the spectral parameters, e.g. u_N and u_{N-1} are free. Thus for even N the exponents of the minors in the final result (31) involve besides of the $N - 2$ independent weights the extra parameter $u_N - u_{N-1}$. The particular case with $N = 4$ will be discussed below.

4. Scattering amplitudes

For applications to scattering amplitudes we introduce a further kind of representation. It is obtained from the uniform representation by partial canonical transformations after dividing the index set labelling the components at each site. $s = 1, ..., n \rightarrow \dot{\alpha} = 1, ..., p;\ \alpha = p + 1, ..., n$. The elementary canonical transformation is applied to the dotted components only as $\mathcal{C}^{-1}(\mathbf{x}_{\dot{\alpha}}, \mathbf{p}_{\dot{\alpha}})\mathcal{C}$. The matrix in the L operator splits into blocks transforming in different ways as

$$\mathbf{p} \otimes \mathbf{x} = \begin{pmatrix} \mathbf{p}_{\dot{\alpha}}\mathbf{x}_{\dot{\alpha}} & \mathbf{p}_{\dot{\alpha}}\mathbf{x}_{\alpha} \\ \mathbf{p}_{\alpha}\mathbf{x}_{\dot{\alpha}} & \mathbf{p}_{\alpha}\mathbf{x}_{\alpha} \end{pmatrix} \rightarrow \begin{pmatrix} -\mathbf{x}_{\dot{\alpha}}\mathbf{p}_{\dot{\alpha}} & -\mathbf{x}_{\dot{\alpha}}\mathbf{x}_{\alpha} \\ \mathbf{p}_{\alpha}\mathbf{p}_{\dot{\alpha}} & \mathbf{p}_{\alpha}\mathbf{x}_{\alpha} \end{pmatrix}.$$

We reconsider the basic symmetry condition (19). By expansion of $T(\mathbf{u} + \Delta)$ we obtain at the $N - 1$st power of Δ the well known result that the sums over the chain sites of the off-diagonal $g\ell_n$ generators annihilate the symmetric correlator. In this way the conserved quantities of the global chain symmetry are expressed.

In the helicity representation a part of the off-diagonal operators (the ones in the upper right block in L) acts multiplicatively on functions of the coordinate variables:

$$\sum_{s=1}^{N} k_{s,\dot{\alpha},\alpha} \Phi = 0, \qquad k_{s,\dot{\alpha},\alpha} = \mathbf{x}_{s,\dot{\alpha}} \mathbf{x}_{s,\alpha}.$$

If $k_{s,\dot{\alpha},\alpha}$ are the components of the momenta of scattering particles the particular conservation law is just the energy-momemtum conservation. It works for massless particles where the mass-shell condition is the one for the factorisability. We recall that a light-like four vector can be represented by the product of Weyl spinor components of left and right chirality. In the following we rename the coordinate components introducing different symbols for the ones in the two index ranges, $\bar{\lambda}_{\dot{\alpha}}$ and λ_α. We conclude that a generic Yangian symmetric correlator in the helicity representation appears with the delta distribution as a factor expressing this conservation law.

$$\Phi(\bar{\lambda}_1, \lambda_1, ..., \bar{\lambda}_N, \lambda_N) = \delta(\sum_1^N k_{s,\dot{\alpha},\alpha}) \times \phi(\bar{\lambda}_1, \lambda_1, ..., \bar{\lambda}_N, \lambda_N). \tag{33}$$

In this representation the scalar product becomes

$$(\mathbf{x}_1\mathbf{p}_2) \to (\lambda_1\pi_2) - (\bar{\pi}_1\bar{\lambda}_2)$$

and the action of the R operator has the form

$$\mathrm{R}_{12}^{++}(u)F(\bar{\lambda}_{1,\dot{\alpha}}, \lambda_{1,\alpha}, \bar{\lambda}_{2,\dot{\alpha}}, \lambda_{2,\alpha}) = \int \frac{dc}{c^{1+u}} F(\bar{\lambda}_{1,\dot{\alpha}} + c\bar{\lambda}_{2,\dot{\alpha}}, \lambda_{1,\alpha}, \bar{\lambda}_{2,\dot{\alpha}}, \lambda_{2,\alpha} - c\lambda_{1,\alpha}).$$

We consider the transformation of the dilatation operator from the uniform to the helicity representation

$$(\mathbf{x}_i \cdot \mathbf{p}_i) \to \lambda_{i,\alpha}\frac{\partial}{\partial\lambda_{i,\alpha}} - \bar{\lambda}_{i,\dot{\alpha}}\frac{\partial}{\partial\bar{\lambda}_{i,\dot{\alpha}}} - 2.$$

The helicity of a scattering particle state h_i is calculated as one half of the difference of the weights in λ and $\bar{\lambda}$, therefore

$$2\ell_i = 2h_i - 2. \tag{34}$$

On the other hand we have $2\ell_i = v_i^+ - v_i$ for the weights of a correlator $\Phi(v_1, v_1^+; v_2, v_2^+, ..., v_N, v_N^+)$. Thus a part of the spectral parameter dependence can be expressed in terms of the particle helicities.

We return to the example of a 4-point correlator (22) and study its application to $2 \to 2$ scattering. We put the parameters back to the standard ordering.

$$\Phi(u_1, u_1^+; u_2, u_2^+; u_3, u_3^+, u_4, u_4^+) =$$

$$\mathrm{R}_{32}(u_2^+ - u_3^+)\mathrm{R}_{12}(u_2 - u_1)\mathrm{R}_{34}(u_4 - u_3)\mathrm{R}_{14}(u_3 - u_2)\ \delta(\mathbf{x}_2)\delta(\mathbf{x}_4) =$$

$$\int \frac{dc_{23}dc_{21}dc_{43}dc_{41}}{c_{23}^{1+u_1-u_4-n}c_{21}^{1+u_2-u_1}c_{43}^{u_4-u_3}c_{41}^{u_3-u_2}} \times$$

$$\delta(\bar{\lambda}_1 + c_{41}\bar{\lambda}_4 + c_{21}\bar{\lambda}_2)\delta(\bar{\lambda}_3 + c_{43}\bar{\lambda}_4 + c_{23}\bar{\lambda}_2)\delta(\lambda_2 - c_{21}\lambda_1 - c_{23}\lambda_3)\delta(\lambda_4 - c_{41}\lambda_1 - c_{43}\lambda_3),$$

$$u_1^+ = u_3, u_2^+ = u_4, u_3^+ = u_1 - n, u_4^+ = u_2 - n. \tag{35}$$

Now we specify to the case $n = 4, p = 2$. We shall transform to the form (33) following [11]. We count 8 delta distributions. We intend to transform the expression in such a way that the 4 deltas of the energy momentum conservation appear explicitly and the remaining 4 are used to do the 4 link integrals. First we address the 4 linear equations related to the deltas involving λ components.

$$\lambda_2 - c_{21}\lambda_1 - c_{23}\lambda_3 = \lambda_2', \quad \lambda_4 - c_{41}\lambda_1 - c_{43}\lambda_3 = \lambda_4'. \tag{36}$$

We do projections by performing (antisymmetric) spinor products,

$$< 1, 2 >=< \lambda_1, \lambda_2 >= \lambda_{1,1}\lambda_{2,2} - \lambda_{1,2}\lambda_{2,1}.$$

Note the difference to the above products (3) denoted by ordinary brackets (..., ...) which are defined in the euclidean way. The conditions (36) with $\lambda'_2 = \lambda'_4 = 0$ are solved by

$$c^*_{21} = \frac{<2,3>}{<1,3>},\ c^*_{23} = \frac{<2,1>}{<3,1>},\ c^*_{41} = \frac{<4,3>}{<1,3>},\ c^*_{43} = \frac{<4,1>}{<3,1>},$$

and the Jacobi determinant of the transformation $\lambda'_{2,\alpha}, \lambda'_{4,\alpha} \to c_{21}, c_{23}, c_{41}, c_{43}$ is

$$\frac{\partial(\lambda'_{2,1}\lambda'_{2,2})}{\partial(c_{21}, c_{23})}\frac{\partial(\lambda'_{4,1}\lambda'_{4,2})}{\partial(c_{41}, c_{43})} =< 1,3 >^2 .$$

In this way we have obtained

$$\delta(\lambda_2 - c_{21}\lambda_1 - c_{23}\lambda_3)\delta(\lambda_4 - c_{41}\lambda_1 - c_{43}\lambda_3) =< 1,3 >^{-2}\ \delta(c_{21} - c^*_{21})\delta(c_{23} - c^*_{23})\delta(c_{41} - c^*_{41})\delta(c_{43} - c^*_{43}).$$

The deltas involving $\bar{\lambda}$ components can be transformed by multiplying the matrix of the involved equations

$$\tilde{C} = \begin{pmatrix} 1 & c_{21} & 0 & c_{41} \\ 0 & c_{23} & 1 & c_{43} \end{pmatrix}$$

from the left by a 2×2 matrix A taking into account the appropriate Jacobian. We choose this matrix in such a way that

$$A\,\tilde{C} = \begin{pmatrix} \lambda_{1,1} & \lambda_{2,1} & \lambda_{3,1} & \lambda_{4,1} \\ \lambda_{1,2} & \lambda_{2,2} & \lambda_{3,2} & \lambda_{4,2} \end{pmatrix}.$$

It works indeed with

$$A = \begin{pmatrix} \lambda_{1,1} & \lambda_{3,1} \\ \lambda_{2,1} & \lambda_{3,2} \end{pmatrix},$$

if we substitute simultaneously the link variables $c_{i,j}$ by the solutions $c^*_{i,j}$ obtained above.

We observe that the Jacobi determinant is the inverse of the one in the first transformation. Finally all factors of $< 1,3 >$ cancel. We obtain the form (33) in our case.

$$\Phi(u_1, u_1^+; u_2, u_2^+; u_3, u_3^+, u_4, u_4^+) = \delta^{(4)}(\sum_1^4 \lambda_{i,\alpha}\bar{\lambda}_{i,\dot{\alpha}})\ \phi(\lambda, \bar{\lambda}),$$

$$\phi(\lambda, \bar{\lambda}) = \frac{<1,2>^4}{<1,2>^{1+u_1-u_4}<2,3>^{1+u_2-u_1}<4,1>^{1+u_4-u_3}<3,4>^{1+u_3-u_2}}.$$

We reconsider the relations for the spectral parameters (35) and calculate the scaling weights as $2\ell_i = u_i^+ - u_i$,

$$2\ell_1 = u_3 - u_1,\ 2\ell_2 = u_4 - u_2,\ 2\ell_3 = u_1 - n - u_3, 2\ell_4 = u_2 - n - u_4. \tag{37}$$

We see that they are pairwise connected, $2\ell_1 = -2\ell_3 - n, 2\ell_2 = -2\ell_4 - n$. The relation between the weights implies for the helicities ($n = 4$) $h_1 = -h_3,\ h_2 = -h_4$. The obtained expression implies the spinor-helicity expressions of $2 \to 2$ helicity amplitudes of several cases by appropriate choices of the helicity values, $h = \pm 1, \pm\frac{1}{2}, 0$.

We can substitute the spectral parameters partially by the helicities.

$$\phi(\lambda, \bar{\lambda}) = \left(\frac{<2,3><4,1>}{<1,2><3,4>}\right)^{u_3-u_4-1} \frac{<1,2>^{2h_1}<3,4>^{-2h_2}}{<2,3>^{2h_1-2h_2}}$$

The amplitude contributions still involve $u_3 - u_4$ as a free parameter. The parameter extension of amplitudes has been proposed in [12]. The analytic structure expected from physical unitarity (dispersion relations) must be used to fix it.

Acknowledgments
The author is grateful to D. Chicherin, S. Derkachov and D. Karakhanyan for useful discussions.

References

[1] Faddeev L D Sklyanin E K and Takhtajan L A 1979 The Quantum Inverse Problem Method 1 *Theor. Math. Phys.* **40** 688 [*Teor. Mat. Fiz.* **40** 194]
[2] Takhtajan L A and Faddeev L D 1979 The Quantum method of the inverse problem and the Heisenberg XYZ model *Russ. Math. Surveys* **34** 11; [*Usp. Mat. Nauk* **34** 13]
[3] Kulish P P Reshetikhin N Y and Sklyanin E K 1981 Yang-Baxter Equation and Representation Theory 1 *Lett. Math. Phys.* **5** 393.
[4] Sklyanin E K 1992 *Quantum Inverse Scattering Method* in *Quantum Groups and Quantum Integrable Systems* (Nankai lectures) editor Mo-Lin Ge pp. 63-97 World Scientific Publ Singapore [hep-th/9211111]
[5] Faddeev L D 1998 How algebraic Bethe ansatz works for integrable model In: Quantum Symmetries/Symetries Quantiques Proc.Les-Houches summer school LXIV Eds. Connes A Kawedzki K and Zinn-Justin J North-Holland 149-211 hep-th/9605187
[6] Drinfeld V G 1985 Hopf algebras and the quantum Yang-Baxter equation *Sov. Math. Dokl.* **32** 254 [*Dokl. Akad. Nauk Ser. Fiz* **283** 1060]; A New realization of Yangians and quantized affine algebras *Sov. Math. Dokl.* **36** (1988) 212
[7] Molev A Nazarov M and Olshansky G 1996 Yangians and classical Lie algebras *Russ. Math. Surveys* **51** 205 [hep-th/9409025]
[8] Drummond J M Henn J Korchemsky G P and Sokatchev E 2010 Dual superconformal symmetry of scattering amplitudes in N=4 super-Yang-Mills theory *Nucl. Phys. B* **828** 317 [arXiv:0807.1095 [hep-th]]
[9] Drummond J M Henn J and Plefka J 2009 Yangian symmetry of scattering amplitudes in N=4 super Yang-Mills theory *JHEP* **0905** 046 [arXiv:0902.2987 [hep-th]]
[10] Arkani-Hamed N Cachazo F Cheung C and Kaplan J 2010 A Duality For The S Matrix *JHEP* **1003** 020 [arXiv:0907.5418 [hep-th]]
[11] Arkani-Hamed N Cachazo F and Cheung C 2010 The Grassmannian Origin Of Dual Superconformal Invariance *JHEP* **1003** 036 [arXiv:0909.0483 [hep-th]]
[12] Ferro L Lukowski T Meneghelli C Plefka J and Staudacher M 2013 Harmonic R-matrices for Scattering Amplitudes and Spectral Regularization *Phys. Rev. Lett.* **110** 121602 [arXiv:1212.0850 [hep-th]]
[13] Ferro L Lukowski T and Staudacher M 2014 N=4 Scattering Amplitudes and the Deformed Grassmannian arXiv:1407.6736 [hep-th]
[14] Derkachov S Karakhanyan D and Kirschner R 2007 Yang-Baxter R operators and parameter permutations *Nucl. Phys. B* **785** 263 [hep-th/0703076]
[15] Karakhanyan D and Kirschner R 2009 Jordan-Schwinger representations and factorised Yang-Baxter operators *SIGMA* **6** 029 [arXiv:0910.5144 [hep-th]]
[16] Chicherin D and Kirschner R 2013 Yangian symmetric correlators *Nucl. Phys. B* **877** 484 [arXiv:1306.0711 [math-ph]]
[17] Chicherin D Derkachov S and Kirschner R 2014 Yang-Baxter operators and scattering amplitudes in N=4 super-Yang-Mills theory *Nucl. Phys. B* **881** 467 [arXiv:1309.5748 [hep-th]]

Quasicrystallography from B_n lattices

M Koca[1], N O Koca[2], A Al-Mukhaini[3] and A Al-Qanabi[4]

[1,2,3,4] Department of Physics, College of Science, Sultan Qaboos University
P.O. Box 36, Al-Khoud, 123 Muscat, Sultanate of Oman

E-mail: [1] kocam@squ.edu.om
[2] nazife@squ.edu.om
[3] m008439@squ.edu.om
[4] m021300@squ.edu.om

Abstract. We present a group theoretical analysis of the hypercubic lattice described by the affine Coxeter-Weyl group $W_a(B_n)$. An *h*-fold symmetric quasicrystal structure follows from the hyperqubic lattice whose point group is described by the Coxeter-Weyl group $W(B_n)$ with the Coxeter number *h=2n*. Higher dimensional cubic lattices are explicitly constructed for $n = 4,5,6$ by identifying their rank-3 Coxeter subgroups and maximal dihedral subgroups. Decomposition of their Voronoi cells under the respective rank-3 subgroups $W(A_3)$, $W(H_2)\times W(A_1)$ and $W(H_3)$ lead to the rhombic dodecahedron, rhombic icosahedron and rhombic triacontahedron respectively. Projection of the lattice B_4 describes a quasicrystal structure with 8-fold symmetry. The B_5 lattice leads to quasicrystals with both 5-fold and 10 fold symmetries. The lattice B_6 projects on a 12-fold symmetric quasicrystal as well as a 3D icosahedral quasicrystal depending on the choice of subspace of projections. The projected sets of lattice points are compatible with the available experimental data.

1. Introduction

The first discovery of the icosahedral quasicrystal by D. Shechtman [1] attracted a lot of interest on the quasicrystallography. The 3D quasicrystals can be described by the Coxeter group $W(H_3)$ representing the icosahedral symmetry of order 120. The 2D quasicrystals exhibit 5-fold, 8-fold, 10-fold, 12-fold, and 18-fold symmetries [2, 3, 4, 5, 6]. Lubin and his collaborators [7] reported recently that a quasicrystallographic structure with 36-fold symmetry is possible. It seems there is no limitation on the order of the planar point symmetry of the quasicrystallography which can be described by the dihedral groups. This reminds us the classification of the Coxeter-Weyl groups with different Coxeter numbers h [8, 9, 10, 11]. Every Coxeter-Weyl group has a dihedral subgroup D_h of order $2h$. We illustrate relations between the affine Coxeter groups $W_a(B_n)$ and the h-fold symmetric quasicrystallography obtained from higher dimensional cubic lattices by orthogonal projections. A general projection technique of the higher dimensional cubic lattice has been proposed by Duneau and Katz [12] but without any discussion on the group theoretical aspects of the lattices.

The five-fold symmetric 2D quasicrystals as well as quasicrystals with icosahedral symmetry have been studied in the references [13, 14, 15, 16]. More elaborate discussions on the quasicrystal structures can be found in the set theoretic approach initiated by Yves Meyer [17,18] and later developed as *Model Set* by Robert V. Moody [19,20].

The Lie group applications derived from the root systems of the Coxeter-Weyl groups are well known by the high energy physicists. The standard model of the High Energy Physics described by the Lie group $SU(3)\times SU(2)\times U(1)$ [21, 22, 23] is based on the Coxeter-Weyl group $W(A_2)\times W(A_1)$. The Grand Unified theories, $SU(5)\approx E_4$ [24], $SO(10)\approx E_5$ [25] and the exceptional group E_6 [26] depend on the respective Coxeter-Weyl groups $W(A_4)$, $W(D_5)$, and $W(E_6)$. One may assume that some of the Coxeter-Weyl groups may play an important role in the study of h-fold symmetric quasicrystallography. The argument goes as follows.

Every Coxeter group with a Coxeter number h has a maximal dihedral subgroup D_h of order $2h$ which acts in the Coxeter plane. In two recent papers [27, 28] it was proposed that a quasicrystallographic structure with h-fold symmetry can be determined by projections of the higher dimensional lattices onto the Coxeter planes. The Coxeter groups are characterized by their Coxeter exponents [8, 11]. This paper will study the general structure of the root lattice of the affine Coxeter group $W_a(B_n)$ which is the simple cubic lattice in nD Euclidean space with the point symmetry determined by the Coxeter-Weyl group $W(B_n)$ of order $2^n n!$. The projections of the 5D cubic lattice onto a plane and the 6D cubic lattice into a 3D subspace have been studied earlier without using Coxeter group techniques [12, 29]. Projections of some 4D root lattices have been also studied earlier [30, 31, 32].

The paper is organized as follows. Section 2 deals with the general structure of the Coxeter-Weyl group $W(B_n)$ by identifying its maximal subgroups which could be useful for the projections of the hypercubic lattices. In Section 3 we study the rank- 3 subgroups of the Coxeter-Weyl groups $W(B_4)$, $W(B_5)$, and $W(B_6)$ and projections of some of their polytopes into 3D Euclidean spaces. The projection of the Voronoi cell of the lattice $W(B_n)$ plays a crucial role in the description of the quasicrystallographic structures in 3D and 2D. Section 4 is devoted to the projection techniques of the lattices onto the Coxeter planes and the projection of the 6D cubic lattice into 3D subspace with icosahedral symmetry. Our predictions are confronted with experimental data in Section 5 and some conclusive remarks are added.

2. The Coxeter-Weyl group $W(B_n)$ and its Maximal Subgroups

The Coxeter-Weyl groups are well known [8, 9,10,11]. The classification includes an infinite series of crystallographic groups A_n, B_n, C_n, D_n, and a finite number of crystallographic exceptional groups G_2, F_4, E_6, E_7, E_8. Moreover, there are an infinite number of noncrystallographic dihedral Coxeter groups $I_2(h)$ $(h\geq 5,\ h\neq 6)$ including two rank-3 and rank-4 noncrystallographic Coxeter groups $W(H_3)$ and $W(H_4)$ respectively. In this section we study the Coxeter-Weyl group $W(B_n)$ of rank n. It is represented by the Coxeter-Dynkin diagram shown in figure 1.

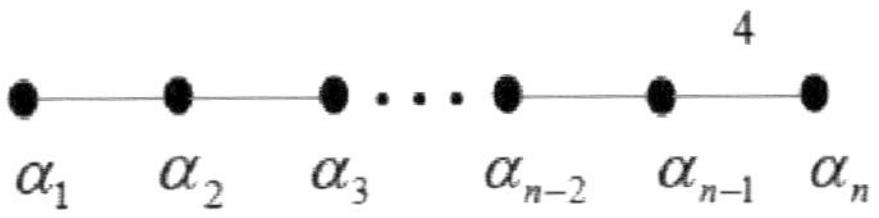

Figure 1. Coxeter-Dynkin diagram of the Coxeter-Weyl group $W(B_n)$.

From left to right the nodes represent the long simple roots α_i $(i=1,2,...,n-1)$ with norm $\sqrt{2}$ and the last simple root α_n is the short root with norm 1. Angle between any two adjacent simple roots with norm $\sqrt{2}$ is 120^0 and the angle between the last two simple roots is 135^0. Any two disconnected roots are orthogonal to each other. The nodes also denote the reflection generators r_i whose action on an arbitrary vector Λ in the n-dimensional Euclidean space is given by

$$r_i\Lambda = \Lambda - \frac{2(\Lambda,\alpha_i)}{(\alpha_i,\alpha_i)}\alpha_i \ . \tag{1}$$

The dual space of the root space is represented by the weight vectors ω_i defined by $(\omega_i, \frac{2\alpha_j}{(\alpha_j,\alpha_j)}) = \delta_{ij}$. The Cartan matrix of the root space (Gram matrix) and the metric tensor in the dual space are defined respectively by the matrix elements

$$A_{ij} = \frac{2(\alpha_i,\alpha_j)}{(\alpha_j,\alpha_j)}, \ G_{ij} = (\omega_i,\omega_j) = (A^{-1})_{ij}\frac{(\alpha_j,\alpha_j)}{2} \ . \tag{2}$$

Let $l_i (i = 1,2,...,n)$ denote the set of orthonormal vectors $(l_i, l_j) = \delta_{ij}$ in n- dimensional Euclidean space. The simple roots of $W(B_n)$ can be written as

$$\alpha_1 = l_1 - l_2, \alpha_2 = l_2 - l_3, ..., \alpha_{n-1} = l_{n-1} - l_n, \alpha_n = l_n \ . \tag{3}$$

The root system consists of $2n$ short roots $\pm l_i$ and $2n(n-1)$ long roots $\pm l_i \pm l_j (i \neq j)$. Reflection generators r_i act on the unit vectors as follows

$$r_1 : l_1 \leftrightarrow l_2, r_2 : l_2 \leftrightarrow l_3, ..., r_{n-1} : l_{n-1} \leftrightarrow l_n, r_n : l_n \rightarrow -l_n. \tag{4}$$

It is implicit that the reflection generators leave the other unit vectors (not shown) invariant. The weight vectors can be determined from $\omega_i = (A^{-1})_{ij}\alpha_j$ as

$$\omega_1 = l_1, \omega_2 = l_1 + l_2, ..., \omega_{n-1} = l_1 + l_2 + \cdots + l_{n-1}, \omega_n = \frac{1}{2}(l_1 + l_2 + \cdots + l_n). \tag{5}$$

The highest weight vector [33] in the weight space can be defined as $\Lambda = a_1\omega_1 + a_2\omega_2 + \cdots + a_n\omega_n \equiv (a_1, a_2, ..., a_n)$ with integer coefficients $a_i \geq 0$. An orbit of Λ under the group $W(B_n)$ will be denoted by $W(B_n)\Lambda \equiv (a_1a_2..a_n)_{B_n}$. With this notation, e.g., the orbits $(100...0)_{B_n} = \pm l_i$ and $(010...0)_{B_n} = \pm l_i \pm l_j, (i \neq j)$ represent the sets of short roots and long roots respectively. The orbit $(00...01)_{B_n} = \frac{1}{2}(\pm l_1 \pm l_2 \pm \cdots \pm l_n)$ represents the vertices of a cube in n-dimensional Euclidean space. We point out that the group $W(B_n)$ represents also the group obtained from the Coxeter-Dynkin diagram C_n where the short and long roots are represented by the roots $\alpha_1 = l_1 - l_2, \alpha_2 = l_2 - l_3, ..., \alpha_{n-1} = l_{n-1} - l_n, \alpha_n = 2l_n$. Certain maximal subgroups of $W(B_n)$ can be useful in the study of quasicrystals. One of the maximal subgroup of the group $W(B_n)$ is the Coxeter-Weyl group $W(D_n)$ with the Coxeter-Dynkin diagram given in figure 2. It consists of only long roots of norm $\sqrt{2}$.

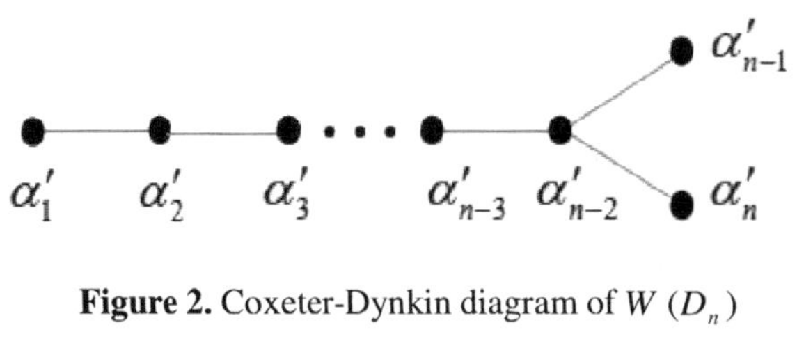

Figure 2. Coxeter-Dynkin diagram of $W(D_n)$

Denote by r'_i the reflection generators of $W(D_n)$ and the simple roots by

$$\alpha_1' = l_1 - l_2, \alpha_2' = l_2 - l_3, ..., \alpha_{n-1}' = l_{n-1} - l_n, \alpha_n' = l_{n-1} + l_n. \tag{6}$$

The first n-1 simple roots are identical to the simple roots of the group $W(B_n)$. The generators of $W(D_n)$ transform the unit vectors as in (4) but the last generator differs in its action

$$r_1' : l_1 \leftrightarrow l_2, r_2' : l_2 \leftrightarrow l_3, ..., r_{n-1}' : l_{n-1} \leftrightarrow l_n, r_n' : l_{n-1} \leftrightarrow -l_n, \tag{7}$$

Then one can identify $r_i' = r_i$, for $(i = 1,2,...,n-1)$ and $r_n' = r_n r_{n-1} r_n$. The Coxeter-Dynkin diagram of $W(D_n)$ has a diagram symmetry $\gamma : \alpha_{n-1}' \leftrightarrow \alpha_n'$ which transforms $\gamma : l_n \rightarrow -l_n$ leaving the other unit vectors invariant and can be identified with r_n. It follows that the automorphism group of D_n is $Aut(D_n) \approx W(D_n) : \mathbf{Z}_2 \approx W(B_n)$ where the group $\mathbf{Z}_2$ is generated by γ. There is one exception to this general result: the automorphism group of D_4 is larger than $W(B_4)$ since its Dynkin diagram symmetry is isomorphic to the symmetric group S_3 of order 6. The automorphism group is then the group $Aut(D_4) \approx W(D_4) : S_3 \approx W(F_4)$. This is known as the triality of the D_4 symmetry. We note that the orbit $(00...01)_{B_n}$ is the union of two orbits of $W(D_n)$,

$$(00...1)_{B_n} = (00...10)_{D_n} \cup (00...01)_{D_n}. \tag{8}$$

The group $W(B_n)$ admits the group $W(B_{n-1})$ and $W(A_{n-1})$ as maximal subgroups. An interesting maximal subgroup of $W(B_n)$ is the dihedral group $W(I_2(h)) \approx D_h$ of order $2h$ which is relevant for the quasicrystallography. The simple roots of $W(B_n)$ decompose in such a way that the corresponding reflection generators $r_1, r_3, ..., r_{n-1}$ commute pair wise as well as do the set $r_2, r_4, ..., r_n$ for even n. A similar decomposition can be done for odd n. Define the generators R_1 and R_2 by $R_1 = r_1 r_3 ... r_{n-1}$ and $R_2 = r_2 r_4 ... r_n$ [34]. The generators R_1 and R_2 act as reflections on the simple roots of the Coxeter diagram $I_2(h)$ and the Coxeter element $R_1 R_2$ represents a rotation of order h in the plane spanned by the simple roots [35] of the diagram $I_2(h)$. The Coxeter exponents $m_i \frac{\pi}{h} (i = 1,2,...,n)$ with $m_i = 1,3,5,...,2n-1$ and the Coxeter number $h = 2n$ are useful to determine the principal planes in which one may have a quasicrystallographic point symmetry of the projected set of points. The eigenvalues of the Cartan matrix in (2) can be written simply $\lambda_i = 2[1 - \cos(m_i \frac{\pi}{h})]$ [8]. The principal planes are determined by the eigenvectors corresponding to the pair of eigenvalues $2[1 - \cos(m_i \frac{\pi}{h})]$, $2[1 - \cos((h - m_i)\frac{\pi}{h})]$. One can choose a convenient set of orthogonal unit vectors obtained from the eigenvectors of the Cartan matrix [27, 28, 36]

$$\hat{x}_i = \frac{1}{\sqrt{h\lambda_i}} \sum_j \frac{2\alpha_j}{(\alpha_j, \alpha_j)} X_{ji} \tag{9}$$

where $\vec{X}_i$ is the eigenvector of the Cartan matrix given in (2) corresponding to the eigenvalue λ_i with a normalization where the last components of the eigenvectors all equal 1.

The root lattice is the simple cubic lattice which is invariant under the affine Coxeter group $W_a(B_n)$ that can be generated by adding a generator r_0 to the set of generators of the group $W(B_n)$. The generator r_0 represents a reflection with respect to the hyperplane bisecting the highest long root $\tilde{\alpha} = \omega_2 = l_1 + l_2$. Its action on an arbitrary vector is a translation.

A general vector of the root lattice then will be given by $p = b_1\alpha_1 + b_2\alpha_2 + \cdots + b_n\alpha_n, b_i \in \mathbf{Z}$. This indicates that a general vector of the lattice is the linear combinations of the unit vectors l_i with integer coefficients. The primitive cell of the lattice can be chosen as the cube with the vertices

$$0, l_i, l_i + l_j (i < j), l_i + l_j + l_k (i < j < k), ..., l_i + l_j + \cdots + l_k (i < j < ... < k). \quad (10)$$

There are 2^n such cubes sharing the origin as a vertex. The Voronoi cells around the lattice points are congruent polytopes tiling the n-dimensional Euclidean space. The Voronoi cells of lattices are important in the theory of coding [37]. Denote by $V(0)$ the Voronoi cell around the origin. The vertices of the Voronoi polytope $V(0)$ can be determined as the intersection of the hyperplanes surrounding the origin. They are the hyperplanes determined as the orbits of the fundamental weights

$$\frac{\omega_1}{2} = \frac{l_1}{2}, \frac{\omega_2}{2} = \frac{l_1 + l_2}{2}, \dots, \frac{\omega_{n-1}}{2} = \frac{l_1 + l_2 + \cdots l_{n-1}}{2}, \omega_n = \frac{1}{2}(l_1 + l_2 + \cdots + l_n). \quad (11)$$

The Voronoi polytope $V(0)$ is then a cube around the origin with the vertices [38, 39]

$$(00...01)_{B_n} = \frac{1}{2}(\pm l_1 \pm l_2 \cdots \pm l_n). \quad (12)$$

We emphasize here that the lattices generated by the affine Coxeter-Weyl group $W_a(C_n)$ are identical to the root and weight lattices of the Coxeter-Weyl group $W_a(D_n)$ [37]. The group $Aut(D_n) \approx W(C_n)$ should be taken into account when the point symmetry of the two lattices is of concern. Therefore in an n-dimensional Euclidean space with $n > 3$ one can construct five different lattices with affine Coxeter-Weyl groups, two for $W_a(A_n)$, two for $W_a(D_n)$, and the other is the simple cubic lattice described by $W_a(B_n)$. When n coincides with the rank of the exceptional groups the number of lattice will be more than 5. The projection of the lattice onto the principal planes is nothing other than determination of the components of lattice vectors in the principal planes defined by the pairs of unit vectors $(\hat{x}_1, \hat{x}_n)$, $(\hat{x}_2, \hat{x}_{n-1})$,.... Here the plane determined by the vectors $(\hat{x}_1, \hat{x}_n)$ is known as the Coxeter plane. Note that for odd n one of the unit vectors is unpaired that represents the direction orthogonal to all principal planes. The representations of the generators R_1 and R_2 of the dihedral subgroup can be put into block-diagonal matrices with 2×2 and/or 1×1 matrix entries. The component of the simple cubic lattice vector in the basis of $\hat{x}_i$ is given by

$$p_i = \frac{1}{\sqrt{h\lambda_i}}(\sum_{j=1}^{n-1} a_j X_{ji} + 2a_n), \; a_i \in \mathbf{Z}. \quad (13)$$

This introduction on the hypercubic lattice will be useful in the following chapters where we study the 8-fold, 5-fold, 10-fold, and 12-fold symmetric quasicrystal structures induced by the projections of the lattices B_4, B_5, and B_6.

3. The Coxeter-Weyl groups $W(B_4)$, $W(B_5)$, $W(B_6)$ and projections of their polytopes into 3D subspaces

The Voronoi cells of the root lattices of the Coxeter-Weyl groups $W(B_4)$, $W(B_5)$, and $W(B_6)$ are projected on the rhombic dodecahedron, rhombic icosahedron, and rhombic triacontahedron respectively. A similar work, not in the context of the affine Coxeter groups, has been studied earlier by [40].

3.1 The Coxeter-Weyl group $W(B_4)$ and projection of its fundamental polytopes into a 3D subspace with octahedral symmetry

The vertices of the fundamental polytopes of the group $W(B_4)$ are given by

$$\begin{aligned}
(1000)_{B_4} &= \pm l_i, \\
(0100)_{B_4} &= \pm l_i \pm l_j \, (i \neq j), \\
(0010)_{B_4} &= \pm l_i \pm l_j \pm l_k \, (i \neq j \neq k), \\
(0001)_{B_4} &= \frac{1}{2}(\pm l_1 \pm l_2 \pm l_3 \pm l_4), \; i,j,k = 1,2,3,4.
\end{aligned} \tag{14}$$

The first one represents the short roots of $W(B_4)$ which constitutes a 4D octahedron with 8 vertices whose facets (3-faces) are tetrahedra. The second polytope consists of 24 long roots as vertices and is known as the 24-cell with octahedral facets. Its full symmetry is the Coxeter-Weyl group $W(F_4)$ which embeds $W(B_4)$ as a subgroup with index 3. The third polytope with 32 vertices has two types of facets: tetrahedra and truncated octahedra. The last one is the 4D cube with 16 vertices consisting of the cubic facets.

The subspace we are interested in here is a 3D Euclidean space with the octahedral symmetry represented by the Coxeter-Weyl group $W(B_3)$ of order 48. Let us recall the isomorphism $W(B_3) \approx Aut(D_3) \approx Aut(A_3) = W(A_3) : \mathbf{Z}_2$. Here one can choose the tetrahedral symmetry as the Coxeter-Weyl group $W(D_3) \approx W(A_3) = \langle r_1, r_2, r_3 \rangle$. The Dynkin diagram symmetry $\mathbf{Z}_2$ generated by γ which permutes the unit vectors as $l_1 \leftrightarrow l_3$ and $l_2 \leftrightarrow l_4$ extends the group to the octahedral symmetry. It is clear that the vector $\pm\frac{1}{2}(l_1 + l_2 + l_3 + l_4)$ is invariant under the octahedral subgroup. It is convenient to introduce a new set of orthonormal vectors defined by

$$\begin{aligned}
t_0 &= \frac{1}{2}(l_1 + l_2 + l_3 + l_4), \; t_1 = \frac{1}{2}(l_1 - l_2 + l_3 - l_4), \\
t_2 &= \frac{1}{2}(-l_1 + l_2 + l_3 - l_4), \; t_3 = \frac{1}{2}(l_1 + l_2 - l_3 - l_4).
\end{aligned} \tag{15}$$

When the vectors in (14) are expressed in terms of the vectors t_0, t_1, t_2, and t_3 it will be simpler to identify the 3D vertices of the 4D polytopes. The 4D octahedron projects onto a cube in 3D with the vertices $\frac{1}{2}(\pm t_1 \pm t_2 \pm t_3)$. The 24-cell projected into the 3D space represents two octahedra as well as one cuboctahedron. The polytope with 32 vertices is projected into one cube and two truncated tetrahedra.

The polytope $(0001)_{B_4}$ is more interesting since it constitutes the Voronoi cell of the 4D cubic lattice. When expressed in terms of the unit vectors t_0, t_1, t_2, and t_3 and the component of an arbitrary vector along t_0 is deleted they will decompose under the octahedral group as two orbits, one with the vertices $(\pm t_1, \pm t_2, \pm t_3)$ representing the vertices of an octahedron and the other with vertices $\frac{1}{2}(\pm t_1 \pm t_2 \pm t_3)$ representing a cube. Union of these two orbits represents a rhombic dodecahedron with 14 vertices as shown in figure 3.

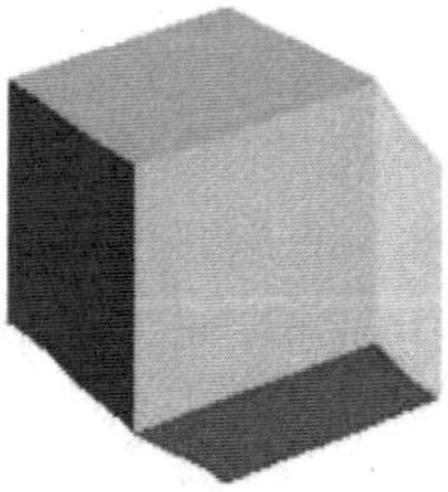

Figure 3. The rhombic dodecahedron.

3.2 Projection of fundamental polytopes of the group $W(B_5)$ into 3D subspace with $W(H_2)\times C_2 \approx D_{5d}$ symmetry

One of the fundamental orbits of the group $W(B_5)$ is the polytope $(10000)_{B_5} = \pm l_i\ (i=1,2,...,5)$ which represents an octahedron in 5D Euclidean space whose facets are 5-cells (4-simplexes). Its projection into a 3D space represents a pentagonal antiprism. To understand this better we have to define a new set of orthonormal vectors $\hat{x}_i\ (i=1,2,...,5)$. The first four unit vectors $\hat{x}_i\ (i=1,2,...,4)$ is obtained by using the eigenvectors of the Cartan matrix of $W(A_4)$ and the fifth vector is chosen to be orthogonal to the rest [36]

$$\begin{aligned}
\hat{x}_1 &= \frac{1}{\sqrt{2(2+\sigma)}}(\alpha_1 + \tau\alpha_2 + \tau\alpha_3 + \alpha_4), \\
\hat{x}_2 &= \frac{1}{(2+\sigma)\sqrt{2}}(\alpha_1 - \sigma\alpha_2 + \sigma\alpha_3 - \alpha_4), \\
\hat{x}_3 &= \frac{1}{\sqrt{2(2+\tau)}}(\alpha_1 + \sigma\alpha_2 + \sigma\alpha_3 + \alpha_4), \\
\hat{x}_4 &= \frac{1}{(2+\tau)\sqrt{2}}(\alpha_1 - \tau\alpha_2 + \tau\alpha_3 - \alpha_4), \\
\hat{x}_5 &= \frac{1}{\sqrt{5}}(\alpha_1 + 2\alpha_2 + 3\alpha_3 + 4\alpha_4 + 5\alpha_5).
\end{aligned} \tag{16}$$

The unit vectors l_i can be expressed as a linear combination $l_i = \sum_{j=i}^{5} b_{ij}\hat{x}_j$ where the matrix B is given by

$$B = \frac{1}{\sqrt{10}}\begin{pmatrix} a & \tau & b & -\sigma & \sqrt{2} \\ b & \sigma & -a & -\tau & \sqrt{2} \\ 0 & -2 & 0 & 2 & \sqrt{2} \\ -b & \sigma & a & -\tau & \sqrt{2} \\ -a & \tau & -b & -\sigma & \sqrt{2} \end{pmatrix}, \tag{17}$$

$$a = \sqrt{2+\tau},\ b = \sqrt{2+\sigma},\ \tau = \frac{1+\sqrt{5}}{2},\ \sigma = \frac{1-\sqrt{5}}{2}.$$

Here we only discuss the projections of two polytopes $(10000)_{B_5} = \pm l_i \,(i = 1,2,...,5)$, and $(00001)_{B_5} = \frac{1}{2}(\pm l_1 \pm l_2 \pm l_3 \pm l_4 \pm l_5)$ into the 3D space described by the symmetry group $W(H_2)\times C_2 \approx D_{5d}$. The group D_{5d} is generated by the group elements $R_1 = r_1 r_3$, $R_2 = r_2 r_4$ and $R_3 = (r_1 r_2 r_3 r_4 r_5)^5$. Here R_1 and R_2 generate the dihedral group $W(H_2)\approx D_5$ of order 10 and $R_3 = -I$ where I is the 5×5 unit matrix in the l_i basis. The center of the group $W(B_5)$ is represented by the elements $C_2 = \{I, -I\}$ and therefore it commutes with all the elements of the group. Let us choose the 3D space defined by the set of unit vectors $(\hat{x}_1, \hat{x}_4, \hat{x}_5)$. The set of vectors $\pm l_i$ form a single orbit under the group D_{5d}. The polytope $(10000)_{B_5} = \pm l_i$ projected into 3D space represents a pentagonal antiprism as shown in figure 4.

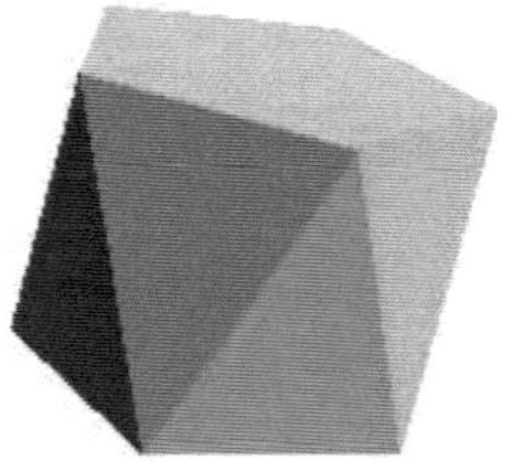

Figure 4. The pentagonal Antiprism.

The polytope $(00001)_{B_5} = \frac{1}{2}(\pm l_1 \pm l_2 \pm l_3 \pm l_4 \pm l_5)$ represents the Voronoi cell of the 5D cubic lattice. The 32 vertices decompose under the group D_{5d} as $32 = 2 + 10 + 20$. The first orbit of size 2 represents the vectors $\pm\frac{1}{2}(l_1 + l_2 + l_3 + l_4 + l_5)$. Each vector is invariant under the dihedral group D_5 but changed to each other under the elements of the center C_2. The next orbit of size 10 consists of the vectors like $\pm\frac{1}{2}(-l + l_2 + l_3 + l_4 + l_5)$ with one or four negative signs. They also constitute a pentagonal antiprism like the one in figure 4. The orbit of size 20 consists of the vectors with two negative and/or three negative signs. The union of two orbits 2+20 constitutes a rhombic icosahedron as shown in figure 5.

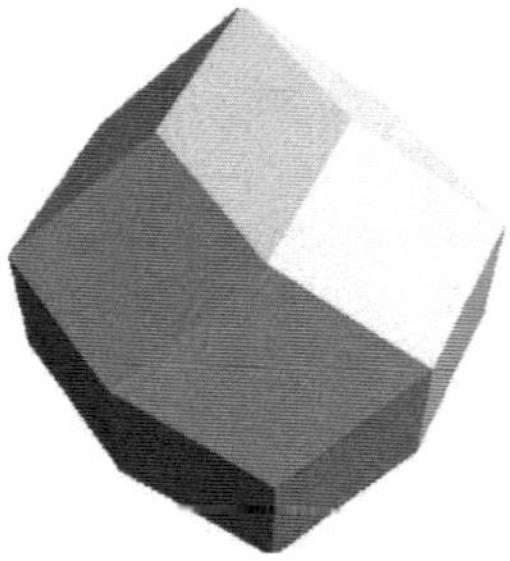

Figure 5. The rhombic icosahedron with the symmetry D_{5d}.

3.3 Projections of the $W(B_6)$ polytopes into 3D subspace with $W(H_3)\approx I_h$ symmetry

The icosahedral symmetry is the one which describes some quasicrystal structures in 3D. Here we first discuss how the icosahedral symmetry, the Coxeter group $W(H_3) \approx I_h$, can be obtained as a subgroup of the Coxeter-Weyl group $W(B_6)$. We have seen in Section 2 that one of the maximal subgroup of the group $W(B_n)$ is the group $W(D_n)$. We introduce the generators $R_1 = r_1'r_5'$, $R_2 = r_2'r_4'$, and $R_3 = r_3'r_6'$. They generate the Coxeter group $W(H_3)$ [41, 42, 43] which in turn can be written as $R_1 = r_1r_5$, $R_2 = r_2r_4$, and $R_3 = r_3r_6r_5r_6$. They satisfy the relations

$$R_1^2 = R_2^2 = R_3^2 = (R_1R_3)^2 = (R_1R_2)^3 = (R_2R_3)^5 = 1 \tag{18}$$

leading to the usual generation relations of the icosahedral group $W(H_3) \approx A_5 \times C_2 = I_h$. The generators of the icosahedral group are 6×6 matrices in the space of orthonormal vectors l_i and constitute a reducible representation of the icosahedral group. They can be transformed into block-diagonal forms of 3×3 matrices which act on two sets of orthonormal vectors $(\hat{x}_1, \hat{x}_2, \hat{x}_3)$ and $(\hat{x}_1', \hat{x}_2', \hat{x}_3')$. Each block represents a different 3×3 irreducible matrix representation.

The block-diagonal form of the generators of the Coxeter group $W(H_3)$ induces the relation

$$\begin{pmatrix} l_1 \\ l_2 \\ l_3 \\ l_4 \\ l_5 \\ l_6 \end{pmatrix} = \frac{1}{\sqrt{2(2+\tau)}} \begin{pmatrix} -1 & -\tau & 0 & -\tau & 1 & 0 \\ 1 & -\tau & 0 & \tau & 1 & 0 \\ 0 & -1 & -\tau & 0 & -\tau & 1 \\ 0 & -1 & \tau & 0 & -\tau & -1 \\ -\tau & 0 & -1 & 1 & 0 & -\tau \\ \tau & 0 & -1 & -1 & 0 & -\tau \end{pmatrix} \begin{pmatrix} \hat{x}_1 \\ \hat{x}_2 \\ \hat{x}_3 \\ \hat{x}_1' \\ \hat{x}_2' \\ \hat{x}_3' \end{pmatrix}. \tag{19}$$

We also note that one can also construct the generators of the Coxeter group $W(H_3)$ as $R_1 = r_1r_6r_5r_6$, $R_2 = r_2r_4$, and $R_3 = r_3r_5$ which is obtained from the Dynkin diagram symmetry of D_6. Therefore two icosahedral groups are conjugate to each other in the group $Aut(D_6) \approx W(B_6)$.

We discuss now the projections of certain $W(B_6)$ polytopes into 3D subspace with a residual icosahedral symmetry. The polytope with 12 vertices $(100000)_{B_6} = \pm l_i$ representing the short roots of the group $W(B_6)$ is projected into an icosahedron represented by the vectors of norm $\frac{1}{\sqrt{2}} \approx 0.707$ as shown in figure 6.

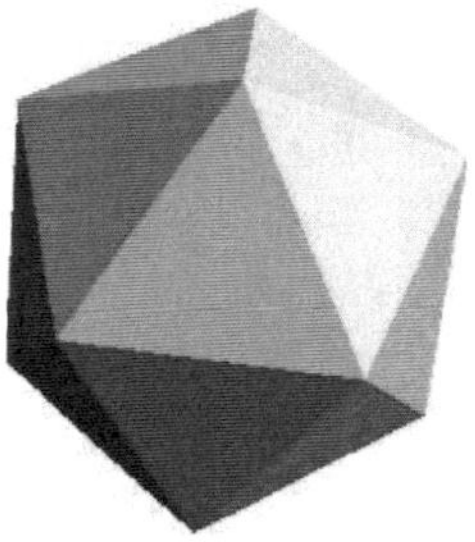

Figure 6. An icosahedron projected from the 6D octahedron $(100000)_{B_6}$.

The long roots of the group $W(B_6)$ are the vertices of the polytope $(010000)_{B_6}$ which are given by the 60 vectors $(010000)_{B_6} = \pm l_i \pm l_j (i \neq j)$. When they are projected into 3D space they represent two copies of icosidodecahedra with 30 vertices each; one copy is expanded with respect to the other by a factor of τ. Their norms in 3D are $\sqrt{\frac{2}{2+\tau}} \approx 0.743$ and $\sqrt{\frac{2}{2+\sigma}} \approx 1.203$. The icosidodecahedron is a polyhedron with 30 vertices, 32 faces (20 triangles+12 pentagons) and 60 edges. One of the icosidodecahedron is depicted in figure 7.

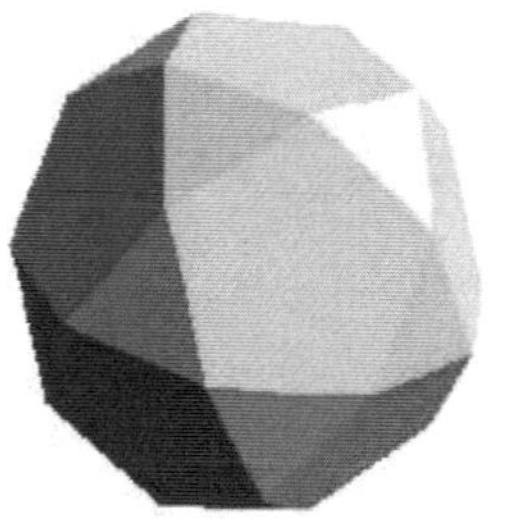

Figure7. Icosidodecahedron projected from the 6D polytope $(010000)_{B_6}$.

The polytope $(000001)_{B_6} = \frac{1}{2}(\pm l_1 \pm l_2 \pm l_3 \pm l_4 \pm l_5 \pm l_6)$ represents the Voronoi cell of the 6D cubic lattice with 64 vertices. They decompose into sets with even $(-)$ sign and odd $(-)$ sign representing two orbits of $W(D_6)$ as mentioned in Section 2. One can demonstrate that each orbit of $W(D_6)$ with 32 vertices decomposes as $32 = 20 + 12$. Normally, the orbits of size 20 and 12 represent a dodecahedron and an icosahedron respectively. Therefore we have two copies of dodecahedron and two copies of classified icosahedron as orbit (I), orbit (II), Orbit (III) , and orbit (IV).However, the situation here is such that the dodecahedron (orbit I) and icosahedron (orbit III) form a rhombic triacontahedron, a Catalan solid dual to the icosidodecahedron [44] as shown in figure 8. The union of the icosahedron (orbit II) and dodecahedron (orbit IV) represents a star dodecahedron which is displayed in figure 9.

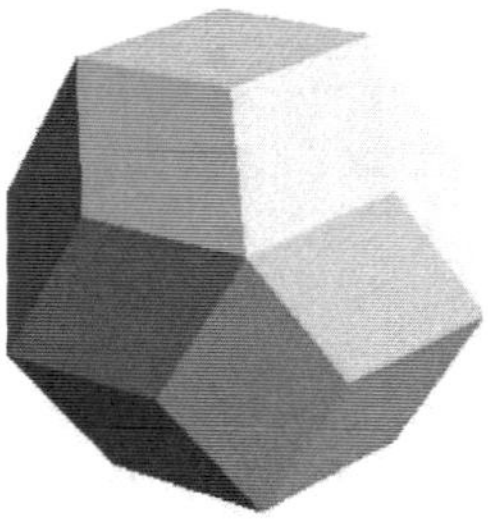

Figure 8. The rhombic triacontahedron obtained as part of the projection of 6D cube.

Figure 9. The dodecahedral star projected from 6D cube.

The vectors representing the vertices of the icosahedra and dodecahedra have the following norms in decreasing order:

$$\begin{aligned} &N(\text{icosahedron III}) = \frac{\tau}{\sqrt{2}} \approx 1.144, \ N(\text{dodecahedron I}) = \sqrt{0.3\,(2+\tau)} \approx 1.042, \\ &N(\text{dodecahedron IV}) = \sqrt{0.3\,(2+\sigma)} \approx 0.644, \ N(\text{icosahedron II}) = \frac{-\sigma}{\sqrt{2}} \approx 0.438. \end{aligned} \tag{20}$$

These are exactly the same results obtained earlier by Conway and Knowles [45]. Note also that

$$\begin{aligned} &N(\text{icosahedron III})/N(\text{icosahedron II}) = \tau^2, \\ &N(\text{dodecahedron I})/N(\text{dodecahedron IV}) = \tau. \end{aligned} \tag{21}$$

Since the icosahedron obtained from the short roots has a norm 0.707 these three icosahedra follow the ratio: $(1:\tau:\tau^2)$. Note also that the vectors $\pm l_i$, when projected into 3D subspace, form rhombohedra when taken in groups of three. For example the sets of vectors (l_1,l_2,l_3) and (l_4,l_5,l_6) form an acute and an obtuse rhombohedra respectively as shown in figure 10.

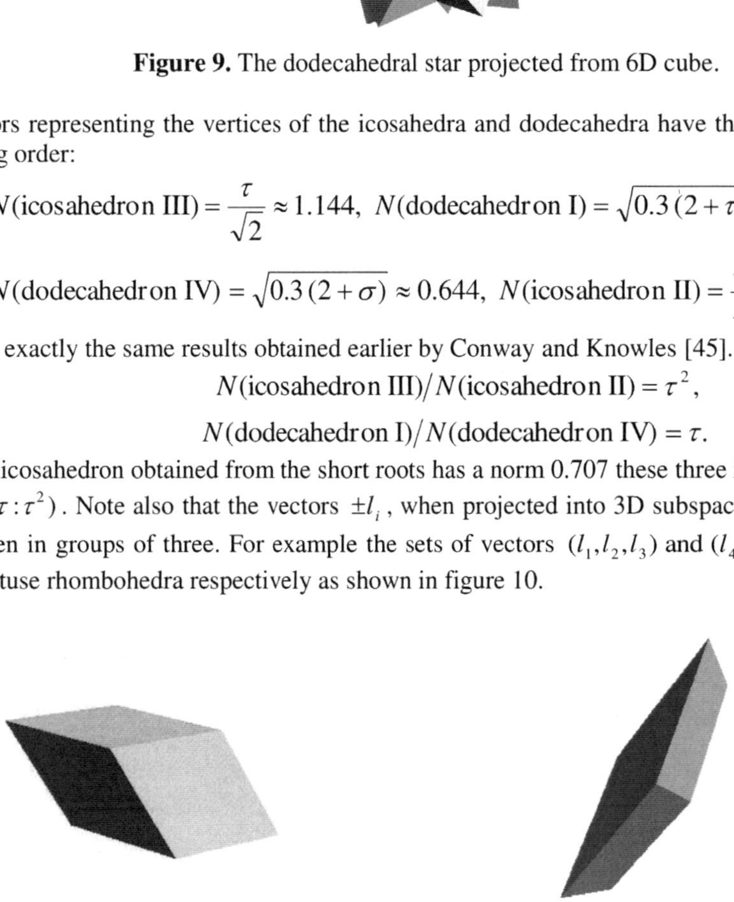

Figure 10. The acute and obtuse rhombohedra generated by the vectors l_i .

4. Projections of the lattices of the Coxeter-Weyl groups $W(B_4)$, $W(B_5)$, and $W(B_6)$ into subspaces

We decompose the nD space into two subspaces where $E_{\parallel}$ represents the subspace into which the lattice points to be projected and the subspace $E_{\perp}$ is the complementary orthogonal subspace. The shift of the Voronoi cell $V(0)$ along the space $E_{\parallel}$ creates an open strip and the projection of the Voronoi polytope into the subspace $E_{\perp}$ determines a region $\boldsymbol{K}$. It has been shown in a number of examples that the set of lattice points projected from the open strip onto the subspace $E_{\parallel}$ determines a quasicrystallographic structure [46, 47, 48]. Now we discuss the projections in the flowing subsections.

4.1 Projection of the lattice points of B_4 into a 2D space

We will show here that the projected set of lattice points display a quasicrystal with 8-fold symmetry. The components of a lattice vector $p = a_1\omega_1 + a_2\omega_2 + a_3\omega_3 + a_4\omega_4, a_i \in \mathbf{Z}$ along the unit vectors $\hat{x}_i$ are given as follows:

$$\begin{aligned} p_1 &= \frac{1}{2\sqrt{2(2-\sqrt{2+\sqrt{2}})}}(\sqrt{2-\sqrt{2}}a_1 + \sqrt{2}a_2 + \sqrt{2+\sqrt{2}}a_3 + 2a_4) \\ p_4 &= \frac{1}{2\sqrt{2(2+\sqrt{2+\sqrt{2}})}}(-\sqrt{2-\sqrt{2}}a_1 + \sqrt{2}a_2 - \sqrt{2+\sqrt{2}}a_3 + 2a_4) \\ p_2 &= \frac{1}{2\sqrt{2(2-\sqrt{2-\sqrt{2}})}}(-\sqrt{2+\sqrt{2}}a_1 - \sqrt{2}a_2 + \sqrt{2-\sqrt{2}}a_3 + 2a_4) \\ p_3 &= \frac{1}{2\sqrt{2(2+\sqrt{2-\sqrt{2}})}}(\sqrt{2+\sqrt{2}}a_1 - \sqrt{2}a_2 - \sqrt{2-\sqrt{2}}a_3 + 2a_4) \end{aligned} \tag{22}$$

Assume now that the subspaces are defined by $E_{II} = (\hat{x}_1, \hat{x}_4)$ and $E_{\perp} = (\hat{x}_2, \hat{x}_3)$. If Voronoi cell $V(0) = (0001)_{B_4}$ is projected onto the plane $E_{\perp} = (\hat{x}_2, \hat{x}_3)$ one obtains a disc of radius $R_0 = Max(p_2^2 + p_3^2)$. When the shifted Voronoi cell $V(\omega_4) = (0001)_{B_4} + \omega_4$ is projected into the subspace $E_{\perp} = (\hat{x}_2, \hat{x}_3)$ the components (p_1, p_4) determines the quasicrystal structure with 8-fold symmetry as shown in figure 11. Similar structures are obtained in a recent paper [49]. The quasicrystal structure with 8-fold symmetry was observed in rapidly solidified $Cr_5Ni_3Si_2$ and $V_{15}Ni_{10}Si$ alloys [50] which can be represented by the quasicrystal structures like in figure 11.

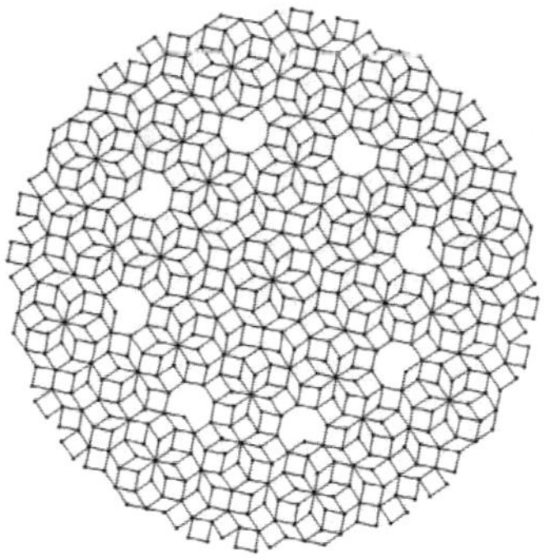

Figure 11. The quasicrystal structure obtained from 4D cubic lattice.

4.2 Projection of the lattice B_5 into a 2D subspace

Projection of the 5D cubic lattice onto a 2D plane with 5-fold symmetry has been discussed earlier [29] with no reference to the Coxeter-Weyl group $W(B_5)$. As we pointed out in Section 2, the Coxeter number of the group $W(B_5)$ is 10 and has a dihedral subgroup D_{10} of order 20. In this section we will study the projections with both symmetries. The 5-fold symmetric projection of the lattice can be obtained by taking the set of orthogonal vectors given by (16). Projection of the Voronoi cell $V(0)$ into the space defined by the unit vectors $(\hat{x}_2, \hat{x}_3, \hat{x}_5)$ determines the window K which constraints the lattice vectors to be projected onto the plane $(\hat{x}_1, \hat{x}_4)$. When the shifted vector $V(\omega_5) = (00001)_{B_5} + \omega_5$

is projected into the subspace $(\hat{x}_2,\hat{x}_3,\hat{x}_5)$ the distribution of the quasicrystallographic lattice structure in the Coxeter plane $(\hat{x}_1,\hat{x}_4)$ is depicted in figure 12. The $\hat{x}_5$ direction in this case preserves the translational invariance.

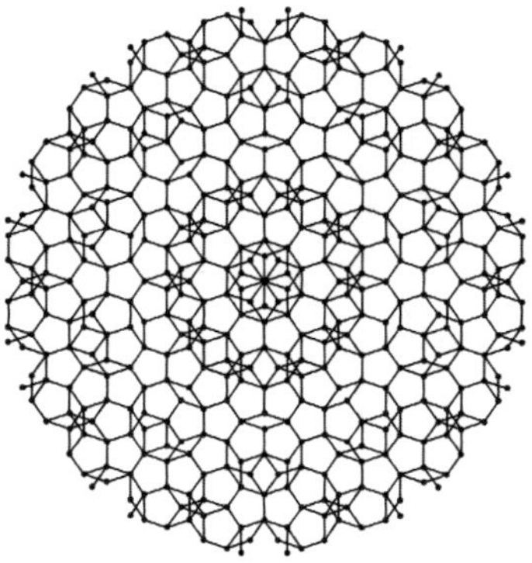

Figure 12. 5-fold symmetric quasicrystal structure from projection of the $W(B_5)$ root lattice.

The 10-fold symmetric quasicrystal structure can be obtained by using the orthonormal vectors in (9). The unit vectors in (9) follow the sequence of the Coxeter exponents $m_1=1, m_2=3, m_3=5, m_4=7, m_5=9$. Now the pairs of unit vectors $(\hat{x}_1,\hat{x}_5)$ and $(\hat{x}_2,\hat{x}_4)$ determine the principal planes with the first pair being the Coxeter plane; the first one can be taken as E_{II} and the second pair as $E_\perp$ and the unit vector $\hat{x}_3$ defines the direction orthogonal to the planes and preserves the translational invariance. The components of the lattice vectors in the planes and the orthogonal direction are given by

$$E_{II}:\left\{\begin{aligned} p_1 &= \frac{1}{\sqrt{10}\sqrt{2-\sqrt{2+\tau}}}(-\sigma a_1+\sqrt{2+\sigma}\,a_2+\tau a_3+\sqrt{2+\tau}\,a_4+2a_5),\\ p_5 &= \frac{1}{\sqrt{10}\sqrt{2+\sqrt{2+\tau}}}(-\sigma a_1-\sqrt{2+\sigma}\,a_2+\tau a_3-\sqrt{2+\tau}\,a_4+2a_5)\end{aligned}\right\},$$

$$E_\perp:\left\{\begin{aligned} p_2 &= \frac{1}{\sqrt{10}\sqrt{2-\sqrt{2+\sigma}}}(-\tau a_1-\sqrt{2+\tau}\,a_2+\sigma a_3+\sqrt{2+\sigma}\,a_4+2a_5),\\ p_4 &= \frac{1}{\sqrt{10}\sqrt{2+\sqrt{2+\sigma}}}(-\tau a_1+\sqrt{2+\tau}\,a_2+\sigma a_3-\sqrt{2+\sigma}\,a_4+2a_5)\\ p_3 &= \frac{1}{\sqrt{5}}(a_1-a_3+a_5)\end{aligned}\right\}. \qquad (23)$$

The quasicrystal structure with 10-fold symmetry from the projection of the root lattice of the Coxeter group $W(B_5)$ is shown in figure 13.

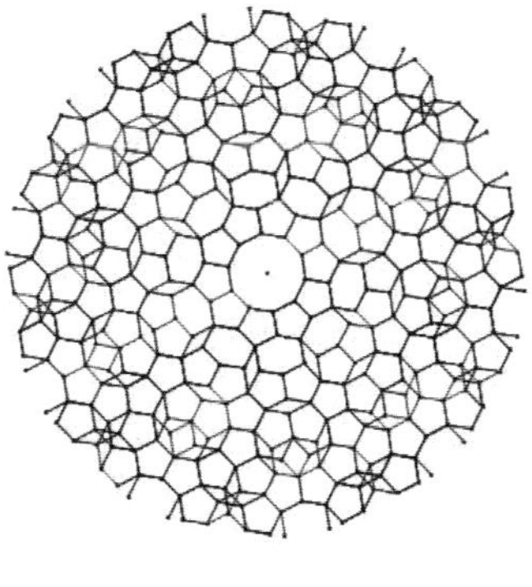

Figure 13. 10-fold symmetric quasicrystal structure from 5D cubic lattice.

4.3 Projection of the lattice of B_6 into 2D and 3D subspaces

The Coxeter number of the group $W(B_6)$ is 12. In two other publications [27, 28] we have studied projection of the 6D cubic lattice using (9) for the group $W(B_6)$ and obtained the following quasicrystal structure in figure 14.

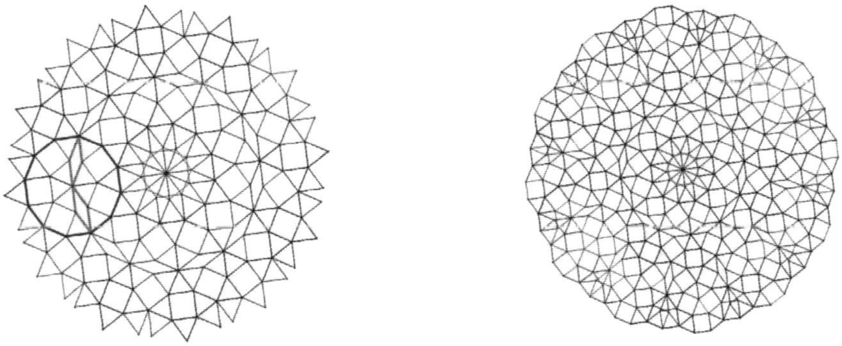

Figure 14. 12-fold symmetric quasicrystal structure from 6D cubic lattice.

The figure 14 shows that the tiling displayed here is different than the usual tiling based on the triangle and square tiles only. It also involves rhombi in addition to square and triangle tiles. Such a structure has been recently observed in the dodecagonal quasicrystal formation of $BaTiO_3$ [51] .

Earlier, projection of the 6D cubic lattice into 3D subspace has been discussed in various papers but with no reference to the detailed structure of the Coxeter group $W(B_6)$ [12, 45, 47]. An explicit construction of the icosahedral subgroup $W(H_3)$ in terms of the generators of the Coxeter group $W(B_6)$ was given in Section 3.3 and the space was decomposed as $E_{\parallel}$ and $E_{\perp}$ defined by the unit vectors $(\hat{x}_1, \hat{x}_2, \hat{x}_3)$ and $(\hat{x}'_1, \hat{x}'_2, \hat{x}'_3)$ respectively. A general root lattice vector can be written as

$$p = \frac{1}{\sqrt{2(2+\tau)}}[(n_1 + n_2\tau)\hat{x}_1 + (n_3 + n_4\tau)\hat{x}_2 + (n_5 + n_6\tau)\hat{x}_3 + \tau(n_1 + n_2\sigma)\hat{x}'_1 + \tau(n_3 + n_4\sigma)\hat{x}'_2 + \tau(n_5 + n_6\sigma)\hat{x}'_3], \quad (24)$$

where the integers n_i are given by

$$\begin{aligned} & n_1=-a_1, n_2=-a_5, n_3=-(a_3+2a_4+2a_5+2a_6), \\ & n_4=-(a_1+2a_2+2a_3+2a_4+2a_5+2a_6), n_5=-(a_5+2a_6), n_6=-a_3. \end{aligned} \tag{25}$$

A similar formula to (24) involving just the coefficients of the unit vectors $\hat{x}_1, \hat{x}_2$, and $\hat{x}_3$ were obtained earlier from another consideration [52].

The root vectors $\beta_i\ (i=1,2,3)$ of the Coxeter diagram of $W(H_3)$ can be determined as

$$\beta_1=-\sqrt{2}\hat{x}_1, \beta_2=\frac{1}{\sqrt{2}}(\hat{x}_1+\sigma\hat{x}_2+\tau\hat{x}_3), \beta_3=-\sqrt{2}\hat{x}_3.$$

The pairs of vectors $(\beta_1,\beta_3), (\beta_1,\beta_2)$, and (β_2,β_3) define 2-fold, 3-fold, and 5-fold symmetry planes respectively. By choosing the vectors in these planes such as $(\hat{x}_1,\hat{x}_3), (\hat{y}_1,\hat{y}_2)$, and $(\hat{z}_1,\hat{z}_2)$ where

$$\begin{aligned} & \hat{y}_1=\frac{1}{2}(-\hat{x}_1+\sigma\hat{x}_2+\tau\hat{x}_3), \hat{y}_2=\frac{-1}{2\sqrt{3}}(3\hat{x}_1+\sigma\hat{x}_2+\tau\hat{x}_3), \\ & \hat{z}_1=\frac{1}{2}(\tau\hat{x}_1-\hat{x}_2+\sigma\hat{x}_3), \hat{z}_2=\frac{1}{2\sqrt{2+\tau}}(\hat{x}_1+\sigma\hat{x}_2+(2+\tau)\hat{x}_3), \end{aligned} \tag{26}$$

one can project the lattice vectors in (24) onto these planes provided the components of the vectors in $E_\perp$ remain in the window $\boldsymbol{K}$.

The distribution of the lattice points are displayed in figures 15, 16 and 17.

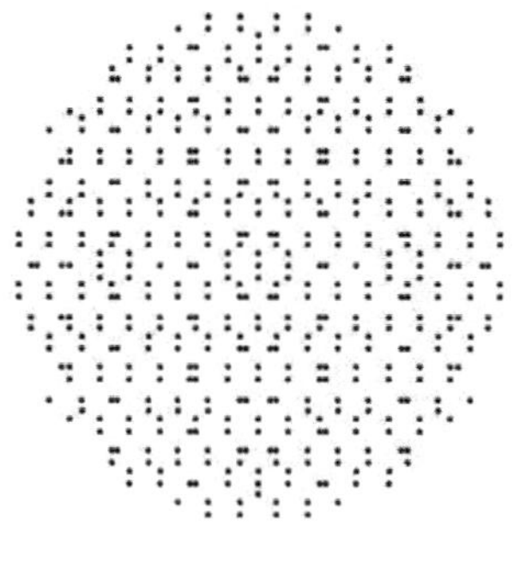

Figure 15. 2-fold symmetry from projection of B_6 lattice.

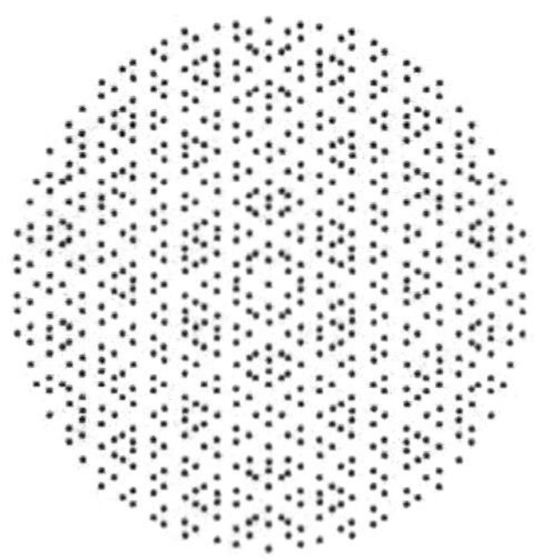

Figure 16. 3-fold symmetry from projection of B_6 lattice.

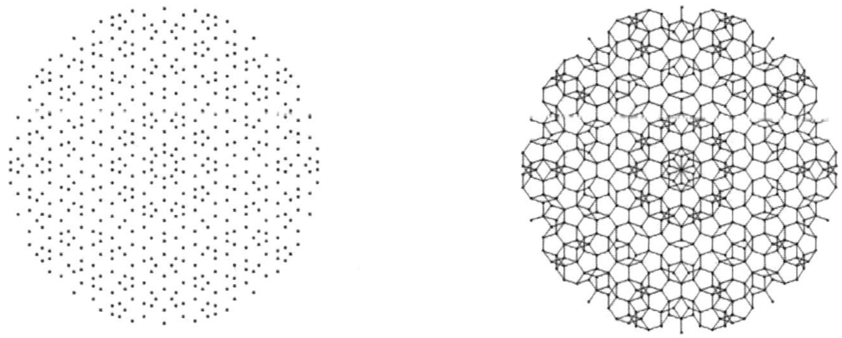

(a) Point distribution (b)The tiles

Figure 17. 5- fold symmetry from projection of B_6 lattice.

5. Conclusions

A systematic analysis of the higher dimensional lattice projection technique has been presented by emphasizing on the group theoretical structure of the nD cubic lattices. We note that the Coxeter number h of the Coxeter-Weyl group $W(B_n)$ plays a crucial role. The dihedral subgroup D_h of the group $W(B_n)$ determines the symmetry of the quasicrystal structure in the Coxeter plane. The eigenvalues and eigenvectors of the Cartan matrix (Gramm matrix) lead to the correct choice of the Coxeter plane onto which the lattice is projected. In two cases we obtained new results: the 10-fold symmetric quasicrystallography from $W(B_5)$ and the 12-fold symmetric quasicrystal structure from $W(B_6)$ lead to some novel structures. In particular, the tiling displayed in figure 14 is compatible with a recent experiment with 12-fold symmetry [51].

References

[1] Shechtman D, Blech I, Gratias D and Cahn J W 1984 *Phys. Rev. Lett.***53** 1951
[2] Janot C 2012 *Quasicrystals: a primer* (Oxford University Press-2nd edition)
[3] DiVincenzo D and Steinhardt P J 1991 *Quasicrystals: the state of the art* (Singapore: World Scientific Publishers)
[4] Senechal M 1995 *Quasicrystals and Geometry* (Cambridge University Press)
[5] Steurer W 2004 *Z. Kristallogr.* **219** 391
[6] Tsai A P 2008 *Sci. Technol. Adv. Mater.***9** 013008
[7] Lubin S M, Zhou W, Hryn A J, Huntington M D and Odom T W 2012 *Nano Lett.* **12** 4948
[8] Coxeter H S M 1952 *Duke Math J.* **18** 765
[9] Coxeter H S M and Moser W O J 1965 *Generators and Relations for Discrete Groups* (Springer-Verlag)
[10] Bourbaki N 1968 *Groupes et Algèbres de Lie. Chap. IV-VI Actualités Scientifiques et Industrielles* (Paris: Hermann) **288**; 2008 English translation: *Lie Groups and Lie algebras* (Springer 2nd printing)
[11] Humphreys J E 1990 *Reflection Groups and Coxeter Groups* (Cambridge University Press)
[12] Duneau M and Katz A 1985 *Phys Rev. Lett.***54** 2688
[13] Moody R V and Patera J 1993 *J. Phys. A: Math. Gen.***26** 2829
[14] Patera J and Twarock R 2002 *J. Phys. A: Math. Gen.***35** 1551

[15] Dechant P P, Boehm C and Twarock R 2012 *J. Phys. A: Math. Theor.***45** 285202
[16] Dechant P P, Boehm C and Twarock R 2013 *J. Math. Phys.* **54** 093508
[17] Meyer Y 1972 *Algebraic Numbers and Harmonic Analysis* (Amsterdam: North-Holland Publishing Company)
[18] Meyer Y 1995 Quasicrystals, Diophantine Approximations and Algebraic Number *Beyond Quasicrystals* eds F Axel and D Gratias Les Edition de Physique (Berlin Heidelberg New York: Springer-Verlag) p 3
[19] Moody R V 1995 Meyer Sets and finite generation of quasicrystals in *Symmetries in Science VIII* ed B Gruber (New York: Plenum)
[20] Moody R V 2000 Meyer Sets: A survey in: *From Quasicrystals to More Complex Systems* eds F Axel, F Denoyer and J P Gazeau (Centre de physique Les Houches France: Springer-Verlag) p 145
[21] Weinberg S 1967 *Phys. Rev. Lett.***19** 1264
[22] Salam A 1968 In Elementary Particle Physics: Relativistic Groups and Analyticity *Proc. eighth Nobel symposium* ed N Svartholm (Stockholm:Almquvist and Wiksell) p 367
[23] Fritzsch H, Gell-Mann M and Leutwyler H 1973 *Phys. Lett.* B **47** 365
[24] Georgi H and Glashow S 1974 *Phys. Rev. Lett.* **32 438**
[25] Fritzsch H and Minkowski P 1975 *Ann. Phys.* **93** 193
[26] Gürsey F, Ramond P and Sikivie P 1976 *Phys. Lett.* B **60** 177
[27] Koca N O, Koca M and Koc R 2014 arXiv:1306.3316 v4 (To be published in *Acta Cryst.* A)
[28] Koca N O, Koca M and Al-Sawafi M 2014 *Symmetry: Culture and Science* **25**(3) 233
[29] de Brujin N G 1981 *Proceedings of Koninklijke Nederlandse Akademie van Wetenschappen Series* A **84** 38 (*Indagationes Mathematicae* **43**)
[30] Kramer P and Neri R 1984 *Acta Cryst.* A **40** 580
[31] Baake M, Kramer P, Schlottmann M and Zeidler D 1990 *Int. J. Mod. Phys.* B **4** 2217
[32] Baake M, Joseph D, Kramer P and Schlottmann M 1990 *J. Phys. A: Math. & Gen.* **3** L1037
[33] Slansky R 1981*Phys. Rep.*C **79** 1
[34] Steinberg R 1951 *Trans. Amer. Math. Soc.* **71** 274
[35] Carter R W 1972 *Simple Groups of Lie Type* (John Wiley & Sons Ltd)
[36] Koca M, Koca N O and Koc R 2014 *Int. J. Geom. Methods Mod. Phys.***11**(4) 1450031
[37] Conway J H and Sloane N J A 1988 *Sphere Packings, Lattices and Groups* (Springer-Verlag New York Inc.)
[38] Conway J H and Sloane N J A 1982 *Voronoi Regions of Lattices, Second Moments of Polytopes and Quantization* (IEE Transactions on Information Theory IT-28, No. 2)
[39] Moody R V and Patera J 1992 *J. Phys. A: Math. Gen.* **25** 5089
[40] Kramer P 1986 *J. de Physique* **47** C3
[41] Shcherbak O P 1988 *Russian Math. Surveys* **43** 149
[42] Koca M, Koca N O and Koc R 1998 *Tr. J. Phys.* **22** 421
[43] Koca M, Koc R and Al- Barwani M 2001 *J. Phys. A: Math. Gen.* **34** 11201
[44] Koca M, Koca N O and Koc R 2010 *J. Math. Phys.* **51** 043501
[45] Conway J H and Knowles K M 1986 *J. Phys. A: Math. Gen.* **19** 3645
[46] Kalugin P A, Kitaev A Yu and Levitov L S 1985 *JETP Lett.* **41**(3)145
[47] Elser V 1985*Phys. Rev.* B **32** 4892
[48] Elser V and Henley CL 1985 *Phys. Rev. Lett.* **55** 2883

[49] Jagannahtan A and Duneau M 2013 *EPL* **104** 66003; 2014 arXiv:1402.7319
[50] Wang N, Chen H and Kuo K H 1987*Phys. Rev. Lett.* **59** 1010
[51] Forster S, Meinel K, Hammer R, Trautmann M and Widdra W 2013 *Nature* **50** 2215
[52] Rokshar D S, Mermin N D and Wright D C 1987 *Phys. Rev.* B **35** 5487

h-fold Symmetric quasicrystallography from affine Coxeter groups

N O Koca[1], **M Koca** [2] **and S Al-Shidhani** [3]

[1, 2,3] Department of Physics, College of Science, Sultan Qaboos University, P.O. Box 36, Al-Khoud 123, Muscat, Sultanate of Oman

E-mail: [1] nazife@squ.edu.om
[2] kocam@squ.edu.om
[3]shidhani7@squ.edu.om

Abstract. One can obtain the quasicrystallographic structures from the projection of the higher dimensional lattices into 2D or 3D subspaces. Here we introduce a general technique applicable to any higher dimensional lattice described by the affine Coxeter groups. It is pointed out that the Coxeter number h and the Coxeter exponents play an important role in determining the principal planes onto which the lattice to be projected. The quasicrystal structures obtained by projection display the dihedral symmetry of order $2h$. The projection subspaces are determined by using the eigenvectors and the corresponding eigenvalues of the Cartan matrix. Examples are given for 12-fold symmetric quasicrystal structures obtained by projections of the lattices determined by the affine Coxeter-Weyl groups $W_a(F_4)$, $W_a(B_6)$, and $W_a(E_6)$. The reflection generators R_1 and R_2 of the dihedral group D_{12} can be obtained as the products of the generators of the Coxeter-Weyl groups. It is noted that the quasicrystal structures obtained from the lattices $W_a(F_4)$ and $W_a(B_6)$ are compatible with the experimental data.

1. Introduction

To describe the symmetry of a natural phenomenon the group theory plays a fundamental role. In this paper we will explore the quasicrystal structures obtained by projections of the lattices described by the Coxeter-Weyl groups. Applications of the Lie groups based on the Coxeter-Weyl groups are well known in High Energy Physics. Extension of the standard model $SU(3)\times SU(2)\times U(1)$ [1, 2, 3] of the High Energy Physics to the grand unified theories $SU(5)$ [4], $SO(10)$ [5] , and E_6 [6] shed light on the unification of fundamental forces. They are intimately related to their Coxeter-Weyl groups which are the point groups of the higher dimensional lattices described by the affine extensions. One wonders whether the Coxeter-Weyl groups $W(A_4)$, $W(D_5)$, and $W(E_6)$ of these Lie groups play any special role in the study of quasicrystallography. Any Coxeter-Weyl group with Coxeter number h has a maximal dihedral subgroup of order $2h$ which acts faithfully in the Coxeter plane. The cut and project technique applied to the lattices described by the point groups $W(A_4)$ [7, 8, 9] and $W(B_5)$ [10, 11] have been already used for the description of the quasicrystallography with 5-fold symmetries. The lattice described by the affine group $W_a(B_5)$ is the simple cubic lattice in 5D and may describe 10-fold symmetric quasicrystallography via the cut and project technique. On the other hand, the lattices generated by the affine extension of the Coxeter-Weyl group $W(D_5):C_2 \approx W(C_5) \approx W(B_5)$ (: means semi direct product) are the root and weight lattices of the group $W_a(D_5)$. This proves that the lattice $W_a(D_5)$ can be used to describe both 5-fold and 10-fold symmetric quasicrystallography. The 3D quasicrystallographic structures with icosahedral symmetry can be described by canonical projections of the 6D lattices determined by the affine extension of the Coxeter-Weyl group $W(D_6)$ which embeds the non-crystallographic group $W(H_3)$ as a maximal subgroup [12, 13, 14, 15, 16]. In a recent paper [17] it was proposed that any quasicrystallographic structure with h-fold symmetry can be determined by projections of the higher dimensional lattices onto the relevant Coxeter planes.

The recent developments after the discovery of the icosahedral quasicrystal by D. Shechtman [18] indicate that metallic alloys or other chemical compositions exhibiting 5-fold, 8-fold, 10-fold, 12-fold, and 18-fold symmetries can be composed. A general exposition can be found in the references [19, 20, 21]. A more recent report on the quasicrystallography predicts the existence of a structure of 36-fold symmetry [22]. It seems that here is no limitation on the order of the dihedral group describing the symmetry of the 2D quasicrystal structures. This reminds us the classification of the Coxeter-Weyl groups with different Coxeter numbers.

The Coxeter-Weyl groups are naturally characterized by some integers known as Coxeter exponents [23, 24, 25] and, in particular, each Coxeter-Weyl group has a Coxeter number h which can be used to describe the dihedral subgroup D_h of order $2h$. It is then quite natural to expect that the projection of the higher dimensional lattice determined by the affine Coxeter-Weyl group with Coxeter number h can describe the quasicrystallography with h-fold symmetry. After introducing a general framework, we study the projections of the lattices described by the affine extensions of the Coxeter-Weyl groups $W_a(F_4)$, $W_a(B_6)$, and $W_a(E_6)$ whose Coxeter numbers are given by $h = 12$. For further study of quasicrystallography a set theoretic approach is initiated by Yves Meyer [26, 27] and later it is developed by Robert V. Moody [28, 29] in the name of *Model Set*.

The paper is organized as follows. Section 2 deals with a general technique for the canonical projections of the lattices determined by the simply-laced Coxeter-Weyl groups. The eigenvalues, eigenvectors and simple roots of the Cartan matrix (Gram matrix) of the Coxeter-Weyl group are used to determine the principal planes and the dihedral subgroup. In Section 3 the canonical projection of the exceptional lattice described by the affine Coxeter-Weyl group $W_a(F_4)$ has been worked out in detail. The Voronoi cell of the group $W(F_4)$ is projected onto the Coxeter plane and the method of canonical projection [30, 11] is used to obtain the 12-fold symmetric quasicrystal structure from the projection of the $W(F_4)$ lattice which is compatible with an experimental observation [31]. Section 4 involves the projections of the hypercubic lattice Z^6 described by the group $W_a(B_6)$ agreeing with the experimental result published recently [32]. In Section 5 the same technique is applied to the root and weight lattices of the group $W_a(E_6)$. Section 6 is devoted for the discussions of the relevant techniques and further extensions of the work.

2. Decomposition of the lattice spaces into principal planes and the canonical projection

Let G be the Coxeter-Dynkin diagram [25] representing the Coxeter-Weyl group of rank n described by the simple roots $\alpha_i, (i = 1, 2, ..., n)$ with norm$(\alpha_i, \alpha_i) = 2$. The Cartan matrix (Gram matrix) is defined by the scalar product

$$C_{ij} = \frac{2(\alpha_i, \alpha_j)}{(\alpha_j, \alpha_j)}. \tag{1}$$

For the simply-laced root systems it is a real and symmetric matrix. The weight vectors ω_i are defined by the relation $(\omega_i, \alpha_j) = \delta_{ij}$ where δ_{ij} is the Kronocker-delta. Weight vectors and the roots are related to each other by the relations

$$\alpha_i = \sum_j C_{ij}\omega_j, \quad \omega_i = \sum_j (C^{-1})_{ij}\alpha_j . \tag{2}$$

The affine extension of the Coxeter-Weyl group $W_a(G)$ can be generated by adjoining a generator r_0 representing the reflection with respect to the hyperplane bisecting the highest root. It can be determined from the extended Coxeter-Dynkin diagram of the Lie group. The root lattice G and the weight lattice G^* are defined respectively by the vectors $p = \sum_{i=1}^{n} b_i \alpha_i$, $b_i \in \mathbf{Z}$ and $q = \sum_{i=1}^{n} a_i \omega_i$, $a_i \in \mathbf{Z}$. A standard notation for the weight vector is $q \equiv (a_1, a_2, ..., a_n)$, with $a_i \in \mathbf{Z}$. Let $r_i, (i = 1, 2, ..., n)$ be the reflection generator with respect to the hyperplane orthogonal to the simple root α_i. The highest weight vector $(a_1, a_2, ..., a_n)$ with $a_i \geq 0$, designates the irreducible representation of a Lie group associated with the Coxeter-Weyl group [33]. Denote by $W(G) = \langle r_1, r_2, ..., r_n \,|\, (r_i r_j)^{m_{ij}} = 1 \rangle$ the Coxeter-Weyl group generated by the reflections. The following notation $W(G)\Lambda \equiv \Lambda_G$ represents the orbit generated from the vector Λ by the Coxeter group $W(G)$. The set of vectors $\Lambda_G = (a_1, a_2, ..., a_n)_G$ describes the vertices of a polytope and should not be confused with the vector $(a_1, a_2, ..., a_n)$, with $a_i \geq 0$ used for the designation of the irreducible representation of the associated Lie algebra. The set of simple roots can be decomposed into two sets in such a way that the corresponding reflection generators $r_1, r_2, ..., r_k$ commute pairwise as well as do the set $r_{k+1}, r_{k+2}, ..., r_n$ for some $k < n$. Define the generators R_1 and R_2 by $R_1 = r_1 r_2 ... r_k$ and $R_2 = r_{k+1} r_{k+2} ... r_n$ [34, 35]. It is easy to show that the generators R_1 and R_2 act as reflections on the simple roots of the Coxeter diagram $I_2(h)$. The Coxeter element $R_1 R_2$ represents a rotation of order h in the plane spanned by the simple roots of the graph $I_2(h)$. The Coxeter exponents m_i, $(i = 1, 2, ..., n)$ [34, 35] are listed in table 1. The eigenvalues of the Cartan matrix can be written simply as $2[1 - \cos(m_i \frac{\pi}{h})]$. The lattice of even dimension n described by the affine Coxeter group of rank-n can be decomposed into principal planes determined by the pair of orthogonal unit vectors obtained from the eigenvectors of the Cartan matrix corresponding to the eigenvalues $2[1 - \cos(m_i \frac{\pi}{h})]$ and $2[1 - \cos(h - m_i)\frac{\pi}{h})]$.

Since the Cartan matrix of a simply-laced Coxeter-Weyl group is real and symmetric it follows that the eigenvalues are real and the corresponding eigenvectors are orthogonal. Let $\lambda_i, (i = 1, 2, ..., n)$ be the eigenvalue and the $\vec{X}_i$ is the corresponding normalized eigenvector satisfying the eigenvalue equation

$$\sum_j C_{ij} X_{jk} = \lambda_k X_{ik} . \tag{3}$$

The above equation then can be written as

$$(X^T C X)_{ij} = \lambda_i \delta_{ij} . \tag{4}$$

One can define the orthogonal set of unit vectors $\hat{x}_i$ by the relation

$$\hat{x}_i - \frac{1}{\sqrt{\lambda_i}}\sum_j \alpha_j X_{ji} \;. \tag{5}$$

Table1. Coxeter exponents and the Coxeter number.

	$m_1, m_2, ..., m_n$	h
A_n	$1,2,...,n$	$n+1$
B_n	$1,3,5,...,2n-1$	$2n$
C_n	$1,3,5,...,2n-1$	$2n$
D_n	$1,3,5,...,2n-3, n-1$	$2(n-1)$
E_6	$1,4,5,7,8,11$	12
E_7	$1,5,7,9,11,13,17$	18
E_8	$1,7,11,13,17,19,23,29$	30
G_2	$1,5$	6
F_4	$1,5,7,11$	12
H_3	$1,5,9$	10
H_4	$1,11,19,29$	30
$I_2(h)$	$1,h-1$	h

The vectors in (5) form an orthonormal set in an n-dimensional Euclidean space. A general proof can be obtained for all Coxeter groups by simple modification of the proof given in the references [36, 34]. A similar relation for the non-simply laced groups can be obtained where $\hat{x}_i = \frac{1}{\sqrt{h\lambda_i}}\sum_j X_{ji}\frac{2\alpha_j}{(\alpha_j,\alpha_j)}$. Here the eigenvectors are such that the last components are all equal 1.

If we choose $m_1 = 1$ and $m_n = h-1$ (in the case of the Coxeter-Weyl group D_n this reads $m_1 = 1$ and $m_{n-1} = h-1$) then the corresponding eigenvalues will be $2[1 \mp \cos(\frac{\pi}{h})]$. Let us denote the corresponding unit vectors by $\hat{x}_1$ and $\hat{x}_n$ ($\hat{x}_{n-1}$ for D_n). Then one can show that the root vectors

$$\beta_1 = \sqrt{2}[\sin(\frac{\pi}{2h})\hat{x}_1 + \cos(\frac{\pi}{2h})\hat{x}_n],\ \beta_n = \sqrt{2}[\sin(\frac{\pi}{2h})\hat{x}_1 - \cos(\frac{\pi}{2h})\hat{x}_n] \tag{6}$$

determine the Coxeter plane and the simple roots of the Coxeter graph $I_2(h)$ satisfy the Cartan matrix

$$\begin{pmatrix} 2 & -2\cos(\frac{\pi}{h}) \\ -2\cos(\frac{\pi}{h}) & 2 \end{pmatrix}. \tag{7}$$

It is easy to prove that the generators R_1 and R_2 act as reflections on the simple roots β_1 and β_n respectively. Consequently, the Coxeter element R_1R_2 acts as a rotation by $\frac{2\pi}{h}$ in the Coxeter plane determined by the vectors β_1 and β_n . All principal planes can be compactly defined by the pair of simple roots β_i, β_{n+1-i} with

$$\beta_i = \sqrt{2}[\sin(\frac{m_i\pi}{2h})\hat{x}_i + \cos(\frac{m_i\pi}{2h})\hat{x}_{n+1-i}],$$
$$\beta_{n+1-i} = \sqrt{2}[\sin(\frac{m_i\pi}{2h})\hat{x}_i - \cos(\frac{m_i\pi}{2h})\hat{x}_{n+1-i}],\ (i = 1,2,...,\frac{n}{2}) \tag{8}$$

for even n. Therefore each pair of simple roots β_i, β_{n+1-i} determines a Coxeter graph $I_2(\frac{h}{m_i})$ where the Coxeter element acts as a rotation by the angle $m_i\frac{2\pi}{h}$. The scalar product $(\beta_i, \beta_{n+1-i}) = -2\cos(\frac{m_i\pi}{h})$ implies that the lattice space is decomposed into orthogonal set of planes each of which is associated with a pair of Coxeter exponents satisfying $m_i + m_{n+1-i} = h$. The Cartan matrix of an arbitrary Coxeter-Weyl group can be transformed into a block-diagonal matrix where each entry is represented by a 2×2 matrix similar to (7) for even n otherwise it includes a 1×1 matrix with entry 2. The Coxeter element can be represented by a block-diagonal matrix

$$R_1R_2 = \begin{pmatrix} M_1 & 0 & 0 & 0 & 0 & 0 \\ 0 & M_2 & 0 & 0 & 0 & 0 \\ 0 & 0 & . & 0 & 0 & 0 \\ 0 & 0 & 0 & . & 0 & 0 \\ 0 & 0 & 0 & 0 & . & 0 \\ 0 & 0 & 0 & 0 & 0 & M_k \end{pmatrix} \tag{9}$$

where each M_k, acting on the set of orthogonal unit vectors $\hat{x}_i$, is either $1, -1$ or a 2×2 matrix of the form [37, 11]

$$\begin{pmatrix} \cos(m_i\frac{2\pi}{h}) & -\sin(m_i\frac{2\pi}{h}) \\ \sin(m_i\frac{2\pi}{h}) & \cos(m_i\frac{2\pi}{h}) \end{pmatrix}. \tag{10}$$

The planes determined by (10) are called the principal planes by Engel [37]. The characteristic equation of the Coxeter element is the product of the characteristic equations obtained from (10). Components of the root and weight vectors $p = \sum_{i=1}^{n} b_i\alpha_i$ and $q = \sum_{i=1}^{n} a_i\omega_i \equiv (a_1, a_2, ..., a_n)$ in the orthonormal basis can be easily determined by using (5) as

$$p_i = \sqrt{\lambda_i}\sum_{j=1}^{n} b_j X_{ji}, \quad q_i = \frac{1}{\sqrt{\lambda_i}}\sum_{j=1}^{n} a_j X_{ji};\ \mathrm{a}_i, b_i \in \mathbf{Z}. \tag{11}$$

The pair of components (p_i, p_{n+1-i}) and (q_i, q_{n+1-i}) in the plane determined by the root and weight vectors represents respectively the orthogonal projections of the respective lattices. An n-dimensional Euclidean space can be decomposed as the space $E_{\parallel}$ determined by the pair of unit vectors $\hat{x}_1$ and $\hat{x}_n$ and the complement $E_{\perp}$. The canonical projection can be defined as follows: Pick up the set of vectors $V(0)$ [38, 39, 40] representing the Voronoi cell of the root lattice [41, 42] or weight lattice [41] around the origin. Projection of the Voronoi cell into the space $E_{\perp}$ determines a ball whose surface is a sphere S^{n-3} of radius R_0 which is also called window $\boldsymbol{K}$ [11]. One projects those lattice points which are restricted by the sphere of radius R_0 onto the plane determined by the unit vectors

$\hat{x}_1$ and $\hat{x}_n$. This is a general technique for every higher dimensional lattice. In the following sections we apply this technique for the projections of the lattices determined by the Coxeter-Weyl groups $W_a(F_4)$, $W_a(B_6)$, and $W_a(E_6)$.

The Voronoi regions of the root lattices are discussed in detail in reference [42] and can be easily summarized in our notations as follows. The Voronoi cell V (0) of A_n is a polytope dual to the root polytope $(1,0,0,...,0,1)_{A_n}$ and can be written as the union of the orbits $(1,0,0,...,0,0)_{A_n} \cup (0,1,0,...,0,0)_{A_n} \cup ... \cup (0,0,0,...,0,1)_{A_n}$. For instance, the Voronoi cell of the fcc lattice of A_3 (the root lattice) is a Catalan solid, the rhombic dodecahedron, described by the union of the polytopes $(1,0,0)_{A_3} \cup (0,1,0)_{A_3} \cup (0,0,1)_{A_3}$ dual to the root polytope $(1,0,1)_{A_3}$ [43]. Similarly the Voronoi cell of A_4 is the union of the polytopes $(1,0,0,0)_{A_4} \cup (0,1,0,0)_{A_4} \cup (0,0,1,0)_{A_4} \cup (0,0,0,1)_{A_4}$ dual to the root polytope $(1,0,0,1)_{A_4}$ [44]. One can check that for all simply laced Coxeter-Weyl groups the Voronoi cell of the root lattice is the dual polytope of the root polytope. The Voronoi cells of the weight lattices have been studied in reference [45] that can be expressed in terms of our notation. The Voronoi cell V (0) of the weight lattice A_n^* [44] can be represented by the polytope $\frac{1}{n+1}(1,1,...,1)_{A_n}$.

3. Canonical projection of the lattice generated by the exceptional group $W_a(F_4)$

The Coxeter-Dynkin diagram of the exceptional group $W(F_4)$ is shown in figure 1. The order of the group is $|W(F_4)| = 1152$.

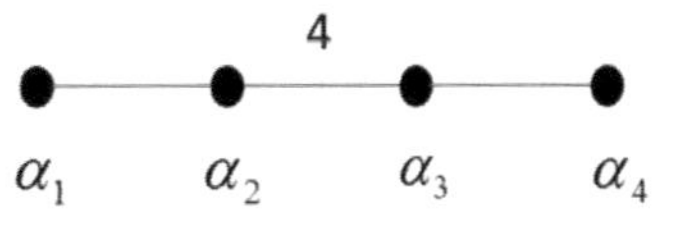

Figure 1. The Coxeter-Dynkin diagram of $W(F_4)$.

It has two long roots (first two from left) and two short roots (last two). The Cartan matrix and its inverse are given as follows

$$C_{F_4} = \begin{pmatrix} 2 & -1 & 0 & 0 \\ -1 & 2 & -2 & 0 \\ 0 & -1 & 2 & -1 \\ 0 & 0 & -1 & 2 \end{pmatrix}, \quad (C_{F_4})^{-1} = \begin{pmatrix} 2 & 3 & 4 & 2 \\ 3 & 6 & 8 & 4 \\ 2 & 4 & 6 & 3 \\ 1 & 2 & 3 & 2 \end{pmatrix}. \tag{12}$$

The dual space basis vectors are defined by the relation $(\omega_i, \frac{2\alpha_j}{(\alpha_j, \alpha_j)}) = \delta_{ij}$ where the scalar product in the dual space is defined by the metric tensor $(\omega_i, \omega_j) = G_{ij}$ which is given in the matrix form as

$$G = \begin{pmatrix} 2 & 3 & 2 & 1 \\ 3 & 6 & 4 & 2 \\ 2 & 4 & 3 & \frac{3}{2} \\ 1 & 2 & \frac{3}{2} & 1 \end{pmatrix}. \tag{13}$$

The F_4 lattice is generated by its short roots and it is self-dual because the inverse matrix in (12) has integer entries only. Therefore the root and the weight lattices coincide up to an action of the Coxeter Weyl group $W_a(F_4)$. We will work with the dual space basis vectors to define a general lattice vector $q=\sum_{i=1}^{4} a_i\omega_i \equiv (a_1,a_2,a_3,a_4)$, $a_i \in \mathbf{Z}$. The F_4 polytopes have been studied in detail in reference [46]. The self- dual regular polytope of F_4 is called the 24-cell and it is either represented by the orbit $(1,0,0,0)_{F_4}$ or $(0,0,0,1)_{F_4}$. In figure 2 we display the orthogonal projection of the Voronoi cell (24-cell) of the group $W(F_4)$.

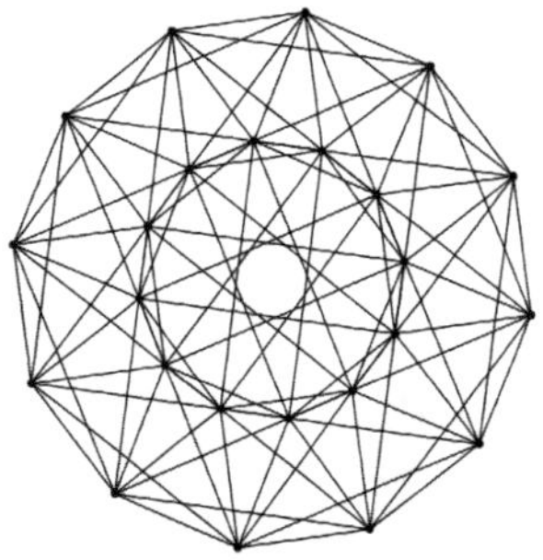

Figure 2. Orthogonal projection of the 24-cell.

The only difference between these two representations of the 24-cell is that norms of the vectors are given by $(\omega_1,\omega_1)=2$ versus $(\omega_4,\omega_4)=1$ and one is rotated with respect to the other within the lattice space. A quaternionic representation of the polytope $(1,0,0,0)_{F_4}$ corresponds to the quaternionic elements of the binary tetrahedral group T [47]. The F_4 lattice is generated by the polytope $(0,0,0,1)_{F_4}$ and the scaled copy of the 24-cell $\frac{1}{2}(0,0,0,1)_{F_4}$ [42] represents the Voronoi cell of the lattice. The four orthonormal unit vectors can be constructed by a similar technique studied in the previous section.

The orthogonal components of a general lattice vector in the principle planes $(\hat{x}_1,\hat{x}_4)$ and $(\hat{x}_2,\hat{x}_3)$ can be written respectively as

$$\begin{aligned} q_1 &= \frac{1}{\sqrt{2}\sqrt{2-\sqrt{2+\sqrt{3}}}\sqrt{3+\sqrt{3}}}[a_1+\frac{\sqrt{3}+1}{\sqrt{2}}a_2+\frac{\sqrt{3}+1}{2}a_3+\frac{1}{\sqrt{2}}a_4], \\ q_4 &= \frac{1}{\sqrt{2}\sqrt{2+\sqrt{2+\sqrt{3}}}\sqrt{3+\sqrt{3}}}[a_1-\frac{\sqrt{3}+1}{\sqrt{2}}a_2+\frac{\sqrt{3}+1}{2}a_3-\frac{1}{\sqrt{2}}a_4], \end{aligned} \quad (14)$$

$$\begin{aligned} q_2 &= \frac{1}{\sqrt{2}\sqrt{2-\sqrt{2-\sqrt{3}}}\sqrt{3-\sqrt{3}}}[a_1+\frac{1-\sqrt{3}}{\sqrt{2}}a_2+\frac{1-\sqrt{3}}{2}a_3+\frac{1}{\sqrt{2}}a_4], \\ q_3 &= \frac{1}{\sqrt{2}\sqrt{2+\sqrt{2-\sqrt{3}}}\sqrt{3-\sqrt{3}}}[a_1-\frac{1-\sqrt{3}}{\sqrt{2}}a_2+\frac{1-\sqrt{3}}{2}a_3-\frac{1}{\sqrt{2}}a_4], \end{aligned} \quad (15)$$

When the Voronoi cell $\frac{1}{2}(0,0,0,1)_{F_4}$ is projected onto the plane $(\hat{x}_2,\hat{x}_3)$ it describes a disc of radius $R_0=\frac{1}{2}(q_2{}^2+q_3{}^2)^{\frac{1}{2}}$. One can select the lattice points which project into this region with their components q_2 and q_3 satisfying the relation $\frac{1}{2}(q_2{}^2+q_3{}^2)^{\frac{1}{2}}\leq R_0$. Now project those points onto the plane $(\hat{x}_1,\hat{x}_4)$. The result of this projection is shown in figure 3. One can repeat the same procedure by interchanging the planes that leads to the same result since the dihedral group acts in both planes faithfully. The quasicrystal structure in figure 3 is strikingly similar observed in the quasicrystals of the Ni-Cr particles [31].

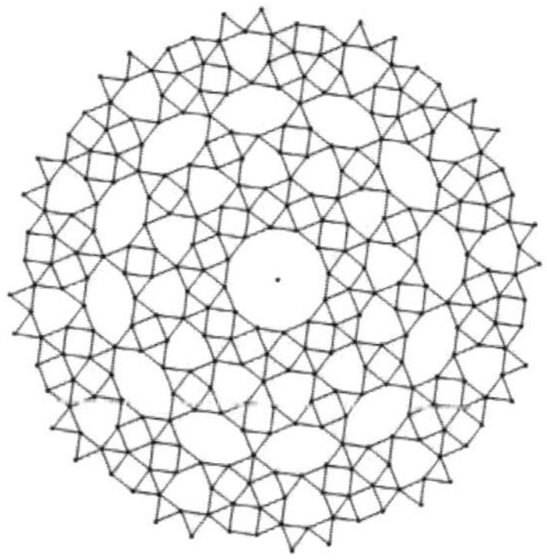

Figure 3. The canonical projection of the F_4 lattice onto the plane $(\hat{x}_1,\hat{x}_4)$.

4. Canonical projection of the lattice generated by the group $W_a(B_6)$

The Coxeter-Dynkin diagram of the group $W(B_6)$ is displayed in figure 4.

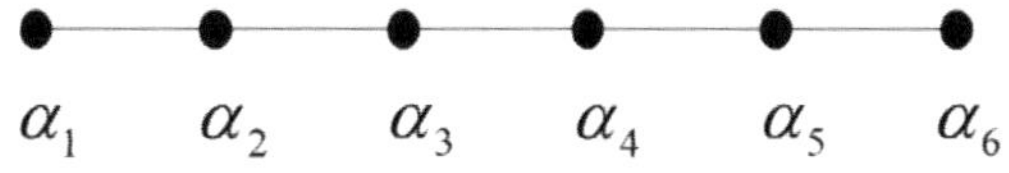

Figure 4. The Coxeter-Dynkin diagram of the Coxeter-Weyl group $W(B_6)$.

Its last root is a short root of norm 1. The Cartan matrix and its inverse are given as follows:

$$C_{B_6}=\begin{pmatrix}2&-1&0&0&0&0\\-1&2&-1&0&0&0\\0&-1&2&-1&0&0\\0&0&-1&2&-1&0\\0&0&0&-1&2&-2\\0&0&0&0&-1&2\end{pmatrix},\quad (C_{B_6})^{-1}=\begin{pmatrix}1&1&1&1&1&1\\1&2&2&2&2&2\\1&2&3&3&3&3\\1&2&3&4&4&4\\1&2&3&4&5&5\\\frac{1}{2}&1&\frac{3}{2}&2&\frac{5}{2}&3\end{pmatrix}. \quad (16)$$

The metric tensor in the dual space is given by the matrix $(\omega_i,\omega_j)=G_{ij}=(C^{-1})_{ij}\frac{(\alpha_j,\alpha_j)}{2}$.

The root lattice is a simple cubic lattice in 6D generated by the short roots of B_6. The lattice can be represented in the weight space with the lattice vectors given by $q=\sum_{i=1}^{5}a_i\omega_i+2a_6\equiv(a_1,a_2,...,2a_6)$, $a_i\in\mathbf{Z}$. For example, the Voronoi cell is the polytope $(0,0,0,0,0,1)_{B_6}$ which represents a 6D cube with 64 vertices. Its fundamental polytopes are represented by the orbits $(1,0,0,0,0,0)_{B_6}$, $(0,1,0,0,0,0)_{B_6}$, $(0,0,1,0,0,0)_{B_6}$, $(0,0,0,1,0,0)_{B_6}$, $(0,0,0,0,1,0)_{B_6}$, and $(0,0,0,0,0,1)_{B_6}$. For example, the two polytopes $(1,0,0,0,0,0)_{B_6}$ and $(0,0,0,0,0,1)_{B_6}$ are dual of each other. Properties of the fundamental polytopes are described in table 2. The vertices of the 6D cube can also be written as $(0,0,0,0,0,1)_{B_6}=\frac{1}{2}(\pm l_1\pm l_2\pm l_3\pm l_4\pm l_5\pm l_6)$ in terms of some suitable orthonormal set of vectors with $(l_i,l_j)=\delta_{ij},(i,j=1,2,...,6)$ where the simple roots are defined as $\alpha_1=l_1-l_2,\alpha_2=l_2-l_3,\alpha_3=l_3-l_4,\alpha_4=l_4-l_5,\alpha_5=l_5-l_6,\alpha_6=l_6$.

Table 2. Numbers of facets of some of the B_6 polytopes.

Polytope	N_0	N_1	N_2	N_3	N_4	N_5
$(1,0,0,0,0,0)_{B_6}$	12	60	160	240	192	64
$(0,1,0,0,0,0)_{B_6}$	60	480	1,120	1,200	576	76
$(0,0,1,0,0,0)_{B_6}$	160	1,440	2,880	2,160	636	76
$(0,0,0,1,0,0)_{B_6}$	240	1,920	3,200	2,080	636	76
$(0,0,0,0,1,0)_{B_6}$	192	960	1,600	1,200	444	76
$(0,0,0,0,0,1)_{B_6}$	64	192	240	160	60	12

The numbers $N_0-N_1+N_2-N_3+N_4-N_5=0$ satisfy the Euler equation.
The projections of some of those polytopes onto the Coxeter plane are depicted in figure 5.

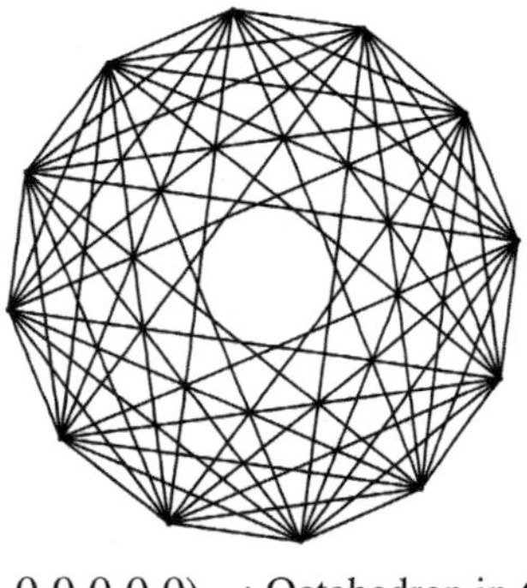

$(1,0,0,0,0,0)_{B_6}$: Octahedron in 6D

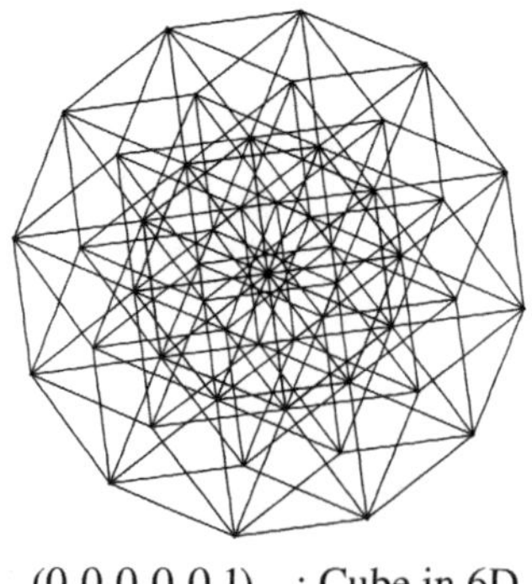

$(0,0,0,0,0,1)_{B_6}$: Cube in 6D

Figure 5. Orthogonal projections of some of the polytopes of $W(B_6)$.

The orthonormal set of basis vectors of the B_6 can be determined as follows

$$\begin{aligned}
\hat{x}_1 &= \frac{1}{\sqrt{12}\sqrt{2-\sqrt{2+\sqrt{3}}}}[\sqrt{2-\sqrt{3}}\alpha_1+\alpha_2+\sqrt{2}\alpha_3+\sqrt{3}\alpha_4+\sqrt{2+\sqrt{3}}\alpha_5+2\alpha_6],\\
\hat{x}_6 &= \frac{1}{\sqrt{12}\sqrt{2+\sqrt{2+\sqrt{3}}}}[-\sqrt{2-\sqrt{3}}\alpha_1+\alpha_2-\sqrt{2}\alpha_3+\sqrt{3}\alpha_4-\sqrt{2+\sqrt{3}}\alpha_5+2\alpha_6],\\
\hat{x}_2 &= \frac{1}{\sqrt{12}\sqrt{2-\sqrt{2}}}[-\sqrt{2}\alpha_1-2\alpha_2-\sqrt{2}\alpha_3+\sqrt{2}\alpha_5+2\alpha_6],\\
\hat{x}_5 &= \frac{1}{\sqrt{12}\sqrt{2+\sqrt{2}}}[\sqrt{2}\alpha_1-2\alpha_2+\sqrt{2}\alpha_3-\sqrt{2}\alpha_5+2\alpha_6],\\
\hat{x}_3 &= \frac{1}{\sqrt{12}\sqrt{2-\sqrt{2-\sqrt{3}}}}[\sqrt{2+\sqrt{3}}\alpha_1+\alpha_2-\sqrt{2}\alpha_3-\sqrt{3}\alpha_4+\sqrt{2-\sqrt{3}}\alpha_5+2\alpha_6]\\
\hat{x}_4 &= \frac{1}{\sqrt{12}\sqrt{2+\sqrt{2-\sqrt{3}}}}[-\sqrt{2+\sqrt{3}}\alpha_1+\alpha_2+\sqrt{2}\alpha_3-\sqrt{3}\alpha_4-\sqrt{2-\sqrt{3}}\alpha_5+2\alpha_6].
\end{aligned} \tag{17}$$

These sets of vectors define three orthogonal planes determined by the pair of vectors $(\hat{x}_1,\hat{x}_6),(\hat{x}_2,\hat{x}_5),(\hat{x}_3,\hat{x}_4)$ in which the Coxeter element acts like rotations with respective angles $\frac{\pi}{6},\frac{\pi}{2},\frac{5\pi}{6}$. Orthogonal projections of a general lattice vector $q=\sum_{i=1}^{6}a_i\omega_i$ onto these planes are given by the following pairs.

The $(\hat{x}_1,\hat{x}_6)$ plane:

$$\begin{aligned}
q_1 &= \frac{1}{\sqrt{12}\sqrt{2-\sqrt{2+\sqrt{3}}}}[\sqrt{2-\sqrt{3}}a_1+a_2+\sqrt{2}a_3+\sqrt{3}a_4+\sqrt{2+\sqrt{3}}a_5+2a_6],\\
q_6 &= \frac{1}{\sqrt{12}\sqrt{2+\sqrt{2+\sqrt{3}}}}[-\sqrt{2-\sqrt{3}}a_1+a_2-\sqrt{2}a_3+\sqrt{3}a_4-\sqrt{2+\sqrt{3}}a_5+2a_6];
\end{aligned} \tag{18}$$

The $(\hat{x}_2,\hat{x}_5)$ plane:

$$\begin{aligned}
q_2 &= \frac{1}{\sqrt{12}\sqrt{2-\sqrt{2}}}[-\sqrt{2}a_1-2a_2-\sqrt{2}a_3+\sqrt{2}a_5+2a_6],\\
q_5 &= \frac{1}{\sqrt{12}\sqrt{2+\sqrt{2}}}[\sqrt{2}a_1-2a_2+\sqrt{2}a_3-\sqrt{2}a_5+2a_6];
\end{aligned} \tag{19}$$

The $(\hat{x}_3,\hat{x}_4)$ plane:

$$\begin{aligned}
q_3 &= \frac{1}{\sqrt{12}\sqrt{2-\sqrt{2-\sqrt{3}}}}[\sqrt{2+\sqrt{3}}a_1+a_2-\sqrt{2}a_3-\sqrt{3}a_4+\sqrt{2-\sqrt{3}}a_5+2a_6],\\
q_4 &= \frac{1}{\sqrt{12}\sqrt{2+\sqrt{2-\sqrt{3}}}}[-\sqrt{2+\sqrt{3}}a_1+a_2+\sqrt{2}a_3-\sqrt{3}a_4-\sqrt{2-\sqrt{3}}a_5+2a_6].
\end{aligned} \tag{20}$$

The Voronoi cell $V(0)$ of the $W_a(B_6)$ lattice, the 6D cube represented by the polytope $(0,0,0,0,0,1)_{B_6}$, is projected into the space $E_\perp$ determined by the 4D sub-space $(\hat{x}_2,\hat{x}_3,\hat{x}_4,\hat{x}_5)$ defining a window $\boldsymbol{K}$. Projection of the strip onto the Coxeter plane $(\hat{x}_1,\hat{x}_6)$ leads to the $W_a(B_6)$ quasicrystal with 12-fold symmetry as shown in figure 6. It is interesting to note that the tiles include rhombi in addition to the usual square and triangle tiles. A recent experimental result [32] confirms this kind of tiling.

Journal of Physics: Conference Series **563** (2014) 012017 doi:10.1088/1742-6596/563/1/012017

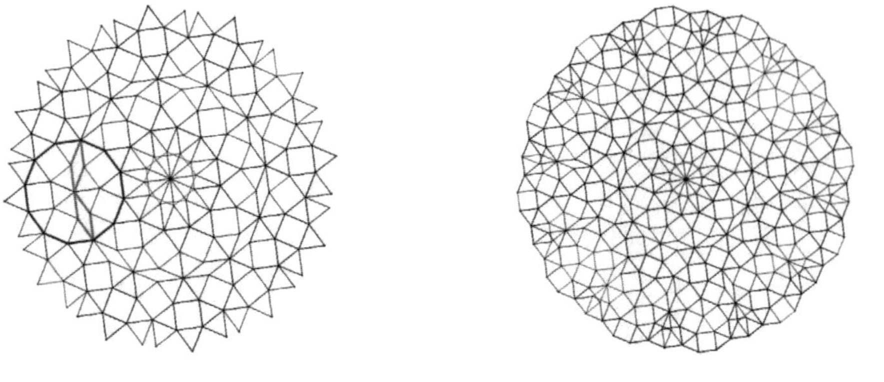

Figure 6. The canonical projection of the lattice $W_a(B_6)$ onto the plane $(\hat{x}_1, \hat{x}_6)$.

5. Canonical projections of the root and weight lattices of $W_a(E_6)$

The Coxeter-Dynkin diagram of the exceptional group $W(E_6)$ is given in figure 7. The Corresponding Cartan matrix and its inverse are given as follows:

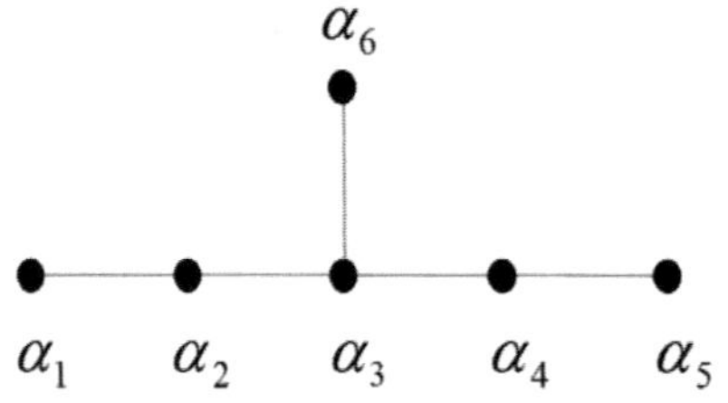

Figure 7. The Coxeter-Dynkin diagram of E_6.

$$C_{E_6} = \begin{pmatrix} 2 & -1 & 0 & 0 & 0 & 0 \\ -1 & 2 & -1 & 0 & 0 & 0 \\ 0 & -1 & 2 & -1 & 0 & -1 \\ 0 & 0 & -1 & 2 & -1 & 0 \\ 0 & 0 & 0 & -1 & 2 & 0 \\ 0 & 0 & -1 & 0 & 0 & 2 \end{pmatrix}, \quad (C_{E_6})^{-1} = \begin{pmatrix} \frac{4}{3} & \frac{5}{3} & 2 & \frac{4}{3} & \frac{2}{3} & 1 \\ \frac{5}{3} & \frac{10}{3} & 4 & \frac{8}{3} & \frac{4}{3} & 2 \\ 2 & 4 & 6 & 4 & 2 & 3 \\ \frac{4}{3} & \frac{8}{3} & 4 & \frac{10}{3} & \frac{5}{3} & 2 \\ \frac{2}{3} & \frac{4}{3} & 2 & \frac{5}{3} & \frac{4}{3} & 1 \\ 1 & 2 & 3 & 2 & 1 & 2 \end{pmatrix}. \quad (21)$$

The order of the Coxeter-Weyl group generated by the 6, 6×6 matrices is $|W(E_6)| = 2^7 3^4 5 = 51{,}840$. An arbitrary vector of the root lattice E_6 is defined by the vector $p = \sum_{i=1}^{6} b_i \alpha_i$, $b_i \in \mathbf{Z}$. Similarly, a weight vector belonging to the weight lattice E_6^* is given by the vector $q = \sum_{i=1}^{6} a_i \omega_i \equiv (a_1, a_2, \ldots, a_6)$, $a_i \in \mathbf{Z}$ and the orbit generated by an arbitrary vector is defined by

$W(E_6)q = q_{E_6} \equiv (a_1, a_2, ..., a_6)_{E_6}$. If we denote the root and weight lattice vectors by 6-tuples $\vec{b} \equiv (b_1, b_2, ..., b_6)$ and $\vec{a} \equiv (a_1, a_2, ..., a_6)$ respectively then we have the relation $\vec{b}C_{E_6} = \vec{a}$. These relations imply that when $\vec{b}$ has integer components it belongs to the root lattice. It also belongs to the weight lattice since $\vec{a}$ would also takes integer components. However, for $\vec{a}$ with integer components, $\vec{b}$ may not have integer components. Therefore it is clear that the weight lattice E_6^* admits the root lattice E_6 as a sub-lattice. The orthogonal unit vectors for E_6 obtained from (5) are given explicitly as follows

$$\begin{aligned}
\hat{x}_1 &= \frac{1}{\sqrt{2-\sqrt{2+\sqrt{3}}}\,(2\sqrt{3}+\sqrt{3})}[\alpha_1 + \sqrt{2+\sqrt{3}}\alpha_2 + (1+\sqrt{3})\alpha_3 + \sqrt{2+\sqrt{3}}\alpha_4 + \alpha_5 + \sqrt{2}\alpha_6],\\
\hat{x}_6 &= \frac{1}{\sqrt{2+\sqrt{2+\sqrt{3}}}\,(2\sqrt{3}+\sqrt{3})}[-\alpha_1 + \sqrt{2+\sqrt{3}}\alpha_2 - (1+\sqrt{3})\alpha_3 + \sqrt{2+\sqrt{3}}\alpha_4 - \alpha_5 + \sqrt{2}\alpha_6],\\
\hat{x}_2 &= \frac{1}{2}[-\alpha_1 - \alpha_2 + \alpha_4 + \alpha_5],\\
\hat{x}_5 &= \frac{1}{2\sqrt{3}}[-\alpha_1 + \alpha_2 - \alpha_4 + \alpha_5],\\
\hat{x}_3 &= \frac{1}{\sqrt{2-\sqrt{2-\sqrt{3}}}\,(2\sqrt{3}-\sqrt{3})}[-\alpha_1 - \sqrt{2-\sqrt{3}}\alpha_2 - (1-\sqrt{3})\alpha_3 - \sqrt{2-\sqrt{3}}\alpha_4 - \alpha_5 + \sqrt{2}\alpha_6],\\
\hat{x}_4 &= \frac{1}{\sqrt{2+\sqrt{2-\sqrt{3}}}\,(2\sqrt{3}-\sqrt{3})}[\alpha_1 - \sqrt{2-\sqrt{3}}\alpha_2 + (1-\sqrt{3})\alpha_3 - \sqrt{2-\sqrt{3}}\alpha_4 + \alpha_5 + \sqrt{2}\alpha_6].
\end{aligned} \tag{22}$$

We define three orthogonal planes $(\hat{x}_1, \hat{x}_6)$, $(\hat{x}_2, \hat{x}_5)$, and $(\hat{x}_3, \hat{x}_4)$ of the graph $I_2(\frac{h}{m_i})$ with $m_1 = 1$, $m_2 = 4$, $m_3 = 5$ respectively. The generators R_1 and R_2 act like reflection generators on these principal planes and the corresponding Coxeter element acts like rotations by angles $\frac{\pi}{6}, \frac{2\pi}{3}$, and $\frac{5\pi}{6}$ respectively.

Before we study the details of the canonical projection of the E_6 lattices we would like to study the projections of certain polytopes of the group $W(E_6)$. The fundamental polytopes are characterized by the orbits $(1,0,0,0,0,0)_{E_6}, (0,1,0,0,0,0)_{E_6}, (0,0,1,0,0,0)_{E_6}, (0,0,0,1,0,0)_{E_6}, (0,0,0,0,1,0)_{E_6}$, and $(0,0,0,0,0,1)_{E_6}$. Their vertices are the vectors of the weight lattice. Two particular polytopes belong to both lattices: the root polytope $(0,0,0,0,0,1)_{E_6}$ and the polytope $(0,0,1,0,0,0)_{E_6}$. These two polytopes have a larger symmetry $Aut(E_6) \approx W(E_6):\gamma$ where γ represents the Dynkin diagram symmetry of order 2. The characteristic properties of these polytopes, such as, numbers of vertices, edges, 2-facets, 3-facets, 4-facets, and 5-facets (denoted respectively by $N_0, N_1, N_2, N_3, N_4, N_5$) are listed in table 3. Their projections onto the plane $(\hat{x}_1, \hat{x}_6)$ are illustrated in figure 9. The dual of the root polytope $(0,0,0,0,0,1)_{E_6}$ is the union of the polytopes $(1,0,0,0,0,0)_{E_6} \cup (0,0,0,0,1,0)_{E_6}$ and it constitutes the Voronoi cell of the root lattice with N_0 =54, N_1 =702, N_2 =2,160, N_3 =2,160, N_4 =720, N_5 =72. The root polytope as well as its dual has a larger symmetry $Aut(E_6) \approx W(E_6):\gamma$. Their projections are shown in figure 8.

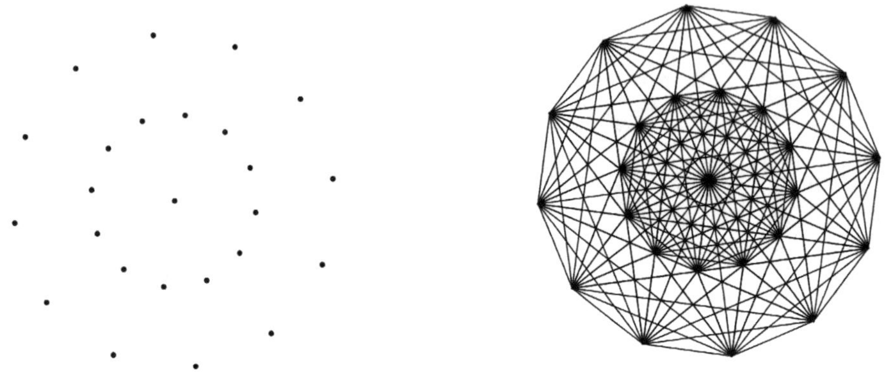

(a) $(1,0,0,0,0,0)_{E_6}$ (b) $(1,0,0,0,0,0)_{E_6}$

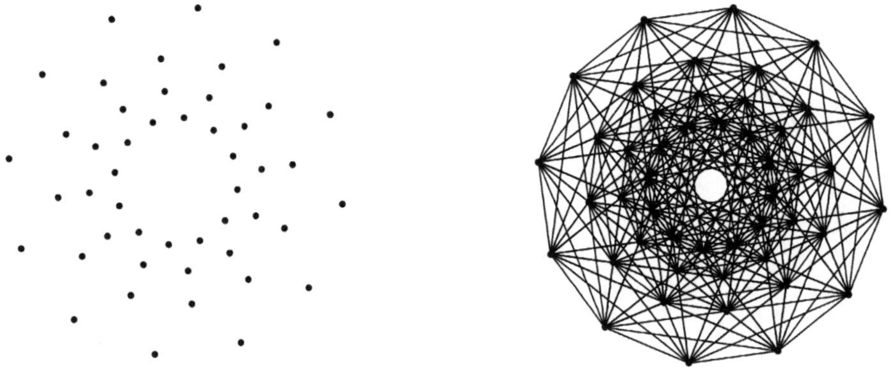

(c) $(0,0,0,0,0,1)_{E_6}$ (d) $(0,0,0,0,0,1)_{E_6}$ root polytope

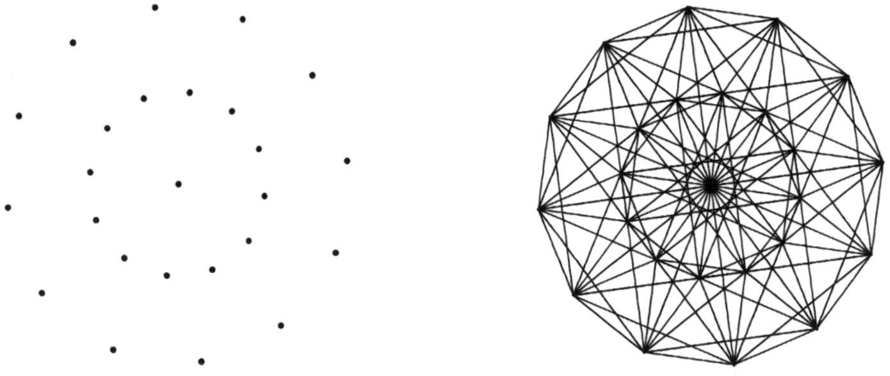

(e) Voronoi cell of the root lattice $(1,0,0,0,0,0)_{E_6} \cup (0,0,0,0,1,0)_{E_6}$

Figure 8. Projections of some of E_6 polytopes onto the Coxeter plane.

Table 3. Numbers of facets of some of the E_6 polytopes.

Polytope	N_0	N_1	N_2	N_3	N_4	N_5
$(1,0,0,0,0,0)_{E_6}$	27	216	720	1,080	648	99
$(0,1,0,0,0,0)_{E_6}$	216	2,160	5,040	4,320	1,350	126
$(0,0,1,0,0,0)_{E_6}$	720	6,480	10,800	6,480	1,566	126
$(0,0,0,1,0,0)_{E_6}$	216	2,160	5,040	4,320	1,350	126
$(0,0,0,0,1,0)_{E_6}$	27	216	720	1,080	648	99
$(0,0,0,0,0,1)_{E_6}$	72	720	2,160	2,160	702	54

Let us choose the plane $(\hat{x}_1,\hat{x}_6)$ as $E_\parallel$ and the rest as $E_\perp$. As we have explained in Section 2 the canonical projection is carried out by projecting the Voronoi cell $V(0)$ of the root/weight lattice into the space $E_\perp$ to obtain a window $\boldsymbol{K}$. The Voronoi cell of the weight lattice E_6^* [45, 48, 49] is a polytope which can be specified as $\frac{1}{3}(0,0,1,0,0,0)_{E_6}$.

The 720 vertices of the polytope are determined by the action of the group elements on the vector $\frac{1}{3}(0,0,1,0,0,0)$ and using the formula

$$R_0 = \frac{1}{3}(q_2^{\,2} + q_3^{\,2} + q_4^{\,2} + q_5^{\,2})^{\frac{1}{2}} \tag{23}$$

one obtains the window $\boldsymbol{K}$ in the form of a 3-sphere. We illustrate the projection of the root lattice in figure 9. They are the same point distributions in which (b) represents the tiling with minimal distance and (c) represents a partial tiling which is exactly the same tiling obtained from the projection of F_4 lattice that is a sublattice in the root lattice E_6.

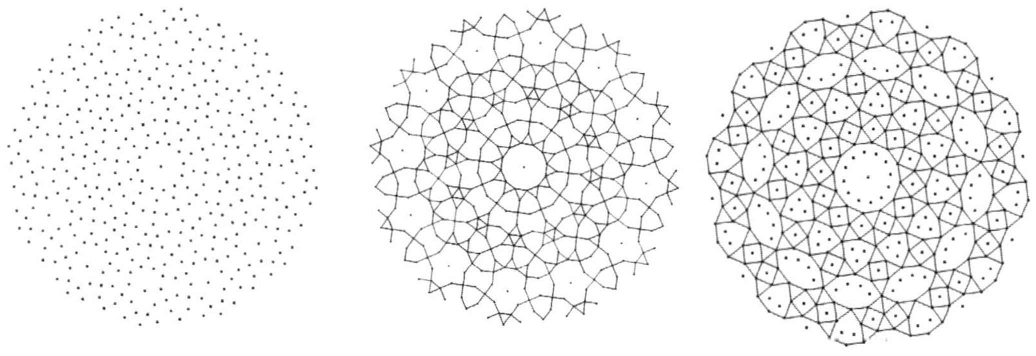

(a) (b) (c)

Figure 9. Projection of root lattice of E_6 .

The projection of the weight lattice is depicted in figure 10.

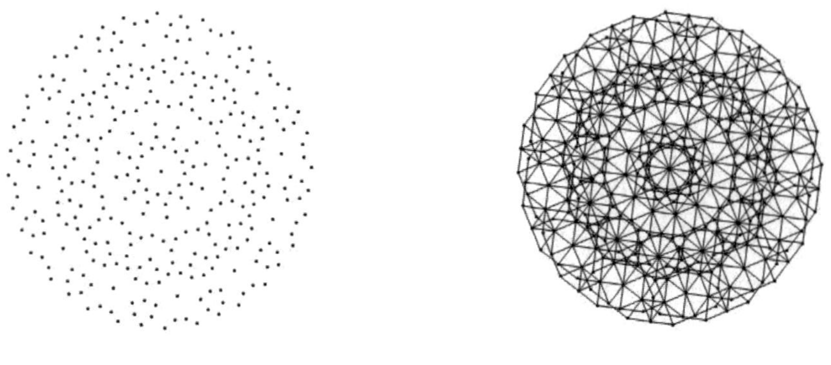

(a) (b)

Figure 10. Canonical projection of the weight lattice of E_6 onto the plane $(\hat{x}_1, \hat{x}_6)$.

Now let us define the $E_{\|}$ plane by the pair of vectors $(\hat{x}_2, \hat{x}_5)$. The representation of the Coxeter group in the plane $(\hat{x}_2, \hat{x}_5)$ is such that $(R_1R_2)^3 = 1$. Therefore the dihedral group is represented by the group $D_3 \approx S_3 \approx W(A_2)$ which is not maximal. Consequently, the 4-dimensional Euclidean sub-space $E_{\perp}$ is determined by the orthogonal planes $(\hat{x}_1, \hat{x}_6)$ and $(\hat{x}_3, \hat{x}_4)$. This time we consider only the weight lattice projection in which the window $\boldsymbol{K}$ is determined by the projection of the Voronoi cell $\frac{1}{3}(0,0,1,0,0,0)_{E_6}$ into the space $E_{\perp}$. After repeating the previous procedure we project the lattice points into the plane determined by the vectors $(\hat{x}_2, \hat{x}_5)$ and obtain the planar lattice as shown in figure 11. It is a honeycomb lattice as expected because the projection has been made onto a plane where the crystallographic subgroup $W(A_2) \subset W(E_6)$ acts. It is interesting to observe that the honeycomb lattice exits in nature as the graphene sheet.

Figure 11. The honeycomb lattice obtained by projection of E_6 lattice.

6. Conclusion

This paper has introduced a general technique applicable to the projections of all higher dimensional lattices generated by the affine Coxeter-Weyl groups. It was shown that the eigenvalues, the corresponding eigenvectors, and the simple roots of the Cartan matrix (Gram matrix) play an important role in the projection technique. The maximal dihedral subgroup of the Coxeter-Weyl group can be generated by two generators whose product defines the Coxeter element of the Coxeter-

Journal of Physics: Conference Series **563** (2014) 012017 doi:10.1088/1742-6596/563/1/012017

Weyl group. It was noted that the Coxeter group acts, in the lattice space spanned by the orthogonal unit vectors, as a reducible representation of the dihedral group in the block diagonal form of 2×2 irreducible representations and/or as 1×1 matrix depending on the rank of the Coxeter-Weyl group. If the dihedral subgroup acting on the principal plane (Coxeter plane) is non-crystallographic the point distribution represents a quasicrystal with *h*-fold symmetry where *h* is the Coxeter number; otherwise it represents a crystallographic point distribution such as square and/or hexagonal lattices. We have studied examples of 12-fold symmetry by projection of the lattices $W_a(F_4)$, $W_a(B_6)$, and $W_a(E_6)$. It turned out that the structures obtained by the projections of the lattices $W_a(F_4)$ and $W_a(B_6)$ are compatible with the 12-fold symmetric quasicrystal structure observed in Ni-Cr particles [31] and in dodecagonal quasicrystal formation $BaTiO_3$ [32] respectively. The projection technique we introduced can be applied to any lattice described by the affine Coxeter group.

References

[1] Weinberg S 1967 *Phys. Rev. Lett.* **19** 1264
[2] Salam A 1968 In Elementary Particle Physics: Relativistic Groups and Analyticity *Proc. eighth Nobel symposium* ed N Svartholm (Stockholm: Almquvist and Wiksell) p 367
[3] Fritzsch H, Gell-Mann M and Leutwyler H 1973 *Phys. Lett.* B **47** 365
[4] Georgi H and Glashow S 1974 *Phys. Rev. Lett.* **32** 438
[5] Fritzsch H and Minkowski P 1975 *Ann. Phys.* **93** 193
[6] Gürsey F, Ramond P and Sikivie P 1976 *Phys. Lett.* B **60** 177
[7] Baake M, Kramer P, Schlottmann M and Zeidler D 1990a *Int. J. Mod. Phys.* B **4** 2217
[8] Baake M, Joseph D, Kramer P and Schlottmann M 1990b *J. Phys. A: Math. & Gen.* **23** L1037
[9] Koca M, Koca N O and Koc R 2014 *Int. J. Geom. Methods Mod. Phys.* **11**(4) 1450031
[10] de Brujin N G 1981 *Proceedings of Koninklijke Nederlandse Akademie van Wetenschappen Series* A **84** 38 (*Indagationes Mathematicae* **43**)
[11] Senechal M 1995 *Quasicrystals and Geometry* (Cambridge University Press)
[12] Kramer P and Neri R 1984 *Acta Cryst.* A **40** 580
[13] Bak P 1986 *Phys. Rev. Lett.* **56** 861
[14] Conway J H and Knowles K M 1986 *J. Phys. A: Math. Gen.* **19** 3645
[15] Shcherbak O P 1988 *Russian Math. Surveys* **43** 149
[16] Koca M, Koc R and Al- Barwani M 2001 *J. Phys. A: Math. Gen.* **34** 11201
[17] Koca M, Koca N O and Koc R 2012 Non-crystallographic Coxeter Groups from Higher Dimensional Lattices and Quasicrystallography, talk presented in *the Conference on Groups and Geometry* 12-16 December, Bangalore, India (Unpublished)
[18] Shechtman D, Blech I, Gratias D and Cahn J W 1984 *Phys. Rev. Lett.* **53** 1951
[19] DiVincenzo D and Steinhardt P J 1991 *Quasicrystals: the state of the art* (Singapore: World Scientific Publishers)
[20] Janot C 2012 *Quasicrystals: a primer* (Oxford University Press-2nd edition)
[21] Suck J B, Screiber M and Haussler P (eds.) (2002) *Quasicrystals: An introduction to structure, physical properties and applications* (Berlin Heidelberg: Springer-Verlag)
[22] Lubin S M, Zhou W, Hryn A J, Huntington M D and Odom T W 2012 *Nano Lett.* **12** 4948
[23] Coxeter H S M 1952 *Duke Math J.* **18** 765
[24] Coxeter H S M and Moser W O J 1965 *Generators and Relations for Discrete Groups* (Springer-Verlag)
[25] Coxeter H S M 1973 *Regular Polytopes* (Third Edition, Dover Publications)
[26] Meyer Y 1972 *Algebraic Numbers and Harmonic Analysis* (Amsterdam: North-Holland Publishing Company)
[27] Meyer Y 1995 Quasicrystals, Diophantine Approximations and Algebraic Number *Beyond Quasicrystals* eds F Axel and D Gratias Les Edition de Physique (Berlin Heidelberg New York: Springer-Verlag) p 3

[28] Moody R V 1995 Meyer Sets and finite generation of quasicrystals in *Symmetries in Science VIII* ed B Gruber (New York: Plenum)
[29] Moody R V 2000 Meyer Sets: A survey in: *From Quasicrystals to More Complex Systems* eds F Axel, F Dénoyer and J P Gazeau (Centre de physique Les Houches France: Springer-Verlag) p 145
[30] Duneau M and Katz A 1985 *Phys Rev. Lett.* **54** 2688
[31] Ishimasa T, Nissen H U and Fukamo Y 1985 *Phys. Rev. Lett.* **55** 511
[32] Forster S, Meinel K, Hammer R, Trautmann M and Widdra W 2013 *Nature* **502** 215
[33] Slansky R 1981 *Phys. Rep.* C **79** 1
[34] Carter R W 1972 *Simple Groups of Lie Type* (John Wiley & Sons Ltd)
[35] Humphreys J E 1990 *Reflection Groups and Coxeter Groups* (Cambridge University Press)
[36] Steinberg R 1951 *Trans. Amer. Math. Soc.* **71** 274
[37] Engel P 1986 *Geometric Crystallography* (Dordrecht: D Reidel Publishing Company)
[38] MacKay A L 1982 *Physica* 114**A** 609
[39] Katz A 1988 *Comm. Math. Phys.* **118** 263
[40] Katz A 1989 *Some local properties of the 3-dimensional Penrose tilings* Introduction to the mathematics of quasicrystals ed M Jaric (San Diego CA 147: Academic Press)
[41] Conway J H and Sloane N J A 1982 *Voronoi Regions of Lattices, Second Moments of Polytopes and Quantization* (IEE Transactions on Information Theory IT-28, No. 2)
[42] Moody R V and Patera J 1992 *J. Phys. A: Math. Gen.* **25** 5089
[43] Koca M, Koca N O and Koc R 2010 *J. Math. Phys.* **51** 043501
[44] Koca M, Koca N O and Al-Ajmi M 2012 *Int. J. Geom. Methods Mod. Phys.* **9** (4) 1250035
[45] Conway J H and Sloane N J A 1988 *Sphere Packings, Lattices and Groups* (Springer-Verlag New York Inc.)
[46] Koca M, Al-Ajmi M and Koca N O 2013 *Int. J. Geom. Methods Mod. Phys.* **10** (5) 1350010
[47] Koca M, Koc R and Al-Barwani M 2006 *J. Math. Phys.* **47** (4) 043507
[48] Worley R T 1987 *J. Austral. Math. Soc.* A **43** 268
[49] Pervin E 1993 The Voronoi Cells of E_6*and E_7*Lattices in *Proc. the Marshall Hall conference on Coding theory, design theory, group theory* (New York: John Wiley & Sons) p 243

Journal of Physics: Conference Series **563** (2014) 012018 doi:10.1088/1742-6596/563/1/012018

A Mathematica package to derive weight-homogeneous decompositions of vector fields

Radosław Antoni Kycia

University of Warsaw
Faculty of Mathematics, Informatics and Mechanics
Banacha 2, Warsaw, 02-097, Poland

E-mail: `rkycia@mimuw.edu.pl`

Abstract. Description of a Mathematica package that performs weight-homogeneous decompositions of polynomial vector fields will be presented. The package has also functions that compute the Kovalevskaya matrix and exponents. Some details of implementation and examples will be also outlined. It can be used in the testing of integrability of dynamical systems described by vector fields as well as in the Painlevé analysis.

1. Introduction

Weight-homogeneous decomposition of a polynomial vector field is base of many methods that allows to obtain some results on the nature of solutions of ordinary differential equations (ODEs), e.g., it is an ingredient of the Painlevé test formulated in [3]. The decomposition can also help to deduce the structure of first integrals. It is a purely combinatorial task. For vector fields that contain small number of monomials in small number of dimensions it can be done by hand, however, when a space dimension is large and a vector field contains many terms it becomes a task in which Computer Algebra Systems (CAS) like Mathematica [6] should be used.

There exist very general packages for Mathematica, Maple or Macsyma CAS to perform the Painlevé analysis (see, [3], [4], [1], [7] and the references therein) which do the first step of weight-homogeneous decomposition, i.e., finding the scale-invariant terms, as a part of algorithm, however, they do not do full decomposition. In this approach the vector structure of the field is preserved which is additional advantage. As far as the author know there is no algorithms which are based on this approach. Therefore, it is important to provide implementation of a method of the decomposition which can serve as a building block for future use.

This paper describes the set of functions in Mathematica CAS [6] language that assist user in weight-homogeneous decomposition. It is organized as follows. In the next section the theory and simple use of the package will be presented. Then details of implementation will be outlined. In Conclusions we provide possible applications and plans for future development.

2. Theory and simple example

The presentation of the theory will be based on [3]. Assume that the space dimension is n, i.e., the vector of coordinates is of the form $x = [x_1, \ldots, x_n]$. Take a vector of integer numbers, the weights $w \in \mathbb{Z}^n$. The following convention about taking powers with respect to the vector simplifies notation $x^w := [x_1^{w_1}, \ldots, x_n^{w_n}]$ for the vector x and $t^w := [t^{w_1}, \ldots, t^{w_n}]$ for a scalar t. A function $f(x)$ has a weight degree d with respect to the weight vector w if

$$f(x_1 t^{w_1}, \ldots, x_n t^{w_n}) = t^d f(x_1, \ldots, x_n). \tag{1}$$

The weight degree is usually denoted by $deg(f, w) := d$. If $\forall_{i \in \{1,\ldots,n\}} \quad w_i = 1$ then the classical definition of the homogeneous function is restored. The definition naturally extends to vector fields, i.e., a field has the weight degree d with respect to the weight vector w if every component of this field vector has this property.

Starting point for a weight-homogeneous decomposition is a scale invariant vector field. It is a weight-homogeneous vector field $f = [f_1, \ldots, f_n]$ which components have weights that fulfils

$$\forall_{i \in \{1,\ldots,n\}} \quad deg(f_i, w) = w_i - 1. \tag{2}$$

The system (2) is a linear system for weights and usually is undetermined which leads to some arbitrary weights. This arbitrariness is the main reason for increase of complexity of computations as it will be explained below. For ODEs $\dot{x}(t) = f(x(t))$ defined by these vector fields there exist power-type solutions

$$x(t) = \alpha t^w, \tag{3}$$

where the values of the vector α can be calculated from the equations

$$\forall_{i \in \{1,\ldots,n\}} \quad \alpha_i w_i = f_i(\alpha). \tag{4}$$

In (3) the multiplication produces a vector and it is understand as a multiplication of corresponding components of the vectors. A homogeneous vector fields with a degree $d \neq 1$ is scale-invariant with the weights $w_i = \frac{1}{1-d}$. Therefore, the vector field defined by a linear system with constant coefficients is homogeneous and cannot be scale-invariant($d = 1$). A solution (3) is used in the Painlevé analysis as a leading-order term around possible singularity if there are nonzero components of α with at least one negative corresponding exponent w_i. The weight of a scale-invariant term and the coefficients α of a scale invariant solution form so called dominant balance $\mathfrak{F} = \{w, \alpha\}$.

For polynomial vector fields there can exist many scale-invariant terms. By fixing one of the possibilities the vector field can be expanded in the following form

$$f = f^{(0)} + f^{(1)} + \ldots + f^{(p)}, \tag{5}$$

where the terms are defined as follows

$$\begin{gathered} f_i^{(j)}(t^w x) = t^{w_i - q^{(j)} - 1} f_i^{(j)}(x), \quad i \in \{1, \ldots, n\}; \quad \forall t \in \mathbf{C}, \\ 0 < q^{(i)} < q^{(j)} \quad \forall i < j, \end{gathered} \tag{6}$$

where $f^{(0)}$ is a scale-invariant term that determines the decomposition. Every scale-invariant term generates one of the weight-homogeneous decomposition.

Determination of a scale-invariant term and selection of one of the corresponding dominant balances defines the Kovalevskaya matrix

$$K := Df^{(0)}(\alpha) - diag(w), \tag{7}$$

which is defined as the Jacobian of the scale-invariant term derived on the scale-invariant solution minus diagonal matrix with weights on its diagonal. If K is semi-simple, i.e., diagonalizable then

Journal of Physics: Conference Series **563** (2014) 012018 doi:10.1088/1742-6596/563/1/012018

its eigenvalues are called the Kovalevskaya exponents. It is a fact [3] that one of the exponents is always -1. It corresponds to arbitrariness of choice of initial 'time'. The values of the Kovalevskaya exponents have important meaning in the Painlevé test and in determining the class of possible first integrals [3].

All above quantities can be calculated using the package. The use of it will be presented below on a simple example. The package can be downloaded from [5].

Consider the planar Lotka-Volterra system of ODEs [3]

$$\begin{cases} \dot{x} = x(ay - x), \\ \dot{y} = y(bx - y), \end{cases} \tag{8}$$

where a, b are real parameters. In order to find all possible scale invariant terms we use $AnalyzeScaleInvariantTerms[equations, variables]$ function. It requires the list of RHS of equations and corresponding variables in order in which the system is arranged. It is useful to define new variables in Mathematica that keep these values, i.e.,

```
equation = {x (a*y - x), y*(b*x - y)};
variables = {x, y};
sol = AnalyzeScaleInvariantTerms[equation, variables];
PrintScaleInvariantTerms[sol];
```

The variable *equation* is substituted by the list of RHS of equations, the variable *varaibles* contains the list of variables in the order indicated by the order of LHS of the equations and *sol* keeps the result of analysis of scale-invariant terms. It can be printed in a user-friendly form by $PrintScaleInvariantTerms[solution]$. As the result three terms will be obtained with the weights

$$\begin{aligned} f^{(0)} &= \begin{bmatrix} -x^2 \\ bxy \end{bmatrix}, \quad w = \begin{bmatrix} w_1 \to -1 \\ w_2 \to \text{anything} \end{bmatrix}; \\ f^{(0)} &= \begin{bmatrix} -x^2 + axy \\ bxy - y^2 \end{bmatrix}, \quad w = \begin{bmatrix} w_1 \to -1 \\ w_2 \to -1 \end{bmatrix}; \\ f^{(0)} &= \begin{bmatrix} axy \\ -y^2 \end{bmatrix}, \quad w = \begin{bmatrix} w_1 \to \text{anything} \\ w_2 \to -1 \end{bmatrix}; \end{aligned} \tag{9}$$

where *anything* means arbitrary value. One can also note that the second solution is a special case of the two remaining ones where arbitrary value of the exponents w_i were set to -1. The problem of arbitrariness of some values of the exponents will be addressed below.

The coefficients α for scale-invariant solutions (3) of (9) can be computed with the help of $CalculateScaleInvariantSolution[solution, variables]$. Continuing the example, if we execute

```
sol = CalculateScaleInvariantSolution[sol, variables];
PrintScaleInvariantSolutions[sol];
```

we obtain the list of all values of α that are solutions of (4). Some of scale-invariant solutions do not describe singular solutions and they can be removed by using $RemoveNonsingular[sol, variables]$. If one is interested in terms which do not contain arbitrary values of weights then these terms can be selected by $RemoveArbitrary[sol, variables]$.

In the next step the Kovalevskaya exponents can be calculated and printed using the following set of instructions

```
km = KovalevskayaMatrix[sol, variables];
PrintScaleInvariantKovalevskaya[km];
```

The final task is to compute remaining terms of the decomposition using $WeightHomogenousDecompose[solution, system, variables, solNumber]$, where $solution$ is previously obtained the set of scale-invariant terms (sol variable in the above example), $system$ and $variables$ are RHS of the system and corresponding variables and $solNumber$ is the index of weight and scale-invariant term which will be treated as the weight and the initial term of the expansion, e.g., the command

```
WeightHomogenousDecompose[sol, equation, variables, 1];
```

produces the remaining terms of the decomposition and corresponding q values for the first scale-invariant term of (9).

The function that performs all the above steps is $WeightHomogenousAnalysis[system, variables]$. For complicated vector fields the output will be long and the step by step method with elimination of unwanted terms may be more preferred by the user.

Application-oriented reader can skip the next section in which details of implementation will be discussed.

3. Details of implementation

This section may be interesting for those who want to alter or adjust the package to their needs.

Firstly, the algorithm for the searching of scale-invariant terms will be discussed. It consists of a few steps:

(i) Extract monomials with respect to the variables in a form of the lists for every components of the vector field and add 0 to every monomial list.

(ii) Create all possible combinations of monomials from different components/lists. Exclude zero vector from the list of combinations.

(iii) Substitute $x \to t^w$ for variables.

(iv) Form equations (4).

(v) Solve the equations.

(vi) Remove terms which have no solutions for exponents.

(vii) Simplify solutions for exponents with arbitrary parameters by back substitution and join the terms which have the same weights.

(viii) Create new terms which are terms with contains arbitrary exponents that were substituted by the solutions from other terms - create special cases of a solution with arbitrary parameters. Then join the terms which have the same weights.

(ix) Join terms that one is a special case of the other.

(x) Repeat vii to join the terms introduced in the previous two steps.

The step vii is a necessary because Mathematica $Solve[]$ function used for solving the system for exponents produce sometimes output of the form

$$\begin{bmatrix} w_1 \to 1 \\ w_2 \to 1 - w_1 \end{bmatrix}, \tag{10}$$

which can be obviously simplified by substituting w_1 form the first rule into the RHS of the second one. In order to explain the step viii consider two terms

$$\begin{bmatrix} w_1 \to -\frac{2}{3} \\ w_2 \to -\frac{2}{3} \\ w_3 \to -\frac{5}{3} \\ w_4 \to -\frac{5}{3} \end{bmatrix}, \quad \begin{bmatrix} w_1 \to \text{anything} \\ w_2 \to -\frac{2}{3} \\ w_3 \to -1 + w_1 \\ w_4 \to -\frac{5}{3} \end{bmatrix}. \tag{11}$$

It is obvious that the second vector of rules is a special case ($w_1 = -\frac{2}{3}$) of the first one and the polynomial associated with this general term can be added to the polynomial associated with the specialized term. It is performed by substitution of weights of special cases to the general ones, which create new terms, and then join distinct components of the vector fields with the same weights. In the step ix the algorithm joins the terms which have weights with the value *anything* and are that are 'orthogonal', which is illustrated by the following two terms

$$\begin{bmatrix} w_1 \to \text{anything} \\ w_2 \to 1 \end{bmatrix}, \quad \begin{bmatrix} w_1 \to 2 \\ w_2 \to \text{anything} \end{bmatrix}. \tag{12}$$

They produce more specific term with $w_1 = 2$, $w_2 = 1$, which generates the polynomial vector which is the sum of the polynomial vectors associated with both more general weights.

All these tasks are performed also when finding the scale-invariant terms by hand. The result of the function is a list that every entity is a list that contains a scale-invariant term as a first element and a vector of weights as a second component. The procedure that calculates the scale invariant solutions uses (4) for every scale invariant term. Likewise, the algorithm for finding the Kovalevskaya exponents is also simple use of (7). It creates a list in which to every scale invariant term the lists that contain the Kovalevskaya matrices and their eigenvalues is appended.

Now let us focus on the algorithm that for a given scale-invariant term produces all higher terms with respect to its weight. The outline of the algorithm is as follows

(i) For each term in every component of a vector field exclude terms from the scale-invariant term and calculate their weights $q^{(i)}$ from (6).

(ii) Add the terms with the same weight for every component.

It is simpler comparing to the previous algorithm as now the weights can be calculated independently for each term in every component.

This section should give the reader general idea of implementation of the algorithms in Mathematica. For more detailed description we refer the reader to the code [5]. The notebook contains a section that provides many examples of analysis of dynamical systems from [3]. It also serves as a benchmark. The next section gives general discussion on possible use of the package.

4. Conclusions

The package presented in the paper can be used as a base to implement more complex packages that automatize the Painlevé test or those that check integrability of dynamical systems. It can be also used to check if a dynamical system has scale invariant solutions. The future development of the package will go in these directions. As it was mentioned in Introduction there is a lot of the Painlevé test packages, however to our knowledge none of them is using the weight-homogeneous decomposition. Large parts of the code are written using functional style of programming, which is very effective approach in symbolic computations.

Acknowledgment
RK is supported by the Warsaw Center of Mathematics and Computer Science from the funds of the Polish Leading National Research Centre (KNOW). RK is grateful to Galina Filipuk(MIM UW, PL) and Rod Halburd(UCL, UK) for enlightening discussions.

References

[1] D. Baldwin and W. Hereman 2006 *Symbolic software for the Painlev test of nonlinear differential ordinary and partial equations*, Journal of Nonlinear Mathematical Physics, 13 (1), 90–110

[2] Conte, R. (editor) 1999, *The Painlev property. One century later* CRM Series in Mathematical Physics, Springer
[3] Goriely, A. 2001 *Integrability and nonintegrability of dynamical systems* Advanced Series on Nonlinear Dynamics, World Scientific
[4] Goriely, A. 1992 *Investigation of Painlevé property under time singularities transformations* J. Math. Phys. 33, 2728;
[5] Kycia, R. A. Web page: `http://www.mimuw.edu.pl/%7erkycia/`
[6] Wolfram Mathematica Web page: `http://www.wolfram.com/mathematica/`
[7] Gui-qiong Xu, Zhi-bin Li 2004 *Symbolic computation of the Painlev test for nonlinear partial differential equations using Maple* Comput. Phys. Commun. 161 65–75

Scattering Amplitudes for Multi-indexed Extensions of Soliton Potential and Extended KdV Integer Solitons

Jen-Chi Lee

Department of Electrophysics,National Chiao-Tung University and
Physics Division, National Center for Theoretical Sciences,
Hsinchu, Taiwan, R.O.C.

E-mail: jcclee@cc.nctu.edu.tw

Abstract. We calculate quantum mechanical scattering problems for multi-indexed extensions of soliton potential by Darboux transformations in terms of pseudo virtual wavefunctions. As an application, we calculate infinite set of higher integer KdV solitons by the inverse scattering transform method of KdV equation.

1. Introduction

The discovery of the exceptional orthogonal polynomials [1, 2, 3, 4, 5], and their applications to exactly solvable 1D quantum mechanical systems [3, 4, 5] have recently attracted many substantial researches in mathematical physics. In this talk, instead of the bound state problems, we will report on new development of 1D exactly solvable quantum mechanical scattering problems. In particular, we will calculate the scattering problems of the deformed soliton potential and apply the results to the KdV solitons.

Our calculation will be based on the multiple Darboux-Crum transformations [6, 7, 8]. Such transformations can generate new solvable quantum systems from the known ones by using certain polynomial type seed solutions. These seed functions are called the virtual and pseudo virtual state wavefunctions [5, 9, 10]. One way to obtain these seed solutions is to perform discrete symmetry operations on the eigenfunctions. The one-indexed [11] and more complete multi-indexed extensions [12] of the known quantum scattering problems [13] were recently calculated.

For the case of the deformed soliton potentials, one interesting application is to use the results of the scattering data to generate higher integer KdV solitons from the lower integer ones. Although the KdV solitons generated by this method are not new soliton solutions, we believe that the method we used is simple and effective, and is closely related to the recent developed multi-indexed extention of solvable quantum mechanical potentials. In this calculation, we will obtain an infinite number of reflectionless potentials, which can be served as the initial profiles of integer KdV solitons. We then use the scattering data to solve the Gel'fand-Levitan-Marchenko (GLM) equation [14, 15] in the inverse scattering transform (IST) method [16, 17], and obtain higher integer KdV solitons.

2. The Deformed Soliton Potential

We begin with a specific example of new solvable potential, namely, the deformed soliton potential under Darboux-Crum transformation. We will calculate both its bound state problem and scattering problem, and relate these two calculations. The bound state problem and scattering problem of the original soliton potential

$$U(x;\lambda) = -\frac{h(h+1)}{\cosh^2 x}, \lambda = h, h > 0, -\infty < x < \infty \tag{2.1}$$

were calculated in [18]. This potential contains finitely many bound states

$$\begin{aligned} \phi_n(x) &= \frac{1}{(\cosh x)^{h-n}} P_n^{(h-n,h-n)}(\tanh x) \\ &\sim \frac{1}{(\cosh x)^{h-n}}\ {}_2F_1\left(-n, 2h-n+1, h-n+1, \frac{1-\tanh x}{2}\right), \\ E_n &= -k^2 = -(h-n)^2; n = 0,1,2...[h]' \end{aligned} \tag{2.2}$$

where $P_n^{(h-n,h-n)}$ is the Jacobi polynomial and ${}_2F_1$ is the hypergeometric function. $[h]'$ denotes the greatest integer not exceeding and not equal to h. The soliton potential contains a discrete symmetry

$$h \to -(h+1), \tag{2.3}$$

which can be used to construct the seed function

$$\varphi_v(x) = (\cosh x)^{h+1+v} P_v^{(-h-1-v,-h-1-v)}(\tanh x), v = 0,1,2,3,4,.. \tag{2.4}$$

with energy $E_v = -(h+v+1)^2$. It turns out that for $v = 1,3,5....$, the deformed potential contains pole at $x = 0$. For example, for $v = 1$,

$$U_1 = U - 2\frac{d^2}{dx^2}\log\varphi_1(x) = U - \frac{2(h+1)}{\cosh^2 x} + \frac{2}{\sinh^2 x} \tag{2.5}$$

which contains pole at $x = 0$. We note that although one can define the asymptotic forms of the scattering state for this potential, the corresponding bound state wavefunctions contain singularities. So for our purpose here, only $v = 2,4,6...$can be used to deform the soliton potential. For simplicity, we will use the seed function for $v = 2$

$$\varphi_2(x) = \frac{h+1}{4}(\cosh x)^{h+3}[1 + (2h+3)\tanh^2 x] \tag{2.6}$$

to illustrate the calculation. The deformed potential is

$$\begin{aligned} U_2 = U - 2\frac{d^2}{dx^2}\log\varphi_2(x) &= U - \frac{4(2h+3)(1-2\sinh^2 x)}{[1+(2h+3)\tanh^2 x](\cosh x)^4} \\ &- 8\left(\frac{(2h+3)\tanh x}{[1+(2h+3)\tanh^2 x](\cosh x)^2}\right)^2, \end{aligned} \tag{2.7}$$

which has no pole for the whole regime of x and approaches 0 asymptotically for $x \to \pm\infty$ as U does. Note that $U_2(x=0) - U(x=0) = -4(2h+3) < 0$, which suggests the existence of a lowest new bound state for the deformed potential U_2.

The bound state wavefunctions of the deformed potential Eq.(2.7) can be calculated through the Darboux-Crum transformation to be

$$\begin{aligned}\psi_b^{(1)}(x) &= \phi_n' - \frac{\varphi_2'}{\varphi_2}\phi_n \\ &= -\left(\frac{(2h-n+3)\tanh x}{(\cosh x)^{h-n}} + \frac{2(2h+3)\tanh x}{[1+(2h+3)\tanh^2 x](\cosh x)^{h-n-2}}\right) \\ &\cdot\ {}_2F_1\left(-n, 2h-n+1, h-n+1, \frac{1-\tanh x}{2}\right) \\ &+ \frac{n(2h-n+1)}{2(h-n-1)(\cosh x)^{h-n-2}} \\ &\cdot\ {}_2F_1\left(-n+1, 2h-n+2, h-n+2, \frac{1-\tanh x}{2}\right)\end{aligned} \tag{2.8}$$

with energy

$$E_n = -k^2 = -(h-n)^2; n = 0, 1, 2...[h]'. \tag{2.9}$$

In calculating Eq.(2.8), we have used the identity

$$\frac{d}{dz}\ {}_2F_1(a, b, c, z) = \ {}_2F_1(a+1, b+1, c+1, z). \tag{2.10}$$

It can be easily shown that there is another bound state

$$\frac{1}{\varphi_2} = \frac{4/(h+1)}{(\cosh x)^{h+3}[1+(2h+3)\tanh^2 x]} \tag{2.11}$$

with lowest energy

$$E = -(h+3)^2 \tag{2.12}$$

as was expected previously.

We next consider the scattering problem of the deformed potential. For this purpose, we introduce the wave vector K such that

$$E = K^2 > 0, K = ik. \tag{2.13}$$

In view of the bound state wavefunction in Eq.(2.8), the scattering state wavefunction for the undeformed potential is

$$\psi_+(x) = \frac{1}{(\cosh x)^{-iK}}\ {}_2F_1\left(-iK-h, -iK+h+1, -iK+1, \frac{1-\tanh x}{2}\right). \tag{2.14}$$

The asymptotic form of the scattering state for the deformed potential can be calculated to be

$$\psi_{+\infty}^{(1)}(x) = \psi_{+\infty}'(x) - \frac{\varphi_2'}{\varphi_2}\psi_{+\infty}'(x) = [iK-(h+3)]\exp iKx \tag{2.15}$$

as $x \to +\infty$. To calculate the asymptotic form of the scattering state as $x \to -\infty$, one needs to include the second solution of the Schrodinger equation. One way to achieve this is to use the identity [18]

$$\begin{aligned}{}_2F_1(a, b, c, z) &= \frac{\Gamma(c)\Gamma(c-a-b)}{\Gamma(c-a)\Gamma(c-b)}\ {}_2F_1(a, b, a+b-c+1, 1-z) \\ &+ \frac{\Gamma(c)\Gamma(a+b-c)}{\Gamma(a)\Gamma(b)}(1-z)^{c-a-b} \\ &\cdot {}_2F_1(c-a, c-b, c-a-b+1, 1-z)\end{aligned} \tag{2.16}$$

in Eq.(2.14) to obtain

$$\begin{aligned}\psi_-(x) &= \frac{1}{(\cosh x)^{-iK}}\frac{\Gamma(c)\Gamma(c-a-b)}{\Gamma(c-a)\Gamma(c-b)}\,{}_2F_1\left(a,b,a+b-c+1,1-z\right)\\ &+ \frac{1}{(\cosh x)^{-iK}}\frac{\Gamma(c)\Gamma(a+b-c)}{\Gamma(a)\Gamma(b)}(1-z)^{c-a-b}\\ &\cdot {}_2F_1\left(c-a,c-b,c-a-b+1,1-z\right)\end{aligned} \tag{2.17}$$

where $a=-iK-h, b=-iK+h+1, c=-iK+1$ and $z=\frac{1-\tanh x}{2}$. One can now calculate the asymptotic form of Eq.(2.17)

$$\psi_{-\infty}(x) = \exp -iKx\,\frac{\Gamma(-iK+1)\Gamma(iK)}{\Gamma(1+h)\Gamma(-h)} + \exp iKx\,\frac{\Gamma(-iK+1)\Gamma(-iK)}{\Gamma(-iK-h)\Gamma(-iK+h+1)} \tag{2.18}$$

as $x\to-\infty$. The asymptotic form of the scattering state for the deformed potential can then be calculated to be

$$\begin{aligned}\psi^{(1)}_{-\infty}(x) &= \psi'_{-\infty}(x) - \frac{\varphi_2'}{\varphi_2}\psi'_{-\infty}(x) = [-iK+(h+3)]\exp -iKx\,\frac{\Gamma(-iK+1)\Gamma(iK)}{\Gamma(1+h)\Gamma(-h)}\\ &+ [iK+(h+3)]\exp iKx\,\frac{\Gamma(-iK+1)\Gamma(-iK)}{\Gamma(-iK-h)\Gamma(-iK+h+1)}\end{aligned} \tag{2.19}$$

as $x\to-\infty$. Finally the transmission and reflection coefficients of the deformed soliton potential can be calculated to be

$$t_D(K) = \frac{iK-(h+3)}{iK+(h+3)}\cdot\frac{\Gamma(-iK-h)\Gamma(-iK+h+1)}{\Gamma(-iK+1)\Gamma(-iK)} = \frac{iK-(h+3)}{iK+(h+3)}\cdot t(K), \tag{2.20}$$

$$\begin{aligned}r_D(K) &= (-)\frac{K+i(h+n+1)}{K-i(h+n+1)}\cdot\frac{\Gamma(1+h-iK)\Gamma(-h-iK)\Gamma(ik)}{\Gamma(-h)\Gamma(1+h)\Gamma(-iK)}\\ &= (-)\frac{K+i(h+n+1)}{K-i(h+n+1)}\cdot r(K)\end{aligned} \tag{2.21}$$

where $t(K)$ and $r(K)$ are the transmission and reflection coefficients of the undeformed soliton potential. It is important to note that, in Eq.(2.20), in addition to the well known second factor for the undeformed soliton potential scattering, an additional ratio $\frac{iK-(h+3)}{iK+(h+3)}$ with a new pole $K=i(h+3)$ corresponding to the new bound state energy given by Eq.(2.12) is added to the transmission coefficient of the deformed potential. Moreover, the additional ratio is of module one so that the conservation of flux is preserved for the deformed potential. In general, It turns out that the M-step deformed transmission amplitude $t_D(K)$ and the reflection amplitude $r_D(K)$ are [12]

$$t_D(K) = \prod_{j=1}^{M}\frac{K+i\Delta^+_{d_j}}{K+i\Delta^-_{d_j}}\cdot t(K), \quad r_D(K) = (-1)^M\prod_{j=1}^{M}\frac{K-i\Delta^-_{d_j}}{K+i\Delta^-_{d_j}}\cdot r(K) \tag{2.22}$$

where $\Delta^+_{d_j}$ and $\Delta^-_{d_j}$ are the asymptotic exponents of seed solutions φ_{d_j}

$$\varphi_{d_j} \approx \begin{cases} e^{x\Delta^+_{d_j}} & x\to+\infty \\ e^{x\Delta^-_{d_j}} & x\to-\infty \end{cases}. \tag{2.23}$$

In view of the multiplicative form of $r_D(k)$ in Eq.(2.21), it is important to note that, for integer $h = 1, 2, 3....$, the scattering of the deformed potential remains reflectionless as the undeformed potential due to the factor $\Gamma(-h)$ in the denominator of $r(k)$. This key observation is important for generating higher integer soliton solutions in the calculation of the inverse transform method to be discussed in the next section.

3. Solvable Higher Integer KdV Solitons

The nontopological KdV solitons are described by the Korteweg-de Vries (KdV) equation, i.e.,

$$u_t - 6uu_x + u_{xxx} = 0 \tag{3.24}$$

in one space $x \in (-\infty, \infty)$ and one time $t > 0$ dimension. One of the important methods to solve nonlinear partial differential equation such as the KdV equation is the method of the inverse scattering transform(IST) [16, 17] invented in 1960's. According to the method of IST, the solution of the KdV equation is converted to the solution of two simpler linear equations, namely, the quantum mechanical Schrodinger equation and the Gel'fand-Levitan-Marchenko (GLM) equation [14, 15]. For KdV soliton solutions the related Schrödinger equation is connected with reflectionless potentials. For such reflectionless potentials, the reflection amplitudes vanish, and the corresponding GLM equation is easy to solve. For the general N-soliton solution, one gets $2N$ continuous parameters, N norming constants $c_n(0)$ and N energy parameters κ_n.

In the following [19], for simplicity and clarity of presentation, we present the result for 1-step deformation and take the seed function with $v = 2$, and parameter in the soliton potential $h = 1$

$$\varphi_2(x)_{h=1} = \frac{1}{2}\cosh^4 x(1 + 5\tanh^2 x) \tag{3.25}$$

to illustrate the calculation. The deformed potential is easily calculated to be

$$U_2(x)_{h=1} = U(x) - 2\frac{d^2}{dx^2}\log\varphi_2(x)_{h=1} = -\frac{30(4\cosh^4 x \quad 8\cosh^2 x + 5)}{\cosh^2 x(36\cosh^4 x - 60\cosh^2 x + 25)} \tag{3.26}$$

which has no pole and no zero for the whole regime of x and approaches 0 asymptotically for $x \to \pm\infty$ as $U(x)_{h=1}$ does. The bound state wavefunctions of the deformed potential Eq.(3.26) can be calculated through the Darboux-Crum transformation to be

$$\psi_0(x) = \phi_0' - \frac{\varphi_2'}{\varphi_2}\phi_0 = -5\,\text{sech}x\tanh x\left(1 + \frac{2\,\text{sech}^2 x}{(1 + 5\tanh^2 x)}\right) \tag{3.27}$$

with energy

$$E_0 = -\kappa_0^2 = -(h - n)^2 = -1. \tag{3.28}$$

It can be easily shown that there is another bound state of the deformed potential

$$\psi_1(x) \sim \frac{1}{\varphi_2} = \frac{2}{\cosh^4 x(1 + 5\tanh^2 x)} \tag{3.29}$$

with a lower energy

$$E_1 = -\kappa_1^2 = -(h + 1 + v)^2 = -4^2 \tag{3.30}$$

as was expected previously. The normalized wavefunctions and their asymptotic forms can be calculated to be

$$\psi_0(x) = \sqrt{\frac{15}{2}}\,\text{sech}x\tanh x\left(1 + \frac{2\,\text{sech}^2 x}{(1 + 5\tanh^2 x)}\right) \to \sqrt{\frac{10}{3}}e^{-x} \text{ as } x \to \infty, \tag{3.31}$$

Journal of Physics: Conference Series **563** (2014) 012019 doi:10.1088/1742-6596/563/1/012019

$$\psi_1(x) = \sqrt{\frac{15}{8}} \frac{2}{\cosh^4 x(1+5\tanh^2 x)} \to \sqrt{\frac{40}{3}} e^{-x} \text{ as } x \to \infty. \tag{3.32}$$

The constants

$$c_0(0) = \sqrt{\frac{10}{3}}, \quad c_1(0) = \sqrt{\frac{40}{3}} \tag{3.33}$$

in equations Eq.(3.31) and Eq.(3.32) are called norming constants. The reflection amplitude of the scattering of the M-step ($M = 1$ for the present case) deformed soliton potential Eq.(3.26) was calculated in Eq.(2.22). In view of the multiplicative form of $r_D(k)$, it is important to note that, for integer $h = 1, 2, 3....$, the scattering of the deformed potential remains reflectionless as the undeformed potential due to the factor $\Gamma(-h)$ in the denominator of $r(k)$.

We can now use the scattering data $\{\kappa_n, c_n, r_D(k)\}$ to solve the KdV equation. For the reflectionless potential, $r_D(k) = 0$, the GLM equation is easy to solve, and the solution $u(x,t)$ is given by [17]

$$u(x,t) = -2\frac{d^2}{dx^2}\log(\det A), \tag{3.34}$$

where A is a $N \times N$ matrix ($N \equiv h+1$) with elements A_{mn} given by

$$A_{mn} = \delta_{mn} + c_n^2(t)\frac{\exp-(\kappa_m+\kappa_n)x}{\kappa_m+\kappa_n}; \quad m,n = 0,1,2....,N-1. \tag{3.35}$$

In Eq.(3.35) $c_n(t) = c_n(0)\exp(4\kappa_n^3 t)$ and is one of the Gardner-Greene-Kruskal-Miura (GGKM) equations [16].

For the present case, $N = h+1 = 2$. The integer 2-soliton solution corresponding to $(\kappa_0,\kappa_1) = (1,4)$ can be calculated to be [19]

$$u(x,t)_{(1,4)} = -\frac{120e^{8t+2x}(e^{1024t}+e^{16x}+16e^{520t+6x}+30e^{512t+8x}+16e^{504t+10x})}{(3e^{520t}+3e^{10x}+5e^{512t+2x}+5e^{8t+8x})^2}. \tag{3.36}$$

By taking $t=0$ in Eq.(3.36), one reproduces the initial profile $u(x,0) = U_2(x)_{h=1}$ calculated in Eq.(3.26). The asymptotic form of the $(\kappa_0,\kappa_1) = (1,4)$ solution is

$$u(x,t)_{(1,4)} \sim -2\sum_{n=0}^{N-1}\kappa_n^2 \sec h^2\{\kappa_n(x-4\kappa_n^2 t) \pm \chi_n\}, t \to \pm\infty, \tag{3.37}$$

where

$$\exp(2\chi_n) = \prod_{\substack{m=0\\m\neq n}}^{N-1}\left|\frac{\kappa_n-\kappa_m}{\kappa_n+\kappa_m}\right|^{sgn(\kappa_n-\kappa_m)}. \tag{3.38}$$

Note that the previous integer 2-soliton solution corresponds to $(\kappa_0,\kappa_1) = (1,2)$. It is interesting to see that the calculation of the $(1,4)$ integer soliton is based on the recently developed multi-indexed extensions of the reflectionless soliton potential.

The calculation can be generalized to other higher integer solitons. The profile for $h=2$, for example, is the extended solvable 3-soliton $(1,2,5)$ [19]

$$U_2(x)_{h=2} = u(x,0)_{(1,2,5)} = -\frac{4(144\cosh^4 x - 280\cosh^2 x + 147)}{\cosh^2 x(64\cosh^4 x - 112\cosh^2 x + 49)}, \tag{3.39}$$

and

$$\begin{aligned} u(x,t)_{(1,2,5)} = & -(16e^{8t+2x}(9e^{2128t} + 9e^{28x} + 1575e^{16(63t+x)} + 882e^{16(66t+x)} \\ & + 3252e^{14(76t+x)} + 175e^{8(142t+x)} + 49e^{8(250t+x)} + 126e^{4(516t+x)} \\ & + 56e^{2072t+2x} + 126e^{2056t+6x} + 1008e^{1128t+10x} + 882e^{1072t+12x} \\ & + 1575e^{1120t+12x} + 1008e^{1000t+18x} + 49e^{128t+20x} \\ & + 175e^{992t+20x} + 126e^{72t+22x} + 126e^{64t+24x} + 56e^{56t+26x}) \\ & /(2e^{1072t} + 2e^{16x} + 14e^{4(252t+x)} + 9e^{2(532t+x)} + 7e^{1000t+6x} \\ & + 7e^{72t+10x} + 14e^{64t+12x} + 9e^{8t+14x})^2. \end{aligned} \tag{3.40}$$

4. Discussion

In this talk we report on the recent calculation of the scattering problems of multi-indexed extensions of the soliton potential. Similar calculation can be performed to other known solvable potentials [12]. As an application [19], we point out an infinite set of higher integer initial profiles of the KdV solitons, which are both exactly solvable for the Schrodinger equation and for the Gel'fand-Levitan-Marchenko equation in the inverse scattering transform method of KdV equation. The calculation of these solutions are based on the multi-indexed extensions of the reflectionless soliton potential based on the Darboux-Crum transformation.

5. Acknowledgments

This talk is based on a work with Choon-Lin Ho and Ryu Sasaki [12], and another work with Choon-Lin Ho [19]. The work is supported in part by
NSC-100-2112-M-009-002-MY3, and S.T. Yau center of NCTU, Taiwan.

References

[1] D.Gomez-Ullate, N.Kamran and R.Milson, J. Math. Anal. Appl. 387 (2012) 410-418, arXiv: 1103.5724 [math-ph].
[2] D.Gomez-Ullate, N.Kamran and R.Milson, J. Approx Theory 162 (2010) 987-1006, arXiv:0805. 3376 [math-ph]; J. Math. Anal. Appl. 359 (2009) 352-367, arXiv:0807.3939 [math-ph].
[3] C.Quesne, J. Phys. A41 (2008) 392001, arXiv:0807.4087 [quant-ph].
[4] C.Quesne, SIGMA 5 (2009) 084, arXiv:0906.2331[mathph].
[5] S.Odake and R. Sasaki, Phys. Lett. B702 (2011) 164-170, arXiv:1105.0508 [math-ph].
[6] G.Darboux, *Theorie generale des surfaces* vol 2 (Gauthier-Villars, Paris, 1888).
[7] M.M.Crum, Quart. J. Math. Oxford Ser. (2) 6 (1955) 121-127, arXiv:physics/9908019.
[8] M.G.Krein, Doklady Acad. Nauk. CCCP 113 (1957) 970-973; V.E. Adler, Theor. Math. Phys. 101 (1994) 1381-1386.
[9] S.Odake and R. Sasaki, J. Phys. A46 (2013) 235205 (15pp), arXiv:1301.3980 [math-ph].
[10] S.Odake and R. Sasaki, J. Phys. A46 (2013) 245201 (24pp), arXiv:1212.6595 [math-ph].
[11] R. K. Yadav, A. Khare and B. P. Mandala, Ann. Phys. 331 (2013) 313-316, arXiv:1212.4251 [math-ph]; "The Scattering amplitude for one parameter family of shape invariant potentials related to Xm Jacobi polynomials," arXiv:1303.3669 [quant-ph].
[12] C.-L. Ho, J.-C. Lee and R. Sasaki, "Scattering Amplitudes for Multi-indexed Extensions of Solvable Potentials", Ann. of Phys. 343 (2014) 115-131. arXiv: 1309.5471 [quant-ph].
[13] A. Khare and U. P. Sukhatme, J. Phys. A21 (1988) L501-L508.
[14] I.M. Gel'fand and B.M. Levitan, Amer.Math. Soc. Trans (2) 1 (1955) 253.
[15] V.A. Marchenko, Dokl. Akad. Nauk SSSR 104 (1955) 695.
[16] C.S. Gardner, J.M. Greene, M.D. Kruskal and R.M. Miura, Phys. Rev. Lett. 19 (1967) 1095.
[17] P.G. Drazin and R.S. Johnson, *Solitons: an Introduction* (Cambridge University Press, London, 1989).
[18] L.D. Landau and E.M. Lifshiftz, *Quantum Mechanics*, 3rd ed. (Elsevier, Singapore, 2007).
[19] C.-L. Ho and J.-C. Lee, "Multi-indexed Extensions of Soliton Potential and Extended Integer Solitons of KdV Equation", arXiv: 1401.1150.

Journal of Physics: Conference Series **563** (2014) 012020 doi:10.1088/1742-6596/563/1/012020

Integrable deformations of the $\mathrm{AdS}_5\times\mathrm{S}^5$ superstring and the classical Yang-Baxter equation

– *Towards the gravity/CYBE correspondence* –

Takuya Matsumoto
Institute for Theoretical Physics and Spinoza Institute, Utrecht University, Leuvenlaan 4, 3854 CE Utrecht, The Netherlands.
E-mail: t.matsumoto@uu.nl

Kentaroh Yoshida
Department of Physics, Kyoto University, Kyoto 606-8502, Japan.
E-mail: kyoshida@gauge.scphys.kyoto-u.ac.jp

Abstract. Based on the formulation of Yang-Baxter sigma models developed by Klimcik and Delduc-Magro-Vicedo, we explain that various deformations of type IIB superstring on $\mathrm{AdS}_5\times\mathrm{S}^5$ can be charactered by classical r-matrices satisfying the classical Yang-Baxter equation (CYBE). The relation may be referred to as *the gravity/CYBE correspondence.* We present non-trivial examples of the correspondence including Lunin-Maldacena backgrounds for β-deformations of the $\mathcal{N}=4$ super Yang-Mills theory and the gravity duals for non-commutative gauge theories. We also discuss non-integrable backgrounds such as $\mathrm{AdS}_5\times T^{1,1}$ as a generalization.

1. Introduction

The AdS/CFT correspondence between type IIB superstring on the $\mathrm{AdS}_5\times\mathrm{S}^5$ background and the $\mathcal{N}=4$ super Yang-Mills theory [1] has a rich integrable structure. On the string-theory side, the Green-Schwarz action on $\mathrm{AdS}_5\times\mathrm{S}^5$ is described as a sigma model on a supercoset $PSU(2,2|4)/[SO(4,1)\times SO(5)]$ [2]. Based on the $\mathbb{Z}_4$-grading of the superconformal group $PSU(2,2|4)$, a novel one-parameter flat connection has been constructed [3] (For an argument in the Roiban-Siegel formulation [4] see [5]). On the gauge-theory side, one-loop dilation operator can be mapped to the hamiltonian of an integrable spin chain [6]. For a comprehensive review, see [7]. However, the fundamental mechanism underlying the duality is still unclear. To understand the duality much deeper, it would be of significance to argue its deformations which preserve the integrability. The deformations allow us to take various limits by adjusting deformation parameters and make our understanding much deeper.

A novel prescription to deform principal chiral models was proposed by Klimcik, which is the so-called *the Yang-Baxter sigma models* [8]. The key ingredient in this prescription is *classical r-matrices* satisfying the *modified classical Yang-Baxter equation (mCYBE).* Here it is assumed that the r-matrices are constant and do not depend on spectral parameters. If the r-matrices are skew-symmetric, the integrability of the sigma models is preserved. Then this prescription was generalized to symmetric coset cases by Delduc-Magro-Vicedo [9]. Just after that, they succeeded to construct a q-deformed action of the $\mathrm{AdS}_5\times\mathrm{S}^5$ superstring [10]. Although in the original formulation, the mCYBE was utilized, the usual CYBE is also available to describe integrable deformations along a similar line [11]. Thus one may consider two kinds of integrable deformations of the $\mathrm{AdS}_5\times\mathrm{S}^5$ superstring[1];

(i) integrable deformations from the mCYBE,

(ii) integrable deformations from the CYBE.

As we will explain later, the difference between them should not be confused.

In the case (i), a typical skew-symmetric constant r-matrix is of the Drinfeld-Jimbo type [14, 15]. The deformed string action [10] exhibits the quantum group symmetry $\mathcal{U}_q(\mathfrak{psu}(2,2|4))$ [16][2]. Thus, this type of deformation might be regarded as a q-deformation of the $\mathrm{AdS}_5\times\mathrm{S}^5$ string action. The metric and NS-NS two-form of the deformed background has been explicitly obtained by [20]. However, completing the full supergravity solutions is still open problem. There are several related works in this direction. Some specific limits of the deformation parameter are investigated in [21] and [22, 23]. Especially, the relation to the $\mathrm{AdS}_5\times\mathrm{S}^5$ mirror models was found in [22, 23]. A fast-moving string limit and its relation to anisotropic Landau-Lifshitz sigma models have been studied by [24]. A deformed Neumann model was presented by [25] with a rotating string ansatz on the deformed $\mathrm{AdS}_5\times\mathrm{S}^5$. Classical string solutions, called giant magnons, on the deformed background are constructed in [23, 26]. A new coordinate system is proposed in [27] and minimal surface solutions have been discussed. Interestingly, it exhibits a BPS-like behavior, while supersymmetries are q-deformed and these are not realized in the usual sense any more.

In the case (ii), on the other hand, an application for the $\mathrm{AdS}_5\times\mathrm{S}^5$ superstring is proposed in [11]. This case contains Jordanian type deformations [28, 29] and often called Jordanian deformations of the $\mathrm{AdS}_5\times\mathrm{S}^5$ superstring. This short article is devoted to give a review of integrable deformations of this type. As a remarkable feature of the deformations based on the CYBE, there are several types of r-matrices such as *abelian*, *Jordanian* and *abelian-Jordanian* types. In addition to varieties of the r-matrices, one can consider partial deformations of the $\mathrm{AdS}_5\times\mathrm{S}^5$ background.

An important question is what the physical meaning of the CYBE-type deformation is. Our conjecture for this question is that there exists a duality chain which gives the identical supergravity background defined by the CYBE-type deformation. In other words, we might interpret the solutions of the CYBE as a moduli space of the deformations of the $\mathrm{AdS}_5\times\mathrm{S}^5$ string background. In fact, based on a series of our works [30, 31, 32, 33, 34], we will present some non-trivial examples including the Lunin-Maldacena backgrounds [35, 36] and the gravity duals of the non-commutative gauge theories [37, 38]. More generally, we expect that there exists a correspondence between the classical r-matrices satisfying the CYBE and the deformed type IIB supergravity backgrounds. We refer to this relation as *the gravity/CYBE correspondence* [30].

This article is organized as follows. After recalling the CYBE and classical r-matrix in Section 2, we introduce a deformed $\mathrm{AdS}_5\times\mathrm{S}^5$ superstring action in Section 3. In Section 4, as a generic property of the CYBE-type deformation, we will show that CYBE-type deformations may be interpreted as non-local gauge transformations of the left-invariant one-form. In Section 5, we explicitly derive the deformed metric and the NS-NS two-form, and present some concrete examples of the gravity/CYBE correspondence. We also demonstrate that our deformation technique works beyond the integrability. As an example, in Section 6, we consider deformations of the $\mathrm{AdS}_5 \times T^{1,1}$ background. Section 7 is devoted to conclusion and discussion.

[1] For another approach to integrable deformations, see [12, 13].

[2] For earlier arguments on a q-deformed $\mathfrak{su}(2)$ in squashed sigma models, see [17, 18, 19].

2. Classical r-matrix

In this section, we will give a short review on some properties of the CYBE and classical r-matrices as solutions of the CYBE.

2.1. Classical Yang-Baxter equation (CYBE)

Let $\mathfrak{g}$ be a Lie algebra. A *classical r-matrix* (in tensorial notation)

$$r = \sum_i a_i \otimes b_i \quad \in \quad \mathfrak{g} \otimes \mathfrak{g}$$

is defined as a solution of *the classical Yang-Baxter equation (CYBE)*;

$$[r_{12}, r_{23}] + [r_{12}, r_{13}] + [r_{13}, r_{23}] = 0\,, \tag{1}$$

where $r_{12} = r \otimes 1, r_{23} = 1 \otimes r$ and $r_{13} = \sum_i a_i \otimes 1 \otimes b_i$ with a schematic notation. The CYBE is regarded as a classical analogue of the quantum Yang-Baxter equation (YBE);

$$\mathcal{R}_{12}\mathcal{R}_{13}\mathcal{R}_{23} = \mathcal{R}_{23}\mathcal{R}_{13}\mathcal{R}_{12}\,. \tag{2}$$

Indeed, plugging an expansion

$$\mathcal{R}_{ij} = 1 + \hbar r_{ij} + \mathcal{O}(\hbar^2)$$

with the YBE, it is easy to derive the CYBE (1) at $\mathcal{O}(\hbar^2)$.

When there exists a non-degenerate symmetric bilinear two-from $\langle\ ,\ \rangle : \mathfrak{g} \otimes \mathfrak{g} \to \mathbb{C}$, one can introduce a linear operator denoted by $R \in \mathrm{End}(\mathfrak{g})$ by

$$R(x) = \langle r, x\rangle_2 = \sum_i a_i \langle b_i, x\rangle \qquad \text{for} \qquad {}^\forall x \in \mathfrak{g}\,. \tag{3}$$

Here the subscript of $\langle r, x\rangle_2$ stands for the inner product over the second entry of the tensor. If the classical r-matrix is skew-symmetric $r_{12} = -r_{21}$, the CYBE (1) is equivalently written as

$$[R(x), R(y)] - R([R(x), y] + [x, R(y)]) = 0 \qquad \text{for} \qquad {}^\forall x, y \in \mathfrak{g} \tag{4}$$

in terms of the linear R-operator.

An important generalization of the CYBE is the *modified classical Yang-Baxter equation (mCYBE)*, that is

$$[R(x), R(y)] - R([R(x), y] + [x, R(y)]) = -c^2[x, y] \qquad \text{for} \qquad {}^\forall x, y \in \mathfrak{g}\,. \tag{5}$$

with a parameter $c \in \mathbb{C}$. The mCYBE is reduced to the CYBE by taking the $c \to 0$ limit. When $\mathfrak{g}$ is a real Lie algebra, the possible non-zero parameters are $c = 1$ or i up to rescaling of the R-operator; $R \to |c|R$. Depending on these values, the associated R-operators are classified into *split* ($c = 1$) and *non-split* ($c = i$) type, respectively. The r-matrices discussed in [10, 16] are of non-split type.

2.2. Solutions of the Classical Yang-Baxter equation

In the following, we will introduce some solutions of the (m)CYBE.

In this subsection, we work with $\mathfrak{g} = \mathfrak{gl}(N)$ generated by E_{ij} with $i, j = 1, \cdots, N$ which satisfy the relations;

$$[E_{ij}, E_{kl}] = \delta_{kj}E_{il} - \delta_{il}E_{kj}\,. \tag{6}$$

For $\mathfrak{gl}(N)$, the bilinear form used in (3) is simply given by the trace of the fundamental representations,

$$\langle E_{ij}, E_{kl}\rangle = \mathrm{Tr}(e_{ij}e_{kl}) = \delta_{kj}\delta_{il}\,, \tag{7}$$

where $(e_{ij})_{kl} = \delta_{ik}\delta_{jl}$.

2.2.1. Solutions of mCYBE Let us first look at a solution of the mCYBE.

Drinfeld-Jimbo r-matrix A typical solution of the *modified* CYBE is the Drinfeld-Jimbo type solution,

$$r_{\rm DJ} = c \sum_{1\le i<j\le N} E_{ij} \wedge E_{ji} \tag{8}$$

with $c \in \mathbb{C}\backslash\{0\}$ and the notation $A \wedge B = A\otimes B - B\otimes A$. The associated linear R-operator is

$$R_{\rm DJ}(E_{ij}) = \begin{cases} +cE_{ij} & i<j \\ 0 & i=j \\ -cE_{ij} & i>j \end{cases} . \tag{9}$$

The r-matrix (8) with $c = i$ (non-split type) is used for a q-deformation of the $\mathrm{AdS}_5\times\mathrm{S}^5$ string action [10, 16]. More generally, the Drinfeld-Jimbo r-matrix is determined by fixing a borel subalgebra of $\mathfrak{g}$. For instance, the other choices of r-matrices for $\mathfrak{psu}(2,2|4)$ are discussed in [16].

2.2.2. Solutions of CYBE The next is to consider solutions of the CYBE (1). For linear R-operators, see (4). As a remarkable feature, it is possible to restrict r-matrices to a subalgebra $\mathfrak{gl}(M) \subset \mathfrak{gl}(N)$ with $M < N$. For the r-matrix of Drinfeld-Jimbo type, such restrictions are not allowed due to non-vanishing c^2-terms on the RHS in (5).

Trivial r-matrix It is trivial that $r = 0$ is a solution of the CYBE (1) and it does not deform the original $\mathrm{AdS}_5\times\mathrm{S}^5$ string action, which will be introduced in next section. However, this fact is a very important feature of the CYBE-type deformations since it allows us to partially deform $\mathrm{AdS}_5\times\mathrm{S}^5$ background. Indeed, $R = 0$ does *not* satisfy the mCYBE (5). It seems quite difficult to discuss the partial deformations based on the r-matrices satisfying the mCYBE.

Abelian r-matrix The simplest non-trivial examples are *abelian* r-matrices, which are composed of Cartan generators like

$$r_{\rm Abe} = E_{ii} \wedge E_{jj} \qquad \text{with} \qquad i \neq j\,. \tag{10}$$

Any linear combinations of them also satisfy the CYBE. In terms of the associated linear R-operator, it only acts on the Cartan generators;

$$R_{\rm Abe}(E_{ii}) = -E_{jj}\,, \qquad R_{\rm Abe}(E_{jj}) = E_{ii}\,. \tag{11}$$

Jordanian r-matrix Slightly non-trivial solutions are the so-called *Jordanian* r-matrices. They have non-zero Cartan charges and thus are nilpotent. The generic expression of them is

$$r_{\rm Jor} = E_{ij} \wedge (\alpha E_{ii} - \beta E_{jj}) - \gamma \sum_{i<k<j} E_{ik} \wedge E_{kj} \quad \text{with} \quad \begin{cases} \gamma = \alpha+\beta & \text{or} \\ \gamma = 0 \end{cases} \tag{12}$$

for $1 \le i < j \le N$. The parameter relations are required to satisfy the CYBE. Eventually it contains an extra deformation parameter in addition to the overall scaling. The non-trivial action of the linear R-operator is written down,

$$R_{\rm Jor}(E_{ji}) = -\alpha E_{ii} + \beta E_{jj}\,, \qquad\qquad R_{\rm Jor}(E_{jk}) = -\gamma E_{ik}\,, \tag{13}$$

$$R_{\rm Jor}(E_{kk}) = (\alpha\delta_{ki} - \beta\delta_{kj})E_{ij}\,, \qquad\qquad R_{\rm Jor}(E_{ki}) = +\gamma E_{kj}\,, \tag{14}$$

where $i < k < j$. Indeed, it is easy to see the nilpotency; $(R_{\rm Jor})^n = 0$ for $n \geq 3$.

When the parameters in (12) take special values $\alpha = \beta = c$ and $\gamma = 2c$, the Jordanian r-matrix is obtained from the Drinfeld-Jimbo r-matrix (8) through the *twisting* given by

$$[\Delta(E_{ij}), r_{\rm DJ}] = r_{\rm Jor} \qquad \text{with} \qquad \Delta(E_{ij}) = E_{ij} \otimes 1 + 1 \otimes E_{ij}\,. \tag{15}$$

Abelian-Jordanian r-matrix One may also consider r-matrices commuting each other and nilpotent. Such r-matrices can be constructed from commuting positive (or negative) generators and they may be referred as to *Abelian-Jordanian type*. A typical example takes the form

$$r_{\rm AJ} = E_{ij} \wedge E_{kl} \qquad \text{with} \qquad i<j\,,\ k<l\,,\ j\neq k\,,\ i\neq l\,. \tag{16}$$

The associated linear R-operator is given by

$$R_{\rm AJ}(E_{lk}) = E_{ij}\,, \qquad R_{\rm AJ}(E_{ji}) = -E_{kl}\,, \tag{17}$$

and acts trivially on the other generators. In this case, the nilpotency is $(R_{\rm AJ})^n = 0$ for $n \geq 2$.

3. Deformed $AdS_5 \times S^5$ string action

In the previous section, we have seen some typical solutions of the CYBE. Then, let us utilize them to construct deformed string actions on $AdS_5 \times S^5$.

3.1. Definition of deformed string action

A method to deform principal chiral models with classical r-matrices was introduced by Klimcik, which is called *the Yang-Baxter sigma models* [8]. This technique has been generalized by Delduc-Magro-Vicedo not only for the bosonic cosets [9] but also the $AdS_5 \times S^5$ string action [10, 16]. The point is that these deformations are based on the Drinfeld-Jimbo type r-matrix satisfying the mCYBE. In this section, we will give a short review of the models deformed by classical r-matrices satisfying the CYBE [11].

The deformed action of the $AdS_5 \times S^5$ superstring is given by

$$S = -\frac{1}{4}(\gamma^{\alpha\beta} - \epsilon^{\alpha\beta}) \int_{-\infty}^{\infty} d\tau \int_0^{2\pi} d\sigma\ {\rm Str}\left(A_\alpha d_+ \frac{1}{1 - \eta R_g \circ d_+} A_\beta\right) \tag{18}$$

where the left-invariant one-form A_α is defined as

$$A_\alpha \equiv g^{-1}\partial_\alpha g\,, \qquad g \in SU(2,2|4)\,. \tag{19}$$

Here we work in the conformal gauge. $\gamma^{\alpha\beta}$ and $\epsilon^{\alpha\beta}$ are the flat metric and the anti-symmetric tensor on the string world-sheet normalized by $\gamma^{\alpha\beta} = {\rm diag}(-1,+1)$ and $\epsilon^{\tau\sigma} = 1$. The dressed operator R_g is defined as

$$R_g(X) \equiv g^{-1}R(gXg^{-1})g\,. \tag{20}$$

Noting that both R- and R_g-operators satisfy the CYBE. The R-operator is related to the tensorial representation of a classical r-matrix through

$$R(X) = {\rm Str}_2[r(1 \otimes X)] = \sum_i \left(a_i {\rm Str}(b_i X) - b_i {\rm Str}(a_i X)\right) \tag{21}$$

$$\text{with} \quad r = \sum_i a_i \wedge b_i \equiv \sum_i (a_i \otimes b_i - b_i \otimes a_i)\,.$$

Since a_i, b_i are $\mathfrak{su}(2,2|4)$-valued, the supertrace Str is needed for $(4|4) \times (4|4)$ supermatrix. The operators $d_\pm$ are given by the following,

$$d_\pm = \pm P_1 + 2P_2 \mp P_3\,, \tag{22}$$

where P_i $(i = 0, 1, 2, 3)$ are the projections to the $\mathbb{Z}_4$-graded components of $\mathfrak{su}(2,2|4)$. $P_0\,, P_2$ and $P_1\,, P_3$ are the projectors to the bosonic and fermionic generators, respectively. In particular, $P_0(\mathfrak{su}(2,2|4))$ is nothing but $\mathfrak{so}(1,4) \oplus \mathfrak{so}(5)$, which is a local symmetry of the action (18).

It is noted that the kappa-symmetry of this model is also proved without specifying the form of classical r-matrix [11] and its derivation is similar to the q-deformed case [10].

3.2. The light-cone expression

For later arguments, we will rewrite the action (18) in terms of the light-cone coordinates; $x^\pm = (\tau \pm \sigma)/2$. Then, the light-cone currents are given by

$$A_\pm = A_\tau \pm A_\sigma\,. \tag{23}$$

It is also convenient to introduce the deformed light-cone currents as follows;

$$J_\pm \equiv \frac{1}{1 \mp \eta R_g \circ d_\pm} A_\pm\,. \tag{24}$$

With these notations, the Lagrangian (18) is now simply written as

$$L = \tfrac{1}{4}\text{Str}\,(A_- d_+(J_+)) = \tfrac{1}{4}\text{Str}\,(A_+ d_-(J_-))\,. \tag{25}$$

It is also noted that, by solving the relation (24) with respect to $A_\pm$, the Lagrangian has *dipole-like* expressions as follows;

$$L = \tfrac{1}{4}\text{Str}\,[J_+ d_-(J_-)] - \tfrac{\eta}{4}\text{Str}\,[d_+(J_+) R_g(d_-(J_-))] \tag{26}$$

$$= \tfrac{1}{4}\text{Str}\,[J_- d_+(J_+)] + \tfrac{\eta}{4}\text{Str}\,[d_+(J_+) R_g(d_-(J_-))]\,. \tag{27}$$

3.3. On-shell flat current

The equation of motion of Lagrangian (25) is given by

$$\mathcal{E} \equiv \partial_- d_+(J_+) + \partial_+ d_-(J_-) + [J_-, d_+(J_+)] + [J_+, d_-(J_-)] = 0\,. \tag{28}$$

By the definition of the current $A_\pm$ in (19), it satisfies the zero-curvature condition,

$$\mathcal{Z} \equiv \partial_+ A_- - \partial_- A_+ + [A_+, A_-] = 0\,. \tag{29}$$

Rewriting $\mathcal{Z}$ in terms of the deformed current (24), we obtain the following expression,

$$\mathcal{Z} = \partial_+ J_- - \partial_- J_+ + [J_+, J_-] + \eta R_g(\mathcal{E}) + \eta^2 \text{CYBE}(d_+(J_+), d_-(J_-))\,, \tag{30}$$

where we have denoted the LHS of the CYBE (4) with the dressed R_g-operator by

$$\text{CYBE}(X, Y) \equiv [R_g(X), R_g(X)] - R_g\big([R_g(X), Y] + [X, R_g(Y)]\big)\,, \tag{31}$$

which vanishes when R satisfies the CYBE[3]. This relation implies that the deformed current $J_\pm$ is also flat current if the equation of motion is satisfied $\mathcal{E} = 0$. In this sense, $J_\pm$ is an *on-shell* flat current.

[3] If the R-operator satisfies the mCYBE rather than the CYBE, the condition (29) is modified as [10]

$$\mathcal{Z} = \partial_+ J_- - \partial_- J_+ + [J_+, J_-] + \eta R_g(\mathcal{E}) + \eta^2 [d_+(J_+), d_-(J_-)]\,. \tag{32}$$

Thus, the deformed current J_μ is no longer flat even if the equation of motion is satisfied.

3.4. Lax pairs

We are ready to introduce the Lax pair for the deformed string action (25). They are given by[4]

$$\mathcal{L}_\pm \equiv J_\pm^{(0)} + \lambda J_\pm^{(1)} + \lambda^{\pm 2} J_\pm^{(2)} + \lambda^{-1} J_\pm^{(3)}\,, \tag{34}$$

with a spectral parameter $\lambda \in \mathbb{C}$, where the $\mathbb{Z}_4$-grading is denoted by $J_\pm^{(k)} \equiv P_k(J_\pm)$ for $k = 0, 1, 2, 3$. Then, the equation of motion (28) and flatness condition (29) are equivalent with the zero-curvature condition of the Lax pair;

$$\partial_+ \mathcal{L}_- - \partial_- \mathcal{L}_+ + [\mathcal{L}_+, \mathcal{L}_-] = 0\,. \tag{35}$$

Thus, the deformed string action (25) is classically integrable in the sense that the Lax pair exists.

4. CYBE deformation as non-local gauge transformation

As a remarkable feature of the CYBE-type deformation, we will show that the deformed current $J_\pm$ in (24) is obtained as a consequence of a non-local gauge transformation from the undeformed current $A_\pm = g^{-1}\partial_\pm g$. More precisely, the deformed current can be expressed as

$$J_\pm = \tilde{g}^{-1} \partial_\pm \tilde{g}\,, \tag{36}$$

where a group element $\tilde{g}$ is related to the undeformed one g by a *twist operator* $\mathcal{F}$ like

$$\tilde{g} = \mathcal{F}^{-1} g\,. \tag{37}$$

As we will see here, such an operator $\mathcal{F}$ does exist when the R-operator satisfies the CYBE.

4.1. Twist operator

To construct the operator $\mathcal{F}$, let us consider the following gauge transformation of $J_\pm$:

$$J_\pm^g \equiv g J_\pm g^{-1} - \partial_\pm g g^{-1} \qquad \Longleftrightarrow \qquad \partial_\pm + J_\pm^g = g(\partial_\pm + J_\pm) g^{-1}\,. \tag{38}$$

The current $J_\pm$ is also flat when equation of motion is satisfied. Hence the resulting current $J_\pm^g$ is obviously flat and explicitly computed as

$$J_\pm^g = g(J_\pm - A_\pm) g^{-1} = g\left(\frac{\pm \eta R_g \circ d_\pm}{1 \mp \eta R_g \circ d_\pm} A_\pm\right) g^{-1} = \pm \eta R\left(g d_\pm (J_\pm) g^{-1}\right)\,. \tag{39}$$

The flatness of $J_\pm^g$ can be seen as a consistency condition of auxiliary linear problem;

$$\begin{aligned} \partial_\tau \mathcal{F} &= -J_\tau^g \mathcal{F}\,, \\ \partial_\sigma \mathcal{F} &= -J_\sigma^g \mathcal{F}\,, \end{aligned} \tag{40}$$

where $\mathcal{F} \equiv \mathcal{F}(\sigma, \tau)$ is a $(4|4) \times (4|4)$ matrix. This problem is formally solved up to world-sheet boundary term $\mathcal{F}(0, \tau)$ as follows;

$$\mathcal{F}(\sigma, \tau) = \mathrm{Pexp}\left[-\int_0^\sigma d\sigma\, J_\sigma^g\right] \mathcal{F}(0, \tau)\,. \tag{41}$$

[4] If the case of the Drinfeld-Jimbo type deformation, the Lax pair is given by [10]

$$\mathcal{L}_\pm \equiv J_\pm^{(0)} + \lambda\sqrt{1+\eta^2} J_\pm^{(1)} + \lambda^{\pm 2} \frac{1+\eta^2}{1-\eta^2} J_\pm^{(2)} + \lambda^{-1} \sqrt{1+\eta^2} J_\pm^{(3)}\,. \tag{33}$$

The equations (40) allow us to express the current $J^g_\pm$ as

$$J^g_\pm = -\partial_\pm \mathcal{F}\mathcal{F}^{-1} \,. \tag{42}$$

Combining the gauge transformation (38) and the above relation, we notice that the deformed current $J_\pm$ could be casted into the following nice form,

$$J_\pm = g^{-1}J^g_\pm g + A_\pm = -g^{-1}\partial_\pm \mathcal{F}\mathcal{F}^{-1}g + g^{-1}\partial_\pm g = (\mathcal{F}^{-1}g)^{-1}\partial_\pm(\mathcal{F}^{-1}g) \,. \tag{43}$$

This expression is nothing but the desired form (36) with $\tilde{g} \equiv \mathcal{F}^{-1}g$. Furthermore, by introducing the difference between g and $\tilde{g}$ as

$$\mathcal{G} \equiv \tilde{g}^{-1}g = g^{-1}\mathcal{F}g \,, \tag{44}$$

we notice that the deformed current $J_\pm$ is directly obtained from $A_\pm$ by the following non-local gauge transformation;

$$J_\pm = \mathcal{G}A_\pm\mathcal{G}^{-1} - \partial_\pm\mathcal{G}\mathcal{G}^{-1} \qquad \Longleftrightarrow \qquad \partial_\pm + J_\pm = \mathcal{G}(\partial_\pm + A_\pm)\mathcal{G}^{-1} \,. \tag{45}$$

4.2. Conserved current

The conserved current for the deformed model is easily obtained by the equation of motion (28). To find out the conserved current, we introduce

$$\Lambda^\alpha \equiv 2\gamma^{\alpha\beta}J^{(2)}_\beta + \epsilon^{\alpha\beta}(J^{(1)}_\beta - J^{(3)}_\beta) \,. \tag{46}$$

Then, by using this quantity, the projected current is expressed as

$$d_\pm(J_\pm) = \Lambda^\tau \pm \Lambda^\sigma \,. \tag{47}$$

The equation of motion (28) is now written as

$$\mathcal{E} = -2\left(\partial_\alpha\Lambda^\alpha + [J_\alpha, \Lambda^\alpha]\right) = 0 \,. \tag{48}$$

This relation leads us to define the following conserved current

$$k^\alpha \equiv \tilde{g}\Lambda^\alpha\tilde{g}^{-1} \,. \tag{49}$$

In fact, the conservation law $\partial_\alpha k^\alpha = 0$ immediately follows from the expression (36).

Here it is worth giving some comments on the symmetry algebra of the conserved charges. Firstly, we should note that the quantities $\oint k^\alpha$ are not conserved in general due to the twisted boundary condition. Thus, to construct the conserved charges, it would be necessary to suppose the infinitely extended world-sheet with boundary conditions;

$$g \to \text{const.} \qquad \text{when} \qquad \sigma \to \pm\infty \,. \tag{50}$$

With these asymptotic boundary conditions, the conserved charges can be defined as

$$Q^\alpha \equiv \int_{-\infty}^{\infty} d\sigma k^\alpha \,. \tag{51}$$

One may expect that the resulting symmetry algebra would be the undeformed $\mathfrak{psu}(2,2|4)$. It is, however, quite sensitive to the choice of the constant matrix K which appears in

$$\mathcal{F}(\sigma,\tau) = \text{Pexp}\left[-\int_{-\infty}^{\sigma} d\sigma\, J^g_\sigma\right] K \,. \tag{52}$$

Indeed, we have a degree of freedom to multiplying such a constant matrix in the linear problem (40). This matrix corresponds to boundary conditions of the string world-sheet. The conserved current is transformed as

$$k^\alpha \to K k^\alpha K^{-1} \qquad \text{under} \qquad \mathcal{F} \to \mathcal{F} K\,. \tag{53}$$

Then, it triggers the mixing of the conserved charges defined in (51). In fact, in the case of three-dimensional Schrödinger spacetimes, a suitable choice of the K-matrix brings a q-deformed Poincaré algebra [39, 40] and the associated infinite dimensional algebra (called an *exotic symmetry* [41]) into a standard $\mathcal{U}(\mathfrak{sl}(2))$ and a Yangian symmetry $\mathcal{Y}(\mathfrak{sl}(2))$ [42].

5. From CYBE to deformations of AdS$_5\times$S^5

The deformed action (18) is rather algebraic and hence it is not so obvious to deduce the background metric. We demonstrate here how to read off the metric and the NS-NS two-form from the deformed action. Then we present some examples of the gravity/CYBE correspondence.

To read off the metric and NS-NS two-form, it is sufficient to see the bosonic part of the deformed action. Firstly, we need to introduce the coordinate system x^i of AdS$_5\times$S^5 via a parameterization of the group element;

$$g = \exp\left(\textstyle\sum_i x^i T_i\right) \quad \in \quad SU(2,2) \otimes SU(4)\,, \tag{54}$$

where T^i belong to the fundamental representation of $\mathfrak{su}(2,2) \oplus \mathfrak{su}(4)$. The coordinate system x^i may be the Poincaré coordinates or the global coordinates, depending on the purpose. In the practical computation, it is often useful to utilize gamma-matrices as the generators T_i.

Secondary, plugging the projector (22) with (18), the bosonic part of the Lagrangian reads

$$L = -\frac{1}{2}(\gamma^{\alpha\beta} - \epsilon^{\alpha\beta})\,\mathrm{Tr}[A_\alpha P_2(J_\beta)] \qquad \text{with} \qquad J_\beta = \frac{1}{1 - 2\eta R_g \circ P_2} A_\beta\,. \tag{55}$$

The metric and NS-NS two-forms are obtained as the symmetric and skew-symmetric part with respect to the world-sheet coordinates in the above expression, respectively. Then, the remaining is to evaluate the projected deformed bosonic current $P_2(J_\beta)$ in (55) explicitly. This is easily done by multiplying $P_2 \circ (1 - 2\eta R_g \circ P_2)$ on J_β and the following form is obtained,

$$(1 - 2\eta P_2 \circ R_g) P_2(J_\beta) = P_2(A_\beta)\,. \tag{56}$$

The above relation may be regarded as the definition of $P_2(J_\beta)$. This relation takes the value on $\mathfrak{su}(2,2) \oplus \mathfrak{su}(4)$, but the projector P_2 reduces the number of the linearly independent relations just to $10 = 5 + 5 = \dim(\mathrm{AdS}_5\times\mathrm{S}^5)$. In particular, when a deformation of either AdS$_5$ or S^5 is considered, only five linear relations are concerned with the analysis.

5.1. Examples

We will present some examples of deformed backgrounds, which are obtained from typical solutions of the CYBE (explained in Section 2)[5].

5.1.1. Three parameter $\hat{\gamma}_i$-deformation of S^5 Firstly, let us consider an abelian classical r-matrix with $\mathfrak{su}(4)$, which consists of the three Cartan generators h_1, h_2, h_3;

$$r_{\mathrm{Abe}} = \frac{1}{8}(\hat{\gamma}_3 h_1 \wedge h_2 + \hat{\gamma}_1 h_2 \wedge h_3 + \hat{\gamma}_2 h_3 \wedge h_1)\,, \tag{57}$$

[5] We set that $\eta = 1$ in (55) here.

Table 1. A catalog of the gravity/CYBE correspondence. Here we list some typical examples of classical r-matrices and the associated gravity backgrounds, which are explained in the Subsection 5.1 and Section 6.

Classical r-matrices	$\rightarrow$	Deformed gravity solutions
Trivial: $r=0$		$\mathrm{AdS}_5\times\mathrm{S}^5$ [2]
Jordanian: $r_{\rm Jor}=E_{24}\wedge(E_{22}-E_{44})$		Melvin twists [43]
Abelian: $r_{\rm Abe}=h_1\wedge h_2$ with $h_i\in\mathfrak{su}(4)$		γ_i-deformations of S^5 [35, 36]
Abelian: $r_{\rm Abe}=h_1\wedge h_2$ with $h_i\in\mathfrak{su}(2,2)$		γ_i-deformations of AdS_5 [44, 45, 46, 47]
Abelian Jordanian: $r_{\rm AJ}=p_1\wedge p_2$		Gravity duals for NC Yang-Mills [37, 38]
Abelian: $r_{\rm Abe}=K_3\wedge L_3$		γ_i-deformations of $T^{1,1}$ [35, 48]

where $\hat{\gamma}_i$ $(i=1,2,3)$ are deformation parameters. This r-matrix deforms only the S^5 part.

The resulting metric and NS-NS two-forms turn out to be [30]

$$ds^2=ds^2_{\mathrm{AdS}_5}+\sum_{i=1}^{3}(d\rho_i^2+G\rho_i^2d\phi_i^2)+G\rho_1^2\rho_2^2\rho_3^2\Big(\sum_{i=1}^{3}\hat{\gamma}_i d\phi_i\Big)^2\,, \tag{58}$$

$$B_2=G\left(\hat{\gamma}_3\rho_1^2\rho_2^2\,d\phi_1\wedge d\phi_2+\hat{\gamma}_1\rho_2^2\rho_3^2\,d\phi_2\wedge d\phi_3+\hat{\gamma}_2\rho_3^2\rho_1^2\,d\phi_3\wedge d\phi_1\right)\,, \tag{59}$$

where ρ_i satisfy the following constraints and the scalar function G is defined by

$$\rho_1^2+\rho_2^2+\rho_3^2=1\,,\qquad G^{-1}\equiv 1+\hat{\gamma}_3^2\rho_1^2\rho_2^2+\hat{\gamma}_1^2\rho_2^2\rho_3^2+\hat{\gamma}_2^2\rho_3^2\rho_1^2\,. \tag{60}$$

These backgrounds agree with the well-known three-parameter $\hat{\gamma}_i$-deformations of S^5 [35, 36]. In particular, it reduces to the Lunin-Maldacena case [35] when $\hat{\gamma}_1=\hat{\gamma}_2=\hat{\gamma}_3$. As pointed out in [36], these deformed backgrounds are reproduced by performing TsT transformations (which is a sequence of two T-dualities and a shift).

5.1.2. Gravity duals for noncommutative gauge theories The next example is an abelian Jordanian classical r-matrix with $\mathfrak{su}(2,2)=\mathfrak{so}(2,4)$;

$$r_{\rm AJ}=\frac{1}{2}\left(a^2p_2\wedge p_3+a'^2p_0\wedge p_1\right). \tag{61}$$

Here p_μ are translation generators of the conformal algebra and commute each other $[p_\mu,p_\nu]=0$. Thus, it trivially satisfies the CYBE. The explicit matrix representations of p_μ are given in [31]. It is noted that this r-matrix only deforms the AdS_5 part.

The resulting metric and NS-NS two-form in the Poincaré coordinates are [31]

$$ds^2=\frac{z^2}{z^4-4a'^4}(-(dx^0)^2+(dx^1)^2)+\frac{z^2}{z^4+4a^4}((dx^2)^2+(dx^3)^2)+\frac{dz^2}{z^2}+ds^2_{\mathrm{S}^5}\,, \tag{62}$$

$$B_2=-\frac{a'^2}{z^4-a'^4}dx^0\wedge dx^1+\frac{a^2}{z^4+a^4}dx^2\wedge dx^3\,. \tag{63}$$

These completely agree with the gravity duals for non-commutative gauge theories [37, 38].

It is worth mentioning that the classical integrability of this background automatically follows from our formulation as a byproduct. It is because the Lax pair has already been constructed in a generic form (34).

One may also consider abelian twists of AdS_5 rather than S^5. For this purpose, one can simply adopt the abelian r-matrix consisting of the Cartan generators in $\mathfrak{su}(2,2)$ instead of (57). The resulting metric and NS-NS two-form agree with [44] and also include [45] as a particular case, which is obtained by means of a Melvin twist.

5.1.3. New backgrounds from Jordanian r-matrix We can also consider new integrable backgrounds which are obtained by Jordanian r-matrices such as (12). For instance, it is easy to see that the following r-matrix satisfies the CYBE,

$$r_{\rm Jor} = E_{24} \wedge (\gamma E_{22} - \gamma^* E_{44}) \qquad \text{with} \qquad \gamma \in \mathbb{C}\,, \tag{64}$$

where $(E_{ij})_{kl} = \delta_{ik}\delta_{jl}$ are the fundamental representations of $\mathfrak{su}(2,2)$ and γ^* denotes the complex conjugate of $\gamma \in \mathbb{C}$.

The resulting metric of AdS_5 and NS-NS two-form turn out to be [33, 11],

$$\begin{aligned} ds^2 &= \frac{-2dx^+dx^- + (dx^1)^2 + (dx^2)^2 + dz^2}{2z^2} - \frac{|\gamma|^2\big((x^1)^2 + (x^2)^2\big) + ({\rm Re}\gamma)^2 z^2}{2z^6}(dx^+)^2\,, \\ B_2 &= -{\rm Re}\gamma\frac{\big(x^1dx^1 + x^2dx^2 + zdz\big)\wedge dx^+}{2z^4} + {\rm Im}\gamma\frac{\big(x^2dx^1 - x^1dx^2\big)\wedge dx^+}{2z^4}\,. \end{aligned} \tag{65}$$

It is noted that both the metric and the NS-NS two-form are real even though the Jordanian r-matrix itself (64) does depend on a complex deformation parameter γ. When the parameter γ is real, the deformed background was computed in [11] and the complete type IIB supergravity solutions are constructed in [32]. In the case that γ is a pure imaginary, the above backgrounds reduce to the one obtained by means of a *null Melvin twist* [43]. Furthermore, for a generic complex parameter γ, the deformed backgrounds (65) are obtained from the undeformed $AdS_5\times S^5$ by a chain of T-dualities and S-duality [33]. This strongly suggests that the Yang-Baxter sigma model with the Jordanian r-matrix (64) gives rise to a consistent string background.

6. An application to a non-integrable background

So far, we have concentrated on deformations of the $AdS_5\times S^5$ case. The $AdS_5\times S^5$ background is known as an integrable background because the metric can be represented by a supercoset with the $\mathbb{Z}_4$-grading [3]. An interesting question is whether or not our deformation technique is applicable to non-integrable backgrounds. In the recent progress, many examples of non-integrable backgrounds have been discovered. Among them, we will consider type IIB string theory on the $AdS_5 \times T^{1,1}$ background. This theory is known to be dual for the $\mathcal{N} = 1$ superconformal field theory in four dimensions [49]. The non-integrability of this background is shown in [50] by presenting the existence of chaotic string solutions. The next issue that we would like to consider is deformations of this background by utilizing the Yang-Baxter sigma model approach. Interestingly, it turns out that it successfully works even for this non-integrable background [34]. The point is that, to apply the Yang-Baxter sigma model description, the parent background needs to be represented by a coset, although it is not necessary to be symmetric.

6.1. A coset construction of $T^{1,1}$

Let us consider the $T^{1,1}$ part of the $AdS_5 \times T^{1,1}$ background. The first step to consider CYBE-type deformations of $T^{1,1}$ is to find out a coset construction. A five-dimensional Sasaki-Einstein manifold $T^{1,1}$ is represented by a $U(1)$-fibration over $S^2 \times S^2$ and it has a coset structure like $[SU(2) \times SU(2)]/U(1)$ as denoted in [51, 52]. This coset description, however, does not leads the correct metric of $T^{1,1}$ without suitable rescalings of vierbeins [53, 54]. For our purpose, we would like to avoid the rescaling because the dressed R-operator (20) depends on a group element g itself. In order to overcome this difficulty, we found another coset description of $T^{1,1}$. With this coset, it is not necessary to perform any rescaling so as to obtain the $T^{1,1}$ metric. The coset description we found in [34] is given by

$$T^{1,1} = \frac{SU(2)\times SU(2)\times U(1)_R}{U(1)_1\times U(1)_2}\,. \tag{66}$$

With this description, the manifold obviously has the $SU(2)\times SU(2)\times U(1)_R$ symmetry, where $SU(2)\times SU(2)$ and $U(1)_R$ correspond to a flavor symmetry and an R-symmetry in the dual gauge theory, respectively. Let T_i $(i=1,2)$ be the generators of two $U(1)_i$'s in the denominator. Then, the two $U(1)_i$'s are embedded into the numerator as

$$T_1 = K_3 + L_3\,, \qquad T_2 = K_3 - L_3 + 4M\,, \tag{67}$$

where K_i and L_i $(i=1,2,3)$ are the generators of the two $\mathfrak{su}(2)$'s, respectively. In particular, K_3 and L_3 are the Cartan generators, and M is the $\mathfrak{u}(1)_R$ generator.

By introducing a parameterization of the group element $g\in SU(2)\times SU(2)\times U(1)_R$ and defining the left-invariant one-form $A=g^{-1}dg$, the metric of $T^{1,1}$ can be reproduced as [34]

$$\begin{aligned} ds^2_{T^{1,1}} &= -\frac{1}{3}\mathrm{Str}[AP(A)] \\ &= \frac{1}{6}\sum_{i=1,2}(d\theta_i^2+\sin^2\theta_i d\phi_i^2) + \frac{1}{9}(d\psi+\cos\theta_1 d\phi_1+\cos\theta_2 d\phi_2)^2\,, \end{aligned} \tag{68}$$

where the coset projector P is defined by

$$P(A) \equiv A - \frac{\mathrm{Str}[T_1A]}{\mathrm{Str}[T_1T_1]}T_1 - \frac{\mathrm{Str}[T_2A]}{\mathrm{Str}[T_2T_2]}T_2\,. \tag{69}$$

One may wonder why the supertrace Str is adopted here rather than the usual trace. Indeed, the superstrace in (68) is assumed to be taken over the $(4|1)\times(4|1)$ supermatrices, which is the fundamental representation of $SU(2)\times SU(2)\times U(1)_R$. This may be justified by interpreting the matrix as a part of a bigger $(8|1)\times(8|1)$ supermatrix including an $\mathcal{N}=1$ superconformal algebra as follows;

$$\left(\begin{array}{cc|c} SU(2) & 0 & 0 \\ 0 & SU(2) & 0 \\ \hline 0 & 0 & U(1)_R \end{array}\right) \hookrightarrow \left(\begin{array}{ccc|c} SU(2,2) & 0 & 0 & \overline{F}^A \\ 0 & SU(2) & 0 & 0 \\ 0 & 0 & SU(2) & 0 \\ \hline F_A & 0 & 0 & U(1)_R \end{array}\right). \tag{70}$$

The lines in the matrices distinguish the $\mathbb{Z}_2$-grading. In the right big supermatrix, an $\mathcal{N}=1$ superconformal algebra $PSU(2,2|1)$ sitting at the four corners together with the supercharges $F_A, \overline{F}^A$ $(A=1,\cdots,4)$. The point is that the $U(1)_R$ symmetry is originated in the superconformal symmetry $PSU(2,2|1)$. In other words, the coset (66) should be regarded as a subsector of the supercoset;

$$\frac{PSU(2,2|1)\times SU(2)\times SU(2)}{SO(1,4)\times U(1)_1\times U(1)_2}\,. \tag{71}$$

6.2. Deformations of $T^{1,1}$ as Yang-Baxter sigma models
We have already revealed the coset structure of $T^{1,1}$. We are now ready to apply the deformation technique based on classical r-matrices. The Lagrangian of the Yang-Baxter sigma model on $T^{1,1}$ is given by

$$L \equiv \frac{1}{3}(\gamma^{\alpha\beta} - \epsilon^{\alpha\beta})\text{Str}\left[A_\alpha P \frac{1}{1-2R_g \circ P} A_\beta\right], \tag{72}$$

with the projector P defined in (69).

As a natural generalization of the S^5 case (57), let us consider an abelian r-matrix which is composed of the three Cartan generators of $\mathfrak{su}(2) \oplus \mathfrak{su}(2) \oplus \mathfrak{u}(1)_R$,

$$r_{\text{Abe}} = \frac{1}{3}\hat{\gamma}_1 L_3 \wedge M + \frac{1}{3}\hat{\gamma}_2 M \wedge K_3 - \frac{1}{6}\hat{\gamma}_3 K_3 \wedge L_3, \tag{73}$$

with deformation parameters $\hat{\gamma}_i$ $(i = 1, 2, 3)$.

After some computations, the resulting metric and NS-NS two-from turn out to be [34]

$$\begin{aligned} ds^2 &= G(\hat{\gamma}_1, \hat{\gamma}_2, \hat{\gamma}_3)\Big[\frac{1}{6}\sum_{i=1,2}(G(\hat{\gamma}_1, \hat{\gamma}_2, \hat{\gamma}_3)^{-1}d\theta_i^2 + \sin^2\theta_i d\phi_i^2) \\ &\quad + \frac{1}{9}(d\psi + \cos\theta_1 d\phi_1 + \cos\theta_2 d\phi_2)^2 + \frac{\sin^2\theta_1\sin^2\theta_2}{324}(\hat{\gamma}_3 d\psi + \hat{\gamma}_1 d\phi_1 + \hat{\gamma}_2 d\phi_2)^2\Big], \end{aligned} \tag{74}$$

$$\begin{aligned} B_2 &= G(\hat{\gamma}_1, \hat{\gamma}_2, \hat{\gamma}_3)\Big[\Big\{\hat{\gamma}_3\Big(\frac{\sin^2\theta_1\sin^2\theta_2}{36} + \frac{\cos^2\theta_1\sin^2\theta_2 + \cos^2\theta_2\sin^2\theta_1}{54}\Big) \\ &\qquad - \hat{\gamma}_2\frac{\cos\theta_2\sin^2\theta_1}{54} - \hat{\gamma}_1\frac{\cos\theta_1\sin^2\theta_2}{54}\Big\}d\phi_1 \wedge d\phi_2 \\ &\quad + \frac{(\hat{\gamma}_3\cos\theta_2 - \hat{\gamma}_2)\sin^2\theta_1}{54}d\phi_1 \wedge d\psi - \frac{(\hat{\gamma}_3\cos\theta_1 - \hat{\gamma}_1)\sin^2\theta_2}{54}d\phi_2 \wedge d\psi\Big], \end{aligned} \tag{75}$$

where the scalar function is defined as

$$\begin{aligned} G(\hat{\gamma}_1, \hat{\gamma}_2, \hat{\gamma}_3)^{-1} &\equiv 1 + \hat{\gamma}_3^2\Big(\frac{\sin^2\theta_1\sin^2\theta_2}{36} + \frac{\cos^2\theta_1\sin^2\theta_2 + \cos^2\theta_2\sin^2\theta_1}{54}\Big) + \hat{\gamma}_2^2\frac{\sin^2\theta_1}{54} \\ &\quad + \hat{\gamma}_1^2\frac{\sin^2\theta_2}{54} - \hat{\gamma}_2\hat{\gamma}_3\frac{\sin^2\theta_1\cos\theta_2}{27} - \hat{\gamma}_3\hat{\gamma}_1\frac{\sin^2\theta_2\cos\theta_1}{27}. \end{aligned} \tag{76}$$

Indeed, these expressions agree with the ones obtained in [48] via TsT-transformations. In particular, when $\hat{\gamma}_1 = \hat{\gamma}_2 = 0$ and $\hat{\gamma}_3$ is non-zero, the above deformed backgrounds reduce to the Lunin-Maldacena case [35].

7. Conclusion and discussion

In this article, we have given a short review of the recent progresses on novel relations between classical r-matrices satisfying the CYBE and deformed string backgrounds [11, 32, 30, 31, 34]. The relations are worthy of being called the *gravity/CYBE correspondence* [30, 31]. According to them, the solution space of the CYBE may be regarded as a moduli space of a certain class of deformed gravitational backgrounds. In particular, we have presented several important examples including three parameter $\hat{\gamma}_i$-deformations of the S^5 part [35, 36] and the gravity duals of non-commutative $\mathcal{N} = 4$ super Yang-Mills theories [37, 38] in Section 5. We have also discussed an application of the deformation technique to a non-integrable background, $AdS_5 \times T^{1,1}$ [34] in Section 6.

There remain, however, several open questions to be answered. Let us list some of them in the following;

- Firstly, we are not sure whether or not any deformed backgrounds obtained by classical r-matrices are consistent with the equations of motion of type IIB supergravity. Even though we proved that the deformed models have kappa-symmetry in general [11] and some important deformed supergravity backgrounds are successfully obtained as demonstrated in Section 5, it is not straightforward to conclude that the resulting backgrounds always satisfy the equations of motion of type IIB supergravity.
- It is not obvious that mutually different r-matrices label pair-wise different backgrounds. It is observed in [32] that two non-equivalent r-matrices lead to the identical metric, up to coordinate transformations. Thus, to establish the one to one correspondence between r-matrices and deformed backgrounds, it would be necessary to take into account an appropriate quotient by algebra automorphisms.
- Can we find the corresponding r-matrix for a given deformed background? More generally, does there alway exist an r-matrix for any analytically deformed $\mathrm{AdS}_5\times\mathrm{S}^5$ background? Even though we know how to compute the deformed back metric and the NS-NS two-forms from the r-matrices, the inverse direction should also be clarified.
- Finally, it is desirable to figure out the composite rule of the r-matrices. On the gravity theory side, some deformed backgrounds are obtained by string dualities such as TsT-transformations and we could think of a chain of them to construct more complicated backgrounds. In this sense, we have a structure of product of the dualities. The question is how it works for the classical r-matrices if the gravity/CYBE correspondence holds. We expect that the twist operator discussed in Section 4 might be a clue to solve this problem.

Even though there are still many issues to establish the gravity/CYBE correspondence, we believe that classical r-matrices are very fascinating objects to characterize deformed gravity solutions. As a final remark, the deformation technique presented here preserves the classical integrability of the undeformed background, but it can describe a non-integrable deformation of a non-integrable background, $\mathrm{AdS}_5 \times T^{1,1}$ [34]. This result indicates that the gravity/CYBE correspondence is not restricted to a class of integrable backgrounds and capture a wider class of gravity solutions. It is an important task to ascertain the limitation of this scenario. We hope to report on some new results along this direction in the near future.

Acknowledgment

We are grateful to Gleb Arutyunov, Riccardo Borsato, Takashi Kameyama and Stefan Vandoren for valuable comments and discussions. We also thank Io Kawaguchi for his collaboration at the beginning of this project. Especially, we would like to appreciate P. Marcos Crichigno for the collaboration concerning with [34] and comments on this manuscript. T.M. is supported by the Netherlands Organization for Scientific Research (NWO) under the VICI grant 680-47-602. This work is also part of the ERC Advanced grant research programme No. 246974, "Supersymmetry: a window to non-perturbative physics" and of the D-ITP consortium, a program of the NWO that is funded by the Dutch Ministry of Education, Culture and Science (OCW).

References

[1] J. M. Maldacena, "The large N limit of superconformal field theories and supergravity," Adv. Theor. Math. Phys. **2** (1998) 231 [Int. J. Theor. Phys. **38** (1999) 1113]. [arXiv:hep-th/9711200].
[2] R. R. Metsaev and A. A. Tseytlin, "Type IIB superstring action in $\mathrm{AdS}_5\times\mathrm{S}^5$ background," Nucl. Phys. B **533** (1998) 109 [hep-th/9805028].
[3] I. Bena, J. Polchinski and R. Roiban, "Hidden symmetries of the $\mathrm{AdS}_5\times\mathrm{S}^5$ superstring," Phys. Rev. D **69** (2004) 046002 [hep-th/0305116].
[4] R. Roiban and W. Siegel, "Superstrings on $\mathrm{AdS}_5\times\mathrm{S}^5$ supertwistor space," JHEP **0011** (2000) 024 [hep-th/0010104].

[5] M. Hatsuda and K. Yoshida, "Classical integrability and super Yangian of superstring on $AdS_5 \times S^5$," Adv. Theor. Math. Phys. **9** (2005) 703 [hep-th/0407044]; "Super Yangian of superstring on $AdS_5 \times S^5$ revisited," Adv. Theor. Math. Phys. **15** (2011) 1485 [arXiv:1107.4673 [hep-th]].
[6] J. A. Minahan and K. Zarembo, "The Bethe ansatz for N=4 superYang-Mills," JHEP **0303** (2003) 013 [hep-th/0212208].
[7] N. Beisert *et al.*, "Review of AdS/CFT Integrability: An Overview," Lett. Math. Phys. **99** (2012) 3 [arXiv:1012.3982 [hep-th]].
[8] C. Klimcik, "Yang-Baxter sigma models and dS/AdS T duality," JHEP **0212** (2002) 051 [hep-th/0210095]; "On integrability of the Yang-Baxter sigma-model," J. Math. Phys. **50** (2009) 043508 [arXiv:0802.3518 [hep-th]]; "Integrability of the bi-Yang-Baxter sigma model," arXiv:1402.2105 [math-ph];
R. Squellari, "Yang-Baxter σ model: Quantum aspects," Nucl. Phys. B **881** (2014) 502 [arXiv:1401.3197 [hep-th]].
[9] F. Delduc, M. Magro and B. Vicedo, "On classical q-deformations of integrable sigma-models," JHEP **1311** (2013) 192 [arXiv:1308.3581 [hep-th]].
[10] F. Delduc, M. Magro and B. Vicedo, "An integrable deformation of the $AdS_5 \times S^5$ superstring action," Phys. Rev. Lett. **112** (2014) 051601 [arXiv:1309.5850 [hep-th]].
[11] I. Kawaguchi, T. Matsumoto and K. Yoshida, "Jordanian deformations of the $AdS_5 \times S^5$ superstring," JHEP **1404** (2014) 153 [arXiv:1401.4855 [hep-th]].
[12] T. J. Hollowood, J. L. Miramontes and D. M. Schmidtt, "Integrable Deformations of Strings on Symmetric Spaces," arXiv:1407.2840 [hep-th];
T. J. Hollowood, J. L. Miramontes and D. Schmidtt, "An Integrable Deformation of the AdS5 x S5 Superstring," arXiv:1409.1538 [hep-th].
[13] K. Sfetsos, "Integrable interpolations: From exact CFTs to non-Abelian T-duals," Nucl. Phys. B **880** (2014) 225 [arXiv:1312.4560 [hep-th]];
G. Itsios, K. Sfetsos, K. Siampos and A. Torrielli, "The classical Yang-Baxter equation and the associated Yangian symmetry of gauged WZW-type theories," arXiv:1409.0554 [hep-th].
[14] V. G. Drinfel'd, "Hopf algebras and the quantum Yang-Baxter equation," Sov. Math. Dokl. **32** (1985) 254; "Quantum groups," J. Sov. Math. **41** (1988) 898 [Zap. Nauchn. Semin. **155**, 18 (1986)].
[15] M. Jimbo, "A q difference analog of $U(g)$ and the Yang-Baxter equation," Lett. Math. Phys. **10** (1985) 63.
[16] F. Delduc, M. Magro and B. Vicedo, "Derivation of the action and symmetries of the q-deformed AdS_5 x S^5 superstring," arXiv:1406.6286 [hep-th].
[17] I. Kawaguchi and K. Yoshida, "Hybrid classical integrability in squashed sigma models," Phys. Lett. B **705** (2011) 251 [arXiv:1107.3662 [hep-th]]; "Hybrid classical integrable structure of squashed sigma models: A short summary," J. Phys. Conf. Ser. **343** (2012) 012055 [arXiv:1110.6748 [hep-th]].
[18] I. Kawaguchi, T. Matsumoto and K. Yoshida, "The classical origin of quantum affine algebra in squashed sigma models," JHEP **1204** (2012) 115 [arXiv:1201.3058 [hep-th]]; "On the classical equivalence of monodromy matrices in squashed sigma model," JHEP **1206** (2012) 082 [arXiv:1203.3400 [hep-th]].
[19] I. Kawaguchi and K. Yoshida, "Hidden Yangian symmetry in sigma model on squashed sphere," JHEP **1011** (2010) 032 [arXiv:1008.0776 [hep-th]];
I. Kawaguchi, D. Orlando and K. Yoshida, "Yangian symmetry in deformed WZNW models on squashed spheres," Phys. Lett. B **701** (2011) 475 [arXiv:1104.0738 [hep-th]];
I. Kawaguchi and K. Yoshida, "A deformation of quantum affine algebra in squashed WZNW models," J. Math. Phys. **55** (2014) 062302 [arXiv:1311.4696 [hep-th]].
[20] G. Arutyunov, R. Borsato and S. Frolov, "S-matrix for strings on η-deformed $AdS_5 \times S^5$," JHEP **1404** (2014) 002 [arXiv:1312.3542 [hep-th]].
[21] B. Hoare, R. Roiban and A. A. Tseytlin, "On deformations of $AdS_n \times S^n$ supercosets," JHEP **1406** (2014) 002 [arXiv:1403.5517 [hep-th]].
[22] G. Arutyunov and S. J. van Tongeren, "The $AdS_5 \times S^5$ mirror model as a string," arXiv:1406.2304 [hep-th].
[23] G. Arutynov, M. de Leeuw and S. J. van Tongeren, "On the exact spectrum and mirror duality of the $(AdS_5 \times S^5)_\eta$ superstring," arXiv:1403.6104 [hep-th].
[24] T. Kameyama and K. Yoshida, "Anisotropic Landau-Lifshitz sigma models from q-deformed $AdS_5 \times S^5$ superstrings," arXiv:1405.4467 [hep-th]; "String theories on warped AdS backgrounds and integrable deformations of spin chains," JHEP **1305** (2013) 146 [arXiv:1304.1286 [hep-th]].
[25] G. Arutyunov and D. Medina-Rincon, "Deformed Neumann model from spinning strings on $(AdS_5 \times S^5)_\eta$," arXiv:1406.2536 [hep-th].
[26] M. Khouchen and J. Kluson, "Giant Magnon on Deformed $AdS_3 \times S^3$," arXiv:1405.5017 [hep-th];
C. Ahn and P. Bozhilov, "Finite-size giant magnons on η-deformed $AdS_5 \times S^5$," arXiv:1406.0628 [hep-th];
A. Banerjee and K. L. Panigrahi, "On the rotating and oscillating strings in $(AdS_3 \times S^3)_\kappa$," JHEP **1409** (2014) 048 [arXiv:1406.3642 [hep-th]].

[27] T. Kameyama and K. Yoshida, "A new coordinate system for q-deformed $AdS_5 \times S^5$ and classical string solutions," arXiv:1408.2189 [hep-th].
[28] A. Stolin and P. P. Kulish, "New rational solutions of Yang-Baxter equation and deformed Yangians," Czech. J. Phys. **47** (1997) 123 [arXiv:q-alg/9608011].
[29] P. P. Kulish, V. D. Lyakhovsky and A. I. Mudrov, "Extended jordanian twists for Lie algebras," J. Math. Phys. **40** (1999) 4569 [math/9806014 [math.QA]].
[30] T. Matsumoto and K. Yoshida, "Lunin-Maldacena backgrounds from the classical Yang-Baxter equation – Towards the gravity/CYBE correspondence," JHEP **1406** (2014) 135 [arXiv:1404.1838 [hep-th]].
[31] T. Matsumoto and K. Yoshida, "Integrability of classical strings dual for noncommutative gauge theories," JHEP **1406** (2014) 163 [arXiv:1404.3657 [hep-th]].
[32] I. Kawaguchi, T. Matsumoto and K. Yoshida, "A Jordanian deformation of AdS space in type IIB supergravity," JHEP **1406** (2014) 146 [arXiv:1402.6147 [hep-th]].
[33] T. Matsumoto and K. Yoshida, in preparation.
[34] P. M. Crichigno, T. Matsumoto and K. Yoshida, "Deformations of $T^{1,1}$ as Yang-Baxter sigma models," arXiv:1406.2249 [hep-th].
[35] O. Lunin and J. M. Maldacena, "Deforming field theories with $U(1) \times U(1)$ global symmetry and their gravity duals," JHEP **0505** (2005) 033 [hep-th/0502086].
[36] S. Frolov, "Lax pair for strings in Lunin-Maldacena background," JHEP **0505** (2005) 069 [hep-th/0503201].
[37] A. Hashimoto and N. Itzhaki, "Noncommutative Yang-Mills and the AdS / CFT correspondence," Phys. Lett. B **465** (1999) 142 [hep-th/9907166].
[38] J. M. Maldacena and J. G. Russo, "Large N limit of noncommutative gauge theories," JHEP **9909** (1999) 025 [hep-th/9908134].
[39] Ch. Ohn, "A $*$-product on $SL(2)$ and the corresponding nonstandard quantum-$U(sl(2))$," Lett. Math. Phys. 25 (1992) 85.
[40] I. Kawaguchi and K. Yoshida, "Classical integrability of Schrödinger sigma models and q-deformed Poincare symmetry," JHEP **1111** (2011) 094 [arXiv:1109.0872 [hep-th]].
[41] I. Kawaguchi and K. Yoshida, "Exotic symmetry and monodromy equivalence in Schrödinger sigma models," JHEP **1302** (2013) 024 [arXiv:1209.4147 [hep-th]].
[42] I. Kawaguchi, T. Matsumoto and K. Yoshida, "Schrödinger sigma models and Jordanian twists," JHEP **1308** (2013) 013 [arXiv:1305.6556 [hep-th]].
[43] V. E. Hubeny, M. Rangamani and S. F. Ross, "Causal structures and holography," JHEP **0507** (2005) 037 [hep-th/0504034].
[44] T. McLoughlin and I. Swanson, "Integrable twists in AdS/CFT," JHEP **0608** (2006) 084 [hep-th/0605018].
[45] D. Dhokarh, S. S. Haque and A. Hashimoto, "Melvin Twists of global AdS(5) x S(5) and their Non-Commutative Field Theory Dual," JHEP **0808** (2008) 084 [arXiv:0801.3812 [hep-th]].
[46] M. Spradlin, T. Takayanagi and A. Volovich, "String theory in beta deformed spacetimes," JHEP **0511** (2005) 039 [hep-th/0509036].
[47] M. Kulaxizi, "On beta-deformations and noncommutativity," hep-th/0610310.
[48] A. Catal-Ozer, "Lunin-Maldacena deformations with three parameters," JHEP **0602** (2006) 026 [hep-th/0512290].
[49] I. R. Klebanov and E. Witten, "Superconformal field theory on three-branes at a Calabi-Yau singularity," Nucl. Phys. B **536** (1998) 199 [hep-th/9807080].
[50] P. Basu and L. A. Pando Zayas, "Chaos Rules out Integrability of Strings in $AdS_5 \times T^{1,1}$," Phys. Lett. B **700** (2011) 243 [arXiv:1103.4107 [hep-th]].
[51] P. Candelas and X. C. de la Ossa, "Comments on Conifolds," Nucl. Phys. B **342** (1990) 246.
[52] J. P. Gauntlett, D. Martelli, J. Sparks and D. Waldram, "Sasaki-Einstein metrics on $S^2 \times S^3$," Adv. Theor. Math. Phys. **8**, 711 (2004) [hep-th/0403002].
[53] L. J. Romans, "New Compactifications of Chiral $N = 2d = 10$ Supergravity," Phys. Lett. B **153** (1985) 392.
[54] L. Castellani, L. J. Romans and N. P. Warner, "Symmetries of Coset Spaces and Kaluza-Klein Supergravity," Annals Phys. **157**, 394 (1984).

Journal of Physics: Conference Series **563** (2014) 012021 doi:10.1088/1742-6596/563/1/012021

Supersymmetry and non-Abelian geometric phase for a free particle on a circle with point-like interactions

Satoshi Ohya[1,2]

[1] Department of Physics, Faculty of Nuclear Sciences and Physical Engineering, Czech Technical University in Prague, Pohraniční 1288/1, 40501 Děčín, Czech Republic

[2] Doppler Institute for Mathematical Physics and Applied Mathematics, Czech Technical University in Prague, Břehová 7, 11519 Prague, Czech Republic

E-mail: `ohyasato@fjfi.cvut.cz`

Abstract. Though not so widely appreciated in the literature, supersymmetric quantum mechanics provides an ideal playground for studying non-Abelian geometric phase, because supersymmetry always guarantees degeneracies in energy levels. In this paper we first present a simple supersymmetric model for a free particle on a circle with point-like interactions that exhibits $\mathscr{N} = 2$ supersymmetry and doubly degenerate energy levels. We then show that Berry's connection in this model is given by the Wu-Yang-like magnetic monopole in $SU(2)$ Yang-Mills gauge theory. This article is largely based on our recent work [1].

1. Introduction

Berry's geometric phase [2] and its non-Abelian generalization [3] are ubiquitous phenomena in quantum physics when model parameters are adiabatically driven along a closed loop on the parameter space. They appear in a wide spectrum of physics and have applications in many disciplines, ranging from condensed matter physics to nuclear and high energy physics.

The purpose of this paper is to introduce a new example of quantum mechanical models that exhibits nontrivial non-Abelian geometric phase. In order to realize non-Abelian geometric phase, we must at least have:

(i) degenerate energy levels; and

(ii) a certain number of (experimentally controllable) parameters.

Let us first look at the criterion (i). As is well-known, degeneracies in energy levels are closely tied to symmetries of the systems. A typical example of such symmetries is time-reversal invariance, which results in the Kramers degeneracy in Fermi systems and plays an important role in geometric phase and/or topological invariants in molecular systems [4, 5, 6]. In this paper we focus on yet another symmetry which induces spectral degeneracy: that is, *supersymmetry*, which also guarantees degenerate energy levels between "bosonic" and "fermionic" degrees of freedom of the system.[1]

[1] This is always true except for the ground state(s): In supersymmetric quantum mechanics the ground state degeneracy is related to the topological index $\mathrm{Tr}_{\mathcal{H}}(-1)^F$ rather than symmetries [7]. In [8, 9] non-Abelian geometric phase over the space of degenerate ground states has been studied in the context of supersymmetric quantum mechanics.

Let us next look at the criterion (ii). A typical example of (experimentally controllable) parameters discussed in the context of geometric phase is electromagnetic field. Here we would like to point out that generic point-like interactions may also serve the purpose, because point-like interactions are all known to be described by $U(2)$ family of boundary conditions [10] and characterized by $\dim U(2) = 4$ independent continuous parameters. Though, as far as we know, $U(2)$ family of point-like interactions have not yet realized in laboratory experiments, they are completely allowed in quantum mechanics and hence will provide a promising ingredient of future nanotechnology.

The aim of the paper is to present a canonical example of quantum mechanical models with point-like interactions that exhibits supersymmetry, spectral degeneracy, and nontrivial non-Abelian geometric phase.[2] As in any study of theoretical physics, it is tempting to construct a minimal model that captures all the essence of the ideas. In this paper we consider a single spinless particle on a circle of circumference 2ℓ that freely propagates in the bulk yet interacts only at the origin $x = 0$ and its antipodal point $x = \ell$, where $x = 0$ and $x = 2\ell$ are identified. The bulk dynamics is governed by the Schrödinger equation for a free particle

$$-\frac{\hbar^2}{2m}\frac{d^2}{dx^2}\psi(x) = E\psi(x), \qquad x \neq 0, \ell. \tag{1}$$

Point-like interactions, on the other hand, are all described by the boundary conditions consistent with *Kirchhoff's law* of probability current at $x = 0$ and ℓ [10]:

$$j(0_+) = j(2\ell_-) \qquad \text{and} \qquad j(\ell_-) = j(\ell_+), \tag{2}$$

where $j(x) = \frac{\hbar}{2im}(\psi^*(x)\psi'(x) - \psi'^*(x)\psi(x))$ is the local probability current (with prime ($'$) indicating the derivative with respect to x) and $j(a_\pm)$ stand for the limits approaching $x = a$ from above and from below. The conditions (2) are quadratic in wavefunctions, but they can be linearized and are known to enjoy the following $U(2)$ family of solutions at each point [10]:

$$(1_2 - U)\begin{pmatrix}\psi(0_+)\\ \psi(2\ell_-)\end{pmatrix} - iL_0(1_2 + U)\begin{pmatrix}\psi'(0_+)\\ -\psi'(2\ell_-)\end{pmatrix} = 0, \qquad U \in U(2), \tag{3a}$$

$$(1_2 - \tilde{U})\begin{pmatrix}\psi(\ell_-)\\ \psi(\ell_+)\end{pmatrix} + iL_0(1_2 + \tilde{U})\begin{pmatrix}\psi'(\ell_-)\\ -\psi'(\ell_+)\end{pmatrix} = 0, \qquad \tilde{U} \in U(2), \tag{3b}$$

where 1_2 stands for the 2×2 unit matrix and L_0 is an arbitrary length scale that needs to be introduced on account of dimensional analysis. Equations (3a) and (3b) describe the most general point-like interactions in quantum mechanics consistent with Kirchhoff's law of probability current (2), or, equivalently, self-adjointness of the free Hamiltonian. The full parameter space of the model is thus $U(2) \times U(2)$. Though energy levels are not degenerate in general, there indeed exists a subspace $\mathcal{M}_{\text{SUSY}} \subset U(2) \times U(2)$ in which $\mathcal{N} = 2$ supersymmetry[3] emerges and energy levels become doubly degenerate (up to the question of ground state degeneracy). As shown in [1], the supersymmetric subspace $\mathcal{M}_{\text{SUSY}}$ is given by the following direct product:

$$\mathcal{M}_{\text{SUSY}} = U(1) \times \frac{U(2)}{U(1) \times U(1)} \cong S^1 \times S^2. \tag{4}$$

It can be shown that only the first factor controls the energy eigenvalues and the second factor does not affect the energy spectrum at all: the coset factor $U(2)/(U(1) \times U(1))$ describes the isospectral family of the system and hence provides an ideal playground for studying geometric

[2] In [11, 12] Abelian geometric phase has been studied in simple models of point-like interactions.

[3] Here $\mathcal{N}$ refers to the number of real (i.e. self-adjoint) supercharges.

Journal of Physics: Conference Series **563** (2014) 012021 doi:10.1088/1742-6596/563/1/012021

phase. The goal of the paper is to report that the Berry connection on $\mathcal{M}_{\text{SUSY}}$ with fixed $U(1)$ parameter is given by a magnetic monopole-like configuration in non-Abelian gauge theory.

The organization of the paper is as follows. In section 2 we construct $\mathcal{N}=2$ supersymmetric quantum mechanics in this system and solve the energy eigenvalue problems. We then compute non-Abelian Berry's connection on the parameter space $U(2)/(U(1)\times U(1))\cong S^2$ in section 3 and show that it is nothing but the Wu-Yang-like magnetic monopole in $SU(2)$ Yang-Mills gauge theory. Section 4 is devoted to conclusions and discussions.

In the rest of the paper we will work in the units $\hbar = 2m = 1$.

2. Supersymmetric point-like interactions

The key to construct $\mathcal{N}=2$ supersymmetric quantum mechanics for a free particle on a circle with two point-like interactions is to find the generic form of fermion parity operator $(-1)^F$, which should satisfy $((-1)^F)^2 = 1$ and have two eigenvalues $+1$ and -1. In this section we first introduce $(-1)^F$ in our model and then discuss point-like interaction invariant under supersymmetry transformations. We also explicitly construct $\mathcal{N}=2$ supersymmetry algebra and solve the superspectrum in this model.

2.1. Folding trick

In the following discussions it is convenient to work in the so-called *folding picture*,[4] which is just a change of viewpoint from "scalar" quantum mechanics (i.e. quantum mechanics with scalar-valued wavefunction) on a circle of circumference 2ℓ to "vector" quantum mechanics (i.e. quantum mechanics with vector-valued wavefunction) on an interval of length ℓ. To do this, let us consider the wavefunction on the upper- and lower-semicircles separately and embed them into a single two-component vector-valued function as follows:

$$\boldsymbol{\Psi}(x) := \begin{pmatrix} \psi(x) \\ \psi(2\ell - x) \end{pmatrix}, \qquad 0 < x < \ell. \tag{5}$$

(Throughout the paper we will use boldface symbols for vectors.) The boundary conditions (3a) and (3b) are then cast into the following forms:

$$(1_2 - U)\boldsymbol{\Psi}(0) - iL_0(1_2 + U)\boldsymbol{\Psi}'(0) = 0, \tag{6a}$$

$$(1_2 - \tilde{U})\boldsymbol{\Psi}(\ell) + iL_0(1_2 + \tilde{U})\boldsymbol{\Psi}'(\ell) = 0. \tag{6b}$$

Here and hereafter we simply write $\boldsymbol{\Psi}(0)$ and $\boldsymbol{\Psi}(\ell)$ for $\boldsymbol{\Psi}(0_+)$ and $\boldsymbol{\Psi}(\ell_-)$. In the folding picture the Hilbert space is $\mathcal{H} = L^2(0,\ell)\otimes\mathbb{C}^2$ and the free Hamiltonian that acts on $\mathcal{H}$ is given by the 2×2 matrix-valued operator, $H = \left(\begin{smallmatrix} h & 0 \\ 0 & h \end{smallmatrix}\right)$, where $h = -d^2/dx^2$. The boundary conditions (6a) and (6b) specify the most general self-adjoint domain of the matrix-valued operator H.

Let us next introduce the fermion parity $(-1)^F$ in this model. As discussed in [15, 1], $(-1)^F$ is given by the following hermitian unitary operator:

$$\mathcal{Z} : \boldsymbol{\Psi}(x) \mapsto (\mathcal{Z}\boldsymbol{\Psi})(x) := Z\boldsymbol{\Psi}(x), \tag{7}$$

where Z is a generic 2×2 hermitian unitary matrix satisfying $Z = Z^\dagger = Z^{-1}$, or $Z^2 = 1_2$. Such hermitian unitary matrix is parameterized as $Z = \boldsymbol{n}\cdot\boldsymbol{\sigma}$, where $\boldsymbol{\sigma} = (\sigma_1, \sigma_2, \sigma_3)$ is the vector of Pauli matrices and $\boldsymbol{n} = (n_1, n_2, n_3)$ is a real unit 3-vector fluffing the condition $n_1^2 + n_2^2 + n_3^2 = 1$. The parameter space of Z is therefore 2-sphere S^2, which turns out to provide the coset factor in the supersymmetric parameter space (4). Notice that the unitary transformation (7) commutes with the Hamiltonian and leaves the Schrödinger equation unchanged. However, $\mathcal{Z}$ does not

[4] Folding trick has been developed in the context of boundary conform field theory [13, 14].

leave the boundary conditions (6a) and (6b) unchanged in general and hence is not a symmetry for generic point-like interactions. It is easy to show that the boundary conditions remain invariant under $\mathcal{Z}$ if and only if both U and $\tilde{U}$ commute with Z, $[U, Z] = [\tilde{U}, Z] = 0$ [16, 1]. The solutions to these conditions are parameterized as follows:

$$U = \mathrm{e}^{i\alpha_+} P_+ + \mathrm{e}^{i\alpha_-} P_-, \tag{8a}$$

$$\tilde{U} = \mathrm{e}^{i\tilde{\alpha}_+} P_+ + \mathrm{e}^{i\tilde{\alpha}_-} P_-, \tag{8b}$$

where $P_\pm = (1_2 \pm Z)/2$ are projection operators and $\{\alpha_+, \alpha_-, \tilde{\alpha}_+, \tilde{\alpha}_-\}$ are eigenphases of U and $\tilde{U}$. The fermion parity $(-1)^F = \mathcal{Z}$ becomes the symmetry of the system if and only if U and $\tilde{U}$ are given by (8a) and (8b). In this case $(-1)^F = \mathcal{Z}$ provides a good grading operator such that the Hilbert space splits into two orthogonal subspaces, $\mathcal{H} = \mathcal{H}_+ \oplus \mathcal{H}_-$, where $\mathcal{H}_\pm = \{\boldsymbol{\Psi} \in \mathcal{H} \mid \mathcal{Z}\boldsymbol{\Psi} = \pm\boldsymbol{\Psi}\}$ are "bosonic" and "fermionic" sectors of the model. Correspondingly, any element $\boldsymbol{\Psi} \in \mathcal{H}$ of the Hilbert space $\mathcal{H}$ is decomposed as follows:

$$\boldsymbol{\Psi}(x) = \boldsymbol{\Psi}_+(x) + \boldsymbol{\Psi}_-(x), \tag{9}$$

where

$$\boldsymbol{\Psi}_\pm(x) = P_\pm \boldsymbol{\Psi}(x) = \Psi_\pm(x) \boldsymbol{e}_\pm. \tag{10}$$

Here $\boldsymbol{e}_\pm$ are orthonormal eigenvectors of Z that satisfy the eigenvalue equations $Z\boldsymbol{e}_\pm = \pm\boldsymbol{e}_\pm$, the orthonormality $\boldsymbol{e}_\alpha^\dagger \boldsymbol{e}_\beta = \delta_{\alpha\beta}$ and the completeness $\boldsymbol{e}_+\boldsymbol{e}_+^\dagger + \boldsymbol{e}_-\boldsymbol{e}_-^\dagger = 1_2$. Notice that the projection operators can also be written as $P_\pm = \boldsymbol{e}_\pm \boldsymbol{e}_\pm^\dagger$. In what follows we will call the set of eigenvectors $\{\boldsymbol{e}_+, \boldsymbol{e}_-\}$ the basis and $\Psi_\pm = \boldsymbol{e}_\pm^\dagger \boldsymbol{\Psi}$ the components. It is easy to check that the $\mathcal{Z}$-invariant boundary conditions completely decouple into the "bosonic" and "fermionic" parts and the components $\Psi_\pm$ satisfy the following Robin boundary conditions:

$$\sin\left(\frac{\alpha_\pm}{2}\right) \Psi_\pm(0) + L_0 \cos\left(\frac{\alpha_\pm}{2}\right) \Psi'_\pm(0) = 0, \tag{11a}$$

$$\sin\left(\frac{\tilde{\alpha}_\pm}{2}\right) \Psi_\pm(\ell) - L_0 \cos\left(\frac{\tilde{\alpha}_\pm}{2}\right) \Psi'_\pm(\ell) = 0. \tag{11b}$$

2.2. *Supersymmetric boundary conditions*

Let us next introduce supersymmetry transformations. To this end, let us first introduce the following one-parameter family of first-order differential operators:

$$A_\alpha^\pm = \pm\frac{d}{dx} + \frac{1}{L(\alpha)}, \tag{12}$$

where $L(\alpha) := L_0 \cot(\alpha/2)$. An important observation is that the free Hamiltonian $h = -d^2/dx^2$ is factorized as $h = A_\alpha^\mp A_\alpha^\pm - 1/L(\alpha)^2$ such that the Schrödinger equations $h\Psi_\pm = E\Psi_\pm$ can be written as $A_\alpha^- A_\alpha^+ \Psi_+ = (E + 1/L(\alpha)^2)\Psi_+$ and $A_\alpha^+ A_\alpha^- \Psi_- = (E + 1/L(\alpha)^2)\Psi_-$, which imply the following relations:

$$A_\alpha^+ \Psi_+(x) = \sqrt{E + \frac{1}{L(\alpha)^2}}\,\Psi_-(x), \tag{13a}$$

$$A_\alpha^- \Psi_-(x) = \sqrt{E + \frac{1}{L(\alpha)^2}}\,\Psi_+(x). \tag{13b}$$

Table 1. Supersymmetric boundary conditions.

	"bosonic" sector	"fermionic" sector
type DD	$\Psi_+(0) = 0 = \Psi_+(\ell)$	$(A_\alpha^-\Psi_-)(0) = 0 = (A_\alpha^-\Psi_-)(\ell)$
type RR	$(A_\alpha^+\Psi_+)(0) = 0 = (A_\alpha^+\Psi_+)(\ell)$	$\Psi_-(0) = 0 = \Psi_+(0)$
type DR	$\Psi_+(0) = 0 = (A_\alpha^+\Psi_+)(\ell)$	$(A_\alpha^-\Psi_-)(0) = 0 = \Psi_-(\ell)$
type RD	$(A_\alpha^+\Psi_+)(0) = 0 = \Psi_+(\ell)$	$\Psi_-(0) = 0 = (A_\alpha^-\Psi_-)(\ell)$

These equations give the supersymmetry transformations that map the "bosonic" state Ψ_+ to the "fermionic" state Ψ_- and vice versa. It should be emphasized that the Robin boundary conditions (11a) and (11b) are not invariant under these transformations for generic values of $\alpha_\pm$ and $\tilde{\alpha}_\pm$. The boundary conditions invariant under the supersymmetry transformations are classified in [1], and it is shown that there are four distinct supersymmetric boundary conditions, which are summarized in table 1. We note that the supersymmetric boundary conditions admit only one parameter α, which is an angle parameter and provides the $U(1)$ factor in (4).

2.3. $\mathscr{N} = 2$ *supersymmetry algebra*

Now we are in a position to introduce $\mathscr{N} = 2$ supersymmetry algebra that consists of a single self-adjoint Hamiltonian H, two self-adjoint supercharges Q_1 and Q_2, and a single fermion parity $(-1)^F$. For the following discussions it is convenient to consider the nilpotent supercharges $Q^\pm$ which are given by the linear combinations $Q^\pm = (Q_1 \pm iQ_2)/2$. Let us first work in the basis in which the fermion parity becomes diagonal. In that basis the wavefunction becomes $\boldsymbol{\Psi} = (\Psi_+, \Psi_-)^T$ such that the operators H, $(-1)^F$ and $Q^\pm$ are given by the following standard forms:

$$H = \begin{pmatrix} h & 0 \\ 0 & h \end{pmatrix}, \tag{14a}$$

$$(-1)^F = \begin{pmatrix} 1 & 0 \\ 0 & -1 \end{pmatrix}, \tag{14b}$$

$$Q^+ = \begin{pmatrix} 0 & 0 \\ A_\alpha^+ & 0 \end{pmatrix}, \tag{14c}$$

$$Q^- = \begin{pmatrix} 0 & A_\alpha^- \\ 0 & 0 \end{pmatrix}. \tag{14d}$$

These operators satisfy the $\mathscr{N} = 2$ supersymmetry algebra with the trivial central extension

$$((-1)^F)^2 = 1, \tag{15a}$$

$$(Q^\pm)^2 = 0, \tag{15b}$$

$$Q^\pm(-1)^F = -(-1)^F Q^\pm, \tag{15c}$$

$$Q^+Q^- + Q^-Q^+ = H + c, \tag{15d}$$

where $c = 1/L(\alpha)^2$ is the trivial center that commutes with all the operators. In a generic basis, these operators that act on the Hilbert space $\mathcal{H} = L^2(0, \ell) \otimes \mathbb{C}^2$ are given by $H = h \otimes 1_2$, $(-1)^F = 1 \otimes Z$, $Q^\pm = A_\alpha^\pm \otimes XP_\pm$, where X is a generic 2×2 hermitian unitary matrix that satisfies the anticommutation relation $XZ = -XZ$ and $XP_\pm X = P_\mp$.

Journal of Physics: Conference Series **563** (2014) 012021 doi:10.1088/1742-6596/563/1/012021

It should be noted that, under the supersymmetric boundary conditions, Q^+ and Q^- are indeed adjoint to each other such that the combination $H + c$ becomes positive semidefinite. In other words, the spectrum of the Hamiltonian Spec(H) is bounded from below, Spec(H) $\geq -c = -1/L(\alpha)^2$, and the lower bound is saturated by the zero-modes of $Q^{\pm}$.

2.4. $\mathcal{N} = 2$ superspectrum

In order to compute Berry's connection explicitly we have to know the spectrum and normalized energy eigenfunctions $\{\boldsymbol{\Psi}_{\pm,n} = \Psi_{\pm,n}\boldsymbol{e}_{\pm}\}_{n=0}^{\infty}$. It is a straightforward exercise to solve the Schrödinger equations $h\Psi_{\pm,n} = E_n\Psi_{\pm,n}$ for the components $\Psi_{\pm}$ with the supersymmetric boundary conditions in table 1. Here are the solutions:

(i) *Type DD.* In this case the positive energy levels are all doubly degenerate and normalized energy eigenfunctions are found to be of the forms

$$\Psi_{+,n}(x) = \sqrt{\frac{2}{\ell}}\sin\left(\sqrt{E_n}x\right), \tag{16a}$$

$$\Psi_{-,n}(x) = \sqrt{\frac{2}{\ell}\frac{1}{1+E_nL(\alpha)^2}}\left[\sin\left(\sqrt{E_n}x\right) + \sqrt{E_n}L(\alpha)\cos\left(\sqrt{E_n}x\right)\right], \tag{16b}$$

where $E_n = (n\pi/\ell)^2$, $n = 1, 2, 3, \cdots$. In addition to these positive energy eigenstates, there exists a single negative energy eigenstate at $E = E_0 = -1/L(\alpha)^2$ in the "fermionic" sector, which is the ground state of the model and takes the following form:

$$\Psi_{-,0}(x) = \sqrt{\frac{2}{L(\alpha)}\frac{1}{\mathrm{e}^{2\ell/L(\alpha)} - 1}}\exp\left(\frac{x}{L(\alpha)}\right). \tag{17}$$

(ii) *Type RR.* As in the previous case the positive energy levels are doubly degenerate too and normalized energy eigenfunctions take the following forms:

$$\Psi_{+,n}(x) = \sqrt{\frac{2}{\ell}\frac{1}{1+E_nL(\alpha)^2}}\left[\sin\left(\sqrt{E_n}x\right) - \sqrt{E_n}L(\alpha)\cos\left(\sqrt{E_n}x\right)\right], \tag{18a}$$

$$\Psi_{-,n}(x) = \sqrt{\frac{2}{\ell}}\sin\left(\sqrt{E_n}x\right), \tag{18b}$$

where $E_n = (n\pi/\ell)^2$, $n = 1, 2, 3, \cdots$. In addition, there is a single negative energy eigenstate at $E = E_0 = -1/L(\alpha)^2$ in the "bosonic" sector, whose eigenfunction is given by

$$\Psi_{+,0}(x) = \sqrt{\frac{2}{L(\alpha)}\frac{1}{1 - \mathrm{e}^{-2\ell/L(\alpha)}}}\exp\left(-\frac{x}{L(\alpha)}\right). \tag{19}$$

(iii) *Type DR.* In this case all the energy levels are doubly degenerate and normalized energy eigenfunctions take the following forms:

$$\Psi_{+,n}(x) = \sqrt{\frac{2}{\ell + \frac{L(\alpha)}{1+E_nL(\alpha)^2}}}\sin\left(\sqrt{E_n}x\right), \tag{20a}$$

$$\Psi_{-,n}(x) = \sqrt{\frac{2}{\ell + \frac{L(\alpha)}{1+E_nL(\alpha)^2}}}\sin\left(\sqrt{E_n}(\ell - x)\right), \tag{20b}$$

where E_n ($E_0 < E_1 < \cdots$) are given by the real roots of the transcendental equation

$$\tan\left(\sqrt{E}\ell\right) = -\sqrt{E}L(\alpha). \tag{21}$$

We note that the ground state energy E_0 becomes negative when $-\ell < L(\alpha) < 0$.

(iv) *Type RD.* As in the previous case all the energy levels are doubly degenerate too and normalized energy eigenfunctions take the following forms:

$$\Psi_{+,n}(x) = \sqrt{\frac{2}{\ell - \frac{L(\alpha)}{1+E_nL(\alpha)^2}}}\sin\left(\sqrt{E_n}(\ell - x)\right), \tag{22a}$$

$$\Psi_{-,n}(x) = \sqrt{\frac{2}{\ell - \frac{L(\alpha)}{1+E_nL(\alpha)^2}}}\sin\left(\sqrt{E_n}x\right), \tag{22b}$$

where E_n ($E_0 < E_1 < \cdots$) are the real roots of the transcendental equation

$$\tan\left(\sqrt{E}\ell\right) = \sqrt{E}L(\alpha). \tag{23}$$

3. Non-Abelian Berry's connection: The Wu-Yang-like monopole

Let us finally study geometric phase in $\mathcal{N} = 2$ supersymmetric quantum mechanics in this system, whose parameter space is $\mathcal{M}_{\text{SUSY}} = U(1) \times U(2)/(U(1) \times U(1))$. To this end, let us consider a simple time-dependent situation in which the $U(1) \cong S^1$ parameter is kept fixed yet the coset $U(2)/(U(1)\times U(1)) \cong S^2$ parameters are adiabatically driven along a closed loop γ on S^2. Then, if the initial state is an element of the subspace $\mathcal{H}_n = \text{span}\{\boldsymbol{\Psi}_{+,n}, \boldsymbol{\Psi}_{-,n}\}$, the final state remains in the subspace $\mathcal{H}_n$ and coincides with the initial states up to the time-dependent trivial dynamical phase $\exp(-iE_nT)$ and the time-independent nontrivial non-Abelian geometric phase $W_\gamma(A)$ given by the formula [3]:

$$W_\gamma(A) = \mathcal{P}\exp\left(i\oint_\gamma A\right), \tag{24}$$

where $A = \begin{pmatrix} A_{++} & A_{+-} \\ A_{-+} & A_{--}\end{pmatrix}$ is the hermitian matrix-valued 1-from on the parameter space $U(2)/(U(1) \times U(1)) \cong S^2$ defined by

$$A_{\alpha\beta} = i\int_0^\ell dx\, \boldsymbol{\Psi}^\dagger_{\alpha,n}(x) d\boldsymbol{\Psi}_{\beta,n}(x), \qquad \alpha,\beta \in \{+,-\}. \tag{25}$$

Here d stands for the exterior derivative on S^2. Substituting the solutions one readily finds that the Berry connection takes the following form:

$$A = \begin{pmatrix} i\boldsymbol{e}_+^\dagger d\boldsymbol{e}_+ & iK_n(\alpha)\boldsymbol{e}_+^\dagger d\boldsymbol{e}_- \\ iK_n(\alpha)\boldsymbol{e}_-^\dagger d\boldsymbol{e}_+ & i\boldsymbol{e}_-^\dagger d\boldsymbol{e}_- \end{pmatrix}, \tag{26}$$

where $K_n(\alpha)$ is the overlap integral between the components

$$K_n(\alpha) = \int_0^\ell dx\, \Psi_{+,n}(x)\Psi_{-,n}(x). \tag{27}$$

Under the local gauge transformation $\boldsymbol{\Psi}_{\alpha,n} \mapsto \tilde{\boldsymbol{\Psi}}_{\alpha,n} = \boldsymbol{\Psi}_{\beta,n} g_{\beta\alpha}$, where $g = (g_{\beta\alpha})$ is a 2×2 unitary matrix that depends on the S^2 parameters, the Berry connection transforms as follows:

$$A \mapsto \tilde{A} = g^\dagger A g + i g^\dagger dg = g^\dagger A g - i(dg^\dagger)g. \tag{28}$$

For the following discussions it is convenient to work in the gauge given by

$$g = \begin{pmatrix} \boldsymbol{e}_+^\dagger \\ \boldsymbol{e}_-^\dagger \end{pmatrix}. \tag{29}$$

In this gauge the Berry connection takes the simple form $\tilde{A} = \frac{i}{2}(1 - K_n(\alpha))ZdZ$ [1]. Let us now parameterize the hermitian unitary matrix Z as $Z = \frac{\boldsymbol{r}}{r} \cdot \boldsymbol{\sigma}$, where $\boldsymbol{r} = (x_1, x_2, x_3) \in \mathbb{R}^3 \setminus \{\boldsymbol{0}\}$ and $r = \sqrt{x_1^2 + x_2^2 + x_3^2}$. A straightforward calculation gives $\tilde{A} = \tilde{A}_i dx_i$, where

$$\tilde{A}_i = \epsilon_{ijk} \frac{x_j}{r^2} \frac{\sigma_k}{2} (1 - K_n(\alpha)). \tag{30}$$

Notice that, when $K_n(\alpha) = 0$, this is nothing the Wu-Yang magnetic monopole which is a classical solution of pure $SU(2)$ Yang-Mills gauge theory. (For the Wu-Yang monopole, see e.g. [17] for review.) Note also that, as shown in [1], (30) is gauge equivalent to the non-Abelian Berry connection discussed in the context of geometric phase in the adiabatic decoupling limit of diatomic molecule [18]. (See also [19, 20].)

4. Conclusions and discussions

In this paper we discussed non-Abelian geometric phase in $\mathscr{N} = 2$ supersymmetric quantum mechanics with point-like interactions. We computed Berry's connection on the parameter space of supersymmetric point-like interactions and showed that it is given by the Wu-Yang-like magnetic monopole in $SU(2)$ Yang-Mills gauge theory. It would be interesting to point out here that, if K_n was a function of r and had the asymptotic behaviors $K_n(r) \to 1$ as $r \to 0$ and $K_n(r) \to 0$ as $r \to \infty$, the Berry connection would be the celebrated singularity-free 't Hooft-Polyakov monopole [21, 22] in $SU(2)$ Yang-Mills-Higgs theory. It would be also quite interesting to point out that, if the zero modes of Q^+ and Q^- coexisted in the spectrum, the overlap integral $K_0(\alpha) = \int_0^\ell dx\, \Psi_{+,0}(x)\Psi_{-,0}(x)$ between the components (17) and (19) would become

$$K_0(\alpha) = \frac{\ell/L(\alpha)}{\sinh(\ell/L(\alpha))}. \tag{31}$$

The Berry connection (30) would then take the form

$$\tilde{A}_i = \epsilon_{ijk} \frac{x_j}{r^2} \frac{\sigma_k}{2} \left(1 - \frac{\ell/L(\alpha)}{\sinh(\ell/L(\alpha))}\right), \tag{32}$$

which exactly coincides with the Bogomolny-Prasad-Sommerfield (BPS) monopole [23, 24] in $SU(2)$ Yang-Mills-Higgs theory under the identification $\ell/L(\alpha) \equiv evr$, where e is the electric charge and v is the vacuum expectation value of Higgs field. As in the model studied in [9], it might be interesting to explore a model of supersymmetric point-like interactions in which the "bosonic" and "fermionic" zero-modes coexist and the Berry connection is given by the 't Hooft-Polyakov monopole that saturates the BPS bound.

Acknowledgments
This work is supported in part by ESF grant CZ.1.07/2.3.00/30.0034.

References

[1] Ohya S 2014 Non-Abelian monopole in the parameter space of point-like interactions *Preprint* arXiv:1406.4857

[2] Berry M V 1984 Quantal phase factors accompanying adiabatic changes *Proc. Roy. Soc. Lond.* A **392** 45–57

[3] Wilczek F and Zee A 1984 Appearance of gauge structure in simple dynamical systems *Phys. Rev. Lett.* **52** 2111–14

[4] Mead C A 1987 Molecular Kramers degeneracy and non-Abelian adiabatic phase factors *Phys. Rev. Lett.* **59** 161–4

[5] Avron J E, Sadun L, Segert J and Simon B 1988 Topological invariants in Fermi systems with time reversal invariance *Phys. Rev. Lett.* **61** 1329–32

[6] Avron J E, Sadun L, Segert J and Simon B 1989 Chern numbers, quaternions, and Berry's phases in Fermi systems *Commun. Math. Phys.* **124** 595–627

[7] Witten E 1982 Constraints on supersymmetry breaking *Nucl. Phys.* B **202** 253–316

[8] Pedder C, Sonner J and Tong D 2008 Geometric phase in supersymmetric quantum mechanics *Phys. Rev.* D **77** 025009 (*Preprint* arXiv:0709.0731)

[9] Sonner J and Tong D 2009 Scheme for building a 't Hooft-Polyakov monopole *Phys. Rev. Lett.* **102** 191801 (*Preprint* arXiv:0809.3783)

[10] Fülöp T and Tsutsui I 2000 A free particle on a circle with point interaction *Phys. Lett.* A **264** 366–74 (*Preprint* quant-ph/9910062)

[11] Exner P and Grosse H 1999 Some properties of the one-dimensional generalized point interactions (a torso) *Preprint* math-ph/9910029

[12] Cheon T, Fülöp T and Tsutsui I 2000 Symmetry, duality, and anholonomy of point interactions in one dimension *Ann. Phys.* **294** 1-23 (*Preprint* quant-ph/0008123)

[13] Oshikawa M and Affleck I 1997 Boundary conformal field theory approach to the critical two-dimensional Ising model with a defect line *Nucl. Phys.* B **495** 533–82 (*Preprint* cond-mat/9612187)

[14] Bachas C, de Boer J, Dijkgraaf R and Ooguri H 2002 Permeable conformal walls and holography *J. High Energy Phys.* JHEP06(2002)027 (*Preprint* hep-th/0111210)

[15] Nagasawa T, Sakamoto M and Takenaga K 2003 Supersymmetry in quantum mechanics with point interactions *Phys. Lett.* B **562** 358–64 (*Preprint* hep-th/0212192)

[16] Fülöp T, Tsutsui I and Cheon T 2003 Spectral properties on a circle with a singularity *J. Phys. Soc. Jpn.* **72** 2737-46 (*Preprint* quant-ph/0307002)

[17] Actor A 1979 Classical solutions of $SU(2)$ Yang-Mills theories *Rev. Mod. Phys.* **51** 461–525

[18] Moody J, Shapere A D and Wilczek F 1986 Realizations of magnetic-monopole gauge fields: diatoms and spin precession *Phys. Rev. Lett.* **56** 893–6

[19] Zee A 1988 Non-Abelian gauge structure in nuclear quadrupole resonance *Phys. Rev.* A **38** 1–6

[20] Chandrasekharan S, Jiang F J, Pepe M and Wiese U J 2008 Rotor spectra, berry phases, and monopole fields: from antiferromagnets to QCD *Phys. Rev.* D **78** 077901 (*Preprint* cond-mat/0612252)

[21] 't Hooft G 1974 Magnetic monopoles in unified gauge theories *Nucl. Phys.* B **79** 276-84

[22] Polyakov A 1974 Particle spectrum in quantum field theory *JETP Lett.* **20** 194-5

[23] Prasad M K and Sommerfield C M 1975 Exact classical solution for the 't Hooft monopole and the Julia-Zee dyon *Phys. Rev. Lett.* **35** 760–2

[24] Bogomolny E B 1976 Stability of classical solutions *Sov. J. Nucl. Phys.* **24** 449-54

Exact Results for the SU(∞) Principal Chiral Model

Peter Orland

1. Baruch College, The City University of New York, 17 Lexington Avenue, New York, NY 10010, U.S.A.
2. The Graduate School and University Center, The City University of New York, 365 Fifth Avenue, New York, NY 10016, U.S.A.

E-mail: `orland@nbi.dk`

Abstract. Exact correlation functions of $N \times N$ matrix-valued quantized fields are not generally known, even as N approaches infinity (the planar limit). This is in stark contrast to isovector field theories, which have straightforward $1/N$-expansions. I review the method by which exact correlation functions of the large-N limit $(1+1)$-dimensional sigma model with SU(N) $\times$ SU(N) symmetry. This field theory is asymptotically free, with a dynamically generated mass gap. The technique is a combination of the $1/N$-expansion of the S matrix and Smirnov's form-factor axioms. I briefly discuss how to extract the short-distance behavior of the exact correlation function, which can be compared with perturbation theory.

Feynman diagrams in the large-N limit of matrix theories are planar [1]. Planar diagrams are a small subclass of all the diagrams generated at finite N. The cardinality of planar diagrams grows only as a power of the number of loops, in contrast to the cardinality of all diagrams, which grows as the factorial of the number of loops. Furthermore, the expansion in $1/N$ is analogous to the perturbative expansion in a string theory, where the diagrams themselves play the role of worldsheets. This analogy has been nearly completely realized in the AdS/CFT correspondence. For field theories which are not conformal, little progress has been made, unless an ultraviolet cut-off is present, with a large bare coupling (in a more realistic approach, the bare coupling is infinitesimal). In contrast, the $1/N$-expansion an effective way to calculate properties of N-component isovector theories (such as the O(N) sigma model); the reason is that a saddle point in a Lagrange multiplier appears. Unfortunately, such saddle-point methods do not work for matrix field theories.

Here I review some recent work on the correlation functions of the SU(N) $\times$ SU(N) principal chiral sigma model (PCSM) in one space and one time dimension. This model is integrable, and its particle spectrum and S matrix have been known for some time [2]. Karowski and Weisz found an exact form factor for the current operator of the $N = 2$ PCSM [3]. For $2 < N < \infty$, the PCSM has bound states, which complicates the determination of exact form factors.

By combining integrability and the $N \to \infty$ limit, exact form factors and exact Green's functions can be found [4], [5]. A key ingredient is Smirnov's axioms for form factors [6].

The action of the SU(N) $\times$ SU(N)-symmetric PCSM in Minkowski spacetime is

$$S = \frac{N}{2g_0^2} \int d^2x \, \eta^{\mu\nu} \, \mathrm{Tr}\, \partial_\mu U(x)^\dagger \partial_\nu U(x), \tag{1}$$

where $U(x) \in$ SU(N) is the bare field, $\mu, \nu = 0, 1$, $U(x) \in$ SU(N) (that is, $U(x)$ is an $N \times N$ unitary matrix of determinant one), and the metric is that of flat Minkowski space, $\eta^{00} = 1$,

$\eta^{11} = -1$, $\eta^{01} = \eta^{10} = 0$. An implicit ultraviolet cut-off Λ is present in (1). The action has the invariance $U(x) \to V_L U(x) V_R$, for two constant matrices V_L, $V_R \in \mathrm{SU}(N)$. The PCSM is asymptotically ultraviolet free. Analytic and numerical evidence strongly indicates that the spectrum has a mass gap m. Particle masses are given by the sine formula:

$$m_r = m \frac{\sin \frac{\pi r}{N}}{\sin \frac{\pi}{N}}, \quad r = 1, \ldots, N-1. \tag{2}$$

If m is fixed, as $N \to \infty$, the only surviving bound state of the elementary particles corresponds to $r = N-1$. This is the antiparticle. The binding energy, for $1 < r < N-1$, vanishes. Furthermore, the residues of the $1 < r < N-1$-bound-state poles of the S matrix also vanish. Thus, in the large-N limit, only the elementary particle and its antiparticle remain in the spectrum.

The S matrix of two elementary particles of the PCSM, with incoming rapidities θ_1 and θ_2 (the components of the momenta are $(p_j)_0 = m\cosh\theta_j$, $(p_j)_1 = m\sinh\theta_j$), outgoing rapidities θ_1' and θ_2' and rapidity difference $\theta = \theta_{12} = \theta_1 - \theta_2$ is

$$\mathcal{S}_{PP} = S_{PP}(|\theta|)\, 4\pi\delta(\theta_1' - \theta_1)\, 4\pi\delta(\theta_2' - \theta_2),$$

where $S_{PP}(|\theta|)$ is a function which acts on the quantum numbers of the particles. The general convention is to refer to $S_{PP}(|\theta|)$ by the term "S matrix", rather than $\mathcal{S}_{PP}$. The $1/N$-expansion of the S matrix is explicitly

$$S_{PP}(\theta) = \left[1 + O(1/N^2)\right] \left[1 - \frac{2\pi \mathrm{i}}{N\theta}(\mathrm{P}_{\mathrm{L}} \otimes 1_{\mathrm{R}} + 1_{\mathrm{L}} \otimes \mathrm{P}_{\mathrm{R}}) - \frac{4\pi^2}{N^2\theta^2}\mathrm{P}_{\mathrm{L}} \otimes \mathrm{P}_{\mathrm{R}}\right], \tag{3}$$

where P_{L} and P_{R} exchange left and right color indices, respectively. The S matrix elements of one particle and one antiparticle $S_{AP}(\theta)$ can be found from (3), using crossing.

The bare field $U(x)$ has vanishing correlations, as g_0 is taken to zero and the ultraviolet cut-off is removed. Thus, $U(x)$ is not the physical field of the renormalized quantum field theory. There is, however, such a physical field, $\Phi(x)$, which is a complex $N \times N$ matrix. This *scaling* field is not directly proportional to the unitary matrix $U(x)$, but there is an equivalence

$$\Phi(x) \sim \mathrm{Z}(g_0, \Lambda)^{-1/2} U(x),$$

in the sense that

$$\frac{1}{N}\langle 0|\mathrm{Tr}\ \Phi(x)\Phi(0)^\dagger|0\rangle = \mathrm{Z}(g_0, \Lambda)^{-1}\frac{1}{N}\langle 0|\mathrm{Tr}\ U(x)U(0)^\dagger|0\rangle. \tag{4}$$

The renormalization factor $\mathrm{Z}(g_0(\Lambda), \Lambda)$ goes to zero as $\Lambda \to \infty$. The coupling is a function of the cut-off $g_0(\Lambda)$, with the property that the mass gap $m(g_0(\Lambda), \Lambda)$ is independent of Λ. The field Φ is normalized by

$$\langle 0|\Phi(0)_{b_0a_0}|P, \theta, a_1, b_1\rangle_{\mathrm{in}} = N^{-1/2}\delta_{a_0a_1}\delta_{b_0b_1}, \tag{5}$$

where the ket on the right is a one particle ($r = 1$) state, with rapidity θ and we implicitly sum over left and right colors a_1 and b_1, respectively. There is no contribution from the one-antiparticle state (with $r = N-1$):

$$\langle 0|\Phi(0)_{b_0a_0}|A, \theta, b_1, a_1\rangle_{\mathrm{in}} = 0.$$

In much of what follows, vacuum expectation values, such as those in (4), are Wightman functions, rather than Schwinger functions. This means that there is no time-ordering of local operators. Towards the end of this article, I will discuss the short-distance behavior of Euclidean Schwinger functions.

Knowledge of form factors, that is matrix elements of local operators, may be used to express Wightman functions through completeness:

$$\mathcal{W}(x) = \frac{1}{N}\langle 0|\mathrm{Tr}\ \Phi(x)\Phi(0)^\dagger|0\rangle = \sum_X \mathrm{Tr}\ \frac{1}{N}\langle 0|\Phi(x)|X\rangle_{\mathrm{in}}\ {}_{\mathrm{in}}\langle X|\Phi(0)^\dagger|0\rangle, \tag{6}$$

where X denotes an arbitrary asymptotic state. A basic form factor is a matrix element of an operator $\mathfrak{B}(x)$ between multi-particle states, that is

$$f^{\mathfrak{B}}(\theta_1,\ldots,\theta_M)_{j_1\cdots j_M} = \langle 0|\,\mathfrak{B}(0)\,|\theta_M, j_M,\ldots,\theta_1, j_1\rangle_{\mathrm{in}}\,, \tag{7}$$

where $j_1, \ldots, j_M$ are the quantum numbers of the excitation (including whether it is a particle or antiparticle). Other matrix elements may be found by crossing and the Lorentz-transformation properties of the operator $\mathfrak{B}(x)$. Smirnov's axioms [6] are:

1. *Scattering Axiom* or Watson's theorem. This is

$$\begin{aligned} &f^{\mathfrak{B}}(\theta_1,\ldots,\theta_{i-1},\theta_{i+1},\theta_i,\theta_{i+2}\ldots,\theta_M)_{j_1\cdots j_{i-1}j_{i+1}j_ij_{i+2}\cdots j_M} \\ &= S^{k_ik_{i+1}}_{j_ij_{i+1}}(\theta_i-\theta_{i+1})f^{\mathfrak{B}}(\theta_1,\ldots,\theta_{i-1},\theta_i,\theta_{i+1},\theta_{i+2},\ldots,\theta_M)_{j_1\cdots j_{i-1}k_ik_{i+1}j_{i+2}\cdots j_M}. \end{aligned} \tag{8}$$

2. *Periodicity Axiom*, or generalized crossing, which states

$$f^{\mathfrak{B}}(\theta_1,\ldots,\theta_M)_{j_1\cdots j_M} = f^{\mathfrak{B}}(\theta_M - 2\pi\mathrm{i}, \theta_1\ldots,\theta_{M-1})_{j_Mj_1\cdots j_{M-1}}\,. \tag{9}$$

3. *Annihilation Pole Axiom.* A justification for this axiom in the context of the SU(∞) PCSM is given in the first of Refs. [4]. The axiom states:

$$\begin{aligned} &\mathrm{i}\ \mathrm{Res} f^{\mathfrak{B}}(\theta_1,\ldots,\theta_M)_{j_1\cdots j_M}\Big|_{\theta_M=\theta_{M-1}+\pi\mathrm{i}} = f^{\mathfrak{B}}(\theta_1,\ldots,\theta_{M-2})_{j_1\cdots j_{M-2}}C_{j_{M-1}j_M} \\ &- S^{k_{M-1}k_1}_{t_1j_1}(\theta_1-\theta_{M-1})S^{t_1k_2}_{t_2j_2}(\theta_2-\theta_{M-1})\cdots S^{t_{M-4}k_{M-3}}_{t_{M-3}j_{M-3}}(\theta_{M-3}-\theta_{M-1}) \\ &\times S^{k_{M-3}k_{M-2}}_{j_{M-1}j_{M-2}}(\theta_{M-2}-\theta_{M-1})f^{\mathfrak{B}}(\theta_1,\ldots,\theta_{M-2})_{k_1\cdots k_{M-2}}C_{k_{M-1}j_M}\,, \end{aligned} \tag{10}$$

where C is the charge-conjugation matrix.

4. *Lorentz-Invariance Axiom.* If an operator $\mathfrak{B}(x)$ carries Lorentz spin s, the form factors must transform under a boost $\theta_j \to \theta_j + \alpha$ for all $j = 1,\ldots,M$ as

$$f^{\mathfrak{B}}(\theta_1+\alpha,\ldots,\theta_M+\alpha) = e^{\mathrm{s}\alpha}f^{\mathfrak{B}}(\theta_1,\ldots,\theta_M)\,, \tag{11}$$

The spin for the scaling field is zero.

5. *Minimality Axiom.* This is the assumption that the form factors consistent with the other axioms have as much analyticity as possible.

Paradoxically, Axioms 1. and 3. appear trivial at large-N, because interactions (see (3)) are of order $1/N$. The resolution of the paradox is that the color indices of the in-state are contracted in such a way as to produce extra factors of N [4], [5]. These contractions produce an *effective* S matrix between particles and and antiparticles. The rules for constructing exact form factors are described below.

The general matrix element of $\Phi(0)$ between the vacuum and an $(M-1)$-antiparticle, M-particle state has many terms. By comparing it to the S-matrix element describing the scattering of M-particles, we can determine the leading part of the form factors in the $1/N$-expansion. This part is proportional to $N^{-M+1/2}$. Left and right permutations (in the permutation group may be written S_M) by σ and τ, respectively and σ and τ take the set of numbers 0, 1, 2, ..., $M-1$ to $\sigma(0)$, $\sigma(1)$, ..., $\sigma(M-1)$ and $\tau(0)$, $\tau(1)$, ..., $\tau(M-1)$, respectively. The most general form factor of the renormalized field is

$$\begin{aligned}&\langle\, 0 \mid \Phi(0)_{b_0a_0}\, |I_1,\theta_1,C_1;I_2,\theta_2,C_2;\ldots;I_{2M-1},\theta_{2M-1},C_{2M-1}\rangle \\ &= \frac{\sqrt{N}}{N^M}\sum_{\sigma,\tau\in S_M} F_{\sigma\tau}(\theta_1,\theta_2,\ldots,\theta_{2M-1})\prod_{j=0}^{M-1}\delta_{a_j\, a_{\sigma(j)+M}}\delta_{b_j\, b_{\tau(j)+M}},\end{aligned} \tag{12}$$

where I_j is A or P and C_j consists of the color pair a_j, b_j. The order $(1/N)^0$ parts of the coefficients of the tensors

$$N^{-M+1/2}\prod_{j=0}^{M-1}\delta_{a_j\, a_{\sigma(j)+M}}\delta_{b_j\, b_{\tau(j)+M}}, \tag{13}$$

that is $F^0_{\sigma\tau}(\theta_1,\theta_2,\ldots,\theta_{2M-1})$, are the same, no matter the order of the particles in the in-state ket of (12), except for a phase, as we explain below.

The function $F_{\sigma\tau}$ can be expanded in powers of $1/N$, *i.e.*

$$F_{\sigma\tau}(\theta_1,\theta_2,\ldots,\theta_{2M-1}) = F^0_{\sigma\tau}(\theta_1,\theta_2,\ldots,\theta_{2M-1}) + \frac{1}{N}F^1_{\sigma\tau}(\theta_1,\theta_2,\ldots,\theta_{2M-1}) + \cdots . \tag{14}$$

Only the leading term on the right-hand side of (14) is known.

Suppose we interchange two adjacent excitations in the left-hand side of (12). Axiom 1 implies that as $N\to\infty$:

(i) If both excitations are antiparticles or both excitations are particles, the result is the interchange of the rapidities of these two excitations, in the function $F^0_{\sigma\tau}$.

(ii) If one excitation is an antiparticle with rapidity θ_j and colors a_j, b_j and the other excitation is a particle with rapidity θ_k and colors a_k, b_k, and $\sigma(j)+M\neq k$, $\tau(j)+M\neq k$, there is no effect on the function $F^0_{\sigma\tau}$.

(iii) If one excitation is an antiparticle with rapidity θ_j and colors a_j, b_j and the other excitation is a particle with rapidity θ_k and colors a_k, b_k, and $\sigma(j)+M=k$, $\tau(j)+M\neq k$, then $F^0_{\sigma\tau}$ is multiplied by the phase $\frac{\theta_{jk}+\pi\mathrm{i}}{\theta_{jk}-\pi\mathrm{i}}$.

(iv) If one excitation is an antiparticle with rapidity θ_j and colors a_j, b_j and the other excitation is a particle with rapidity θ_k and colors a_k, b_k, and $\sigma(j)+M\neq k$, $\tau(j)+M=k$, then $F^0_{\sigma\tau}$ is multiplied by the phase $\frac{\theta_{jk}+\pi\mathrm{i}}{\theta_{jk}-\pi\mathrm{i}}$.

(v) If one one excitation is an antiparticle with rapidity θ_j and colors a_j, b_j and the other excitation is a particle with rapidity θ_k and colors a_k, b_k, and $\sigma(j)+M=k$, $\tau(j)+M=k$, then $F^0_{\sigma\tau}$ is multiplied by the phase $\left(\frac{\theta_{jk}+\pi\mathrm{i}}{\theta_{jk}-\pi\mathrm{i}}\right)^2$.

These conditions yield a choice of $F^0_{\sigma\tau}(\theta_1,\theta_2,\ldots,\theta_{2M-1})$, satisfying the annihilation-pole axiom and having as much analyticity as possible:

$$F^0_{\sigma\tau}(\theta_1,\theta_2,\ldots,\theta_{2M-1}) = \frac{(-4\pi)^{M-1}K_{\sigma\tau}}{\prod_{j=1}^{M-1}[\theta_j-\theta_{\sigma(j)+M}+\pi\mathrm{i}][\theta_j-\theta_{\tau(j)+M}+\pi\mathrm{i}]}, \tag{15}$$

where

$$K_{\sigma\tau} = \begin{cases} 1\,, & \sigma(j) \neq \tau(j), \text{ for all } j \\ 0\,, & \text{otherwise} \end{cases} \quad . \tag{16}$$

Notice that the expression for $K_{\sigma\tau}$ insures the absence of double poles. Thus $K_{\sigma\tau}$ is unity if and only if the composite permutation $\sigma \circ \tau^{-1}$ has no fixed points, *i.e.* has the smallest possible fundamental character in S_M. The number of pairs σ and τ in S_M satisfying this condition is $(M-1)!M!$. Equations (15) and (16) completely determine the form factors in the large-N limit.

Now that the form factors are determined, the Wightman function can be found. It is convenient to write (6) as

$$\begin{aligned} \mathcal{W}(x) &= \frac{1}{N}\sum_{a_0,b_0}\sum_X \langle 0|\Phi(0)_{b_0a_0}|X\rangle_{\text{in}}\ {}_{\text{in}}\langle X|\ [\Phi(0)_{b_0a_0}]^*\ |0\rangle e^{\mathrm{i}p_X\cdot x} \\ &= \frac{1}{N}\sum_{a_0,b_0}\sum_X |\langle 0|\Phi(0)_{b_0a_0}|X\rangle_{\text{in}}|^2\ e^{\mathrm{i}p_X\cdot x}, \end{aligned}$$

where p_X is the momentum eigenvalue of the state $|X\rangle$. After some work, the result is the remarkably simple expression:

$$\mathcal{W}(x) = \int \frac{d\theta}{4\pi} e^{\mathrm{i}x\cdot p} + \frac{1}{4\pi}\sum_{l=1}^{\infty}\int d\theta_1\cdots d\theta_{2l+1}\, e^{\mathrm{i}\sum_{j=1}^{2l+1} x_j\cdot p_j}\prod_{j=1}^{2l}\frac{1}{(\theta_j \quad \theta_{j+1})^2+\pi^2}\,. \tag{17}$$

For $m|x| \gg 1$, the expression (17) decays exponentially, as expected. On the other hand, if (17) is exact for all choices of x, then for $m|x| \ll 1$, the Wick-rotated time-ordered product of two scaling field operators should behave as

$$\frac{1}{N}\langle 0|\ \mathcal{T}\ \text{Tr}\ \Phi(x)\Phi(0)^\dagger|0\rangle = C_2(\ln m|x|)^2 + C_1\ln m|x| + C_0 + O(1/\ln m|x|), \tag{18}$$

for some constants C_2, C_1, etc.

Here is the standard argument for (18) (see for example, Ref. [7]). After a Wick rotation $x^0 \to \mathrm{i}x^0$, the regularized Euclidean correlation function is $G(|x|,\Lambda) = N^{-1}\langle 0|\ \mathcal{T}\ \text{Tr}\ \Phi(x)\Phi(0)^\dagger|0\rangle$. This function and the coupling $g_0(\Lambda)$ satisfy the renormalization group equations

$$\frac{\partial \ln G(R,\Lambda)}{\partial \ln \Lambda} = \gamma(g_0) = \gamma_1 g_0^2 + \cdots\,, \qquad \frac{\partial g_0(\Lambda)}{\partial \ln \Lambda} = \beta(g_0) = -\beta_1 g_0^3 + \cdots\,, \tag{19}$$

respectively. The coefficients of the anomalous dimension $\gamma(g_0)$ and the beta function $\beta(g_0)$ are $\gamma_1 = (N^2-1)/(2\pi N^2)$ and $\beta_1 = 1/(4\pi)$. For large Λ, $G(R,\Lambda)$ becomes a function of the product of the two variables $G(R\Lambda)$. Integrating (19) yields the leading behavior

$$G(R,\Lambda) \sim C[\ln(R\Lambda)]^{\gamma_1/\beta_1}\,. \tag{20}$$

As $N \to \infty$, the power γ_1/β_1 becomes 2.

Performing the Wick-rotation on (17) is done by setting $x^1 = 0$ and replacing x^0 by $\mathrm{i}R$, with $R > 0$. This changes the phases in (17) by $\exp \mathrm{i}p_j\cdot x \to \exp -mR\cosh\theta_j$. It is convenient to introduce $L = \ln\frac{1}{mR}$. As mR becomes small, $\exp -mR\cosh\theta_j$ becomes approximately the characteristic function of $(-L,L)$, equal to unity for $-L < \theta < L$ and zero everywhere else (the same argument was used to explain how walls form in the Feynman-Wilson gas [8]). This basic idea was used to find the scaling behavior of Ising-model correlation functions [9] from the exact form factors [10]. The short-distance approximation to the Euclidean two-point function is

$$G(mR) = \frac{L}{2\pi} + \frac{1}{4\pi}\sum_{l=1}^{\infty}\int_{-L}^{L} d\theta_1 \cdots \int_{-L}^{L} d\theta_{2l+1} \prod_{j=1}^{2l} \frac{1}{(\theta_j - \theta_{j+1})^2 + \pi^2}\,. \tag{21}$$

This series (21) is the partition function of a polymer in a box of size $2L$. The $j^{\rm th}$ atom in the polymer chain is located at θ_j. The long-range potential energy is $\ln[(\theta_j - \theta_{j+1})^2 + \pi^2]$, between connected pairs of atoms in the chain.

Redefining the variables by $\theta_j = Lu_j$, (21) becomes

$$G(mR) = \frac{L}{2\pi} + \frac{L}{4\pi}\sum_{l=1}^{\infty}\int_{-1}^{1} du_1 \cdots \int_{-1}^{1} du_{2l+1} \prod_{j=1}^{2l} \frac{1}{L[(u_j - u_{j+1})^2 + (\pi/L)^2]}\,. \tag{22}$$

The factors in the series (22) are closely related to the generators of the Poisson semigroup. Consider the transfer operators $P(a)$, whose matrix elements are defined by $\langle u'|P(a)|u\rangle = a[(u'-u)^2 + a^2]^{-1}\pi^{-1}$, where u' and u are arbitrary real numbers. These operators form the Poisson semigroup, with the composition law $P(a)P(b) = P(a+b)$. Specifically, $P(a) = \exp -a\Delta^{1/2}$, where $\Delta^{1/2} = \sqrt{-d^2/du^2}$ is the fractional Laplacian in one dimension.

As I will discuss in a future publication, Eq. (18) follows from the existence of a self-adjoint extension of $\Delta^{1/2}$ on the Hilbert space of functions which vanish outside the open interval $u \in (-1,1)$. This operator has an infinite set of discrete eigenvalues λ_n, of the eigenfunctions $\varphi_n(u)$, $n = 1, 2, \ldots$, with $0 < \lambda_1 < \lambda_2 < \cdots$, with $\varphi_n(\pm 1) = 0$. Discussion of the spectrum of $\Delta^{\alpha/2}$, with real $\alpha \in (0,2)$, may be found in [11]. For $u, u' \in (-1,1)$,

$$\frac{1}{L[(u-u')^2 + (\pi/L)^2]} = \langle u'|e^{-\frac{\pi}{L}H(L)}|u\rangle,$$

where the operator in the exponent is

$$H(L) = \Delta^{1/2} + O(L^{-1}). \tag{23}$$

Summing over l in Eq. (22) yields

$$G(mR) = \frac{L}{4\pi}\sum_{n=1}^{\infty}\left|\int_{-1}^{1} du\,\varphi_n(u)\right|^2 \frac{1}{1 - \exp[-2\pi\lambda_n/L + O(L^{-2})]}. \tag{24}$$

The expression (24) yields the first two coefficients in (18):

$$C_2 = \frac{1}{8\pi^2}\sum_{n=1}^{\infty}\left|\int_{-1}^{1} du\,\varphi_n(u)\right|^2 \lambda_n^{-1},\;\; C_1 = \frac{1}{4\pi} = 0.0796\,. \tag{25}$$

An upper bound on the leading coefficient C_2 is obtained by replacing λ_n in (25) by λ_1. This yields $C_2 < \frac{1}{4\pi^2\lambda_1} = 0.0219$, from the numerical result $\lambda_1 = 1.1577$, found in the second of Ref. [11].

To summarize, the short-distance behavior of the exact correlation function agrees with that predicted by perturbation theory. This, in my opinion, is convincing support of the validity of the large-N form factors, found in Refs. [4], [5].

References

[1] G. 't Hooft, Nucl. Phys. **B72** (1974) 461.

[2] A.M. Polyakov and P.B. Wiegmann, Phys. Lett. **131B** (1983) 121; E. Abdalla, M.C.B. Abdalla and M. Lima-Santos, Phys. Lett. **140B** (1984) 71; P.B. Wiegmann, Phys. Lett. **142B** (1984) 173; L.D. Faddeev, N.Yu. Reshetikhin, Ann. Phys. **167** (1986) 227.

[3] M. Karowski and P. Wiesz, Nucl. Phys. **B139** (1978) 455.

[4] P. Orland, Phys. Rev. **D 84** (2011) 105005, (2011); Phys. Rev. **D 86** (2012) 045023.

[5] A. Cortés Cubero, Phys. Rev. **D 86** (2012) 025025; A. Cortés Cubero and P. Orland, Phys. Rev. **D 88** (2013) 025044. An error in some of the higher-point form factors in these papers is corrected in : A. Cortés Cubero, Phys. Rev. **D 90** (2014) 065002.

[6] F.A. Smirnov, **Form Factors in Completely Integrable Models of Quantum Field Theory**, Adv. Series in Math. Phys. **14**, World Scientific, Singapore (1992).

[7] A.M. Polyakov, **Gauge Fields and Strings**, Section 2.1, Harwood Academic Pulishers, Chur (1987).

[8] R.P. Feynman, Phys. Rev. Lett. **23** (1969) 1415; K.G. Wilson, Cornell preprint CLNS-131 (1970) in: **Proc. Fourteenth Scottish Universities Summer School in Physics** (1973), eds. R.L. Crawford and R. Jennings, Academic Press, New York, 1974.

[9] J.L. Cardy and G. Mussardo, Nucl. Phys. **B 340** (1990) 387; V.P. Yurov and Al.B. Zamolodchikov, Int. J. Mod. Phys. **A 6** (1991) 3419.

[10] B. Berg, M. Karowski and P. Weisz, Phys. Rev. **D 19** (1979) 2477; T.T. Wu, B.M. McCoy, C.A. Tracy and E. Barouch, Phys. Rev. **B 13** (1976) 316; M. Sato, T. Miwa and M. Jimbo, Proc. Japan Acad. **53 A** (1977) 6; Publ. RIMS, Kyoto Univ. **16** (1980), 531; B. Schroer and T.T. Truong, Nucl. Phys. **B 144** (1978) 80.

[11] T. Kulczycki, M. Kwaśnicki, J.Małecki, A. Stós, Proc. London Math. Soc. **101** (2) (2010) 589; M. Kwaśnicki, J. Funct. Anal. **262** (2012) 2379.

Journal of Physics: Conference Series **563** (2014) 012023 doi:10.1088/1742-6596/563/1/012023

Generalized symmetries generating Noether currents and canonical conserved quantities

Marcella Palese and Ekkehart Winterroth

Department of Mathematics, University of Torino, via C. A.lberto 10, I-10124 Torino , Italy

E-mail: marcella.palese@unito.it, ekkehart.winterroth@unito.it

Abstract. We determine the condition for a Noether–Bessel-Hagen current, associated with a generalized symmetry, to be variationally equivalent to a Noether current for an invariant Lagrangian. We show that, if it exists, this Noether current is exact on-shell and generates a canonical conserved quantity.

2000 MSC: 58A20,58E30,46M18.
keywords: jet bundle, generalized symmetry, Noether current, conserved quantity, cohomology.

1. Introduction

As it is well known, invariance properties of field dynamics are an effective tool to understand a physical system without solving the equations themselves: the existence of conservation laws associated with symmetries of equations strongly simplifies their study and corresponding conserved currents along solutions (on-shell) appear to be the most significant for the description of a system.

The Noether Theorems [15] stated that, for the description of systems the equations of which arise from a variational problem, off-shell conserved currents are even more relevant since, for a large class of field theories, potentials (also called superpotentials in natural and gauge-natural theories) of such currents provide conserved quantities such as energy, momenta and charges. Therefore, it turns out to be fundamental to understand whether conserved currents associated with invariance of equations could be identified with Noether conserved currents for a certain Lagrangian; in fact, a symmetry of a Lagrangian is also a symmetry of its Euler-Lagrange form, but the converse in general is not true. We are interested to this converse problem which belongs to aspects of inverse problems the calculus of variations. The expression for the so-called canonical Noether current associated with a Lagrangian is the outcome of an integration by parts, related with the precise form taken by the Lie derivative of a Lagrangian (*i.e.* the Lie derivative of an horizontal n-form, n being the dimension of the base manifold). Of course, the particular type of inverse problem we shall consider will strictly interplay with the formula for a variational Lie derivative; analogously to what happens in the study of variationality of equations, the solution to this problem involves a kind of variational integrating factor.

Consider then conserved currents associated with invariance properties of *(locally) variational global* field equations, *i.e.* with so-called generalized or Bessel-Hagen symmetries [1]. Noether currents for different local Lagrangian presentations and corresponding *conserved currents associated with each local presentation* have been characterized in [3, 4, 5, 17]. There exist cohomological obstructions for such local currents be globalized and such obstructions are also related with the *existence of global solutions* for a given global field equation [6].

We refer to the geometric formulation of the calculus of variations as a subsequence of the de Rham sequence of differential forms on finite order prologations of fibered manifolds. We assume the r-th order prolongation of a fibered manifold $\pi : \mathbf{Y} \to \mathbf{X}$, with $\dim \mathbf{X} = n$ and $\dim \mathbf{Y} = n + m$, to be the configuration space; this means that *fields are assumed to be (local) sections* of $\pi^r : J_r\mathbf{Y} \to \mathbf{X}$. Due to the affine bundle structure of $\pi^{r+1}_r : J_{r+1}\mathbf{Y} \to J_r\mathbf{Y}$, we have a natural splitting $J_r\mathbf{Y} \times_{J_{r-1}\mathbf{Y}} T^*J_{r-1}\mathbf{Y} = J_r\mathbf{Y} \times_{J_{r-1}\mathbf{Y}} (T^*\mathbf{X} \oplus V^*J_{r-1}\mathbf{Y})$, which induces natural *spittings* in horizontal and vertical parts of vector fields, forms and of the exterior differential on $J_r\mathbf{Y}$. Let ρ be a q-form on $J_r\mathbf{Y}$; in particular we obtain a natural decomposition of the pull-back by the affine projections of a q-form ρ, as $(\pi^{r+1}_r)^*\rho = \sum_{i=0}^{q} p_i\rho$, where $p_i\rho$ is the i-contact component of ρ (by definition a contact form is zero along any holonomic section of $J_r\mathbf{Y}$). For $1 \leq q \leq n$, the representation mapping is just given by the horizontalization $p_0\rho = h\rho$. For $q > n$, say, $q = n + k$, it is clear that, in this case, $p_k\rho$ denotes the component of ρ with the lowest degree of contactness. Starting from this splitting one can define sheaves of contact forms Θ^*_r, suitably characterized by the kernel of p_i [9]; the sheaves Θ^*_r form an exact subsequence of the de Rham sequence on $J_r\mathbf{Y}$ and one can define the quotient sequence

$$0 \to I\!R_{\mathbf{Y}} \to \ldots \to^{\mathcal{E}_{n-1}} \Lambda^n_r/\Theta^n_r \to^{\mathcal{E}_n} \Lambda^{n+1}_r/\Theta^{n+1}_r \to^{\mathcal{E}_{n+1}} \Lambda^{n+2}_r/\Theta^{n+2}_r \to^{\mathcal{E}_{n+2}} \ldots \to 0$$

the r–th order *variational sequence* on $\mathbf{Y} \to \mathbf{X}$ which is an acyclic resolution of the constant sheaf $I\!R_{\mathbf{Y}}$; see [9].

In the following, if $\rho \in \Lambda^q_r$, $[\rho] \in \mathcal{V}^r_q$ denotes the equivalence class of ρ modulo $ker\, p_i$, $i = 0, 1, \ldots, m$, $q = n + i$. The quotient sheaves $\mathcal{V}^k_r \equiv \Lambda^k_r/\Theta^k_r$ in the variational sequence can be represented as sheaves of k-forms on jet spaces of higher order by the *interior Euler operator* which is uniquely intrinsically defined by the decomposition $p_k\rho = \mathcal{I}(\rho) + p_k dp_k \mathcal{R}(\rho)$, together with the properties

$$(\pi^{2r+1}_r)^*\rho - \mathcal{I}(\rho) \in \Theta^{n+k}_{2r+1}\,; \quad \mathcal{I}(p_k dp_k \mathcal{R}(\rho)) = 0\,; \quad \mathcal{I}^2(\rho) = (\pi^{4r+3}_{2r+1})^*\mathcal{I}(\rho)\,; \quad ker\mathcal{I} \doteq \Theta^{n+k}_r\,;$$

one can define a representation mapping $R_q : \mathcal{V}^q_r \to \Lambda^q_s, : [\rho] \mapsto R_q([\rho])$, with $R_q([\rho]) = p_0\rho \equiv h\rho$ for $0 \leq q \leq n$, $s = r + 1$; $R_q([\rho]) = \mathcal{I}(\rho)$ for $n + 1 \leq q \leq P$, $s = 2r + 1$; $R_q([\rho]) = \rho$ for $P + 1 \leq q \leq N$, $s = r$, where $N = dim\, J_r\mathbf{Y}$ and P is the maximal degree of non trivial contact forms on $J_r\mathbf{Y}$ (see *e.g.* [7, 8, 9, 12, 18], whereby also local coordinate expressions can be found). The representation sequence $\{0 \to R_*(\mathcal{V}^*_r), E_*\}$, is also exact and we have $E_q \circ R_q([\rho]) = R_{q+1} \circ \mathcal{E}_q([\rho]) = R_{q+1}([d\rho])$. Currents are sheaf sections ϵ of $\mathcal{V}^{n-1}_r$ and $\mathcal{E}_{n-1} = d_H$ is the total divergence; Lagrangians are sections λ of $\mathcal{V}^n_r$, while $\mathcal{E}_n$ is called the Euler-Lagrange morphism; Sections η of $\mathcal{V}^{n+1}_r$ are called *source forms* or also *dynamical forms*, $\mathcal{E}_{n+1}$ is called the Helmholtz morphism.

2. Currents associated with (locally) variational dynamical forms

In the following for the sake of generality, we will consider locally variational dynamical forms. We will denote by a subscript i the fact that in general a sheaf section is defined locally. A *variational Lie derivative* operator $\mathcal{L}_{j_r\Xi}$ is well defined acting on the sections of sheaves in the variational sequence: it can be characterized as a *quotient Cartan formula* (which in fact provides a *variation formula*); the basic idea is to factorize modulo contact structures [10, 11, 12, 13]. This enable us to define symmetries of classes of forms of any degree in the variational sequence and corresponding conservation theorems; see [2] and [14] for details. We shall study the interplay of such an operator with the representation by forms. Let $j_r\Xi$ be a projectable vector field on $J_r\mathbf{Y}$, ρ a q-form defined (locally) on $J_r\mathbf{Y}$.

We have a commutative diagram defined by $\hat{R}_q(\mathcal{L}_{j_r\Xi}[\rho]) \doteq \hat{R}_q([L_{j_r\Xi}\rho]) = L_{j_r\Xi}R_q([\rho])$, where $\hat{R}_q(\mathcal{L}_{j_r\Xi}[\rho]) = L_{j_s\Xi}h\rho\,, 0 \leq q \leq n\,, s = r+1$; $\hat{R}_q(\mathcal{L}_{j_r\Xi}[\rho]) = L_{j_s\Xi}\mathcal{I}(\rho)\,, n+1 \leq q \leq P\,, s = 2r+1$; $\hat{R}_q(\mathcal{L}_{j_r\Xi}[\rho]) = L_{j_s\Xi}\rho\,, P+1 \leq q \leq N\,, s = r$. This enables us to deal with ordinary Lie derivatives of forms on Λ^q_s, then apply the Cartan formula for differential forms, therefore return back to the classes of forms to obtain the following *variational Cartan formulae* [14].

Proposition 1 *Let $0 \leq q \leq n$. Let $j_s\Xi$ be a projectable vector field on $J_s\mathbf{Y}$. We have*

$$\mathcal{L}_{j_s\Xi}[\rho] = \Xi\rfloor\mathcal{E}_q([\rho]) + d_H(j_s\Xi_V\rfloor p_{d_V[\rho]} + \Xi_H\rfloor[\rho])\,.$$

Note that, for $q = n-1$, the formula above defines a 'momentum' associated with a current, we shall denote such a momentum by $\tilde{p}$; it is clear that, for $q < n$, $\mathcal{E}_q = d_H$.

Proposition 2 *Let $q = n+k$, with $k \geq 1$. Let $j_s\Xi$ be a projectable vector field on $J_s\mathbf{Y}$. We have*

$$\mathcal{L}_{j_s\Xi}[\rho] = j_s\Xi_V\rfloor\mathcal{E}_q([\rho]) + \mathcal{E}_{q-1}(j_{s-1}\Xi_V\rfloor[\rho]))\,.$$

Let for simplicity η_{λ_i} denote a global Euler–Lagrange class of forms for a (local) variational problem represented by (local) sheaf sections λ_i.

Proposition 1 (Noether Theorem I) reads $\mathcal{L}_\Xi\lambda_i = \Xi_V\rfloor\eta_{\lambda_i} + d_H\epsilon_i$; where $\epsilon_i = j_s\Xi_V\rfloor p_{d_V\lambda_i} + \Xi_H\rfloor\lambda_i$ is the Noether current associated with it.

Definition 1 A generalized symmetry of a (locally variational) dynamical form η_{λ_i} is a projectable vector field $j_s\Xi$ on $J_s\mathbf{Y}$ such that $\mathcal{L}_{j_s\Xi}\eta_{\lambda_i} = 0$.

Since we assume η_{λ_i} to be closed, Proposition 2 reduces (case $q = n+1$) to $\mathcal{L}_{j_s\Xi}\eta_\lambda = \mathcal{E}_n(\Xi_V\rfloor\eta_{\lambda_i})$, and if $j_s\Xi$ is such that $\mathcal{L}_{j_s\Xi}\eta_{\lambda_i} = 0$, then $\mathcal{E}_n(\Xi_V\rfloor\eta_{\lambda_i}) = 0$; therefore, *locally* we have $\Xi_V\rfloor\eta_\lambda = d_H\nu_i$. Notice that, although $\Xi_V\rfloor\eta_\lambda$ is global, in general it defines a non trivial cohomology class [3]; it is clear that ν_i is a (local) current which is conserved on-shell (*i.e.* along critical sections). On the other hand, *and independently* (see [15]), we get *locally* $\mathcal{L}_\Xi\lambda_i = d_H\beta_i$ thus we can write $\Xi_V\rfloor\eta_\lambda + d_H(\epsilon_i - \beta_i) = 0$, where ϵ_i is the usual *canonical* Noether current.

Definition 2 We call the (local) current $\epsilon_i - \beta_i$ a *Noether-Bessel-Hagen current.*

A Noether–Bessel-Hagen current $\epsilon_i - \beta_i$ is a current associated with a generalized symmetry (conserved along critical sections); in [5, 6] we proved that a Noether–Bessel-Hagen current is variationally equivalent to a global (conserved) current if and only if $0 = [\Xi_V\rfloor\mathcal{E}_n(\lambda_i)] \in H^n_{dR}(\mathbf{Y})$.

3. Generalized symmetries generating Noether currents

In the following we investigate under which conditions a Noether–Bessel-Hagen current is variationally equivalent to a Noether conserved current for a suitable Lagrangian. We shall see that this is involved with the existence of a variationally trivial local Lagrangian $d_H\mu_i$, and with a condition on the current associated with it.

Proposition 3 *A Noether-Bessel-Hagen current $\epsilon_{\lambda_i} - \beta_i$ associated with a generalized symmetry of η_{λ_i} is a Noether conserved current if and only if is of the form $\epsilon_{\lambda_i} - \mathcal{L}_{j_s\Xi}\mu_i$, with μ_i a current satisfying $\mathcal{L}_{j_s\Xi}(\lambda_i - d_H\mu_i) = 0$.*

Proof. From $\mathcal{L}_{j_s\Xi}\eta_{\lambda_i} = 0$, we get $\mathcal{L}_{j_s\Xi}\lambda_i = d_H\beta_i$. It is easy to see that the current $\epsilon_{\lambda_i} - \beta_i$ is a Noether conserved current if and only if there exists μ_i such that $\beta_i = \mathcal{L}_{j_s\Xi}\mu_i$, *i.e.* $d_H\beta_i = d_H(j_{s-1}\Xi_V\rfloor p_{d_V d_H\mu_i} + \Xi_H\rfloor d_H\mu_i)$. On the other hand $d_H\beta_i = d_H\mathcal{L}_{j_s\Xi}\mu_i = d_H(\Xi_H\rfloor d_H\mu_i)$. Comparing the two expression we get $d_H(j_{s-1}\Xi_V\rfloor p_{d_V d_H\mu_i}) \equiv 0$. This identity is a consequence of the fact that $\mathcal{L}_{j_s\Xi}$ commutes with d_H; note that it can be obtained also by the fact that $d_H(j^{s-1}\Xi_V\rfloor p_{d_V hd\mu_i}) = p_0 d(j_{s-1}\Xi_V\rfloor p_1\mathcal{R}(d(d\mu_i)))$ (see [7]).

Proposition 4 *The Noether current $\epsilon_{\lambda_i - d_H\mu_i}$ is exact on-shell and it is equal to* $d_H(j_{s-1}\Xi_V \rfloor \tilde{p}_{d_V\mu_i} + \Xi_H \rfloor \mu_i)$.

PROOF. As it is well known, along any section pulling back to zero $\Xi_V \rfloor \eta_{\lambda_i}$ we get the on-shell conservation law $d_H(\epsilon_{\lambda_i} - \beta_i) = 0$. If there exists a current μ_i such that $\beta_i = \mathcal{L}_{j_s\Xi}\mu_i \equiv \Xi_H \rfloor d_H\mu_i + d_H(j_{s-1}\Xi_V \rfloor \tilde{p}_{d_V\mu} + \Xi_H \rfloor \mu_i$, therefore $j_{s-1}\Xi_V \rfloor p_{d_V\lambda_i} + \Xi_H \rfloor (\lambda_i - d_H\mu_i) - d_H(j_{s-1}\Xi_V \rfloor \tilde{p}_{d_V\mu_i} + \Xi_H \rfloor \mu_i)$ is closed on-shell. By an uniqueness argument, we see that the latter expression must be equal to $j_{s-1}\Xi_V \rfloor p_{d_V d_H\mu_i}$; therefore $d_H(j_{s-1}\Xi_V \rfloor \tilde{p}_{d_V\mu_i} + \Xi_H \rfloor \mu_i) = \epsilon_{\lambda_i - d_H\mu_i}$ on-shell.

Remark 1 It turns out that, on-shell, a canonical potential of the Noether current $\epsilon_{\lambda_i - d_H\mu_i}$, then a corresponding canonical conserved quantity, is defined.

Remark 2 An *off-shell* exact Noether current associated with the invariance of $\lambda_i - d_H\mu_i$ would be generated by a generalized symmetry $j_s\Xi$ such that $\Xi_V \rfloor \eta_{\lambda_i} = 0$; the corresponding cohomology class would be, therefore, trivial (see the discussion in [5, 6]).

Acknowledgement
Research supported by Department of Mathematics, University of Torino, research project *Geometric methods in mathematical physics and applications 2013.* The first author (M.P.) gratefully acknowledges J. Janyška for the kind invitation at DGA2013 University of Brno; thanks are due to J. Musilová and O. Rossi for the useful discussions held there.

References

[1] E. Bessel-Hagen: Über die Erhaltungssätze der Elektrodynamik, *Math. Ann.* **84** (1921) 258–276.
[2] M. Francaviglia, M. Palese, R. Vitolo: Symmetries in finite order variational sequences, *Czech. Math. J.* **52(127)** (1) (2002) 197–213.
[3] M. Ferraris, M. Palese, E. Winterroth: Local variational problems and conservation laws, *Diff. Geom. Appl* **29** (2011) S80–S85.
[4] M. Francaviglia, M. Palese, E. Winterroth: Locally variational invariant field equations and global currents: Chern-Simons theories, *Commun. Math.* **20** (1) (2012)13–22.
[5] M. Francaviglia, M. Palese, E. Winterroth: Variationally equivalent problems and variations of Noether currents, *Int. J. Geom. Meth. Mod. Phys.* **10**(1) (2013) art. no. 1220024.
[6] M. Francaviglia, M. Palese, E. Winterroth: Cohomological obstructions in locally variational field theories, *Jour. Phys. Conf. Series* **474** (2013) art. no. 012017.
[7] M. Krbek: The Representation of the Variational Sequence by Forms, PdhThesis (2002).
[8] M. Krbek, J. Musilová: Representation of the variational sequence by differential forms, *Acta Appl. Math.***88** (2) (2005) 177–199.
[9] D. Krupka: Variational Sequences on Finite Order Jet Spaces, *Proc. Diff. Geom. Appl.*; J. Janyška, D. Krupka eds., World Sci. (Singapore, 1990) 236–254.
[10] D. Krupka: Global variational theory in fibred spaces, in *Handbook of global analysis*, 773–836, 1215, Elsevier Sci. B. V., Amsterdam, 2008.
[11] D. Krupka, O. Krupková, D. Saunders: The Cartan form and its generalizations in the calculus of variations, *Int. J. Geom. Methods Mod. Phys.***7** (4) (2010) 631–654.
[12] D. Krupka, J. Šeděnková: Variational sequences and Lepage forms, in *Differential geometry and its applications*, 617–627, Matfyzpress, Prague, 2005.
[13] O. Krupková: Lepage forms in the calculus of variations, in *Variations, geometry and physics*, 27–55, Nova Sci. Publ., New York, 2009.
[14] J. Musilová, M. Palese, O. Rossi, E. Winterroth: Representation of variational derivatives by forms and applications, in preparation.
[15] E. Noether: Invariante Variationsprobleme, *Nachr. Ges. Wiss. Gött., Math. Phys. Kl.* **II** (1918) 235–257.
[16] M. Palese, E. Winterroth: Symmetries of Helmholtz forms and globally variational dynamical forms, *Jour. Phys. Conf. Series* **343** (2012) art. no. 012129.
[17] M. Palese, E. Winterroth, E. Garrone: Second variational derivative of local variational problems and conservation laws, *Arch. Math. (Brno)* **47**(5) (2011) 395–403.
[18] J. Volna, Z. Urban: The interior Euler-Lagrange operator in field theory, *Lepage Inst. Preprint Ser.* **1** (2013) 14pp; avaible at http://www.lepageri.eu/publications/preprint-series (to appear in *Math. Slovaca*).

Consistency and renormalization of the $\lambda = \frac{1}{3}$ soft breaking model of the complete Hořava theory

Alvaro Restuccia[1,2], **Jorge Bellorín** [2], **Adrián Sotomayor**[3]

[1] Physics Department, Antofagasta University, Antofagasta, Chile
[2] Physics Department, Simón Bolívar University, Caracas, Venezuela
[3] Departament of Mathematics, Antofagasta University, Antofagasta, Chile

E-mail: `arestu@usb.ve,jorgebellorin@usb.ve,adrian.sotomayor@uantof.cl`

Abstract. We discuss the $\lambda = \frac{1}{3}$ soft breaking of the complete Hořava theory we have recently introduced. We show the consistency of the theory from the point of view of the algebra of the constraints and of the existence and uniqueness of their solutions. The theory, in distinction to the $\lambda \neq \frac{1}{3}$ theory, has exactly the same physical degrees of freedom as General Relativity. We discuss the structure of the propagator and its behaviour at low and high energies. We comment on the spherically symmetric solutions of the theory.

1. Introduction

The Hořava proposal for a renormalizable theory [1] of gravity give rise to several models of gravity. All of them are in principle renormalizable although a proof of it has to be explicitly given since the models are restricted by second class constraints and one has to show that the quantization procedure may be performed without the introduction of dangerous non-local terms. The main property of Hořava proposal is to consider an anisotropic behaviour of space and time at high energies, the scaling being of the form

$$t \sim [b]^z$$
$$x \sim [b]^1.$$

The scaling breaks the relativistic symmetry with the idea of recovering it, at least in an approximated sense without violating the newtonian limit, at low energies.

The main benefit would be to obtain a renormalizable theory of gravity, which is now possible since higher order spatial derivatives terms may be added to the potential without breaking the spatial diffeomorphism invariance of the action. In a relativistic theory together with the higher order spatial derivatives terms one has to include higher order time derivatives terms which introduces ghosts into the formulation and break the unitarity property of the theory. At high energies the propagator is dominated by the higher spatial derivatives terms which improve the ultraviolet properties of the theory. The final goal of obtaining a renormalizable theory may be achieved by considering a suitable scaling z in order to have a dimensionless coupling on the action. For a four dimensional gravity theory the value of z should be $z = 3$.

In the Hořava proposal the space-time breaks into a spatial foliation parametrized by a preferred time, the foliation preserving diffeomorphism of the three dimensional spatial leaves are the gauge symmetries of the theory. The natural formulation of the theory is then in terms of the ADM variables g_{ij}, N, N_i. The action is given by

$$S = \int dtd^3x\sqrt{g}N\left(G^{ijkl}K_{ij}K_{kl} - \nu\right) \tag{1}$$

where

$$K_{ij} = \frac{1}{2N}\left(\dot{g}_{ij} - 2\nabla_{(iN_j)}\right) \tag{2}$$

$$G^{ijkl} = \frac{1}{2}\left(g^{ik}g^{jl} + g^{il}g^{jk}\right) - \lambda g^{ij}g^{kl}. \tag{3}$$

The dimensionless coupling constant of the action has been absorbed by a redefinition of the fields. The coupling λ is dimensionless and the potential density $\nu = \nu(g_{ij}, a_i, \ldots)$ is the most general combination of the spatial metric, its curvature tensor, the vector $a_i \equiv \frac{N_{,i}}{N}$ and covariant spatial derivatives of these objects transforming as a scalar under spatial diffeomorphisms. In order to have power-counting renormalizability the potential must include terms up to sixth order in spatial derivatives. The potential may include a cosmological constant but we do not consider it here. The highest order derivatives term in the potential is $\sqrt{g}C_{ij}C^{ij}$, it was introduced by Hořava and it is the unique interacting term of that order arising from a detailed balance principle. C_{ij} is the Cotton tensor constructed for the three dimensional metric on the spatial leaves of the foliation. This interacting term has its origin on the three dimensional Chern-Simon theory constructed from the Levi-Civita connection in terms of the three dimensional metric. We will consider as in [1] a detailed balance principle at high energies but we break it at low energies bye introducing the most general quadratic terms on spatial derivatives. It is a soft breaking mechanism which preserves the renormalizability properties of the theory. The potential becomes then

$$\nu = -R - \alpha a_i a^i + wC_{ij}C^{ij} \tag{4}$$

where α and w are coupling constants. The second term on the right hand member was introduced in [2].

The Hořava gravities have been developed for projectable theories where $N = N(t)$ and for non-projectable ones when $N = N(x, t)$. They are inequivalent theories, there is no way to transform by using the gauge freedom of the theory, from one theory to the other. We consider the non-projectable case where the full symmetries of the theory may be completely realized. Among the non-projectable theories most of the work has been performed for $\lambda \neq \frac{1}{3}$. The tensor G^{ijkl} in (1) is not invertible for $\lambda = \frac{1}{3}$, hence two different theories with different physical degrees of freedom arises from the two cases. Most of the work has been done for $\lambda \neq \frac{1}{3}$ [3, 4, 5, 6, 7, 8, 9, 10, 11]. In particular the strong coupling problem has been extensively analyzed. In this case the theory has one additional physical degree of freedom compared to General Relativity [3, 4]. The case we consider corresponds to $\lambda = \frac{1}{3}$, the theory contains exactly the physical degrees of freedom as General Relativity, they may be realized in terms of the tranverse-traceless components of the three dimensional metric.

It is interesting to notice that the particular theory with a potential density $-R$ describes for $\lambda = 1$ General Relativity while for $\lambda \neq 1$ General Relativity on a particular gauge $\pi = 0$ (in the space-time region where this is admissible) [7]

2. The $\lambda = \frac{1}{3}$ soft breaking theory

The hamiltonian of the theory is [12]

$$\int d^3x\left(\frac{N}{\sqrt{g}}\pi^{ij}\pi_{ij} + \sqrt{g}N\nu + N_i\mathcal{H}^i + \sigma\phi + \mu\pi\right) + E_{ADM} \tag{5}$$

where

$$\mathcal{H}^i \equiv -2\nabla_j \pi^{ij} + \phi \partial^i N. \tag{6}$$

ϕ is the conjugate momentum associated to N and σ, μ are Lagrange multipliers.

E_{ADM} denotes the ADM energy

$$E_{ADM} = \oint d\Sigma_i \left(\partial_j g_{ij} - \partial_i g_{jj}\right). \tag{7}$$

It has to be included in the hamiltonian, following [13], in order to obtain the equations of motion compatible with the boundary conditions from the most general variations of the hamiltonian.

The preservation of the primary constraints give rise to two new constraints, the hamiltonian constraint

$$\mathcal{H} \equiv \frac{1}{\sqrt{g}} \pi^{ij}\pi_{ij} - \sqrt{g}R + \alpha\sqrt{g}\left(2\nabla_i a^i + a_i a^i\right) + w\sqrt{g}C_{ij}C^{ij} = 0 \tag{8}$$

and

$$C \equiv \frac{3N}{2\sqrt{g}} \pi^{ij}\pi_{ij} + \frac{1}{2}\sqrt{g}NR - \sqrt{g}N\left(2\nabla_i a^i + \left(2 - \frac{\alpha}{2}\right)a_i a^i\right) + \frac{3w}{2}\sqrt{g}NC_{ij}C^{ij} = 0. \tag{9}$$

The conservation of these new constraints provides two elliptic partial differential equations for the lagrange multipliers σ and μ. The Dirac algorithm ends up at this stage.

From (8) and (9) we obtain

$$\left(1 - \frac{\alpha}{2}\right)\nabla^2 N = N\left(\frac{1}{g}\pi^{ij}\pi_{ij} + wC_{ij}C^{ij}\right). \tag{10}$$

The constraint $\mathcal{H}^i = 0$ is a first class constraint while $\phi = 0, \pi = 0, \mathcal{H} = 0$ and $C = 0$ are second class ones. The primary constraint

$$\pi = 0 \tag{11}$$

is a particular property of the Hořava theory formulated for $\lambda = \frac{1}{3}$.

One way to analyze these constraints is to solve N from (10) in terms of the metric and its conjugate momentum and then go back to the remaining constraints and solve them in terms of the trace of the transverse components of the metric. The longitudinal components of the metric and its conjugate momentum are eliminated, as in General Relativity, from the first class constraints and a gauge fixing condition. It is then crucial to show existence and uniqueness of the solution for N from (10).

We finally remark that the value $\lambda = \frac{1}{3}$ is protected against quantum corrections by the second class constraints $\pi = 0$ which may be imposed fro every time.

3. The existence and the uniqueness of the solution for N and the physical degrees of freedom

Given a Riemannian metric g_{ij} its conjugate momentum π^{ij}, the equation (10) is a strongly elliptic equation for N. Let us consider the Dirichlet problem for this partial differential equation. We consider an open bounded region Ω with smooth boundary $\partial\Omega$. The Dirichlet form

$$D(u,v) \equiv \left\langle \left(1 - \frac{\alpha}{2}\right)\nabla u \nabla v + u\left(\frac{1}{g}\pi^{ij}\pi_{ij} + wC_{ij}C^{ij}\right)v \right\rangle$$

defined for all u and v on the Sobolev space $H_1(\Omega)$ is selfadjoint, bounded and coercive for $\alpha < 2$.

Without loss of generality we may consider $w > 0$, this condition arises from the Chern-Simon action from which the interacting term $\sqrt{g}C_{ij}C^{ij}$ is obtained via a detailed balance principle. The change of sign of w corresponds to a change of orientability of the three dimensional manifold on which the Chern-Simon action is constructed. We can always impose then $w > 0$.

The potential of the Dirichlet form is thus positive. Consequently the operator

$$-\left(1-\frac{\alpha}{2}\right)\nabla^2+\frac{1}{g}\pi^{ij}\pi_{ij}+wC_{ij}C^{ij}$$

has trivial kernel on Ω.

We now impose a boundary value on $\partial\Omega$ for the solution of (10). We consider a function f defined on Ω and satisfying the boundary value condition. The weak solution of (10) with N satisfying the given boundary value is obtained from the solution on $H_1(\Omega)$ to

$$D(u,v)=\langle G,v\rangle \tag{12}$$

for all $v \in H_1(\Omega)$, where $G=\left(1-\frac{\alpha}{2}\right)\nabla^2 f-\left(\frac{1}{g}\pi^{ij}\pi_{ij}+wC_{ij}C^{ij}\right)f$ and $N=u+f$. We now applied general results for Dirichlet forms with the above properties to show that there exist a unique $u \in H_1(\Omega)$, with zero boundary value on $\partial\Omega$, solution of (12). Hence $N=u+f$ is the unique solution of (10) satisfying the specified boundary value on $\partial\Omega$.

Using regularity theorems for elliptic operators we then conclude that the solution for N exist, is unique and it is C^∞ if the metric and its conjugate momentum are C^∞ on the three dimensional spatial leaf of the foliation.

We then have a consistent formulation for the gravity theory obtained for a soft breaking of the Hořava original proposal. In fact, the Poisson bracket of the constraints closed on the submanifold of constraints for suitable values of the Lagrange multipliers and the solution for N exist and it is unique.

The number of physical degrees of freedom is obtained directly by noting that after solving N we are left with the second class constraints $\pi = 0$ and the hamiltonian constraint together with the first class constraint $\mathcal{H}_i = 0$. This counting is analogous to the case of General Relativity, where the hamiltonian constraints is of first class. In fact, a gauge fixing for this constraint, which may even be $\pi = 0$ (on regions of space-time where it is admissible), leave us with exactly the same number of physical degrees of freedom on both theories. Moreover if we consider a perturbative analysis around a Minkowski background, which is an exact solution of the field equations of the action (1), we obtain that the hamiltonian at quadratic level and at low energies is exactly the same as the hamiltonian of linearized General Relativity.

4. The propagator of the theory

The hamiltonian may also be rewritten as a sum of constraints plus boundary terms

$$H=\int d^3x\left(N\mathcal{H}+N_i\mathcal{H}^i+\sigma\phi+\mu\pi\right)+E_{ADM}-2\alpha\Phi_N \tag{13}$$

where

$$\Phi_N \equiv \int d\Sigma_i\partial_i N. \tag{14}$$

The role of the boundary terms ensure the existence of the Gateaux derivative of the Hamiltonian under variations of N that behave as $\delta N=\mathcal{O}(r^{-1})$ at infinity. The energy of the theory is given by the value of the Hamiltonian on the constrained submanifold and it is thus equal to $[E_{ADM}-2\alpha\Phi_N]$.

We now consider the quadratic contribution to the hamiltonian, subject to the linearized contraints. We take

$$g_{ij} = \delta_{ij} + h_{ij} \ , \ \pi^{ij} = p_{ij} \ , \ N = 1 + n \tag{15}$$

and obtain from the linearized contraints

$$h^T = n = 0. \tag{16}$$

C^{ij} only depends on the transverse-traceless part of h_{ij} and it appears quadratically in the costraints. The final expression of the hamiltonian at the quadratic level, from which we directly obtain the propagator of the theory, is

$$H = \int d^3x \left(p_{ij}^{TT} p_{ij}^{TT} + \frac{1}{4} \partial_i h_{jk}^{TT} \partial_i h_{jk}^{TT} + w \partial_i h_{jk}^{TT} \Delta^2 \partial_i h_{jk}^{TT} \right). \tag{17}$$

We then notice that at high energies the anisotropic third term dominates over the relativistic second one, the propagator acquires additional powers of the spatial momentum $\left(|k|^2\right)^3$, compared to $|k|^2$ in the relativistic propagator. This additional powers of momentum improve the UV properties of the theory. At low energies the theory flows to the $z = 1$ theory which exactly concides with General Relativity at the quadratic level. The relativistic scaling for the propagator is then restored at low energies. There is no dependence on the α coupling in the expression of the propagator.

5. Conclusions

The soft breaking model of Hořava gravity we are considering is a consistent theory which propagates the same number of physical degrees of freedom as General Relativity. It is in principle a renormalizable gravity theory. In order to achieve this final goal one should include also the $z = 2$ interacting term in the theory. Although these terms do not contribute to the low energy nor to the high energy behaviour of the theory they have to be included in the action and it must be shown their consistency at any energy level.

We expect due to the structure of the second class contraints that no problem will arise from their insertion in the quantization process.

From the point of view of the classical solutions of the theory, we have recently obtained the complete space of spherically symmetric solutions [14] see also [15]. They depend only on the α coupling. The "external" solutions for small values of α are "small" perturbations of the external Schwarzschild solution corresponding to $\alpha = 0$. Depending on the value of α the solutions presents wormholes or naked singularities.

Acknowledgments

A. R. and A. S. are partially supported by Projects Fondecyt 1121103 and Mecesup ANT398 (Chile).

References

[1] P. Hořava, Phys. Rev. D 79, 084008 (2009).
[2] D. Blas, O. Pujolas, and S. Sibiryakov, Phys. Rev. Lett. 104, 181302 (2010).
[3] C. Charmousis, G. Niz, A. Padilla, and P. M. Saffin, J. High Energy Phys. 08 (2009) 070; X. Gao, Y. Wang, R. Brandenberger, and A. Riotto, Phys. Rev. D 81, 083508 (2010); J.-O. Gong, S. Koh, and M. Sasaki, Phys. Rev. D 81, 084053 (2010); A. Kehagias and K. Sfetsos, Phys. Lett. B 678, 123 (2009); Y.W. Kim, H.W. Lee, and Y. S. Myung, Phys. Lett. B 682, 246 (2009); M.-i. Park, Classical Quantum Gravity 28, 015004 (2011); Gen. Relativ. Gravit. 43, 2979 (2011).
[4] D. Blas, O. Pujolas, and S. Sibiryakov, J. High Energy Phys. 10 (2009) 029.

[5] M. Li and Y. Pang, J. High Energy Phys. 08 (2009) 015.
[6] A. Papazoglou and T. P. Sotiriou, Phys. Lett. B 685, 197 (2010); D. Blas, O. Pujolas, and S. Sibiryakov, Phys. Lett. B 688, 350 (2010); I. Kimpton and A. Padilla, J. High Energy Phys. 07 (2010) 014; A. Cerioni and R. H. Brandenberger, arXiv:1008.3589; R.-G. Cai, B. Hu, and H.-B. Zhang, Phys. Rev. D 83, 084009 (2011).
[7] J. Bellorín and A. Restuccia, Int. J. Mod. Phys. D 21, 1250029-1 (2012).
[8] Bellorín and A. Restuccia, Phys. Rev. D 84, 104037 (2011).
[9] J. Bellorín and A. Restuccia, Phys. Rev. D 83, 044003 (2011).
[10] W. Donnelly and T. Jacobson, Phys. Rev. D 84, 104019 (2011).
[11] J. Bellorín, A. Restuccia and A. Sotomayor, Phys. Rev. D 85, 124060 (2012).
[12] J. Bellorín, A. Restuccia and A. Sotomayor, Phys. Rev. D 87, 084020 (2013).
[13] T. Regge and C. Teitelboim, Ann. Phys. (N.Y.) 88, 286 (1974).
[14] J. Bellorín, A. Restuccia and A. Sotomayor, Phys. Rev. D 90, 044009 (2014).
[15] E. Kiritsis, Phys. Rev. D 81:044009, (2010).

Journal of Physics: Conference Series **563** (2014) 012025 doi:10.1088/1742-6596/563/1/012025

On discrete and continuous nonlinear Fourier transforms

Pavle Saksida

Faculty of Mathematics and Physics, University of Ljubljana
Institute of Mathematics, Physics and Mechanics, Ljubljana

E-mail: pavle.saksida@fmf.uni-lj.si

Abstract. We study a discretization $\mathcal{DF}$ of the nonlinear Fourier transform $\mathcal{F}$ associated with the integrable partial differential equations of AKNS-ZS type. We construct an iterative procedure by means of which we can calculate the inverse transform $\mathcal{DF}(h)$ for an arbitrary small enough argument h to any desired accuracy. The construction of this procedure is based on the analogous construction for the continuous transform $\mathcal{F}$.

1. Introduction

Integrable nonlinear partial differential equations can be analyzed by means of the inverse scattering method. From its beginnings, this method has been known to be a nonlinear analogue of the well-known procedure for solving the linear partial differential equations by means of the Fourier transform. The first similarity between the two methods that springs to mind lies in the fact that in both methods the time evolution of the transformed data is easier that the evolution of the original (physical) data. More concretely, suppose the time evolution of a linear system is given by a curve $t \mapsto q(t)$ in some phase space. The problem is to find this curve if the initial condition is given by $q(0)$. When using the Fourier analysis, we find the solution $q(t)$ in the following way. We first perform the Fourier transform $F(q(0))$ of the initial state $q(0)$, then we find the time evolution $E_t[F(q(0))]$ of the transformed initial state. Finally, we compute the inverse Fourier transform $F^{-1}(E_t[F(q(0))])$ of the time-evolved transformed data. This is precisely the solution of our problem,

$$q(t) = F^{-1}(E_t[F(q(0))]).$$

When using the inverse scattering method to search for the solutions of nonlinear integrable systems, the strategy is the same and can be summarized by essentially the same formula

$$q(t) = \mathcal{F}^{-1}(E_t[\mathcal{F}(q(0))])$$

in which the linear Fourier transform F is replaced by the nonlinear operator $\mathcal{F}$ called the inverse scattering transformation. Finding the inverse scattering transform $\mathcal{F}$ is a difficult inverse problem. In this paper we address this issue.

The analogy between the two methods goes beyond this diagramatic description, and the inverse scattering transform is often called the nonlinear Fourier transform. Deep relations

Journal of Physics: Conference Series **563** (2014) 012025 doi:10.1088/1742-6596/563/1/012025

between the two theories were already studied in the papers [1], [2]. Later important developments were presented in, e.g. [6], [7], [8]. Deep analytical problems of the nonlinear Fourier transform were studied in [3], [4], [5], [9], [10], [11], and in many other sources. In some important cases, the derivative of $\mathcal{F}$ at the origin is the linear Fourier transform. This fact can provide an insight into the Poisson geometric aspect of the problems (see, e.g. [12] and [13]). Apart from the theory of integrable systems, the nonlinear Fourier transform has interesting applications in many other areas, ranging from the theory of orthogonal polynomials to random matrices. For an interesting introduction, see [15].

In the present paper we shall study the nonlinear Fourier transform associated to the AKNS-ZS systems. Let $L^2[0,1]$ be the Hilbert space of square-integrable complex functions on the interval $[0,1]$ and let $l^2_{\mathbb{Z}}$ be the Hilbert space of the bi-infinite square-integrable complex sequences. We define the nonlinear Fourier transform to be the nonlinear operator

$$\mathcal{F} : L^2[0,1] \longrightarrow l^2_{\mathbb{Z}}$$

given as follows. Let $f(x) \in L^2[0,1]$ and let

$$L_f(x,z) = \begin{pmatrix} 0 & f(x)\, e^{-2\pi i x z} \\ -\overline{f(x)}\, e^{2\pi i x z} & 0 \end{pmatrix}, \quad x \in [0,1],\ z \in \mathbb{Z}.$$

The holonomy of L_f is given by

$$\mathrm{Hol}_f(z) = \begin{pmatrix} h^d(z) & h(z) \\ -\overline{h(z)} & \overline{h^d(z)} \end{pmatrix} = \Phi(1,z),$$

where $\Phi(x,z)$ is the solution of the initial problem

$$\Phi_x(x,z) = L_f(x,z)\cdot\Phi(x,z), \quad \Phi(0,z) = I.$$

Let P denote the projection which assigns the upper anti-diagonal term to a 2×2 matrix.

Definition 1 *The nonlinear Fourier transform of the a function $f \in L^2[0,1]$ is given by*

$$\mathcal{F}(f)(z) = P(\mathrm{Hol}_f(z)) = h(z).$$

The holonomy can be given by Dyson's series, thus we can define $\mathcal{F}$ by

$$\mathcal{F}(f)(z) = P\Big[I + \sum_{k=1}^{\infty}\int_{\triangle_k} L_f(x_1,z)L_f(x_2,z)\cdots L_f(x_k,z)\,\mathrm{d}\vec{x}\Big], \tag{1}$$

where

$$\triangle_k = \{(x_1,x_2,\ldots,x_k);\, 1 \ge x_1 \ge x_2 \ge \ldots \ge x_k \ge 0\}$$

is the so-called ordered k-dimensional simplex.

In [14], the fundamental analytic properties of $\mathcal{F}$ are studied. In particular, it is proved that the target space of $\mathcal{F}$ is indeed $l^2_{\mathbb{Z}}$, and that $\mathcal{F}$ is a *real analytic operator* between the L^2-spaces $L^2[0,1]$ and $l^2_{\mathbb{Z}}$. Analyticity turns out to be important in calculating the inverse transform $\mathcal{F}^{-1}$. Finding the inverse $\mathcal{F}^{-1}$ is a difficult inverse problem which is equivalent to a suitable Riemann-Hilbert problem. Recasting the inverse problem in the Riemann-Hilbert form is conceptually very important, but it only yields explicit solutions for some very special elements in the target space of $\mathcal{F}$. In [14], we describe an iterative scheme which yields approximations for $\mathcal{F}^{-1}(h)$ to any desired degree of accuracy and *for any (small enough) element* h in the target space $l^2_{\mathbb{Z}}$. To prove the convergence of this scheme, one needs analyticity of $\mathcal{F}$.

The present paper is primarily devoted to the study of a discretization of $\mathcal{F}$. In section 2, we define the discrete nonlinear transformation $\mathcal{DF}$. Our definition of the discretization is perhaps not a very obvious one, therefore one has to show that it makes sense. Let f_N denote the usual step function approximation of an element $f \in L^2[0,1]$ such that $\lim_{N\to\infty} f_N = f$ in $L^2[0,1]$. We show that $\lim_{N\to\infty} \mathcal{DF}(f_N) = \mathcal{F}(f)$. We exhibit the relation between $\mathcal{DF}$ in two ways. In section 3, we study the inverse nonlinear Fourier transform. First, we consider the continuous case because the idea of the perturbational iterative calculation of $\mathcal{F}^{-1}$ can be more clearly explained in this case. Then we adapt this procedure to obtain an iterative calculation of the inverse $\mathcal{DF}^{-1}$. The formula turns out to be rather complicated, but its calculation demands only calculations of finite sums and of linear inverse discrete Fourier transforms.

2. Discrete nonlinear Fourier transform

The most common way of defining the continuous nonlinear Fourier transform is via the linear system of ordinary differential equations, as we have done above. Clearly, one could define the discrete nonlinear Fourier transform by means of a difference equation. Here we shall propose another definition which is perhaps less obvious, but is in a sense closer to the continuous original, as we shall see later in remark 1.

Let $a \in \mathbb{C}$, $b \in \mathbb{R}$ and let us denote

$$R(a) = \mathrm{Exp}[\begin{pmatrix} 0 & a \\ -\bar{a} & 0 \end{pmatrix}] = \begin{pmatrix} \cos|a| & \frac{a}{|a|}\sin|a| \\ -\frac{a}{|a|}\sin|a| & \cos|a| \end{pmatrix}$$

and

$$E(b) = \mathrm{Exp}[\begin{pmatrix} -ib & 0 \\ 0 & ib \end{pmatrix}) = \begin{pmatrix} e^{-\pi ib} & 0 \\ 0 & e^{\pi ib} \end{pmatrix}.$$

Let now

$$f : \mathbb{Z}_N \longrightarrow \mathbb{C}$$

be a function of discrete variable $k \in \mathbb{Z} = \mathbb{Z}/N\mathbb{Z}$. Denote by $C(\mathbb{Z}_N)$ the vector space of all such functions. Clearly $C(\mathbb{Z}) = \mathbb{C}^N$.

Definition 2 *The discrete nonlinear Fourier transform*

$$\mathcal{DF} : C(\mathbb{Z}_N) \longrightarrow C(\mathbb{Z}_N)$$

is given by the formula

$$\mathcal{DF}(f)(z) = P(\mathrm{Hol}_f^D(z)) = P\Big(\prod_{k=N-1}^{0} \mathrm{Ad}_{E(\frac{zk}{N})} R(\frac{1}{N}f(k)\Big), \quad z = 0,1,\ldots,N-1. \qquad (2)$$

As before, P denotes the operation of taking the upper right element of the 2×2 matrix in the bracket.

The matrix in the above definition can also be written in the form

$$\mathrm{Hol}_f^D(z) = R(\frac{f(N-1)}{N})E(\frac{z}{N})\cdots R(\frac{f(k)}{N})E(\frac{z}{N})\cdots E(\frac{z}{N})R(\frac{f(0)}{N}).$$

Despite the unusual appearance of (2), the transformation $\mathcal{DF}$ is related to the linear discrete Fourier transform

$$DF(f)(z) = \sum_{k=0}^{N-1} f(k)e^{-2\pi i\frac{kz}{N}}.$$

Journal of Physics: Conference Series **563** (2014) 012025 doi:10.1088/1742-6596/563/1/012025

For the derivative of $\mathcal{DF}$ at the origin we have

$$D_0\mathcal{DF}(t)(z) = \lim_{s\to 0}\mathcal{DF}(st)(z) = DF(t)(z), \quad \text{for every } t(k).$$

Taking into account the relation

$$\mathrm{Ad}_{E(\frac{zk}{N})}R(\frac{1}{N}f(k)) = R(\frac{1}{N}f(k)e^{-2\pi i\frac{zk}{N}}),$$

we get the following interesting rewriting of $\mathcal{DF}$:

$$\mathcal{DF}(f)(z) = P\Big(R(\frac{1}{N}f(N-1)e^{-2\pi i\frac{z(N-1)}{N}})\cdots R(\frac{1}{N}f(k)e^{-2\pi i\frac{zk}{N}})\cdots R(\frac{1}{N}f(0)e^{-2\pi i\frac{z0}{N}})\Big). \quad (3)$$

This formula suggests the definitions of the discrete nonlinear cosine and sine transforms given by

$$\mathcal{DS}(f)(z) = P\Big(\prod_{k=N-1}^{0} R(\frac{1}{N}f(k)\cos(\frac{kz}{N}))\Big) \quad \text{and} \quad \mathcal{DC}(f)(z) = P\Big(\prod_{k=N-1}^{0} R(\frac{1}{N}f(k)\sin(\frac{kz}{N})\Big),$$

respectively.

Proposition 1 *Discrete nonlinear sine and cosine transforms of a* real valued *function give essentially the same information as the discrete linear sine and cosine transforms, provided that the norm of the function is small enough.*

Proof: The arguments of R in the definitions of $\mathcal{DC}$ and $\mathcal{DS}$ are real. Therefore, for different values of k, matrices of the form

$$R(\frac{1}{N}f(k)\cos(\frac{zk}{N})) = \mathrm{Exp}(\begin{pmatrix} 0 & \frac{1}{N}f(k)\cos(\frac{zk}{N}) \\ -\frac{1}{N}f(k)\cos(\frac{zk}{N}) & 0 \end{pmatrix})$$

commute. Thus we have

$$\mathcal{DC}(f) = P\Big(R(\frac{1}{N}\sum_{k=0}^{N-1} f(k)\cos(\frac{zk}{N}))\Big),$$

or more explicitly

$$\mathcal{DC}(f)(z) = \sin\Big(\frac{1}{N}\sum_{k=0}^{N-1} f(k)\cos(\frac{zk}{N})\Big).$$

In other words, we have

$$\mathcal{DC}(f)(z) = \sin[DC(f)(z)] \quad \text{and} \quad \mathcal{DS}(f)(z) = \sin[DS(f)(z)].$$

□

From the above we see that the essential difference between the linear and the nonlinear Fourier transforms has its source in the fact that the matrices $R(\frac{1}{N}f(k)\exp(-2\pi i\frac{zk}{N}))$ do *not* commute for different values of k. It is also clear that the above proposition is not true for complex valued functions f, since the matrices $R(\frac{1}{N}f(k)\cos(-2\pi\frac{zk}{N}))$ for different values of k do *not* commute in this case.

We shall now show that the transformation $\mathcal{DF}$ is indeed a sensible discretization of the continuous nonlinear Fourier transformation $\mathcal{F}$ defined in the introduction. Let

$$f : [0,1] \longrightarrow \mathbb{C}$$

be a continuous function, defined on the interval. For every integer N, we shall denote by f_N the discretization

$$f_N : \mathbb{Z}_N \longrightarrow \mathbb{C}$$

of f, given by

$$f_N(k) = f(\frac{k}{N}).$$

We shall prove the following proposition.

Proposition 2 *Let f be a continuous function defined on the interval $[0,1]$. Then we have*

$$\lim_{N\to\infty} \mathcal{DF}(f_N)(z) = \mathcal{F}(f)(z),$$

for every $z \in \mathbb{N}$.

Proof: Consider formula (3) for $\mathcal{DF}$. Expanding

$$R(\frac{1}{N}f_N(k)e^{-2\pi i\frac{zk}{N}}) = \text{Exp}\Big[\begin{pmatrix} 0 & \frac{1}{N}f_N(k)e^{-2\pi i\frac{zk}{N}} \\ -\frac{1}{N}\overline{f_N(k)}e^{2\pi i\frac{zk}{N}} & 0 \end{pmatrix}\Big] = \text{Exp}(\frac{1}{N}\widehat{L}(k,z))$$

into power series and putting the expansions into (3) yields

$$\mathcal{DF}(f_N)(z) = P\Big[\prod_{k=N-1}^{0}\Big(I + \frac{1}{N}\widehat{L}(k,z) + (\frac{1}{N})^2\frac{1}{2!}\widehat{L}(k,z)^2 + \ldots (\frac{1}{N})^n\frac{1}{n!}\widehat{L}(k,z)^n + \ldots\Big)\Big].$$

Multiplying and arranging with respect to the powers of $(\frac{1}{N})$ yields

$$\mathcal{DF}(f_N)(z) = P\Big[I + \frac{1}{N}\sum_{k=0}^{N-1}\widehat{L}(k,z) + (\frac{1}{N})^2\Big(\sum_{k_1>k_2}\widehat{L}(k_1,z)\widehat{L}(k_2,z) + \frac{1}{2!}\sum_{k=0}^{N-1}\widehat{L}(k,z)^2\Big) + \ldots\Big]. \quad (4)$$

The above sum can be rewritten as follows. Let Ω_N be the set of all functions

$$\omega : \{0,1,\ldots,N-1\} \longrightarrow \mathbb{N}\cup\{0\}$$

and let

$$|\omega| = \sum_{k=0}^{N-1}\omega(k).$$

Then we have

$$\mathcal{DF}(f_N)(z) = P\Big[\sum_{\omega\in\Omega_N}(\frac{1}{N})^{|\omega|}\frac{1}{\omega(0)!\cdots\omega(N-1)!}\widehat{L}(N-1,z)^{\omega(N-1)}\ldots\widehat{L}(0,z)^{\omega(0)}\Big].$$

The above sum can be refined into the following expression

$$\mathcal{DF}(f_N)(z) = P\Big[\sum_{n=0}^{\infty}\Big(\sum_{|\omega|=n}(\frac{1}{N})^{|\omega|}\frac{1}{\omega(0)!\cdots\omega(N-1)!}\widehat{L}(N-1,z)^{\omega(N-1)}\ldots\widehat{L}(0,z)^{\omega(0)}\Big)\Big].$$

It is now not difficult to see that for $n \leq N$ the sums

$$\sum_{|\omega|=n} (\frac{1}{N})^{|\omega|} \frac{1}{\omega(0)! \cdots \omega(N-1)!} \widehat{L}(\omega(N-1), z) \cdots \widehat{L}(\omega(0), z) \tag{5}$$

are precisely the integral sums for the integrals

$$\int_{\triangle_n} L(x_1, z) L(x_2, z) \cdots L(x_n, z) \, \mathrm{d}\vec{x}.$$

This shows that the limit, when N goes to the infinity, of the expression (4) is equal to the sum

$$P\Big[I + \int_{\triangle_1} L(x, z) \, \mathrm{d}x + \ldots \int_{\triangle_n} L(x_1, z) \cdots L(x_n, z) \, \mathrm{d}\vec{x} + \ldots\Big].$$

The sum in the bracket is the Dyson sum for the solution of the initial problem

$$\Phi_x(x, z) = L(x, z) \cdot \Phi(x, z), \quad \Phi(0, z) = I,$$

evaluated at $x = 1$. This proves our proposition. □

Remark 1 *The expression (5) is complicated because it takes into account the reduction of volumes in the approximation grid at the boundary of the simplex $\triangle_n$. For this reason, $\mathcal{DF}$ is a better approximation of the continuous $\mathcal{F}$ than a discretization, obtained by replacing the differential equation for Φ by a suitable difference equation. Such discretization does not take into account the geometry of the boundary $\partial\triangle_n$.*

In order to further exhibit the relation between the discrete transform $\mathcal{DF}$ and the continuous transform $\mathcal{F}$, one can also derive formula (2) from the definition of $\mathcal{F}$ given in the introduction. Let $\chi_{[x_0,a]}$ denote the characteristic function of the interval $[x_0, x_0 + a]$. If we apply the Dyson series expression for the nonlinear Fourier transform and the mean value theorem, we get for any complex number g

$$\mathcal{F}(g\,\chi_{[x_0, \frac{1}{N}]})(z) = \frac{g}{|g|} \sum_{n=1}^{\infty} (-1)^{n+1} \frac{1}{(2n+1)!} (\frac{|g|}{N})^{2n+1} \, e^{-2\pi i z (x_1^n - x_2^n + \ldots + x_{2n+1}^n)}.$$

Here $(x_1^n, x_2^n, \ldots x_{2n+1}^n)$ is some point in the ordered standard simplex

$$\widetilde{\triangle}_{2n+1} = \{(x_1, \ldots x_{2n+1}); x_0 + \frac{1}{N} \geq x_1 \geq \ldots \geq x_{2n+1} \geq x_0\}.$$

If we replace the points $(x_1^n, x_2^n, \ldots, x_{2n+1}^n)$ by $(x_0, x_0, \ldots, x_0)$ in the above series, we commit only a small mistake, and we get a perturbation $\widetilde{\mathcal{F}}(\chi_{[x_0, \frac{1}{N}]})(z)$ given by

$$\widetilde{\mathcal{F}}(g\,\chi_{[x_0, \frac{1}{N}]}) = \frac{g}{|g|} \sum_{n=1}^{\infty} (-1)^{n+1} \frac{1}{(2n+1)!} (\frac{|g|}{N})^{2n+1} \, e^{-2\pi i z x_0} = \frac{g}{|g|} \sin(\frac{|g|}{N}) \, e^{-2\pi i z x_0}.$$

Clearly, we have

$$\lim_{N \to \infty} \Big(\mathcal{F}(g\,\chi_{[x_0, \frac{1}{N}]})(z) - \widetilde{\mathcal{F}}(g\,\chi_{[x_0, \frac{1}{N}]})(z)\Big) = 0, \quad \text{for every } z \in \mathbb{Z}.$$

Similar reasoning tells us that the upper diagonal terms $\mathcal{E}(g\,\chi_{[x_0,\frac{1}{N}]})(z)$ of the holonomies $\mathrm{Hol}_{(g\,\chi_{[x_0,\frac{1}{N}]})}(z)$ are given by

$$\mathcal{E}(g\,\chi_{[x_0,\frac{1}{N}]})(z) = \sum_{n=0}^{\infty}(-1)^{n+1}\frac{1}{2n!}(\frac{|g|}{N})^{2n}e^{-2\pi i(x_1^n - x_2^n + \ldots - x_{2n}^n)},$$

and are well approximated by the values

$$\widetilde{\mathcal{E}}(g\,\chi_{[x_0,\frac{1}{N}]})(z) = \cos(\frac{|g|}{N}).$$

If we replace the points $(x_1^n, x_2^n, \ldots x_{2n}^n)$ by $(x_0, x_0, \ldots x_0)$, the sum in the exponent vanishes. So, the holonomies $\mathrm{Hol}_{(g\,\chi_{[x_0,\frac{1}{N}]})}(z)$ are well approximated by the matrices

$$\widetilde{\mathrm{H}}_{g\,\chi_{[x_0,\frac{1}{N}]}}(z) = \begin{pmatrix} \cos(\frac{|g|}{N}) & \frac{g}{|g|}\sin(\frac{|g|}{N})\;e^{-2\pi i z x_0} \\ -\frac{g}{|g|}\sin(\frac{|g|}{N})\;e^{2\pi i z x_0} & \cos(\frac{|g|}{N}) \end{pmatrix} = \mathrm{Ad}_{E(zx_0)}R(\frac{g}{N}).$$

Let f_1 and f_2 be two functions with disjoint supports, and suppose that the support of f_1 lies on the left of the support of f_2. Then we have

$$\mathrm{Hol}_{f_1+f_2}(z) = \mathrm{Hol}_{f_2}(z)\cdot \mathrm{Hol}_{f_1}(z).$$

Perhaps the simplest way to see this is to consider the relevant Dyson series. In the context of the scattering transform, this property is often called the superposition law. Let us now approximate a continuous function f by the step functions

$$\widetilde{f}_N\colon [0,1] \longrightarrow \mathbb{R},$$

given by

$$\widetilde{f}_N(x) = \sum_{k=0}^{N-1} f(\frac{k}{N})\cdot \chi_{[\frac{k}{N},\frac{1}{N}]}.$$

The nonlinear Fourier transform of $\widetilde{f}_N(x)$ is then well approximated by the expression

$$\mathcal{F}(\widetilde{f}_N)(z) \cong P\Big[\prod_{k=N-1}^{0}\widetilde{H}_{f(\frac{k}{N})\chi_{[\frac{k}{N},\frac{1}{N}]}}(z)\Big] = P\Big[\prod_{k=N-1}^{0}\mathrm{Ad}_{E(\frac{kz}{N})}R(\frac{1}{N}f(\frac{k}{N}))\Big].$$

The expression on the right is precisely the discrete nonlinear Fourier transform of the function f_N,

$$\mathcal{DF}(f_N)(z) = P\Big[\prod_{k=N-1}^{0}\mathrm{Ad}_{E(\frac{kz}{N})}R(\frac{1}{N}f(\frac{k}{N}))\Big].$$

From the above it is not difficult to show that we have

$$\lim_{N\to\infty}\mathcal{DF}(f_N)(z) = \mathcal{F}(f)(z), \quad \text{for every } z\in\mathbb{N},$$

but this was already proved more rigorously and in more detail above.

3. The inverse nonlinear Fourier transform

In this section, we shall describe an algorithm for computing arbitrarily good approximations for the inverse nonlinear Fourier transform of arbitrary function $h(z)$, provided that it belongs to a suitable space and lies close enough to the origin. We shall first give the construction for the continuous case because it illustrates the problem better. The adaptation of the construction to the case of the discrete nonlinear Fourier transform is rather straightforward.

In order to obtain a perturbational formula, we consider the action of $\mathcal{F}$ on real analytic curves in the Hilbert space $L^2[0,1]$. We shall consider curves

$$f(x,s) : (-\epsilon, \epsilon) \longrightarrow L^2[0,1]$$

which start at the origin, i.e., $f(x,0) = 0 \in L^2[0,1]$ and are real analytic. We can expand $f(x,s)$ around zero with respect to s. We get the power series

$$f(x,s) = \sum_{n=1}^{\infty} \frac{s^n}{n!} f_n(x)$$

and we assume that the convergence radius is $\alpha > 0$. The following theorem, proved in [14], establishes an essential fact about the nonlinear Fourier transform.

Theorem 1 *Nonlinear Fourier transform*

$$\mathcal{F} : L^2[0,1] \longrightarrow l^2_{\mathbb{Z}}$$

is a real analytic map between the two Hilbert spaces equipped with the appropriate L^2-norms.

The real analyticity ensures that the image $\mathcal{F}(f(x,s)$ is a real analytic curve in $l^2_{\mathbb{Z}}$. Since $\mathcal{F}(0) = 0$, this curve also starts at the origin. If we expand it in $l^2_{\mathbb{Z}}$ around $s = 0$, we obtain the power series

$$\mathcal{F}(f(x,s)) = \mathcal{F}\Big(\sum_{n-1}^{\infty} \frac{s^n}{n!} f_n(x)\Big) = \sum_{n=1}^{\infty} \frac{s^n}{n!} h_n(z). \tag{6}$$

It is not difficult to see that the convergence radius of the above series is also equal to α.

The second essential fact about the nonlinear Fourier transform is given by the following proposition.

Proposition 3 *The derivative of $\mathcal{F}$ at the origin of $L^2[0,1]$ is the linear Fourier transform*

$$F : L^2[0,1] \longrightarrow l^2_{\mathbb{Z}}.$$

Proof: If we evaluate the Dyson series for $\mathrm{Hol}_{f+st}(z)$, where $t \in L^2[0,1]$ and s is real number, we get

$$\mathrm{Hol}_{f+st}(z) = I + \int_{\triangle_1} L_{f+st}(x)\, \mathrm{d}x + \sum_{k=2}^{\infty} \int_{\triangle_k} L_{f+st}(x_1) \cdots L_{f+st}(x_k)\, \mathrm{d}\vec{x}.$$

Deriving with respect to s and evaluating at $s = 0$ and at $f = 0$, we get

$$D_0\mathcal{F}(t) = P[\int_{\triangle_1} L_t(\xi, z)\, \mathrm{d}\xi] = \int_0^1 t(\xi) e^{-2\pi i \xi z}\, \mathrm{d}\xi.$$

□

Above we have proved that F is the weak (directional) derivative of $\mathcal{F}$ at the origin. It is actually also the strong derivative (differential). This fact is also proved in [14]. Now, the linear Fourier transform is an isometry of the L^2-spaces $L^2[0,1]$ and $l^2_{\mathbb{Z}}$. Therefore, by the inverse function theorem for Hilbert spaces, the map $\mathcal{F}$ has a local inverse in a neighbourhood of the origin. Moreover, the analytical version of this theorem tells us that this inverse is also a real analytic map. We can summarize this in the following theorem:

Theorem 2 *There exist neighbourhoods $U \subset L^2[0,1]$ and $V \subset l^2_{\mathbb{Z}}$ of the respective origins and a real analytic map*

$$\mathcal{G} : V \longrightarrow U$$

which is the local inverse of the nonlinear Fourier transform $\mathcal{F}$.

Our next goal is to find an effective way of computing $\mathcal{G}(h)$ of any h close enough to the origin. First, we shall give explicit formulae for the terms h_n of series (6). To this end, let us consider the upper-triangular block-Toeplitz matrix

$$L^n_f(x,z) = \begin{pmatrix} 0 & L_1 & L_2 & \cdots & \frac{1}{n!}L_n \\ 0 & 0 & L_1 & \cdots & \frac{1}{(n-1)!}L_{n-1} \\ 0 & 0 & 0 & \cdots & \frac{1}{(n-2)}L_{n-2} \\ \vdots & \vdots & \vdots & \vdots & \vdots \\ 0 & 0 & 0 & \cdots & L_1 \\ 0 & 0 & 0 & \cdots & 0 \end{pmatrix}$$

with blocks given by

$$L_k = \begin{pmatrix} 0 & e^{-2\pi i z x} f_k(x) \\ -e^{2\pi i z x}\overline{f_k(x)} & 0 \end{pmatrix}.$$

Let $\mathrm{Hol}^n_{f(s)}(z) = \Phi^n(1,z)$, where $\Phi^n(x,z)$ is the solution of the initial problem

$$(\Phi^n)_x = L^n_f \cdot \Phi^n, \quad \Phi^n(0;z) = I.$$

The coefficient matrix L^n_f is strictly upper-triangular, therefore for every $m > n$ and for any choice of $(x_1, x_2, \ldots, x_m)$ we have

$$L^n_f(x_1,z)\cdot L^n_f(x_2,z)\cdots L^n_f(x_m,z) = 0.$$

For this reason the Dyson series for $\mathrm{Hol}^n_{f(s)}(z)$ is finite;

$$\mathrm{Hol}^n_{f(s)} = I + \sum_{k=1}^{n} \int_{\triangle_k} L^n_f(x_1,z)\cdot L^n_f(x_2,z)\cdots L^n_f(x_k,z)\, \mathrm{d}\vec{x}. \tag{7}$$

Finiteness of the above sum is essential for the derivation of the closed expressions of the terms $h_n(z)$. The holonomy matrix is again an upper triangular block-Toeplitz matrix of the form

$$\mathrm{Hol}^n_{f(s)}(z) = \begin{pmatrix} I & H_1(z) & \cdots & \frac{1}{n!}H_n(z) \\ 0 & I & \cdots & \frac{1}{(n-1)!}H_{n-1}(z) \\ 0 & 0 & \cdots & \frac{1}{(n-2)!}H_{n-2}(z) \\ \vdots & \vdots & \cdots & \vdots \\ 0 & 0 & \cdots & I \end{pmatrix}.$$

On the other hand, the analyticity of $\mathcal{F}$ and of the diagonal term of the holonomy Hol_f gives

$$\mathrm{Hol}_{f(s)}(z) = I + \sum_{k=1}^{\infty} \frac{s^k}{k!}\widehat{H_k}(z).$$

It is not difficult to see that the terms of this expansion are precisely the blocks of $\mathrm{Hol}^n_{f(s)}(z)$,

$$H_n(z) \equiv H_n(z), \quad \text{for every } z \in \mathbb{Z} \text{ and } n \in \mathbb{N}. \tag{8}$$

Let us denote

$$\mathrm{Hol}_{f(s)} = I + \sum_{n=1}^{\infty} \frac{s^n}{n!} H_n(z) = I + \sum_{n=0}^{\infty} \frac{s^n}{n!} \begin{pmatrix} h_n^d(z) & h_n(z) \\ -\overline{h_n(z)} & \overline{h_n^d(z)} \end{pmatrix}.$$

A lengthy but straightforward calculation in which we use (7) and (8) gives the following formula

$$h_n(z) = \sum_{k=1}^{\widehat{n}} (-1)^{k-1} [\sum_{|J|=n} \alpha^J \int_{\triangle_{2k-1}} e^{-2\pi i(x_1 - x_2 + \ldots + x_{2k-1})} f_{j_1}(x_1) \overline{f_{j_2}}(x_2) \cdots f_{j_{2k-1}}(x_{2k-1})], \quad (9)$$

where $J = (j_1, j_2, \ldots, j_{2k-1})$ are multi-indices with $j_r \geq 1$ and

$$\alpha^J = \binom{2n+1}{j_1, \ldots, j_{2k-1}}$$

are the multinomial coefficients. The integer $\widehat{n}$ is equal to $n/2$ when n is even and to $(n+1)/2$ when n is odd. If we separate the term at $k = 1$ and the rest in formula (9), we get

$$h_1(z) = \int_{\triangle_1} \phi_1(\xi)\, \mathrm{d}\xi = \int_0^1 f_1(\xi)\, e^{-2\pi i \xi z} \mathrm{d}\xi = F(f_1)(z) \quad (10)$$

for $n = 1$ and

$$h_n(z) = F(f_n)(z) + \mathcal{H}_n(f_1, f_2, \ldots, f_{n-2})(z) \quad (11)$$

for $n > 1$. The term $\mathcal{H}_n$ is given by

$$\mathcal{H}_n(f_1, f_2, \ldots, f_{n-2})(z) = \sum_{k=2}^{\widehat{n}} (-1)^{k-1} \sum_{|J|=n} \alpha^J \int_{\triangle_{2k-1}} \phi_{j_1}(x_1) \overline{\phi_{j_2}}(x_2) \cdots \phi_{j_{2k-1}}(x_{2k-1})\, \mathrm{d}\vec{x}, \quad (12)$$

where

$$\phi_{j_r}(x_r) = f_{j_r} e^{-2\pi i x_r z}.$$

The term $\mathcal{H}_n$ depends only on $f_1, \ldots, f_{n-2}$. This follows from the fact that we have $j_r \geq 1$ for all indices j_r. The products of terms ϕ_j which contain the highest index are of the form $\phi_1(x_{\sigma(1)}) \overline{\phi_1(x_{\sigma(2)})} \phi_{n-2}(x_{\sigma(3)})$ for any permutation $\sigma \in S_3$.

Let G denote the inverse linear Fourier transform

$$G(c_n) = \sum_{n=-\infty}^{\infty} c_n e^{2\pi i n z}.$$

Inversion of formulae (10) and (11) yields the following theorem

Theorem 3 *Let*

$$h(z, s) = \sum_{n=1}^{\infty} \frac{s^n}{n!} h_n(z)$$

be a real analytic curve lying in a small enough neighbourhood of the origin of the space $l^2_{\mathbb{Z}}$. *Then the inverse nonlinear Fourier transform* $\mathcal{G}$ *of this curve is the real analytic curve, given by the convergent series*

$$\mathcal{G}[h(z, s)](x) = \sum_{n=1}^{\infty} \frac{s^n}{n!} f_n(x)$$

in $L^2[0,1]$ *whose terms are given by formulae*

$$\begin{aligned} f_1(x) &= G[h_1(z)](x) &(13)\\ f_n(x) &= G[h_n(z)] - G[\mathcal{H}_n(f_1,\dots,f_{n-2})(z)](x). &(14)\end{aligned}$$

Let now $h(z)$ be an arbitrary element in $l^2_{\mathbb{Z}}$ close enough to the origin.

Theorem 4 *The inverse nonlinear Fourier transform* $\mathcal{G}(h)$ *is given by the convergent series*

$$\mathcal{G}[h(z)](x) = \sum_{n=1}^{\infty} \frac{1}{n!} f_n(x)$$

in $L^2[0,1]$, *whose terms can be computed successively by formulae*

$$\begin{aligned} f_1(x) &= G[h(z)](x)\\ f_{2n}(x) &= 0\\ f_{2n+1}(x) &= -G[\mathcal{H}_{2n+1}(f_1,\dots,f_{2n-1})(z)](x).\end{aligned}$$

Proof: Consider the simplest analytic curve $h(z,s)$ in $l^2_{\mathbb{Z}}$, given by

$$h(z,s) = s \cdot h(z).$$

In terms of the notation in the previous theorem, we have

$$h_1 = h, \quad h_n = 0, \quad \text{for } n \geq 2.$$

Provided the norm of $h(z)$ is small enough, theorem 3 ensures the existence of the convergent series

$$\mathcal{G}(h)(x) = \sum_{n=1}^{\infty} \frac{s^n}{n!} f_n(x)$$

whose terms are given by equations (13) and (14). If we set $s = 1$, we obtain the expressions for the terms f_n of our series. It only remains to be shown that the even terms f_{2n} are equal to zero. We can prove this by induction. Fix an $n_0 \in \mathbb{N}$. Suppose all $f_{2m} = 0$ for $m < n_0$. We claim that $f_{2n_0} = 0$. If this were not so, then by formula (12) we should have

$$j_1 + j_2 + \ldots + j_{2k-1} = 2n_0.$$

But by the induction hypothesis, all the indices j_r in the above sum are odd, therefore this sum cannot be even. This shows that $f_{2n} = 0$ for all n. □

It is not difficult to adapt the result of theorems 3 and 4 to the discrete nonlinear Fourier transform $\mathcal{DF}$. First, we recall that $D_0\mathcal{DF} = DF$. By the inverse function theorem, $\mathcal{DF}$ is locally invertible in the vicinity of the origin. Recall also the the continuous nonlinear Fourier transform of a real analytic curve $f(x,s) = \sum_{n=1}^{\infty}(s^n/n!)\, f_n(x)$ can be written in the form

$$\mathcal{F}\Big(\sum_{n=1}^{\infty} \frac{s^n}{n!} f_n(x)\Big)(z) = \sum_{n=1}^{\infty} \frac{s^n}{n!} h_n(z),$$

where h_n are given by (9), or equivalently, by (10) and (11). Formulae (10) and (11) can be adapted to the discrete case. Let f be again a function of the discrete variable

$$f : \mathbb{Z}_N \longrightarrow \mathbb{C}.$$

After some inspection and bookkeeping we find that the discrete versions of (10) and (11) are given by

$$\begin{aligned} Dh_1(z) &= \frac{1}{N}\sum_{k=N-1}^{0} f_1(k)\, e^{-2\pi i \frac{zk}{N}} = DF(f_1)(z), \\ Dh_n(z) &= DF(f_n)(z) + \mathcal{DH}_n(f_1, f_2, \dots f_{n-2})(z), \quad \text{for } n > 1. \end{aligned}$$

Here the terms $\mathcal{DH}_n$ are given as follows. Denote

$$L_m(k) = \begin{pmatrix} 0 & \frac{1}{N} f_m(k) e^{-2\pi i \frac{kz}{N}} \\ -\frac{1}{N} f_m(k) e^{2\pi i \frac{kz}{N}} & 0 \end{pmatrix}.$$

Then

$$\mathcal{DH}_n(f_1, f_2, \dots f_{n-2})(z) = P\Big[\sum_{\sum_{k=N-1}^{0} n(k)=n} \; \prod_{k=N-1}^{0} \frac{1}{n(k)!} \mathcal{P}_{n(k)}(f_1, \dots, f_{n-2}) \Big] \tag{15}$$

and

$$\mathcal{P}_{n(k)}(f_1, \dots, f_{n-2}) = \sum_{r=1}^{n(k)} \frac{1}{r!} \sum_{m_1^r + \dots + m_r^r = n(k)} \frac{1}{m_1^1! m_2^r! \cdots m_r^r!} L_{m_1^r}(k) L_{m_2^r}(k) \cdots L_{m_r^r}(k). \tag{16}$$

We note that all the above expressions are *finite* sums.

Let now

$$DG(h(z))(k) = \sum_{z=0}^{N-1} h(z)\, e^{2\pi i \frac{zk}{N}}$$

be the linear discrete inverse Fourier transform. We have the following result.

Theorem 5 *Let*

$$h(z, s) = \sum_{n=1}^{\infty} \frac{s^n}{n!} h_n(z) : (-\epsilon, \epsilon) \longrightarrow C(\mathbb{Z}_N) = \mathbb{C}^N$$

be a real analytic curve in $C(\mathbb{Z}_N)$ starting at the origin. Then its inverse image $\mathcal{DG}(h(z,s))(k)$ with respect to $\mathcal{DF}$ is given by the convergent power series

$$\mathcal{DG}(h(z,s)(x) = \sum_{n=1}^{\infty} \frac{s^n}{n!} f_n(k)$$

whose terms are given by

$$\begin{aligned} f_1(k) &= DG[h_1(z)](k) \quad (17) \\ f_n(k) &= DG[h_n(z)](k) - DG[\mathcal{DH}_n(z)](k), \quad (18) \end{aligned}$$

and $\mathcal{DH}_n$ are given by (15)and (16).

From this we get:

Theorem 6 *Let $h(z) \in C(\mathbb{Z})$ be an arbitrary element, close enough to the origin. Then the inverse discrete nonlinear transform $\mathcal{DG}(h)$ is given by the convergent series*

$$\mathcal{DG}(h)(k) = \sum_{n=1}^{\infty} \frac{s^n}{n!} f_n(k),$$

where the terms f_n are given by

$$\begin{aligned} f_1(k) &= DG[h(z)](k) \\ f_n(k) &= -DG(\mathcal{DH}_n). \end{aligned}$$

The above theorems are proved in the same way as in the continuous case. The problem of convergence does not really arise here. Transformation $\mathcal{DF}$ is a map between two finite dimensional spaces and its analyticity is seen rather easily.

References

[1] Ablowitz M J, Kaup D J, Newell A C and Segur H 1973 Nonlinear-evolution equations of physical significance *Phys. Rev. Lett.* **31** 125 - 27

[2] Ablowitz M J, Kaup D J, Newell A C and Segur H 1974 The inverse scattering transform - Fourier analysis for nonlinear problems *Stud. Appl. Math.* **53** 249-315

[3] Christ M and Kiselev A 1998 Absolutely continuous spectrum for one-dimensional Schrödinger operators with slowly decaying potentials: some optimal results *J. Amer. Math. Soc.* **11**, 771 - 97

[4] Christ M and Kiselev A 2001 Maximal functions associated to filtrations *J. Funct. Anal.* **179**, 409 - 25

[5] Frayer C, Hryniv R O, Mykytyuk Ya V and Perry P A 2009 Inverse scattering for Schrödinger operators with Miura potentials: I. Unique Riccati representatives and ZS-AKNS systems *Inverse Problems*, **25** 115007- 32

[6] Fokas A S 2002 Integrable nonlinear evolution equations on the half-line *Commun. Math. Phys.* **230** 1 - 39

[7] Fokas A S and Gelfand I M 1994 Integrability of linear and lonlinear evolution equations and the associated nonlinear Fourier transform *Lett. Math. Phys.* **32** 189 - 210

[8] Fokas A S and Sung L Y 2005 Generalized Fourier transforms, their nonlinearization and the imaging of the Brain *Notices Amer. Math. Soc.*, **52** 1178 - 92

[9] Hryniv R O, Mykytyuk Ya V, Perry P A 2010 Inverse scattering for Schrödinger operators with Miura potentials: II. Different Riccati representatives *Comm. Partial Differential Equations*, **36** 1587-623

[10] Muscalu C, Tao T and Thiele C 2003 A Carleson type theorem for a Cantor group model of the scattering transform *Nonlinearity*, **16**, 219 - 46

[11] Muscalu C, Tao T and Thiele C 2003 A counterexample to a multilinear end-point question by Christ and Kiselev *Math. Res. Lett.* **10**, 237 - 46

[12] Saksida P 2011 On zero-curvature condition and Fourier analysis *J. Phys. A: Math. Gen.* **44** 85203-85222

[13] Saksida P 2012 Lax pairs and Fourier analysis: The case of sine-Gordon and Klein-Gordon equations *J. Phys.: Conf.* Ser. 343 012109, (10pp)

[14] Saksida P, 2013 On the nonlinear Fourier transform associated with periodic AKNS-ZS system and itsinverse *J. Phys. A: Math. Theor.* **46** 465204 (22pp)

[15] Tao T, Thiele C 2012 Nonlinear Fourier analysis, . arXiv:1201.5129, 1 - 94

Journal of Physics: Conference Series **563** (2014) 012026 doi:10.1088/1742-6596/563/1/012026

From noncommutative spacetimes to Lorentz symmetry violation

M Schreck

Indiana University Center for Spacetime Symmetries, Indiana University, Bloomington, Indiana 47405-7105, U.S.A.

E-mail: mschreck@indiana.edu

Abstract. The current article provides a brief review on noncommutative spacetimes in light of Lorentz invariance violation and especially on how these models are linked to the Lorentz-violating Standard-Model Extension. By embedding a general noncommutative spacetime with a constant background tensor into the Standard-Model Extension, a couple of implications are drawn on Lorentz violation in such models.

1. Introduction

Continuous symmetries and their description by Lie groups play a role in a plethora of topics in physics starting with classical mechanics and reaching as far as to the Standard Model of elementary particle physics. For this reason any physicist has come in touch with this highly interesting and powerful concept. Lie groups themselves are linked to homogeneous spaces, which provide examples in classical, commutative differential geometry.

Besides Lie groups there is another class of groups that are important in particular areas of physics; these are known as quantum groups. In principle quantum groups can occur as generalized symmetry groups in certain lattice models [1]. The analogy to Lie groups being connected to homogenous spaces is that quantum groups are linked to quantum homogeneous spaces. These have established the relatively new field of noncommutative differential geometry. Noncommutative spaces may be characterized by preferred directions in their tangent spaces where the existence of such directions leads to a violation of active rotational invariance. Generalizing these concepts to a noncommutative spacetime shows that this is linked to a violation of particle Lorentz invariance.

Lorentz invariance violation is motivated by the existence of a small-scale structure of spacetime. The latter is expected to occur at the Planck length due to quantum-theoretical energy fluctuations and their influence on spacetime geometry. The concept of such a spacetime foam was introduced by Wheeler in the context of his quantum geometrodynamics [2] and also Hawking worked on it a couple of years later [3, 4]. Successively particular spacetime foam models were considered in [5, 6, 7, 8, 9]. Since any direct way of describing a small-scale spacetime structure seems to be very complicated, an effective approach is to introduce noncommutative spacetime coordinates. Such a noncommutativity must necessarily imply a tensorial object, which gives rise to a fixed background and therefore leads to particle Lorentz violation.

The Standard-Model Extension (SME) is a framework, which contains all Lorentz-violating field operators constructed from Standard-Model fields or the Riemann curvature tensor

[10, 11, 12, 13, 14, 15]. The matter sector is invariant under $\mathrm{SU}(3)_c \times \mathrm{SU}(2)_L \times \mathrm{U}(1)_Y$-transformations as well. Each term is characterized by a field operator and particular controlling coefficients transforming as usual under observer Lorentz transformations and general coordinate transformations, respectively. However they are fixed backgrounds under particle Lorentz transformations and diffeomorphisms, respectively. The minimal SME [11] is restricted to all field operators of mass dimension 4 or lower, whereas the nonminimal SME [13, 14, 15] contains all field operators to arbitrary dimension. This is one of the reasons why the SME is a highly powerful framework for the search of Lorentz violation in nature.

The current paper shall briefly review how a noncommutative spacetime can be mapped to the SME whereby existing experimental bounds on certain controlling coefficients translate to bounds on tensor components, which describe noncommutative spacetimes. The article is organized as follows. In Sec. 2 noncommutative spacetimes will be reviewed where we will elaborate on one particular example. Section 3 will give a brief introduction to the SME and Sec. 4 is dedicated to discussing how to link a particular class of noncommutative spacetimes with the SME. Last but not least, Sec. 5 provides a summary of the most important aspects.

2. Noncommutative spacetimes

A noncommutative spacetime can be considered as an effective approach to modeling a small-scale spacetime structure. Spacetime coordinates are assumed to not commute any more, i.e., $[x^\mu, x^\nu] = \mathrm{i}\theta^{\mu\nu}$. The nonvanishing right-hand side of the latter equation must have the same tensor structure as the commutator, which means that an antisymmetric object $\theta^{\mu\nu}$ with two Lorentz indices has to be introduced. Since any particle Lorentz transformation does not transform the coordinates, but physical entities such as particles or fields, $\theta^{\mu\nu}$ is considered as a fixed background under particle Lorentz transformations. This is the reason why any such object manifestly violates particle Lorentz invariance. For extensive, general reviews on noncommutative spacetimes we refer to [16, 17].

To see what implications a noncommutative spacetime may have, a particular model, which was first proposed by Majid and Ruegg, shall be considered [18, 19]:

$$[x^\mu, x^\nu] = \mathrm{i}\lambda\theta^{\mu\nu}\,, \quad \theta^{\mu\nu} = \zeta^\mu\xi^\nu - \xi^\mu\zeta^\nu\,, \tag{1a}$$

$$(\zeta^\mu) = \begin{pmatrix} 0 \\ x^1 \\ x^2 \\ x^3 \end{pmatrix}, \quad (\xi^\mu) = \begin{pmatrix} 1 \\ 0 \\ 0 \\ 0 \end{pmatrix}. \tag{1b}$$

First of all, λ is a length scale, which is assumed to lie in the order of magnitude of the Planck length. This is reasonable, since a noncommutative spacetime is supposed to describe Planck-scale effects. Furthermore for this particular setup the tensor $\theta^{\mu\nu}$ is decomposed into two four-vectors ζ^μ and ξ^μ. The latter can be interpreted as preferred spacetime directions originating from a background field, which penetrates the vacuum. The background may be a low-energy description of physics at the Planck scale such as a spacetime foam. Note that the mass dimension of the equation is consistent since the vector ξ^μ only involves dimensionless components.

For this particular model, the tensor $\theta^{\mu\nu}$ also depends on the spacetime point considered due to the four-vector ζ^μ. The preferred spacetime direction ξ^μ is the same, whereas ζ^μ is different at each point. Additionally this leads to a violation of translation invariance, since the physics supposedly differs at each spacetime point.

Now any noncommutative spacetime gives rise to both a modified calculus and modified Fourier transforms, which imply a modified momentum group. In particular for the Majid-Ruegg model the conventional addition of two four momenta and the inverse four momentum are modified as follows [19] where the corresponding group operation will be denoted by $\oplus$:

$$\begin{pmatrix} E \\ \mathbf{p} \end{pmatrix} \oplus \begin{pmatrix} E' \\ \mathbf{p}' \end{pmatrix} = \begin{pmatrix} E + E' \\ \mathbf{p} + \exp(\lambda E)\mathbf{p}' \end{pmatrix}, \quad \begin{pmatrix} E \\ \mathbf{p} \end{pmatrix}^{-1} = \begin{pmatrix} -E \\ -\exp(-\lambda E)\mathbf{p} \end{pmatrix}. \tag{2}$$

Here E, E' are energies and $\mathbf{p}$, $\mathbf{p}'$ spatial momenta. The constant λ is the length scale, which appears in the definition of the model in Eq. (1). Besides, both the boost and rotation generators of the Lorentz algebra are altered as well resulting in a modified dispersion law [19] for particles of mass m:

$$\frac{\exp(\lambda E) + \exp(-\lambda E) - 2}{\lambda^2} - \mathbf{p}^2 \exp(-\lambda E) = m^2. \tag{3}$$

For particle energies lying well below the Planck scale, i.e., $\lambda E \ll 1$ the dispersion relation can be expanded giving the more illuminating result

$$E \approx \sqrt{\mathbf{p}^2 + m^2} - \lambda \frac{\mathbf{p}^2}{2}. \tag{4}$$

Besides the standard relativistic particle energy there is an additional contribution, which grows quadratically with the momentum but is suppressed by the small length scale (or large inverse energy scale) λ. The latter result is only valid if the energy or the momentum is sufficiently small. If λ is assumed to be larger than zero, the particle energy could become negative for a sufficiently high momentum otherwise.

Any noncommutative spacetime model gives rise to a noncommutative field theory by introducing noncommutative fields and a new product for two of these fields called the Moyal product [16, 17]. In [20] unitarity is investigated for such noncommutative field theories. The outcome is that theories with $\theta^{0i} \neq 0$, $i = 1 \dots 3$ may have issues with perturbative unitarity. Based on this result a model of noncommutative spacetime has no difficulties with perturbative unitarity when the following two conditions hold [21]:

$$\theta_{\mu\nu}\theta^{\mu\nu} > 0, \quad \theta_{\mu\nu}\widetilde{\theta}^{\mu\nu} = 0, \quad \widetilde{\theta}^{\mu\nu} = \frac{1}{2}\varepsilon^{\mu\nu\varrho\sigma}\theta_{\varrho\sigma}, \tag{5}$$

where $\widetilde{\theta}^{\mu\nu}$ is the dual of $\theta^{\mu\nu}$ and $\varepsilon^{\mu\nu\varrho\sigma}$ is the four-dimensional Levi-Civita symbol with $\varepsilon^{0123} = 1$. For the Majid-Ruegg model of Eq. (1) one can check that $\theta_{\mu\nu}\widetilde{\theta}^{\mu\nu} = 0$, but $\theta_{\mu\nu}\theta^{\mu\nu} = -2\mathbf{x}^2 < 0$ with the vector $\mathbf{x} = (x^1, x^2, x^3)$ of the spatial coordinates. Therefore the fate of perturbative unitarity for this particular noncommutative spacetime remains unclear at first.

3. Standard-Model Extension

It is reasonable to consider models of noncommutative spacetimes as effective descriptions of Planck-scale physics. However there are certain drawbacks with such setups. First, they describe the physics only partially since, e.g., Eq. (1) itself does not provide any modified field equations based on commutative field theory. This makes phenomenological investigations very involved. Second, there may be no observable physical effects even if there is a modified particle dispersion relation. These issues are based on the lack of an underlying effective field theory.

The reasons stated, amongst others, led to the development of the SME, which is an effective field theory comprising all Lorentz-violating operators constructed within the ordinary Standard Model and General Relativity. The great advantages of the SME are that all Lorentz-violating terms can be well organized according to their symmetry properties and the particle sector they belong to. Each contribution leads to modified field equations, Feynman rules, and all the tools available that are needed to consider (quantum) field theory and particle physics processes. For example, only within such a framework it can be understood why certain modifications of dispersion relations do not have any observable effects. This is because some Lorentz-violating coefficients can be absorbed by a redefinition of the physical fields at all orders in Lorentz violation, which makes them unphysical.

The SME contains all Lorentz-violating modifications of the free Standard-Model fields and General Relativity plus all modified interaction terms to an operator dimension of less or equal than four. Hence this concerns leptons, quarks, photons, gluons, the mediators of the weak interaction, and finally the recently discovered Higgs particle.

Lorentz violation leads to interesting, unusual physical effects such as a dispersive and/or birefringent vacuum [11, 22, 13] and modified particle decays [23, 24, 25, 26, 27, 28, 29, 30]. Within the SME the properties of modified quantum field theories can be considered by obtaining modified photon polarization vectors, spinors, and particle propagators. They are then used to check the validity of various theoretical properties that must be fulfilled for a reasonable and consistent quantum field theory, which was done in a series of articles [31, 32, 33, 34, 35, 36, 37, 38, 39, 40, 41, 42, 43].

4. Embedding noncommutative spacetimes into the Standard-Model Extension

A very helpful approach to investigating the physical implications of a noncommutative spacetime, i.e., $[x^\mu, x^\nu] = \mathrm{i}\theta^{\mu\nu}$ is to embed it into the SME. To get a first idea which sector of the SME a noncommutative spacetime corresponds to, one may be tempted to map modified dispersion relations to each other. First the expanded dispersion relation of the Majid-Ruegg model given by Eq. (4) shall be considered and be identified with a subset of the SME fermion sector coefficients. The dispersion relation is isotropic and since there is a single one, particles and antiparticles do not differ from each other. Therefore only the CPT-even operators $\widehat{m}$, $\widehat{m}_5$, $\widehat{c}^\mu$, $\widehat{d}^\mu$, and $\widehat{H}^{\mu\nu}$ come into consideration. From these already a large number is excluded for several reasons. The operator $\widehat{m}_5$ can be removed by a chiral transformation in most cases, which is why it will be omitted. Since λ is considered as the Lorentz-violating coefficient in Eq. (4), we must look for coefficients with mass dimension -1. Thus $\widehat{c}^\mu$ and $\widehat{d}^\mu$ are discarded as well, as the corresponding coefficients have even mass dimension. What finally remains are the operators $\widehat{m}$ and $\widehat{H}^{\mu\nu}$ where our interest is based on the isotropic cases. In [15] all isotropic sectors are discussed including operators to dimension 6. On the one hand, the isotropic subspace of the coefficients associated with the dimension-5 operator $\widehat{m}$ is given by:

$$(m^{(5)\mu\nu}) = \mathrm{diag}(m^{(5)00}, m^{(5)33}, m^{(5)33}, m^{(5)33})\,, \tag{6}$$

i.e., there are two distinct nonzero coefficients. The effective coefficient $c_{\mathrm{eff}}^{(6)0000}$ in Eq. (98) of [15] involves $m^{(5)00}$ via the second of their Eq. (27). Furthermore, $c_{\mathrm{eff}}^{(6)00jj}$ and $c_{\mathrm{eff}}^{(6)jjkk}$ contains the sum $m^{(5)jj}$ of the coefficients with two spatial indices set equal. Here $m^{(5)11} = m^{(5)22} \equiv m^{(5)33}$ is considered instead, which is equivalent. With the choice of Eq. (6) the (physical) isotropic fermion dispersion relation reads as follows:

$$\begin{aligned} p^0 &= \sqrt{\frac{1 - 2m^{(5)00}\left(m_\psi + \mathbf{p}^2 m^{(5)33}\right) - \sqrt{1 - 4m_\psi m^{(5)00} - 4\mathbf{p}^2 m^{(5)00}\left(m^{(5)00} + m^{(5)33}\right)}}{2(m^{(5)00})^2}} \\ &\approx \sqrt{\mathbf{p}^2 + m_\psi^2}\left(1 + m^{(5)00} m_\psi + m^{(5)33}\frac{m_\psi \mathbf{p}^2}{\mathbf{p}^2 + m_\psi^2}\right)\,, \end{aligned} \tag{7}$$

where m_ψ is the fermion mass. Note that $m^{(5)00}$, $m^{(5)33}$ have mass dimension -1, which renders the terms in parentheses dimensionless. Furthermore there are spurious dispersion relations, which do not have a limit for vanishing Lorentz-violating coefficients [43]. However they are considered as unphysical for momenta much smaller than the Planck scale where the SME is an effective framework. On the other hand, the isotropic subspace of the dimenion-5 operator $\widehat{H}^{\mu\nu}$ is spanned by the nonvanishing coefficients

$$H^{(5)1203} = H^{(5)2301} = H^{(5)3102} = -H^{(5)1302} = -H^{(5)2103} = -H^{(5)3201} \equiv H\,. \tag{8}$$

These are the coefficients $\widetilde{H}^{(5)0j0j}$, $j = 1\ldots3$ of the dual of $H^{(5)\mu\nu\varrho\sigma}$ as indicated in Eq. (97) in [15]. In that case the modified fermion dispersion law is given by:

$$\begin{aligned} p^0 &= \sqrt{\frac{\mathbf{p}^2 + m_\psi^2(1+H^2\mathbf{p}^2) - H^2\mathbf{p}^4 - 2m_\psi|\mathbf{p}|\sqrt{H^2\left[m_\psi^2 + \mathbf{p}^2(1-H^2\mathbf{p}^2)\right]}}{(1-H^2\mathbf{p}^2)^2}} \\ &\approx \sqrt{\mathbf{p}^2 + m_\psi^2} - Hm_\psi|\mathbf{p}|\,. \end{aligned} \tag{9}$$

The expansions stated hold as long as $m^{(5)00}m_\psi \ll 1$, $m^{(5)33}|\mathbf{p}| \ll 1$, and $H|\mathbf{p}| \ll 1$, i.e., only if the Lorentz-violating coefficient and the momentum are sufficiently small, which will be assumed in what follows. All three Lorentz-violating modifications grow linearly with increasing momentum for $\mathbf{p}^2 \gg m_\psi^2$. However for $\mathbf{p}^2 \ll m_\psi^2$ their momentum dependences differ from each other. The $m^{(5)00}$-term is then independent of the momentum, the H-term still grows linearly and the $m^{(5)33}$-term even grows quadratically. Hence in this case the momentum dependences of the $m^{(5)33}$-term and Eq. (4) would coincide. However if the SME-coefficient $m^{(5)33}$ was identified with the Lorentz-violating parameter λ of Eq. (4) both dispersion laws would be expected to correspond to each other not only when the particle momentum is much smaller than the particle mass. Therefore it seems that there is no operator in the SME, which produces a dispersion relation similar to Eq. (4) for a reasonably large range of the particle momentum. As a matter of fact it is challenging to identify the SME fermion sector corresponding to the Majid-Ruegg model by such an analysis. This does not mean that are is no possibility of performing this identification. However further studies of this issue are beyond the scope of the paper. That is why we will turn to another class of noncommutative spacetimes: models for which $\theta^{\mu\nu}$ itself is spacetime-independent.

These particular models form the basis of [21]. In the latter article a correspondence between the constant $\theta^{\mu\nu}$ and the SME is established. Since the SME is a commutative field theory, the Seiberg-Witten map [44] has to be used to match $\theta^{\mu\nu}$ to SME coefficients. By doing so, the noncommutative photon field $\widehat{A}_\mu$ and the noncommutative spinor field $\widehat{\psi}$ are mapped to the commutative photon and spinor field A_μ and ψ, respectively. At next-to-leading order in the commutative fields this map is given by:

$$\widehat{A}_\mu = A_\mu - \frac{1}{2}\theta^{\alpha\beta}A_\alpha(\partial_\beta A_\mu + F_{\beta\mu}) + \ldots\,, \quad \widehat{\psi} = \psi - \frac{1}{2}\theta^{\alpha\beta}A_\alpha\partial_\beta\psi + \ldots\,. \tag{10}$$

The procedure outlined leads to a modified commutative field theory where two characteristic additional terms will be given as an example:

$$\mathcal{L} = \overline{\psi}\left(\frac{\mathrm{i}}{2}\gamma^\mu \overleftrightarrow{D}_\mu - m_\psi\right)\psi - \frac{1}{4}F^{\mu\nu}F_{\mu\nu} + \frac{1}{4}mq\theta^{\alpha\beta}F_{\alpha\beta}\overline{\psi}\psi - \frac{1}{2}q\theta^{\alpha\beta}F_{\alpha\mu}F_{\beta\nu}F^{\mu\nu} + \ldots\,. \tag{11}$$

Here $F_{\alpha\beta} = \partial_\alpha A_\beta - \partial_\beta A_\alpha$ is the electromagnetic field strength tensor, $\overline{\psi} \equiv \psi^\dagger\gamma^0$ the Dirac-conjugate spinor, γ^μ are the standard Dirac matrices, D_μ is the covariant derivative, m_ψ the fermion mass, and q its charge. The first three terms are standard and the remaining two emerge from the noncommutative field theory by applying the Seiberg-Witten map. Note that both of them involve the noncommutativity tensor $\theta^{\alpha\beta}$. The first describes a photon-fermion interaction and the second is a photon interaction term.

The great advantage of the effective field theory gained from the noncommutative one is that $\theta^{\alpha\beta}$ can be identified with Lorentz-violating coefficients from the SME. Therefore fermions with charge q and photons are considered and the electromagnetic field strength tensor is understood as a sum over a constant background field strength tensor $f_{\mu\nu}$ and a small dynamical fluctuation $F_{\mu\nu}$. Since any realistic noncommutative theory is CPT-even [21], the identification reveals relationships between the noncommutativity tensor and the CPT-even coefficients $c^{\mu\nu}$ of the minimal SME fermion sector and the $(k_F)^{\mu\nu\varrho\sigma}$ of the minimal photon sector. For the fermion sector it reads as follows [21]:

$$c^{\mu\nu} = -\frac{q}{2}f^{\mu}{}_{\lambda}\theta^{\lambda\nu}\,. \tag{12}$$

Using this relation and looking for sidereal variations, e.g., in clock-comparison experiments provides conservative bounds on certain coefficients of $\theta^{\mu\nu}$ in a particular frame lying in the order of magnitude of $(10\,\mathrm{TeV})^{-2}$ [21]. It is impressive that via low-energy experiments such tight constraints on Planck-scale physics can be derived, which demonstrates the phenomenological power of the SME.

Now the correspondence of Eq. (12) shall be further investigated from a theoretical point of view. The properties of the SME reveal that $c^{\mu\nu}$ is symmetric and traceless: $c^{\mu\nu} = c^{\nu\mu}$ and $c^{\mu}{}_{\mu} = 0$. The latter property is valid, since the trace of $c^{\mu\nu}$ does not contain any physics, but it can be absorbed by a redefinition of the physical fields. Therefore these conditions shall be applied to the correspondence stated above. If $c^{\mu\nu}$ is claimed to be symmetric, four of the six components of the antisymmetric $\theta^{\mu\nu}$ can already be eliminated. If the electric and magnetic field vectors $\mathbf{E} = (E^1, E^2, E^3)$, $\mathbf{B} = (B^1, B^2, B^3)$ are introduced according to the definition of the contravariant field strength tensor,

$$(f^{\mu\nu}) = \begin{pmatrix} 0 & -E^1 & -E^2 & -E^3 \\ E^1 & 0 & -B^3 & B^2 \\ E^2 & B^3 & 0 & -B^1 \\ E^3 & -B^2 & B^1 & 0 \end{pmatrix}, \tag{13}$$

the components θ^{0i} for $i = 1\dots 3$ and θ^{23} can be expressed via θ^{12}, θ^{13} as follows:

$$\theta^{0i} = \frac{\sum_{j=2,3}(E^iE^j + B^iB^j)\theta^{1j}}{(\mathbf{E}\times\mathbf{B})^1}\,,\quad \theta^{23} = -\frac{\sum_{i=2,3}(E^1B^i - E^iB^1)\theta^{1i}}{(\mathbf{E}\times\mathbf{B})^1}\,. \tag{14}$$

With the tracelessness condition of $c^{\mu\nu}$ we can eliminate θ^{13} as well:

$$\theta^{13} = -\frac{E^2(\mathbf{E}^2 - \mathbf{B}^2) + 2B^2\,\mathbf{E}\cdot\mathbf{B}}{E^3(\mathbf{E}^2 - \mathbf{B}^2) + 2B^3\,\mathbf{E}\cdot\mathbf{B}}\theta^{12}\,. \tag{15}$$

By using the relationships of Eqs. (14), (15) the set of coefficients $c^{\mu\nu}$ must read as

$$(c^{\mu\nu}) = q\theta^{12}\frac{\mathbf{E}\cdot\mathbf{B}}{E^3(\mathbf{E}^2 - \mathbf{B}^2) + 2B^3\,\mathbf{E}\cdot\mathbf{B}}\begin{pmatrix} \widehat{c}^{00} & \widehat{c}^{01} & \widehat{c}^{02} & \widehat{c}^{03} \\ \widehat{c}^{01} & \widehat{c}^{11} & \widehat{c}^{12} & \widehat{c}^{13} \\ \widehat{c}^{02} & \widehat{c}^{12} & \widehat{c}^{22} & \widehat{c}^{23} \\ \widehat{c}^{03} & \widehat{c}^{13} & \widehat{c}^{23} & \widehat{c}^{33} \end{pmatrix}, \tag{16a}$$

$$\widehat{c}^{00} = (\mathbf{E}^2 + \mathbf{B}^2)/2\,, \tag{16b}$$

$$\widehat{c}^{ij} = \widehat{c}^{00}\delta^{ij} - (E^iE^j + B^iB^j)\,, \tag{16c}$$

$$\widehat{c}^{0i} = (\mathbf{E}\times\mathbf{B})^i\,, \tag{16d}$$

where $i,\ j \in \{1,2,3\}$. There are at least two interesting special cases. First, for orthogonal electric and magnetic fields the coefficients $c^{\mu\nu}$ vanish as long as the denominator in Eq. (16a) remains nonzero. Hence such a field configuration in a noncommutative spacetime does not imply Lorentz violation in the fermion sector. Second, for the choice $E^1 = E^2 = B^1 = B^2 = 0$ there results a fairly simple set of $c^{\mu\nu}$, which is given by a diagonal matrix:

$$(c^{\mu\nu}) = c^{00}\,\mathrm{diag}(1,1,1,-1)\,, \quad c^{00} = \frac{q}{2}\theta^{12}B^3\,. \tag{17}$$

The z-component of the electric field strength vector is completely arbitrary (but nonzero); it does not contribute to $c^{\mu\nu}$ in this case. The latter choice of Lorentz-violating coefficients gives rise to the following modified fermion dispersion relation:

$$p^0 = \sqrt{\left(\frac{1-c^{00}}{1+c^{00}}\right)^2 [(p^1)^2 + (p^2)^2] + (p^3)^2 + \widetilde{m}_\psi}\,, \quad \widetilde{m}_\psi \equiv \frac{m_\psi}{|1+c^{00}|}\,. \tag{18}$$

Here the fermion mass can be redefined as indicated. The resulting dispersion law is modified in the x-y-plane and it has a remaining isotropy in this plane. Therefore for a field configuration with parallel electric and magnetic field vectors in a noncommutative spacetime a Lorentz-violating fermion dispersion relation may emerge, which is characterized by a residual isotropy. Attempts to construct a field configuration giving rise to an isotropic Lorentz-violating dispersion relation, did not produce any results. Hence it seems that a noncommutative spacetime with constant tensor components $\theta^{\mu\nu}$ cannot lead to isotropic Lorentz violating modifications.

5. Conclusions

To summarize, noncommutative spacetimes provide effective models for a microscopic spacetime structure, which may emerge at the Planck scale. The noncommutativity is characterized by an antisymmetric two-tensor $\theta^{\mu\nu}$ with six independent components, which may even be spacetime-dependent. As an example a particular model with spacetime-dependent components $\theta^{\mu\nu}$ was considered with the goal of identifying it with a particular sector of the SME. However no corresponding modified particle dispersion relation was found, which makes a possible identification challenging. In contrast, it has been well-known for over a decade that models with a spacetime-independent $\theta^{\mu\nu}$ can be embedded into the SME by considering photons and fermions propagating in constant electromagnetic fields.

The identification leads to nonvanishing CPT-even coefficients $c^{\mu\nu}$ of the minimal fermion sector and $(k_F)^{\mu\nu\varrho\sigma}$ of the minimal photon sector [21]. In the current paper it was demonstrated that five of the six components of $\theta^{\mu\nu}$ can be eliminated due to the symmetry properties of $c^{\mu\nu}$. Therefore any Lorentz-violating dispersion relation in the fermion sector only depends on one of these components and the electric and magnetic background fields. For orthogonal electric and magnetic fields no Lorentz violation in the fermion sector can be observed. For parallel fields there exists a Lorentz-violating dispersion relation having a residual isotropy in a plane.

6. Acknowledgments

It is a pleasure to thank the organizers of the *International Conference on Integrable Systems and Quantum symmetries 22* (2014) in Prague, Czech Republic for their hospitality and the opportunity of giving a talk about this exciting topic. Furthermore the author is indebted to V. A. Kostelecký for reading the manuscript and giving helpful suggestions. This work was performed with financial support from the *Deutsche Akademie der Naturforscher Leopoldina* within Grant No. LPDS 2012-17.

Journal of Physics: Conference Series **563** (2014) 012026 doi:10.1088/1742-6596/563/1/012026

References

[1] Majid S 2000 *J. Math. Phys.* **41** 3892 (*Preprint* hep-th/0006167)
[2] Wheeler J A 1957 *Ann. Phys.* (N.Y.) **2** 604
[3] Hawking S W 1978 *Nucl. Phys.* B **144** 349
[4] Hawking S W, Page D N and Pope C N 1980 *Nucl. Phys.* B **170** 283
[5] Klinkhamer F R and Rupp C 2004 *Phys. Rev.* D **70** 045020 (*Preprint* hep-th/0312032)
[6] Bernadotte S and Klinkhamer F R 2007 *Phys. Rev.* D **75** 024028 (*Preprint* hep-ph/0610216)
[7] Schreck M, Sorba F and Thambyahpillai S 2013 *Phys. Rev.* D **88** 125011 (*Preprint* arXiv:1211.0084)
[8] Hossenfelder S 2013 *Phys. Rev.* D **88** 124030 (*Preprint* arXiv:1309.0311); Hossenfelder S 2013 *Phys. Rev.* D **88** 124031 (*Preprint* arXiv:1309.0314)
[9] Hossenfelder S 2014 *Adv. High Energy Phys.* **2014** 950672 (*Preprint* arXiv:1401.0276)
[10] Colladay D and Kostelecký V A 1997 *Phys. Rev.* D **55** 6760 (*Preprint* hep-ph/9703464)
[11] Colladay D and Kostelecký V A 1998 *Phys. Rev.* D **58** 116002 (*Preprint* hep-ph/9809521)
[12] Kostelecký V A 2004 *Phys. Rev.* D **69** 105009 (*Preprint* hep-th/0312310)
[13] Kostelecký V A and Mewes M 2009 *Phys. Rev.* D **80** 015020 (*Preprint* arXiv:0905.0031)
[14] Kostelecký V A and Mewes M 2012 *Phys. Rev.* D **85** 096005 (*Preprint* arXiv:1112.6395)
[15] Kostelecký V A and Mewes M 2013 *Phys. Rev.*D **88** 096006 (*Preprint* arXiv:1308.4973)
[16] Douglas M R and Nekrasov N A 2001 *Rev. Mod. Phys.* **73** 977 (*Preprint* hep-th/0106048)
[17] Szabo R J 2003 *Phys. Rept.* **378** 207 (*Preprint* hep-th/0109162)
[18] Majid S and Ruegg H 1994 *Phys. Lett.* B **334** 348 (*Preprint* hep-th/9405107)
[19] Amelino-Camelia G and Majid S 2000 *Int. J. Mod. Phys.* A **15** 4301 (*Preprint* hep-th/9907110)
[20] Gomis J and Mehen T 2000 *Nucl. Phys.* B **591** 265 (*Preprint* hep-th/0005129)
[21] Carroll S M, Harvey J A, Kostelecký V A, Lane C D, and Okamoto T 2001 *Phys. Rev. Lett.* **87** 141601 (*Preprint* hep-th/0105082)
[22] Kostelecký V A and Mewes M 2002 *Phys. Rev.* D **66** 056005 (*Preprint* arXiv:hep-ph/0205211)
[23] Coleman S R and Glashow S L 1999 *Phys. Rev.* D **59** 116008 (*Preprint* hep-ph/9812418)
[24] Kostelecký V A and Pickering A G M 2003 *Phys. Rev. Lett.* **91** 031801 (*Preprint* hep-ph/0212382)
[25] Lehnert R and Potting R 2004 *Phys. Rev. Lett.* **93** 110402 (*Preprint* hep-ph/0406128)
[26] Lehnert R and Potting R 2004 *Phys. Rev.* D **70** 125010; Erratum-ibid. D **70** 129906 (*Preprint* hep-ph/0408285)
[27] Jacobson T, Liberati S, and Mattingly D 2006 *Annals Phys.* **321** 150 (*Preprint* astro-ph/0505267)
[28] Kaufhold C and Klinkhamer F R 2006 *Nucl. Phys.* B **734** 1 (*Preprint* hep-th/0508074)
[29] Kaufhold C and Klinkhamer F R 2007 *Phys. Rev.* D **76** 025024 (*Preprint* arXiv:0704.3255)
[30] Klinkhamer F R and Schreck M 2008 *Phys. Rev.* D **78** 085026 (*Preprint* arXiv:0809.3217)
[31] Kostelecký V A and Lehnert R 2001 *Phys. Rev.* D **63** 065008 (*Preprint* hep-th/0012060)
[32] Adam C and Klinkhamer F R 2001 *Nucl. Phys.* B **607** 247 (*Preprint* hep-ph/0101087)
[33] Casana R, Ferreira M M, Gomes A R, and Pinheiro P R D 2009 *Phys. Rev.* D **80** 125040 (*Preprint* arXiv:0909.0544)
[34] Casana R, Ferreira M M, Gomes A R, and dos Santos F E P 2010 *Phys. Rev.* D **82** 125006 (*Preprint* arXiv:1010.2776)
[35] Klinkhamer F R and Schreck M 2011 *Nucl. Phys.* B **848** 90 (*Preprint* arXiv:1011.4258); Klinkhamer F R and Schreck M 2012 *Nucl. Phys.* B **856** 666 (*Preprint* arXiv:1110.4101)
[36] Schreck M 2012 *Phys. Rev.* D **86** 065038 (*Preprint* arXiv:1111.4182)
[37] Cambiaso M, Lehnert R, and Potting R 2012 *Phys. Rev.* D **85** 085023 (*Preprint* arXiv:1201.3045)
[38] Colladay D, McDonald P, and Potting R 2014 *Phys. Rev.* D **89** 085014 (*Preprint* arXiv:1401.1173)
[39] Maniatis M and Reyes C M 2014 *Phys. Rev.* D **89** 056009 (*Preprint* arXiv:1401.3752)
[40] Schreck M 2014 *Phys. Rev.* D **89** 085013 (*Preprint* arXiv:1311.0032)
[41] Schreck M 2014 *Phys. Rev.* D **89** 105019 (*Preprint* arXiv:1312.4916)
[42] Cambiaso M, Lehnert R, and Potting R 2014 *Phys. Rev.* D **90** (2014) 065003 (*Preprint* arXiv:1401.7317)
[43] Schreck M 2014 (*Preprint* arXiv:1403.6766)
[44] Seiberg N and Witten E 1999 *JHEP* **9909** 032 (*Preprint* hep-th/9908142)

Journal of Physics: Conference Series **563** (2014) 012027 doi:10.1088/1742-6596/563/1/012027

The chiral massive fermions in the graphitic wormhole

Jan Smotlacha

Bogoliubov Laboratory of Theoretical Physics, Joint Institute for Nuclear Research, 141980 Dubna, Moscow region, Russia

E-mail: smota@centrum.cz

Abstract. The graphitic wormhole appears at the center of physical interest because of its interesting properties which can remotely resemble the concept of the space wormhole. Apart from the usual applications of the carbon nanostructures like the electronic computer devices, it seems to be a good material for the accumulation of the electric charge and different kinds of molecules, e.g. the hydrogen molecules which enables using this material for the storage of the new kinds of fuel. Here, we present the electronic structure of this material and its possibly significant derivate - the perturbed wormhole. Next, the influence of some additional factors on the electronic structure like the changes of the Fermi velocity close to the wormhole bridge as well as the the spin-orbit interaction will be investigated. On this base, we predict the massive chiral fermions in the connecting wormhole nanotube.

1. Introduction

In the passed years, the basic forms of the nanostructures were investigated: the fullerene, graphene, nanotubes, nanohorns, nanocones, nanowires etc. Now, the steps are implemented for the preparation of more complicated forms like the wormhole [1, 2] and for the description of the electronic peculiarities of such a structure. Because of the repulsion forces between the carbon atoms on the different graphene sheets, it seemed to be very difficult to pursue this problem on another than the theoretical base. However, the progress of the laboratory methods promises to get good results in this field. The presence of the wormholes in the synthesized structures could be proven by the detection of the zero energy modes of the fermions in the places of the wormhole bridges.

The geometry of the wormhole could also cause a shift of the Fermi energy at different distances from the wormhole center which could direct the electron flux to this center, and in this way, the electric charge could be accumulated. The extraordinary deformation of the wormhole structures could highly influence the character of the Fermi velocity - relativistic effects could appear in these conditions. First of all, it could significantly change the mass of fermions which are usually considered to be massless. Another effect which could significantly influence the character of fermions is the spin-orbit coupling or interaction (SOC) for the fermions located in the carbon nanotube. We will show that the strength of this interaction strongly depends on the radius of the nanotube, it should be inversely proportional to this radius. As a result, the chiral (massive) fermions should be detected around the wormhole bridge.

We investigate the graphitic wormhole: we calculate its electronic structure and look through the corrections coming from the relativistic effects and the spin-orbit interaction. Similar calculations of the electronic structure will be carried out for the case of the perturbed wormhole. Furthermore, some derivates of these exotic structures and their properties will be mentioned.

2. The graphitic and perturbed wormhole

The wormhole (Fig. 1) is usually understood as a form which arises when two graphene sheets are connected together with the help of the connecting nanotube [1, 2]. The border part of the sheets with the nanotube is called the wormhole bridge. This can be achieved by the supply of 2 sets of 6 heptagonal defects onto both sides of the given nanotube, i.e. 12 heptagons altogether. Of course, this brings the restrictions on the form of the nanotube - the chirality must be $(6n, 6n)$ armchair or $(6n, 0)$ zig-zag. Furthermore, because of the physical limitations, the radius of the nanotube must be much larger than its length.

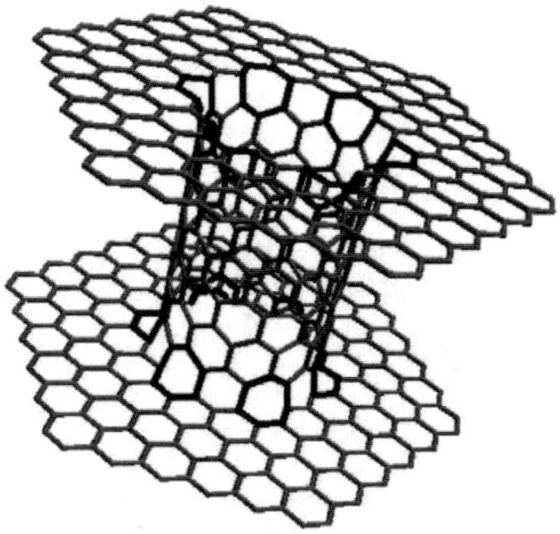

Figure 1. Wormhole structure.

The number of the defects can and then the resulting structure we call the perturbed wormhole. Then, the cases 0 defects (the nanotube) and 12 defects (the wormhole) will be the limiting possibilities. Possible forms of the perturbed wormhole (with the defects located in the middle) can be seen in Fig. 2. Due to symmetry preservation, we consider only the even numbers of the defects, i.e. 2, 4, 6, 8 or 10. The defects can be located in the middle as well as at the edge of the connecting structure - in the former we can say that the hight of the connecting nanotube is negligible (similarly to the case of the above draught structure of the wormhole containing 12 defects). We will be concerned with the case when the defects are located at the edge of the connecting nanotube.

3. Electronic structure

The electronic structure can be characterized by the local density of states ($LDoS$) which gives the number of the electronic states per the unit interval of energies and per the unit area of the molecular surface. The reason is an intuitive assumption that there exists a direct connection between this quantity and the electronic conductivity. For its calculation we use the Dirac-like equation

$$H\psi = E\psi. \tag{1}$$

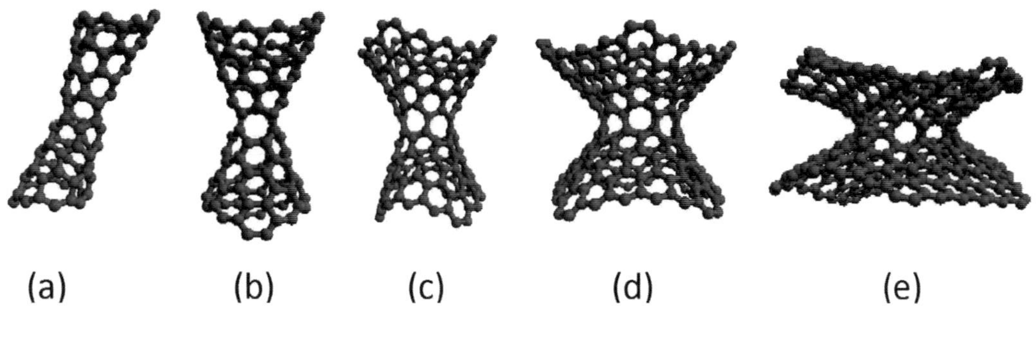

Figure 2. Different forms of the perturbed wormhole: (a) 2 defects, (b) 4 defects, (c) 6 defects, (d) 8 defects, (e) 10 defects.

The wave function ψ, which gives the solution, has two components:

$$\psi = \begin{pmatrix} \psi_A \\ \psi_B \end{pmatrix}, \tag{2}$$

where A, B correspond to inequivalent sublattices of the hexagonal plane. The solution is found by using the substitution

$$\psi = \begin{pmatrix} \psi_A \\ \psi_B \end{pmatrix} = \frac{1}{\sqrt[4]{g_{\varphi\varphi}}} \begin{pmatrix} u_j(\xi)e^{i\varphi j} \\ v_j(\xi)e^{i\varphi(j+1)} \end{pmatrix}, \quad j = 0, \pm1, ..., \tag{3}$$

so

$$\frac{\partial_\xi u_j}{\sqrt{g_{\xi\xi}}} - \frac{\tilde{j}}{\sqrt{g_{\varphi\varphi}}} u_j = E v_j, \qquad -\frac{\partial_\xi v_j}{\sqrt{g_{\xi\xi}}} - \frac{\tilde{j}}{\sqrt{g_{\varphi\varphi}}} v_j = E u_j, \tag{4}$$

where

$$\tilde{j} = j + 1/2 - a_\varphi - a_\varphi^W - A_\varphi, \tag{5}$$

$g_{\xi\xi}, g_{\varphi\varphi}$ are the metric coefficients and j is the angular momentum. Next, the gauge fields a_μ, a_μ^W ensure the continuation of the wave function with respect to the angular coordinate, and they are caused by the presence of the defects and by the rotational symmetry, respectively. The gauge field A_μ characterizes the possible magnetic field.Then, if u, v are the normalized solutions, the local density of states can be calculated as

$$LDoS(E, \xi_0) = u^2(E, \xi_0) + v^2(E, \xi_0). \tag{6}$$

3.1. Case of wormhole

In the solution of the Dirac equation for the case of 12 defects, we distinct 2 possibilities: the first corresponds to the case when the chiral vector has the form $(6n, 6n)$ or $(6n, 0)$ with n divisible by 3, the second corresponds to the case when the chiral vector has the form $(6n, 0)$ with n divisible by 3. The corresponding solution has the form

$$\psi^\pm = \begin{pmatrix} \psi_A^\pm(r,\varphi) \\ \psi_B^\pm(r,\varphi) \end{pmatrix} = c_1 \begin{pmatrix} J_{n\mp a_\varphi - 1/2}(kr) \\ -i\,\mathrm{sgn}\,\varepsilon J_{n\mp a_\varphi + 1/2}(kr) \end{pmatrix} + c_2 \begin{pmatrix} Y_{n\mp a_\varphi - 1/2}(kr) \\ -i\,\mathrm{sgn}\,\varepsilon Y_{n\mp a_\varphi + 1/2}(kr) \end{pmatrix}, \tag{7}$$

where the values of the gauge field $a_\varphi = \frac{3}{2}$ and $a_\varphi = \frac{1}{2}$ correspond to the particular possibilities, $J_n(x)$ and $Y_n(x)$ are the Bessel functions and the energy $\varepsilon = \pm v_F k$. The local density of states is seen in Fig. 3.

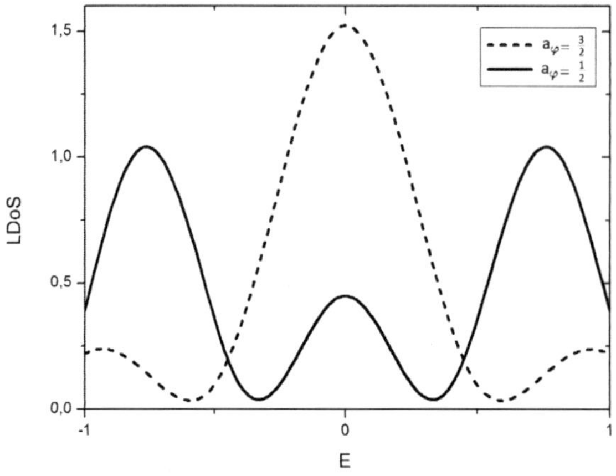

Figure 3. Local density of states on the bridge of the graphitic wormhole for different values of a_φ.

3.2. Case of chiral massive fermions

In our investigations, we suppose that the fermions appearing in the Dirac equation have the zero mass, or more precisely, the mass is very small in comparison with the energy. But it was shown in [7, 8] that the Fermi velocity changes and needs to be renormalized due to the elasticity and the deformations in graphene. In the investigated case of the graphitic wormhole which is very specific because of the big deformations, the velocity of the fermions close to the wormhole bridge could achieve such values that the relativistic effects can appear or break off the symmetry. The result is that the mass of fermions would be non-negligible. (The same situation is also for the graphene bilayer where especially massive particles are predicted, see [9, 10].) Moreover, the radius of the wormhole and its bridge is very small in comparison with the size of the upper and the lower graphene sheet, and by folding the sheet into a tube they acquire non-zero effective mass as they move along the tube axis.

As a result, we supply a mass term into the Dirac equation. First, we transform the system (4) into the differential equation of the second order

$$\left(\partial_{\xi\xi} - \frac{1}{2g_{\xi\xi}}\partial_\xi g_{\xi\xi} + \frac{\tilde{j}}{2}\sqrt{\frac{g_{\xi\xi}}{g_{\varphi\varphi}^3}}\partial_\xi g_{\varphi\varphi} - \tilde{j}^2\frac{g_{\xi\xi}}{g_{\varphi\varphi}} + E^2 g_{\xi\xi}\right)u_j = 0. \tag{8}$$

Now, we will do a substitution $E^2 \to E^2 - M^2$, where M is the mass of the corresponding fermion. Then, resulting equation acquires the form

$$\left(\partial_{\xi\xi} - \frac{1}{2g_{\xi\xi}}\partial_\xi g_{\xi\xi} + \frac{\tilde{j}}{2}\sqrt{\frac{g_{\xi\xi}}{g_{\varphi\varphi}^3}}\partial_\xi g_{\varphi\varphi} - \tilde{j}^2\frac{g_{\xi\xi}}{g_{\varphi\varphi}} + (E^2 - M^2) g_{\xi\xi}\right)u_j = 0. \tag{9}$$

Then, for different values of M, we can find the corrections of the *LDoS* for the graphitic wormhole. For different distances from the wormhole bridge, we can see these corrections in Fig. 4.

An important measurable quantity in the carbon nanostructures (including the nanotubular part of the graphitic wormhole) is the spin-orbit coupling (SOC) [3, 4]. Considering this influence, the 2-component Dirac equation is changed into the usual 4-component form. As a consequence, the chiral fermions should be detected close to the wormhole bridge.

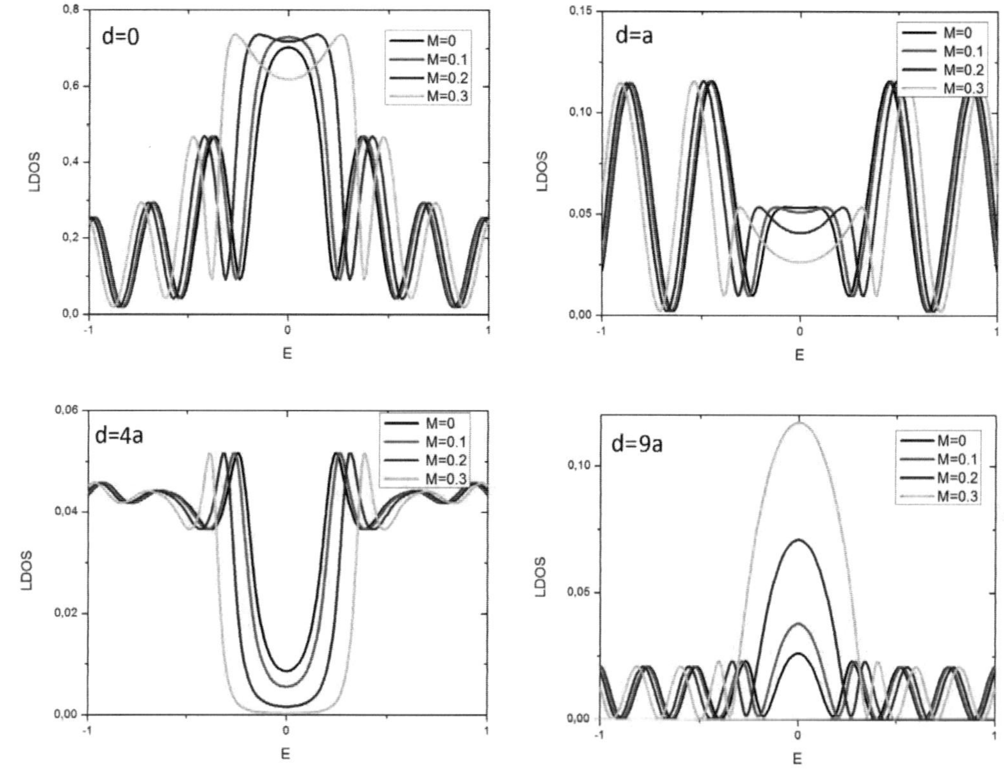

Figure 4. Comparison of *LDoS* for different masses of fermions at different distances d from the wormhole bridge.

We have 2 sources of the SOC: first, the inter-atomic which preserves the z-component of the spin and second, the so-called Rashba-type coming from the external electric field which conserves the z-component of the angular momentum J_z. In both cases, the strength of the SOC is influenced by the nonzero curvature. Here, we introduce in the first, inter-atomic source of the SOC.

It transforms the initial Hamiltonian into the form

$$\hat{H} = \hat{H}_{kin} + \hat{H}_{SOC}, \tag{10}$$

$$\hat{H}_{kin} = -i\gamma \left(\partial_y \mathrm{Id}_2 \otimes \hat{s}_y + \frac{1}{R} \partial_\theta \mathrm{Id}_2 \otimes \hat{s}_x \right), \tag{11}$$

$$\hat{H}_{SOC} = \lambda_y \hat{\sigma}_x \otimes \hat{s}_y - \lambda_x \hat{\sigma}_y \otimes \hat{s}_x. \tag{12}$$

Here, $\hat{H}_{SOC}$ corresponds to the spin-orbit coupling. The operators $\hat{s}_{x,y,z}$ are the Pauli matrices which transform the wave function of the A sublattice into the wave function of the B sublattice and vice versa.

The SOC is described with the aid of two strength parameters λ_x and λ_y which have, in the case of single wall carbon nanotube with different magnitude, the form

$$\lambda_x = \frac{\gamma}{R} \left(\frac{1}{2} + 2\delta p \right), \quad \lambda_y = -\frac{\delta \gamma'}{4R}. \tag{13}$$

Here, $|\lambda_y| \ll |\lambda_x|$ and for $R \to 0$, both strengths go to the infinity, as we require.
So, the chiral massive fermions should be detected around the wormhole bridge.

3.3. Case of perturbed wormhole

In Fig. 5, we see the comparison of the *LDoS* for different kinds of the perturbed wormhole. From the plots follows that the intensity is rising with the increasing number of the defects and it is closer and closer approaching the results in Fig. 3, where the case of 12 defects is shown. The localized states for zero energy is also in coincidence with the results presented for the bilayer and trilayer in [5, 6].

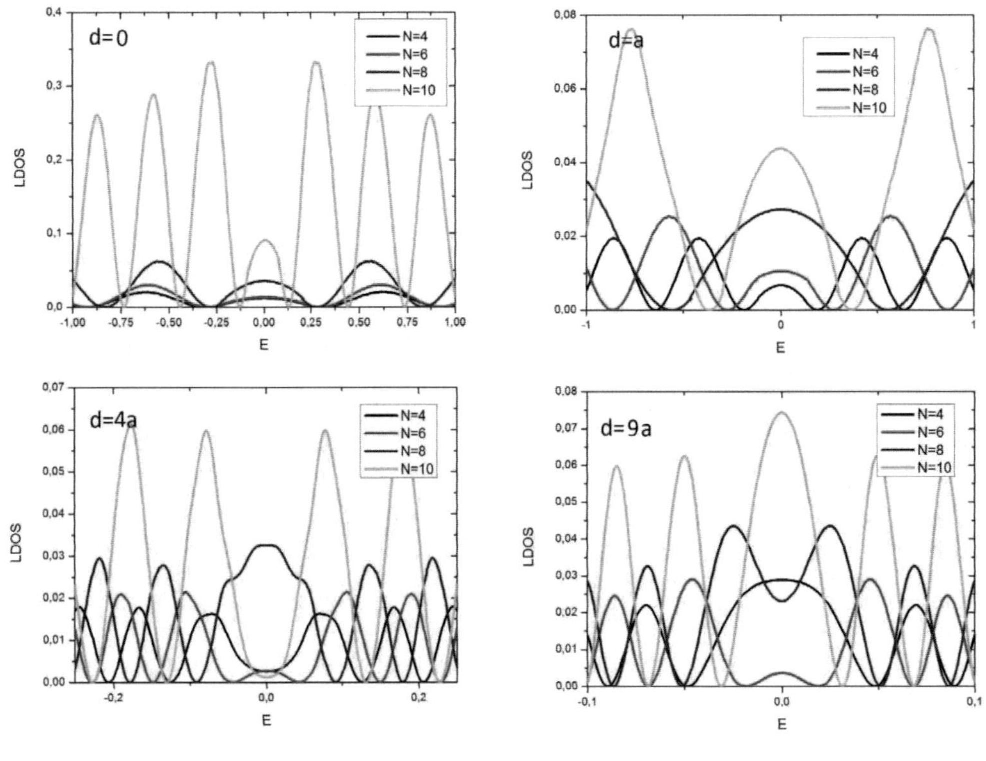

Figure 5. Comparison of *LDoS* for different numbers of the defects in the perturbed wormhole at different distances *d* from the wormhole bridge.

4. Conclusion

The wormhole allotropes could be prepaired by the procedure which would be composed of two steps: first, doping the graphene sheets by the reactive molecules and second, pressing these sheets against each other by achieving their connection through the mentioned molecules. The graphitic wormhole could also be created spontaneously in the bilayer graphene. The created wormholes in the graphene structure could then be indicated by the presence of the corresponding zero modes. We use the continuum gauge field-theory for their calculation. It was shown in the calculations of the *LDoS* for the perturbed wormhole that the closer is the number of the defects to the border values 0 a 12, the more the *LDoS* coincides with the calculations carried out for these 2 values as can be verified by the comparison with [1] and [11].

Next possibilities as to identify the wormhole structure in the graphene bilayer is via spin-orbit interaction or coupling (SOC) in connecting nanotubes [12]. The SOC in graphene could be induced as well by the nonzero curvature [13, 14, 15, 16]. In the case of the perturbed wormhole with negative curvature the chiral fermions penetrating through the connecting nanotube in the wormhole structure could be created. This effect should the higher, the lower is the radius of the connecting wormhole. Moreover, the chiral fermions should prefer permanently only one direction of the massive or massless fermionic current [17], e.g., from the upper graphene sheet to the lower in the graphene bilayer through the connecting nanotube depending on the wormhole curvature. This permanently oriented flow could be detected by the experimental observations.

More complicated wormhole structures can be established. One of them is the pillared graphene (Fig. 6). This kind of nanostructures will have a wide application in the branch of storage of the hydrogen fuel [18]. Initially, the carbon nanotubes were considered as a suitable candidate, but for different reasons, e.g. the insufficiently large molecular surface area, it was shown that these structures are not suitable and the pillared graphene could be much better for this purpose. One of the possible methods of the synthesis of this kind of nanostructure was described in [19]. Here the procedure is suggested in which the layer of the fullerene molecules is inserted between the graphene sheets and the reactive molecules are put to the attachment points. Then, the required chemical structure is achieved via the thermal and radiative processes.

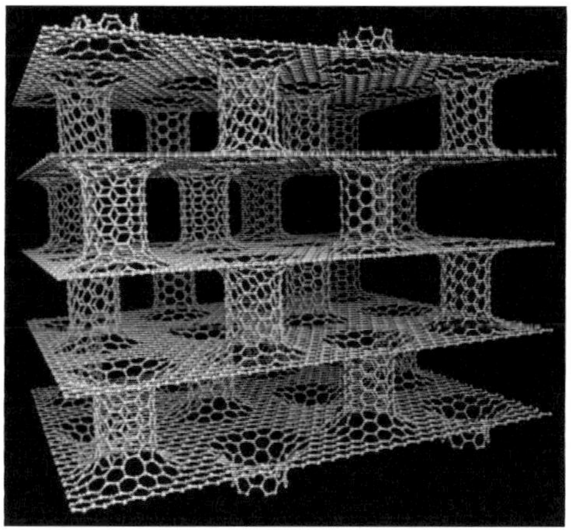

Figure 6. Fragment of pillared graphene.

Summarize, the presented paper is a real connection between nanoscale physics, condensed matter physics, chemical physics, computational physics, relativistic physics and experimental physics. We hope that it reflects a lot of main ideas of the 21st century of scientific research.

References

[1] J. Gonzalez, F. Guinea and J. Herrero, Phys. Rev. B **79**, 165434 (2009).
[2] J. Gonzalez, J. Herrero, Nucl. Phys. B **825**, 426443 (2010).
[3] C. L. Kane, E. J. Mele, Phys. Rev. Lett. **95**, 226801 (2005).
[4] T. Ando, Journal of the Physical Society of Japan **69**, 1757 (2000).
[5] E. V. Castro, M. P. Lopez-Sancho, M. A. H. Vozmediano, Phys. Rev. Lett. **104**, 036802 (2010).
[6] E. V. Castro, M. P. Lopez-Sancho, M. A. H. Vozmediano, Solid State Communications **152**, 1483 (2012).
[7] E.A. Kochetov, V.A. Osipov and R. Pincak, Journal of Physics: Condens. Matter **22**, 395502 (2010).
[8] H. Rostami, R. Asgari, Phys. Rev. B **86** , 155435 (2012).
[9] G. Borghi, M. Polini, R. Asgari and A.H. MacDonald, Solid State Commun. **149**, 1117 (2009).
[10] M. I. Katsnelson, K. S. Novoselov and A. K. Geim, Nature Physics **2**, 620 (2006).

[11] R. Pincak, J. Smotlacha, M. Pudlak, NanoMMTA, **2**, 81 (2013).
[12] F. Kuemmeth, S. Ilani, D. C. Ralph, P. L. McEuen, Nature **452**, 448 (2008).
[13] M. V. Entin, L. I. Magaril, Phys. Rev. B **64**, 085330 (2001).
[14] A. De Martino, R. Egger, K. Hallberg, C. A. Balseiro, Phys. Rev. Lett. **88**, 206402 (2002).
[15] J. S. Jeong, H. W. Lee, Phys. Rev. B **80**, 075409 (2009).
[16] M. P. Lopez-Sancho, M. C. Munoz, Phys. Rev. B **83**, 075406 (2011).
[17] S. Bellucci, A. A. Saharian, V. M. Bardeghyan, Phys. Rev. D **82**, 065011 (2010).
[18] Dmitrakakis, G et all, SPIE newsroom, (2009).
[19] D. L. Barker, W. R. Owens, J. W. Beck: Fabrication of pillared graphene (Patent application), http://www.faqs.org/patents/app/20120152725 (2012).

XXII International Conference on Integrable Systems and Quantum Symmetries (ISQS-22) IOP Publishing
Journal of Physics: Conference Series **563** (2014) 012028 doi:10.1088/1742-6596/563/1/012028

Hamiltonian structure for a class of parametric coupled systems of the Korteweg-de Vries type

Adrián Sotomayor[1], **Alvaro Restuccia**[2]

[1] Departament of Mathematics, Antofagasta University, Antofagasta, Chile
[2] Physics Department, Antofagasta University, Antofagasta, Chile
[2] Physics Department, Simón Bolívar University, Caracas, Venezuela

E-mail: `adrian.sotomayor@uantof.cl,arestu@usb.ve`

Abstract. We obtain from a lagrangian action describing a class of coupled parametric systems of KdV type its hamiltonian structure. The Poisson algebra arises from second class constraints of the theory and the use of Dirac brackets. The coupled system has relevant applications in physics.

1. Introduction

Extensions of the Korteweg-de Vries equation have been object of an extensive study due to its intrinsic mathematical properties and the wide range of applications in which they appear. In particular, the coupled Korteweg-de Vries (KDV) systems are a special class of such extensions, widely used and still deserving attention in view of the relevant open problems related to them [1, 2, 3, 4, 5].

It is known, for example, that coupled KdV systems have, as the KdV equation, many interesting properties, such as associated Bäcklund transformations, multisolitonic solutions, Painlevé property, Lax pairs and hamiltonian formulations, among others.

We remind that the first hamiltonian formulation of KdV [6] was formulated in terms of the Fourier coefficients in the expansion of the defining field. It that sense we have an infinite dimensional hamiltonian equation. It was however necessary, to check the fulfillment of the Jacobi identity, for the Poisson bracket candidate. Later, the method of Dirac was used to obtain the same hamiltonian formulation [7, 8, 9]. The starting point here was a lagrangian formulation in terms of a prepotential of the field which defines the KdV equation and then, via a Legendre transformation and using the Dirac method for singular lagrangians [10], to obtain the second class constraints which induce the Poisson structure. To obtain the second hamiltonian structure it was necessary to consider the associated Miura transformation and again the Dirac method allows to obtain the other hamiltonian structure of the KdV equation. In both cases the Jacobi identity is a direct consequence of the formulation and the hamiltonian structures are the most general ones, because they are defined in terms of the prepotential.

In this work we obtain for a parametric coupled KdV system its hamiltonian structure via a lagrangian formulation and an associated Miura transformation with, of course, the associated coupled Miura system. We follow the method of Dirac for constrained systems. On the way we obtain the natural Casimirs of the theory arising directly from the Dirac construction. We notice that the above results are not a direct consequence of the complexification (or "deformation"

[11]) of the KdV equation. We also notice that this construction is the first step to construct associated pencils of Poisson structures for the given coupled KdV system.

2. The parametric coupled KdV system and it hamiltonian structures

The coupled KdV system to be considered is given by

$$u_t + uu_x + u_{xxx} + \lambda v v_x = 0 \tag{1}$$

$$v_t + u_x v + v_x u + v_{xxx} = 0 \tag{2}$$

where λ is a real parameter. For the value $\lambda > 0$ the system is equivalent to two decoupled KdV equations. The $\lambda < 0$ case is equivalent to the complex version of KdV equation and finally, for $\lambda = 0$ we have a particular and interesting Hirota-Satsuma type system in the classification given in [12, 13].

In [14] a Bäcklund and a Gardner transformation for the system (1),(2) were obtained , also a permutability theorem was shown, which enables to generate new solutions from old ones. From it an infinite family of multisolitonic and periodic solutions were found.

The Miura transformation together with the Miura system associated to the coupled system (1),(2) are given by

$$\begin{aligned} u &= \mu_x - \tfrac{1}{6}\mu^2 - \tfrac{\lambda}{6}\nu^2 \\ v &= \nu_x - \tfrac{1}{3}\mu\nu \end{aligned} \tag{3}$$

and

$$\begin{aligned} \mu_t + \mu_{xxx} - \tfrac{1}{6}\mu^2\mu_x - \tfrac{\lambda}{6}\nu^2\mu_x - \tfrac{\lambda}{3}\mu\nu\nu_x = 0 \\ \nu_t + \nu_{xxx} - \tfrac{1}{6}\mu^2\nu_x - \tfrac{\lambda}{6}\nu^2\nu_x - \tfrac{1}{3}\mu\nu\mu_x = 0. \end{aligned} \tag{4}$$

The fundamental property of (3) is that it maps solutions of (4) into solutions of (1),(2) for every value of the parameter λ.

Now we construct the hamiltonian structure of the given coupled KdV system. We consider the two lagrangians $L_1 = \int_0^T dt \int_{-\infty}^{+\infty} dx\, \mathcal{L}_1$, $L_2 = \int_0^T dt \int_{-\infty}^{+\infty} dx\, \mathcal{L}_2$ where

$$\mathcal{L}_1 = -\frac{1}{2}w_x w_t - \frac{1}{6}{w_x}^3 + \frac{1}{2}{w_{xx}}^2 - \frac{\lambda}{2}w_x {y_x}^2 - \frac{\lambda}{2}y_x y_t + \frac{\lambda}{2}{y_{xx}}^2, \tag{5}$$

$$\mathcal{L}_2 = -\frac{1}{2}w_x y_t - \frac{1}{2}w_t y_x - \frac{1}{2}w_x^2 y_x - y_x w_{xxx} - \frac{\lambda}{6}y_x^3 \tag{6}$$

and

$$\begin{aligned} u(x,t) &= w_x(x,t) \\ v(x,t) &= y_x(x,t) \end{aligned}$$

where w and y are the prepotentials for u and v respectively. For the lagrangian defined by (5) it is necessary to impose $\lambda \neq 0$.

By taking independent variations of L_1 and L_2 with respect to w and to y we obtain the field equations, they are given by (1) and (2).

We start our construction with lagrangian L_1. We introduce the conjugate momenta associated to w and y, we denote them p and q respectively, we have

$$p = \frac{\delta \mathcal{L}_1}{\delta w_t} = -\frac{1}{2}w_x \quad , \quad q = \frac{\delta \mathcal{L}_1}{\delta y_t} = -\frac{\lambda}{2}y_x.$$

We define

$$\phi_1 \equiv p + \frac{1}{2}w_x \quad , \quad \phi_2 = q + \frac{\lambda}{2}y_x.$$

We notice that ϕ_1 and ϕ_2 do not have any w_t nor any y_t dependence, hence $\phi_1 = \phi_2 = 0$ are constraints on the phase space. It turns out that these are the only constraints on the phase space. They are second class contraints. The hamiltonian may be obtain directly from $\mathcal{L}_1$ by performing a Legendre transformation,

$$\mathcal{H}_1 = pw_t + qy_t - \mathcal{L}_1.$$

We obtain

$$\mathcal{H}_1 = \frac{1}{6}w_x^3 - \frac{1}{2}w_{xx}^2 + \frac{\lambda}{2}w_x y_x^2 - \frac{\lambda}{2}y_{xx}^2$$

and the corresponding hamiltonian $H_1 = \int_{-\infty}^{+\infty} dx\, \mathcal{H}_1$.

We introduce a Poisson structure on the phase space defined by

$$\begin{aligned} \{w(x), p(\hat{x})\}_{PB} &= \delta(x - \hat{x}) \\ \{y(x), q(\hat{x})\}_{PB} &= \delta(x - \hat{x}) \end{aligned}$$

with all other brackets between these variables being zero.

The resulting phase space is a constrained phase space and for that reason it is necessary to introduce Dirac brackets. The Dirac brackets between two functionals F and G on phase space is defined as

$$\{F, G\}_{DB} = \{F, G\}_{PB} - \left\langle \left\langle \{F, \phi_i(x')\}_{PB}\, \mathbb{C}_{ij}(x', x'')\, \{\phi_j(x''), G\}_{PB} \right\rangle_{x'} \right\rangle_{x''} \tag{7}$$

where $<>_{x'}$ denotes integration on x' from $-\infty$ to $+\infty$. It turns out, after some calculations, that

$$\begin{aligned} &\{u(x), u(\hat{x})\}_{DB} = -\partial_x \delta(x - \hat{x}) \ , \ \{v(x), v(\hat{x})\}_{DB} = -\frac{1}{\lambda}\partial_x \delta(x - \hat{x}) \\ &\{u(x), v(\hat{x})\}_{DB} = 0. \end{aligned}$$

We notice that this Poisson bracket is not well defined for $\lambda = 0$. We have already assume $\lambda \neq 0$.

From them we obtain the Hamilton equations, which are of course the same as (1),(2):

$$\begin{aligned} u_t = \{u, H_1\}_{DB} &= -uu_x - u_{xxx} - \lambda v v_x \\ v_t = \{v, H_1\}_{DB} &= -u_x v - v_x u - v_{xxx}. \end{aligned} \tag{8}$$

Moreover, we may obtain directly the Dirac bracket of any two functionals $F(u, v)$ and $G(u, v)$ from (7) using the above bracket relations for u and v. We notice that the observables F and G in (7) may be functionals of w, y, p and q, not only of u and v. In this sense the phase space approach for singular lagrangians provides the most general space of observables. The same comment will be valid for the phase space construction using lagrangians L_2 and L_1^M, L_2^M in the following.

We notice that that by construction ϕ_1 and ϕ_2 as well as any functional of them, in all the cases we have considered, are Casimirs of the Poisson structure defined in terms of the Dirac brackets. In fact,

$$\begin{aligned} \{F, \phi_1)\}_{DB} &= 0 \\ \{F, \phi_2)\}_{DB} &= 0 \end{aligned}$$

for any functional F on phase space. This is a general property of the Dirac bracket.

We now consider the lagrangian L_2 and its associated hamiltonian structure. In this case we denote the conjugate momenta to w and y by $\hat{p}$ and $\hat{q}$ respectively. We have

$$\hat{p} = -\frac{1}{2}y_x \quad , \quad \hat{q} = -\frac{1}{2}w_x.$$

The constraints become in this case

$$\widehat{\phi_1} = \hat{p} + \frac{1}{2}y_x = 0 \quad , \quad \widehat{\phi_2} = \hat{q} + \frac{1}{2}w_x = 0.$$

The corresponding Poisson brackets between $\hat{\phi}_i$ and $\hat{\phi}_j, i, j = 1, 2$, are given by

$$\left\{\widehat{\phi_1}(x), \widehat{\phi_1}(x')\right\}_{PB} = 0 \quad , \quad \left\{\widehat{\phi_2}(x), \widehat{\phi_2}(x')\right\}_{PB} = 0,$$
$$\left\{\widehat{\phi_1}(x), \widehat{\phi_2}(x')\right\}_{PB} = \partial_x\delta(x - x').$$

The corresponding construction of the Dirac brackets yields

$$\{u(x), u(\hat{x})\}_{DB} = 0 \quad , \quad \{v(x), v(\hat{x})\}_{DB} = 0,$$
$$\{u(x), v(\hat{x})\}_{DB} = -\partial_x\delta(x - \hat{x}).$$

The hamiltonian $H_2 = \int_{-\infty}^{+\infty} dx\, \mathcal{H}_2$ is given by the hamiltonian density

$$\mathcal{H}_2 = \frac{1}{2}w_x^2 y_x + y_x w_{xxx} + \frac{\lambda}{6}y_x^3.$$

The Hamilton equations

$$u_t(x) = \{u(x), H_2\}_{DB} \quad , \quad v_t(x) = \{v(x), H_2\}_{DB}$$

now using the corresponding Dirac brackets yield the same fields equations (1),(2) for any λ. We have thus constructed two hamiltonian functionals and associated Poisson bracket structures. These two hamiltonian structures arise directly from the basic lagrangians L_1 and L_2. We will now construct two additional hamiltonian structures by considering the Miura transformation.

The hamiltonians H_1, H_2 and H_1^M, H_2^M in the following, were presented in [11].

For the Miura system given by (4) we have the lagrangians L_1^M , L_2^M defined in terms of the lagrangians densities

$$\mathcal{L}_1^M = -\frac{1}{2}\sigma_t\sigma_x - \frac{\lambda}{2}\rho_t\rho_x - \frac{1}{2}\sigma_x\sigma_{xxx} - \frac{\lambda}{2}\rho_x\rho_{xxx} + \frac{1}{72}{\sigma_x}^4 - \frac{\lambda^2}{72}{\rho_x}^4 + \frac{\lambda}{12}\rho_x^2\sigma_x^2 \tag{9}$$

and

$$\mathcal{L}_2^M = -\frac{1}{2}\sigma_t\rho_x - \frac{1}{2}\sigma_x\rho_t - \sigma_{xxx}\rho_x + \frac{1}{18}{\sigma_x}^3\rho_x + \frac{\lambda}{18}{\rho_x}^3\sigma_x \tag{10}$$

where $\mu = \sigma_x, \nu = \rho_x$. As is previous case it is necessary to impose $\lambda \neq 0$ for the lagrangian defined by (9). Using the method of Dirac for singular lagrangians we can deduce the two additional hamiltonian structures corresponding to (9) and (10) for the coupled KdV system given by (1) and (2). They are given by:

$$\begin{aligned}
\{u(x), u(\hat{x})\}_{DB} &= \partial_{xxx}\delta(x,\hat{x}) + \frac{1}{3}u_x\delta(x,\hat{x}) + \frac{2}{3}u\partial_x\delta(x,\hat{x}) \\
\{v(x), v(\hat{x})\}_{DB} &= \frac{1}{\lambda}\partial_{xxx}\delta(x,\hat{x}) + \frac{1}{3\lambda}u_x\delta(x,\hat{x}) + \frac{2}{3\lambda}u\partial_x\delta(x,\hat{x}) \\
\{u(x), v(\hat{x})\}_{DB} &= \frac{1}{3}v_x\delta(x,\hat{x}) + \frac{2}{3}v\partial_x\delta(x,\hat{x})
\end{aligned}$$

and

$$\begin{aligned}
\{u(x), u(\hat{x})\}_{DB} &= \frac{\lambda}{3}v_x\delta(x,\hat{x}) + \frac{2\lambda}{3}v\partial_x\delta(x,\hat{x}) \\
\{v(x), v(\hat{x})\}_{DB} &= \frac{1}{3}v_x\delta(x,\hat{x}) + \frac{2}{3}v\partial_x\delta(x,\hat{x}) \\
\{u(x), v(\hat{x})\}_{DB} &= \partial_{xxx}\delta(x,\hat{x}) + \frac{1}{3}u_x\delta(x,\hat{x}) + \frac{2}{3}u\partial_x\delta(x,\hat{x})
\end{aligned}$$

and the corresponding hamiltonians are given by

$$\begin{aligned}
H_1^M &= \int_{-\infty}^{+\infty} \left(v^2 - u^2\right) dx \\
H_2^M &= \int_{-\infty}^{+\infty} (-uv).
\end{aligned}$$

We recall that it is necessary to assume $\lambda \neq 0$ to consider the hamiltonian structure related with L_1^M.

We have then obtained four self-adjoint operators which define the hamiltonian basic structures of the coupled KdV system given by (1),(2). We mention that this is the starting construction in order to obtain a more general formulation in terms of pencils of Poisson structures, that is, families of Poisson brackets depending on a given parameter. Another interesting related problem is to consider supersymmetric extensions of the given coupled KdV system.

The functions on phase space ϕ_1 and ϕ_2 defining the second class for each hamiltonian structure are Casimirs of the Poisson algebra since they commute with all observables, this is a direct consequence of the Dirac bracket construction.

3. Conclusions

We obtained four basic self-adjoint hamiltonians for the given coupled KdV system and its corresponding hamiltonian stuctures, using the method of Dirac for singular lagrangians. We also commented that this construction is the first step in order to obtain a full hamiltonian structure for the given system in terms of pencils of Poisson structures.

Acknowledgments

A. R. and A. S. are partially supported by Projects Fondecyt 1121103 and Mecesup ANT398 (Chile).

References

[1] Hirota R and Satsuma J 1981 *Phys. Lett.* A **85** 407
[2] Gear J A and Grimshaw R 1984 *Stud. Appl. Math.* **70**, 235 ; Gear J A 1985 *Stud. Appl. Math.* **72**, 95
[3] Lou S Y, Tong B, Hu H C and Tang X Y 2006 *J. Phys. A: Math. Gen.* **39** 513-527
[4] Grimshaw R 2013 *Without Bounds: A Scientific Canvas of Nonlinearity and Complex Dynamics, Understanding Complex Systems* Rubio R G et al (eds.) (Springer-Verlag Berlin Heidelberg)
[5] Bona J L, Ponce G, Saut J C and Tom M N 1992 *Commun. Math. Phys.* **143** 287-313
[6] Gardner C. S. 1971 *J. Math. Phys.***12** 1548-1551
[7] Nutku Y 1984 *J. Math. Phys.* **25** 2007-2008
[8] Kentwell G W 1988 *J. Math. Phys.* **29** 46-48
[9] Restuccia A and Sotomayor A 2013 *J. Math. Phys.* **54** 113510
[10] P. A. M. Dirac 1964, *Lectures on Quantum Mechanics* Belfer Graduate School Monograph Series No.2, Yeshiva University, New York
[11] Zuo D 2014 *Preprint* arXiv: math/ph 1403.0027 v1
[12] Sakovich S Yu 1999 *J. Nonlin. Math. Phys.* **6** N3 255-262
[13] Casati P and Ortenzi G 2006 *J. Geom. and Phys.* **56** 418-449
[14] Cortés Vega L , Restuccia A and Sotomayor A 2014 *Preprint* arXiv: math/ph 1407.7743 v1

Journal of Physics: Conference Series **563** (2014) 012029 doi:10.1088/1742-6596/563/1/012029

Gamma distributed random variables and their semi–quantum operators

Gabriela Popa[1] **and Aurel I. Stan**[2]

[1] Department of Physics, College of Arts and Sciences, 240 Elton Hall, Ohio University, Zanesville, OH 43701, USA

[2] Department of Mathematics, Ohio State University at Marion, 1465 Mount Vernon Avenue, Marion, OH 43302, USA

E-mail: popag@ohio.edu, stan.7@osu.edu

Abstract. We first introduce the joint semi–quantum operators of a finite family of random variables having finite moments of all orders. We then use the semi–quantum operators to characterize the one–dimensional Gamma and Gaussian distributions in terms of their commutators.

1. Introduction

A series of papers, see [1], [2], [9], and [10], have been written by various authors to characterize certain classes of random variables or properties of their probability distributions in terms of their quantum operators. A special attention has been given to the Meixner class because of the Lie Algebra structure enjoyed by the vector space spanned by the identity and quantum operators. Using the commutators of the quantum operators, a recursive way of computing the moments have been used in [8] and [9]. Moreover, one paper has been written about the q–commutators of the quantum operators of some class of random variables, see [4].

In this paper we propose a study of random variables in terms of their semi–quantum operators, which are obtained by splitting the preservation operators in half and adding one half to the creation operators and the other half to the annihilation operators. A particular subclass of the Meixner random variables, namely the Gamma distributed and Gaussian (which can be viewed as limits of the Gamma distributed) random variables, can be characterized in an elegant way using the semi–quantum operators. We observe a Lie Algebra structure for the vector space spanned by the identity and semi–quantum operators of these random variables.

The paper is structured as follows:

In section 2 we present a minimal background on the quantum operators, and use them to define the semi–quantum operators. In section 3 we characterize the one–dimensional Gamma and Gaussian random variables in terms of the commutators of their semi–quantum operators. The proof is done via the Szegö–Jacobi parameters. Finally, in section 4 we show how the moments, of a random variable, can be recovered, in the one–dimensional case, from the commutator of the semi–quantum operators, and use these moments to compute the distribution of the random variable, via the Laplace transform.

2. Background

Let $X_1, X_2, \ldots, X_d$ be d–random variables defined on the same probability space $(\Omega, \mathcal{F}, P)$ and having finite moments of all orders, i.e., for all $i \in \{1, 2, \ldots, d\}$, and all $p > 0$,

$$E\left[|X_i|^p\right] \quad < \quad \infty, \tag{1}$$

where E denotes the expectation.
For each non–negative integer n, we define the space F_n of all polynomial random variables of the form $P(X_1, X_2, \ldots, X_d)$, where P is a polynomial of d variables, with complex coefficients, of total degree less than or equal to n. Since $X_1, X_2, \ldots, X_d$ have finite moments of all orders, each space F_n, for $n \geq 0$, is contained in $L^2(\Omega, \mathcal{F}, P)$. Therefore, we have:

$$\mathbb{C} = F_0 \subset F_1 \subset F_2 \subset \cdots \subset L^2(\Omega, \mathcal{F}, P). \tag{2}$$

We also define the space $F_{-1} := \{0\}$ (the null space). Moreover, since for all $n \geq 0$, F_n is finite dimensional, F_n is a closed subspace of $L^2(\Omega, \mathcal{F}, P)$. Therefore, we can orthogonalize the spaces $\{F_n\}_{n\geq 0}$, and define, for all $n \geq 0$:

$$G_n \quad := \quad F_n \ominus F_{n-1}, \tag{3}$$

that means G_n is the orthogonal complement of F_{n-1} into F_n. We also define $G_{-1} := \{0\}$.
We introduce now the spaces:

$$F \quad = \quad \cup_{n=0}^{\infty} F_n \tag{4}$$

$$\quad = \quad \sum_{n=0}^{\infty} G_n \tag{5}$$

and

$$H \quad = \quad \oplus_{n=0}^{\infty} G_n. \tag{6}$$

The distinction between F and H is the following: F is the algebraic sum of the vector spaces $\{G_n\}_{n\geq 0}$, that means every element in F is a sum of only finitely many non–zero elements of $\cup_{n\geq 0} G_n$, while H is the orthogonal direct sum (a Hilbert space) of the orthogonal closed vector subspaces $\{G_n\}_{n\geq 0}$ of $L^2(\Omega, \mathcal{F}, P)$. We call F the *space of polynomial random variables generated by* $X_1, X_2, \ldots, X_d$, and H the *chaos space generated by* $X_1, X_2, \ldots, X_d$. For each $n \geq 0$, we call G_n *the* n*–th homogenous chaos space generated by* $X_1, X_2, \ldots, X_d$, and each element $P(X_1, X_2, \ldots, X_d)$ a *homogenous polynomial of degree* n.
We are going to view now each random variable $X_1, X_2, \ldots, X_d$ as a multiplication linear operator. For each $i \in \{1, 2, \ldots, d\}$, we consider the multiplication operator: $M_{X_i} : F \to F$ defined by:

$$M_{X_i} P(X_1, X_2, \ldots, X_d) \quad := \quad X_i P(X_1, X_2, \ldots, X_d). \tag{7}$$

To keep the notation simple we denote M_{X_i} by X_i.
The following lemma can be easily checked using the symmetry of the multiplication operators: $X_1, X_2, \ldots, X_d$.

Lemma 2.1 *For all* $i \in \{1, 2, \ldots, d\}$ *and all* $n \geq 0$, *we have:*

$$X_i G_n \quad \perp \quad G_k, \tag{8}$$

for all $k \neq n-1$, n, *and* $n+1$, *where "*$\perp$*" means "orthogonal to".*

Let $i \in \{1, 2, \ldots, d\}$ be fixed. Since, for all $n \geq 0$,

$$\begin{aligned} X_i G_n &\subset H \\ &= G_0 \oplus G_1 \oplus \cdots , \end{aligned}$$

but $X_i G_n \perp G_k$, for all $k \notin \{n-1, n, n+1\}$, we have:

$$X_i G_n \subset G_{n-1} \oplus G_n \oplus G_{n+1}. \tag{9}$$

Thus, for all $f \in G_n$, there exists and are unique $f_{n-1} \in G_{n-1}$, $f_n \in G_n$, and $f_{n+1} \in G_{n+1}$, such that:

$$X_i f := f_{n-1} + f_n + f_{n+1}. \tag{10}$$

We define the following three linear operators:

$$D_n^-(i) : G_n \to G_{n-1}, \tag{11}$$

$$D_n^-(i) f := f_{n-1}, \tag{12}$$

$$D_n^0(i) : G_n \to G_n, \tag{13}$$

$$D_n^0(i) f := f_n, \tag{14}$$

and

$$D_n^+(i) : G_n \to G_{n+1}, \tag{15}$$

$$D_n^+(i) f := f_{n+1}. \tag{16}$$

Since $D_n^-(i)$ maps G_n into G_{n-1}, it associates to a homogenous polynomial of degree n a homogenous polynomial of degree $n-1$ and so, it looses one unit in the degree. $D_n^-(i)$ is called an *annihilation operator*. Similarly, since $D_n^0(i)$ maps G_n into G_n, it preserves the degree of a homogenous polynomial. $D_n^0(i)$ is called a *preservation operator*. Finally, since $D_n^+(i)$ maps G_n into G_{n+1}, it increases the degree of a homogenous polynomial by one unit. $D_n^+(i)$ is called a *creation operator*.
Lemma 2.1 becomes now:

$$X_i|_{G_n} = D_n^-(i) + D_n^0(i) + D_n^+(i). \tag{17}$$

So far, we have defined the annihilation, preservation, and creation operators only on each individual space G_n, $n \geq 0$. Since $F = \sum_{n \geq 0} G_n$, we can extend the definition of these operators, in a linear way, to the whole space F. Namely, let:

$$a^-(i), a^0(i), a^+(i) : F \to F, \tag{18}$$

be defined in the following way: if $f \in F$, there exist and are unique $f_0 \in G_0$, $f_1 \in G_1$, $f_2 \in G_2$, $\cdots$, with only finitely many of them being non–zero, such that:

$$f := f_0 + f_1 + f_2 + \cdots . \tag{19}$$

For such an f, we define:

$$a^-(i)f \;:=\; D_0^-(i)f_0 + D_1^-(i)f_1 + D_2^-(i)f_2 + \cdots, \tag{20}$$

$$a^0(i)f \;:=\; D_0^0(i)f_0 + D_1^0(i)f_1 + D_2^0(i)f_2 + \cdots, \tag{21}$$

and

$$a^+(i)f \;:=\; D_0^+(i)f_0 + D_1^+(i)f_1 + D_2^+(i)f_2 + \cdots. \tag{22}$$

We call $a^-(i)$ an *annihilation operator*, $a^0(i)$ a *preservation operator*, and $a^+(i)$ a *creation operator*.
We call $\{a^-(i)\}_{1\le i\le d}$, $\{a^0(i)\}_{1\le i\le d}$, and $\{a^+(i)\}_{1\le i\le d}$ the *joint quantum operators* generated by X_1, X_2, ..., X_d.
Lemma 2.1 becomes now:

Theorem 2.2 *For all $i \in \{1, 2, \ldots, d\}$, we have:*

$$X_i \;=\; a^-(i) + a^0(i) + a^+(i), \tag{23}$$

where the domain of the linear operators X_i, $a^-(i)$, $a^0(i)$, and $a^+(i)$ is understood to be the space F of al polynomial random variables.

For all $i \in \{1, 2, \ldots, d\}$, $a^0(i)$ is a symmetric operator, while $(a^-(i))^* = a^+(i)$, meaning, that for all f and g in F, we have:

$$\langle a^0(i)f, g\rangle \;:=\; \langle f, a^0(i)g\rangle \tag{24}$$

and

$$\langle a^-(i)f, g\rangle \;:=\; \langle f, a^+(i)g\rangle, \tag{25}$$

where, for all u and v in F, we define:

$$\langle u, v\rangle \;:=\; E\left[u(X_1, X_2, \ldots, X_d)\overline{v(X_1, X_2, \ldots, X_d)}\right]. \tag{26}$$

For every $i \in \{1, 2, \ldots, d\}$, we define the *semi–annihilation* operator:

$$U_i \;:=\; a^-(i) + \frac{1}{2}a^0(i) \tag{27}$$

and the *semi–creation* operator:

$$V_i \;:=\; a^+(i) + \frac{1}{2}a^0(i). \tag{28}$$

We can see that for all $i \in \{1, 2, \ldots, d\}$, we have:

$$V_i \;=\; U_i^* \tag{29}$$

and

$$X_i \;=\; U_i + V_i. \tag{30}$$

We call $\{U_i\}_{1\le i\le d}$ and $\{V_i\}_{1\le i\le d}$ the *joint semi–quantum operators* of X_1, X_2, ..., X_d.

Proposition 2.3 *For all i and j in $\{1, 2, \ldots, d\}$, we have:*

(i) $[U_i, X_j]$ *is a self-adjoint operator.*

(ii)

$$[U_i, X_j] = [U_j, X_i]. \tag{31}$$

(iii)

$$[X_j, V_i] = [X_i, V_j]. \tag{32}$$

Proof.
(i) We have:

$$\begin{aligned} [U_i, X_j]^* &= [X_j, V_i] \\ &= [X_j, X_i - U_i] \\ &= 0 - [X_j, U_i] \\ &= [U_i, X_j]. \end{aligned}$$

(ii) Since, as it was shown in [1] and [2], we have:

$$\left[a^-(i), a^-(j)\right] = 0, \tag{33}$$

$$\left[a^-(i), a^0(j)\right] = \left[a^-(j), a^0(i)\right], \tag{34}$$

$$\left[a^-(i), a^+(j)\right] + \left[a^0(i), a^0(j)\right] + \left[a^+(i), a^-(j)\right] = 0, \tag{35}$$

and

$$\left[a^0(i), a^+(j)\right] = \left[a^0(j), a^+(i)\right], \tag{36}$$

we have:

$$\begin{aligned} & [U_i, X_j] - [U_j, X_i] \\ = & \left[a^-(i) + \frac{1}{2}a^0(i), a^-(j) + a^0(j) + a^+(j)\right] - \left[a^-(j) + \frac{1}{2}a^0(j), a^-(i) + a^0(i) + a^+(i)\right] \\ = & \left[a^-(i), a^-(j)\right] - \left[a^-(j), a^-(i)\right] \\ & + \frac{3}{2}\left(\left[a^-(i), a^0(j)\right] - \left[a^-(j), a^0(i)\right]\right) \\ & + \left[a^-(i), a^+(j)\right] + \left[a^0(i), a^0(j)\right] + \left[a^+(i), a^-(j)\right] \\ & + \frac{1}{2}\left(\left[a^0(i), a^+(j)\right] - \left[a^0(j), a^+(i)\right]\right) \\ = & \ 0. \end{aligned}$$

(iii) follows by taking the adjoint in both sides of (ii). □

In the particular case, when $d = 1$, we have only one random variable X_1, one annihilation operator $a^-(1)$, one preservation operator $a^0(1)$, and one creation operator $a^+(1)$, which we can call simply X, a^-, a^0, and a^+. Since for each $n \geq 0$, F_n is spanned by: $1, X, X^2, \ldots, X^n$, the codimension of F_{n-1} into F_n is at most 1. Thus, for each $n \geq 0$, the dimension of the n–th chaos space G_n is at most 1. More precisely, if the support of the probability distribution μ of the random variable X is not a finite set, then for all $n \geq 0$, the dimension of G_n is equal to 1. On the other hand, if X takes on only finitely many values $x_1, x_2, \ldots, x_k$, with positive probability, then using a Lagrange interpolation polynomial, one can see that for every function $f : \mathbb{R} \to \mathbb{C}$, $f(X)$ is equal almost surely with a polynomial random variable $P(X)$, where P has degree at most $k-1$. Thus, we have:

$$F_{n-1} = F_n = F_{n+1} = \cdots . \tag{37}$$

Therefore, in this case, if "dim" denotes the dimension, he have:

$$\dim(G_n) = \begin{cases} 1 & \text{if} \quad n \leq k-1 \\ 0 & \text{if} \quad n \geq k \end{cases} . \tag{38}$$

For every $n \geq 0$, if $G_n \neq \{0\}$, since $\dim(G_n) = 1$, there exists a unique polynomial f_n of degree n with leading coefficient 1, such that $f_n(X) \in G_n$. We call f_n the *n–th orthogonal polynomial generated by X*. Since $XG_n \subset G_{n-1} \oplus G_n \oplus G_{n+1}$, there exist α_n and ω_n real numbers, such that:

$$Xf_n(X) = f_{n+1}(X) + \alpha_n f_n(X) + \omega_n f_{n-1}. \tag{39}$$

For $n = 0$, we can define $f_{-1} := 0$ and $\omega_0 := 0$.
In the above equation, the coefficient of f_{n+1} is 1, since the coefficient of x^{n+1} must be 1 in both sides of it. The numbers $\{\alpha_n\}_{n\geq 0}$ and $\{\omega_n\}_{n\geq 1}$ are called the *Szegö–Jacobi parameters* of the random variable X.
In this case, for all $n \geq 0$, we have:

$$a^- f_n(X) = \omega_n f_{n-1}(X), \tag{40}$$

$$a^0 f_n(X) = \alpha_n f_n(X), \tag{41}$$

and

$$a^+ f_n(X) = f_{n+1}(X). \tag{42}$$

3. Gamma distributed random variables

In this section we present a characterization, in the one–dimensional case, of the Gamma distributions in terms of the semi–quantum operators. The Gamma distributed random variables, form a sub–class of the Meixner class, see [6].
It is known that a random variable X, having finite moments of all orders, whose Szegö–Jacobi parameters are of the form:

$$\alpha_n = \alpha n + \alpha_0 \tag{43}$$

and

$$\omega_n = \beta n^2 + (t - \beta)n, \tag{44}$$

for all $n \geq 1$, where α, β, t, and α_0 are fixed real numbers, such that: $\alpha^2 = 4\beta$ and $t > 0$, is a shifted *Gamma distributed* random variable with shift parameter $2t/\alpha$ and scaling parameter

$\alpha/2$. That means, up to a shift, X is a continuous distributed random variable, whose density function is:

$$f(x) = \frac{2^{2t/\alpha}}{\alpha^{2t/\alpha}\Gamma(2t/\alpha)} x^{(2t/\alpha)-1} e^{-2x/\alpha} 1_{(0,\infty)},$$

if $\alpha \neq 0$.
If $\alpha = 0$, then X is a Gaussian random variable. Since the case $\alpha = 0$, can be understood as a limit of the case $\alpha \neq 0$, as $\alpha \to 0$, we will call the Gaussian random variables also Gamma random variables, but we should keep in mind that this is true only in a limit sense.
We have the following proposition:

Proposition 3.1 *Let X be a random variable having finite moments of all orders. Then X is a Gamma (or Gaussian) random distributed random variable if and only if the real vector space spanned by its semi-quantum operators U and V, and the identity operator I, equipped with the bracket given by the commutator $[\cdot, \cdot]$ form a Lie Algebra.*

Proof. ($\Rightarrow$) Let us suppose that X is a Gamma distributed random variable. Then, there exists α, α_0, and t real numbers, such that its Szegö–Jacobi parameters are: $\alpha_n = \alpha n + \alpha_0$ and $\omega_n = (1/4)\alpha^2 n^2 + [t - (\alpha^2/4)]n$. If, for all $n \geq 0$, we denote by f_n, the orthogonal monic polynomial of degree n, generated by X, then we have:

$$\begin{aligned}
[U, V] f_n &= \left[a^- + \frac{1}{2}a^0, a^+ + \frac{1}{2}a^0\right] f_n \\
&= \left[a^-, a^+\right] f_n + \frac{1}{2}\left[a^-, a^0\right] f_n + \frac{1}{2}\left[a^0, a^+\right] f_n \\
&= (\omega_{n+1} - \omega_n) f_n + \frac{1}{2}(\alpha_n - \alpha_{n-1})\omega_n f_{n-1} + \frac{1}{2}(\alpha_{n+1} - \alpha_n) f_{n+1} \\
&- \left(\frac{\alpha^2}{2}n + t\right) f_n + \frac{1}{2}\alpha\omega_n f_{n-1} + \frac{1}{2}\alpha f_{n+1} \\
&= \frac{\alpha}{2}\left[\omega_n f_{n-1} + \frac{1}{2}(\alpha n + \alpha_0) f_n\right] + \frac{\alpha}{2}\left[f_{n+1} + \frac{1}{2}(\alpha n + \alpha_0) f_n\right] + \left(t - \frac{\alpha\alpha_0}{2}\right) f_n \\
&= \frac{\alpha}{2}\left(a^- + \frac{1}{2}a^0\right) f_n + \frac{\alpha}{2}\left(a^+ + \frac{1}{2}a^0\right) f_n + \left(t - \frac{\alpha\alpha_0}{2}\right) f_n \\
&= \frac{\alpha}{2}U f_n + \frac{\alpha}{2}V f_n + \left(t - \frac{\alpha\alpha_0}{2}\right) I f_n.
\end{aligned}$$

Thus, we have:

$$[U, V] = \frac{\alpha}{2}U + \frac{\alpha}{2}V + \left(t - \frac{\alpha\alpha_0}{2}\right) I.$$

Therefore, the commutator $[U, V]$ can be expressed as a linear combination of U, V, and I, and so the space $W = \mathbb{R}U + \mathbb{R}V + \mathbb{R}I$ is a Lie Algebra.
($\Leftarrow$) Let us suppose now that the space $W = \mathbb{R}U + \mathbb{R}V + \mathbb{R}I$, equipped with the commutator $[\cdot, \cdot]$ is a Lie Algebra. That means, there exist c, d, and e real numbers, such that:

$$[U, V] = cU + dV + eI. \tag{45}$$

Since we have:

$$\begin{aligned}
[U, X] &= [U, U + V] \\
&= [U, V]
\end{aligned}$$

and we know from Proposition 2.3 that $[U, X]$ is self–adjoint, we conclude that:

$$c = d. \tag{46}$$

Thus, we have:

$$\begin{aligned}[U, V]\, f_n &= \frac{1}{2}\left(\alpha_{n+1} - \alpha_n\right) f_{n+1} + \left(\omega_{n+1} - \omega_n\right) f_n + \frac{1}{2}\left(\alpha_n - \alpha_{n-1}\right) \omega_n f_{n-1} \\ &= cX f_n + eI f_n \\ &= c f_{n+1} + \left(c\alpha_n + e\right) f_n + c\omega_n f_{n-1}.\end{aligned}$$

Since f_{n-1}, f_n, and f_{n+1} are linearly independent, being orthogonal, from the above equality, we conclude that, for all $n \geq 0$, we have:

$$\frac{1}{2}\left(\alpha_{n+1} - \alpha_n\right) = c \tag{47}$$

and

$$\omega_{n+1} - \omega_n = c\alpha_n + e. \tag{48}$$

Thus, for all $n \geq 1$, we have:

$$\begin{aligned}\alpha_n &= \alpha_0 + \sum_{k=0}^{n-1}\left(\alpha_{k+1} - \alpha_k\right) \\ &= \alpha_0 + \sum_{k=0}^{n-1} 2c \\ &= 2cn + \alpha_0.\end{aligned}$$

This implies that since $\omega_0 := 0$, for all $n \geq 1$, we have:

$$\begin{aligned}\omega_n &= \sum_{k=0}^{n-1}\left(\omega_{k+1} - \omega_k\right) \\ &= \sum_{k=0}^{n-1}\left(c\alpha_k + e\right) \\ &= c\sum_{k=0}^{n-1}\left(2ck + \alpha_0\right) + ne \\ &= 2c^2\frac{(n-1)n}{2} + c\alpha_0 n + ne \\ &= c^2 n^2 + \left(c\alpha_0 + e - c^2\right) n.\end{aligned}$$

Thus X is a Gamma distributed random variables with $\alpha := 2c$, $\beta := c^2$, and $t = c\alpha_0 + e$, since $\alpha^2 = 4\beta$. □

Observe, that formula (45) can be written as:

$$[U, X] = cX + eI \tag{49}$$

4. Deriving the Gamma distributions from the semi–quantum operators

Let X be a random variable having finite moments of all orders, such that, if U and V denotes its semi–annihilation and semi–creation operators, the real vector space spanned by U, V, and I, where I is the identity operator, equipped with the commutator bracket $[\cdot,\,\cdot]$, form a Lie Algebra. Since as a multiplication operator $X = U + V$, and $[U,\,X]$ is a symmetric operator, we have seen in the previous section that there exist two real numbers b and α, such that:

$$[U, X] \quad = \quad bX + \alpha I. \tag{50}$$

Let $\phi := 1$ be the constant polynomial equal to 1.
For all non–negative integers n, using Leibniz rule for commutators, we have:

$$\begin{aligned} E\left[X^{n+1}\right] &= \langle (U+V)X^n\phi, \phi\rangle \\ &= \langle UX^n\phi, \phi\rangle + \langle VX^n\phi, \phi\rangle \\ &= \langle [U, X^n]\,\phi, \phi\rangle + \langle X^nU\phi, \phi\rangle + \langle X^n\phi, U\phi\rangle \\ &= \sum_{i=0}^{n-1}\langle X^i\,[U,X]\,X^{n-1-i}\phi, \phi\rangle + \langle X^n[a^- + (1/2)a^0]\phi, \phi\rangle \\ &\quad + \langle X^n\phi, [a^- + (1/2)a^0]\phi\rangle. \end{aligned}$$

Since $a^-\phi = 0$ and $a^0\phi = E[X]\phi$, we obtain:

$$\begin{aligned} E\left[X^{n+1}\right] &= \sum_{i=0}^{n-1}\langle X^i(bX+\alpha I)X^{n-1-i}\phi, \phi\rangle + \frac{1}{2}\langle X^nE[X]\phi, \phi\rangle + \frac{1}{2}\langle X^n\phi, E[X]\phi\rangle \\ &= nbE\left[X^n\right] + n\alpha E\left[X^{n-1}\right] + E[X]E\left[X^n\right]. \end{aligned}$$

If we define:

$$c \quad := \quad E[X], \tag{51}$$

then we obtain the following recursive formula for the moments of X:

$$E[X^{n+1}] \quad = \quad nbE\left[X^n\right] + n\alpha E\left[X^{n-1}\right] + cE\left[X^n\right], \tag{52}$$

for all $n \geq 0$.
Let $K := b + \alpha + c$. We can see from (52), by induction on n, that for all $n \geq 0$, we have:

$$|E\left[X^n\right]| \quad \leq \quad K^n n!. \tag{53}$$

The growth condition (53) allows us to define the Laplace transform of X:

$$\varphi(s) \quad = \quad E\left[\exp(sX)\right], \tag{54}$$

and differentiate its Taylor series term by term, for any s in a neighborhood of 0.
Multiplying both sides of (52) by $s^n/n!$ and summing up from $n = 0$ to ∞, we obtain:

$$\sum_{n=0}^{\infty}\frac{s^n}{n!}E[X^{n+1}] \quad = \quad bs\sum_{n=1}^{\infty}\frac{s^{n-1}}{(n-1)!}E\left[X^n\right] + \alpha s\sum_{n=1}^{\infty}\frac{s^{n-1}}{(n-1)!}E\left[X^{n-1}\right] + c\sum_{n=0}^{\infty}\frac{s^n}{n!}E\left[X^n\right].$$

This is equivalent to:

$$\varphi'(s) \quad = \quad bs\varphi'(s) + \alpha s\varphi(s) + c\varphi(s). \tag{55}$$

Therefore, for all s in a small neighborhood V of 0, such that $1 - bs > 0$ and $\varphi(s) > 0$ (since φ is continuous and $\varphi(0) = 1 > 0$), we have:

$$\frac{\varphi'(s)}{\varphi(s)} = \frac{\alpha s + c}{-bs + 1}. \tag{56}$$

We distinguish now between two cases:
Case 1: If $b \neq 0$, then we can rewrite the previous equation as:

$$\frac{\varphi'(s)}{\varphi(s)} = -\frac{\alpha}{b} + \left(\frac{\alpha}{b} + c\right) \cdot \frac{1}{-bs + 1}. \tag{57}$$

Introducing now the constants: $\beta := -\alpha/b$ and $\gamma := -\beta + c$, we obtain:

$$\frac{d}{ds} \ln(\varphi(s)) = \frac{d}{ds}\left[\beta s - \frac{\gamma}{b}\ln(-bs + 1)\right], \tag{58}$$

for all s in V. Thus, for all $s \in V$, we obtain:

$$E[\exp(sX)] := A\exp(\beta s)(-bs + 1)^{-\gamma/\beta}, \tag{59}$$

where A is a positive real number.
It is not hard to see that the random variables that have a Laplace transform of the form (59) are exactly the shifted and re–scaled Gamma distributed random variables.
Case 2. If $b = 0$, then equation (56) becomes:

$$\frac{\varphi'(s)}{\varphi(s)} = \alpha s + c. \tag{60}$$

It is clear now that:

$$E[\exp(sX)] = B\exp\left(\frac{\alpha^2}{2}s^2 + cs\right), \tag{61}$$

where B is a positive real number.
It is easy to see that the random variables, having a Laplace transform of the form (61), are exactly the Gaussian random variables.

References

[1] Accardi L, Kuo H-H and Stan A I 2004 Characterization of probability measures through the canonically associated interacting Fock spaces *Infin. Dimens. Anal. Quantum Probab. Relat. Top.* **7** No 4 485-505
[2] Accardi L, Kuo H-H and Stan A I 2007 Moments and commutators of probability measures *Infin. Dimens. Anal. Quantum Probab. Relat. Top.* **10** No 4 591-612
[3] Chihara T S 1978 *An Introduction to Orthogonal Polynomials* (New York: Gordon & Breach)
[4] Drożdżewicz K and Matysiak W 2011 Moments and q-commutators of noncommutative random vectors *Infin. Dimens. Anal. Quantum Probab. Relat. Top.* **14** No 4 629-45
[5] Kuo H-H 1996 *White Noise Distribution Theory* (Boca Raton, FL: CRC Press)
[6] Meixner J Orthogonale 1934 Polynomsysteme mit einer besonderen Gestalt der erzeugenden Funktion *J. London Math. Soc.* **9** 6-13
[7] Obata N 1994 *White Noise Calculus and Fock Space* (Berlin Heidelberg: Springer-Verlag)
[8] Popa G and Stan A I 2013 Moments of two noncommutative random variables in terms of their joint quantum operators *J. Phys.: Conf. Ser.* 474 012027 doi:10.1088/1742-6596/474/1/012027
[9] Stan A I and Whitaker J J 2009 A study of probability measures through commutators J. Theoret. Probab. **22** No 1 123-45
[10] Stan A I 2011 Two-dimensional Meixner random vectors of class $\mathcal{M}_L$ J. Theoret. Probab. **24** No 1 39-65
[11] Szegö M 1975 *Orthogonal Polynomials* Coll. Publ. **23** Amer. Math. Soc.
[12] Voiculescu D V, Dykema KJ and Nica A 1982 *Free Random Variables* **1** (Providence, RI: CRM Series, Amer. Math. Soc.)

Belavin–Drinfeld cohomologies and introduction to classification of quantum groups

Alexander Stolin and Iulia Pop

Department of Mathematical Sciences, University of Göteborg, 41296 Göteborg, Sweden

E-mail: astolin@chalmers.se; iulia@chalmers.se

Abstract. In the present article we discuss the classification of quantum groups whose quasi-classical limit is a given simple complex Lie algebra $\mathfrak{g}$. This problem reduces to the classification of all Lie bialgebra structures on $\mathfrak{g}(\mathbb{K})$, where $\mathbb{K} = \mathbb{C}((\hbar))$. The associated classical double is of the form $\mathfrak{g}(\mathbb{K}) \otimes_{\mathbb{K}} A$, where A is one of the following: $\mathbb{K}[\varepsilon]$, where $\varepsilon^2 = 0$, $\mathbb{K} \oplus \mathbb{K}$ or $\mathbb{K}[j]$, where $j^2 = \hbar$. The first case relates to quasi-Frobenius Lie algebras. In the second and third cases we introduce a theory of Belavin–Drinfeld cohomology associated to any non-skewsymmetric r-matrix from the Belavin–Drinfeld list [1]. We prove a one-to-one correspondence between gauge equivalence classes of Lie bialgebra structures on $\mathfrak{g}(\mathbb{K})$ and cohomology classes (in case II) and twisted cohomology classes (in case III) associated to any non-skewsymmetric r-matrix.

Mathematics Subject Classification (2010): 17B37, 17B62.
Keywords: Quantum groups, Lie bialgebras, classical double, r-matrix.

1. Introduction

Let k be a field of characteristic 0. According to [3], a quantized universal enveloping algebra (or a quantum group) is a topologically free topological Hopf algebra H over the formal power series ring $k[[\hbar]]$ such that $H/\hbar H$ is isomorphic to the universal enveloping algebra of a Lie algebra $\mathfrak{g}$ over k.

The quasi-classical limit of a quantum group is a Lie bialgebra. By definition, a Lie bialgebra is a Lie algebra $\mathfrak{g}$ together with a cobracket δ which is compatible with the Lie bracket. Given a quantum group H, with comultiplication Δ, the quasi-classical limit of H is the Lie bialgebra $\mathfrak{g}$ of primitive elements of $H/\hbar H$ and the cobracket is the restriction of the map $(\Delta - \Delta^{21})/\hbar(\mathrm{mod}\hbar)$ to $\mathfrak{g}$.

The operation of taking the semiclassical limit is a functor $SC : QUE \to LBA$ between categories of quantum groups and Lie bialgebras over k. The quantization problem raised by Drinfeld aims at finding a quantization functor, i.e. a functor $Q : LBA \to QUE$ such that $SC \circ Q$ is isomorphic to the identity. Moreover, a quantization functor is required to be universal, in the sense of props.

The existence of universal quantization functors was proved by Etingof and Kazhdan [5, 6]. They used Drinfeld's theory of associators to construct quantization functors for any field k of characteristic zero. Drinfeld introduced the notion of associator in relation to the theory of quasi-triangular quasi-Hopf algebras and showed that associators exist over any field k of characteristic zero. Etingof and Kazhdan proved that for any fixed associator over k one can construct a universal quantization functor. More precisely, let $(\mathfrak{g}, \delta)$ be a Lie bialgebra over k. Then one can associate a Lie bialgebra $\mathfrak{g}_\hbar$ over $k[[\hbar]]$ defined as $(\mathfrak{g} \otimes_k k[[\hbar]], \hbar\delta)$. According to Theorem 2.1 of [6] there exists an equivalence $\widehat{Q}$ between the category $LBA_0(k[[\hbar]])$ of topologically free over $k[[\hbar]]$ Lie bialgebras with $\delta = 0(\text{mod}\hbar)$ and the category $HA_0(k[[\hbar]])$ of topologically free Hopf algebras cocommutative modulo $\hbar$. Moreover, for any $(\mathfrak{g}, \delta)$ over k, one has the following: $\widehat{Q}(\mathfrak{g}_\hbar) = U_\hbar(\mathfrak{g})$.

The aim of the present article is the classification of quantum groups whose quasi-classical limit is a given simple complex Lie algebra $\mathfrak{g}$. Due to the equivalence between $HA_0(\mathbb{C}[[\hbar]])$ and $LBA_0(\mathbb{C}[[\hbar]])$, this problem is equivalent to classification of Lie bialgebra structures on $\mathfrak{g}\otimes_{\mathbb{C}}\mathbb{C}[[\hbar]]$. For simplicity, denote $\mathbb{O} := \mathbb{C}[[\hbar]]$, $\mathbb{K} := \mathbb{C}((\hbar))$, $\mathfrak{g}(\mathbb{O}) := \mathfrak{g} \otimes_{\mathbb{C}} \mathbb{O}$ and $\mathfrak{g}(\mathbb{K}) := \mathfrak{g} \otimes_{\mathbb{C}} \mathbb{K}$.

On the other hand, in order to classify cobrackets on $\mathfrak{g}(\mathbb{O})$ it is enough to classify cobrackets on $\mathfrak{g}(\mathbb{K})$. Indeed, if δ is a Lie bialgebra structure on $\mathfrak{g}(\mathbb{O})$, then it can be naturally extended to $\mathfrak{g}(\mathbb{K})$. Conversely, given a Lie bialgebra structure $\bar{\delta}$ on $\mathfrak{g}(\mathbb{K})$, then by multiplying $\bar{\delta}$ by an appropriate power of $\hbar$, the restriction of $\bar{\delta}$ to $\mathfrak{g}(\mathbb{O})$ is a Lie bialgebra structure on $\mathfrak{g}(\mathbb{O})$.

Now, from the general theory of Lie bialgebras it is known that for each Lie bialgebra structure δ on a fixed Lie algebra L one can construct the corresponding classical double $D(L, \delta)$ which is the vector space $L \oplus L^*$ together with a bracket which is induced by the bracket and cobracket of L, and a non-degenerate invariant bilinear form, see [4]. We consider $L = \mathfrak{g}(\mathbb{K})$ and prove Prop. 2.1 which states that there exists an associative, unital, commutative algebra A, of dimension 2 over $\mathbb{K}$, such that $D(\mathfrak{g}(\mathbb{K}), \delta) \cong \mathfrak{g}(\mathbb{K}) \otimes_{\mathbb{K}} A$. In Prop. 2.3 we show that there are three possibilities for A: $A = \mathbb{K}[\varepsilon]$, where $\varepsilon^2 = 0$, $A = \mathbb{K} \oplus \mathbb{K}$ or $A = \mathbb{K}[j]$, where $j^2 = \hbar$.

Due to the correspondence Lie bialgebras–Manin triples, to any Lie bialgebra structure δ on L one can associate a certain Lagrangian subalgebra W of $D(L, \delta)$ which is complementary to L and conversely, any such W produces a Lie cobracket on L. The main problem is to obtain a classification of all such subalgebras W for the three choices of A as above. We investigate each choice of A separately.

For $A = \mathbb{K}[\varepsilon]$, where $\varepsilon^2 = 0$, it turns out that the classification problem is related to that of quasi-Frobenius Lie subalgebras over $\mathbb{K}$.

In case $A = \mathbb{K} \oplus \mathbb{K}$, we introduce Belavin–Drinfeld cohomologies. Namely, for any non-skewsymmetric constant r-matrix r_{BD} from the Belavin–Drinfeld list [1], we associate a cohomology set $H^1_{BD}(r_{BD})$. We prove that there exists a one-to-one correspondence between any Belavin–Drinfeld cohomology and gauge equivalence classes of Lie bialgebra structures on $\mathfrak{g}(\mathbb{K})$. We first consider $\mathfrak{g} = sl(n)$ and show that all cohomologies are trivial. We then discuss the case of orthogonal algebras $\mathfrak{g} = o(n)$, where it turns out that the cohomology associated to the Drinfeld–Jimbo r-matrix is also trivial. We illustrate an example where the cohomology corresponding to another non-skewsymmetric constant r-matrix for $o(2n)$ is non-trivial.

We finally treat the case $A = \mathbb{K}[j]$, where $j^2 = \hbar$. We restrict our analysis to $\mathfrak{g} = sl(n)$ and we show that in this case a cohomology theory can be introduced too. Our result states that there exists a one-to-one correspondence between Belavin–Drinfeld twisted cohomology and gauge equivalence classes of Lie bialgebra structures on $\mathfrak{g}(\mathbb{K})$. We prove that the twisted cohomology corresponding to the Drinfeld–Jimbo r-matrix is trivial.

In the last section of the article we formulate a conjecture stating that the Belavin–Drinfeld cohomology associated to the Drinfeld–Jimbo r-matrix is trivial for any simple complex Lie algebra $\mathfrak{g}$. We also define the quantum Belavin–Drinfeld cohomology and formulate a second conjecture about the existence of a natural correspondence between classical and quantum cohomologies.

2. Lie bialgebra structures on $\mathfrak{g}(\mathbb{K})$

Let $\mathfrak{g}$ be a simple complex finite-dimensional Lie algebra. Consider the Lie algebras $\mathfrak{g}(\mathbb{O}) = \mathfrak{g}\otimes_{\mathbb{C}}\mathbb{O}$ and $\mathfrak{g}(\mathbb{K}) = \mathfrak{g} \otimes_{\mathbb{C}} \mathbb{K}$.

We have seen that the classification of quantum groups with quasi-classical limit $\mathfrak{g}$ is equivalent to the classification of all Lie bialgebra structures on $\mathfrak{g}(\mathbb{O})$. Moreover, as explained in the introduction, in order to classify Lie bialgebra structures on $\mathfrak{g}(\mathbb{O})$, it is enough to classify them on $\mathfrak{g}(\mathbb{K})$.

Let us assume that $\bar{\delta}$ is a Lie bialgebra structure on $\mathfrak{g}(\mathbb{K})$. This cobracket endows the dual of $\mathfrak{g}(\mathbb{K})$ with a Lie bracket. Then one can construct the corresponding classical double $D(\mathfrak{g}(\mathbb{K}), \bar{\delta})$. As a vector space, $D(\mathfrak{g}(\mathbb{K}), \bar{\delta}) = \mathfrak{g}(\mathbb{K}) \oplus \mathfrak{g}(\mathbb{K})^*$. As a Lie algebra, it is endowed with a bracket which is induced by the bracket and cobracket of $\mathfrak{g}(\mathbb{K})$. Moreover the canonical symmetric non-degenerate bilinear form on this space is invariant.

Similarly to Lemma 2.1 from [8], one can prove that $D(\mathfrak{g}(\mathbb{K}), \bar{\delta})$ is a direct sum of regular adjoint $\mathfrak{g}$-modules. Combining this result with Prop. 2.2 from [2], one obtains the following

Proposition 2.1. *There exists an associative, unital, commutative algebra A, of dimension 2 over $\mathbb{K}$, such that $D(\mathfrak{g}(\mathbb{K}), \bar{\delta}) \cong \mathfrak{g}(\mathbb{K}) \otimes_{\mathbb{K}} A$.*

Remark 2.2. The symmetric invariant non-degenerate bilinear form Q on $\mathfrak{g}(\mathbb{K}) \otimes_{\mathbb{K}} A$ is given in the following way. For arbitrary elements $f_1, f_2 \in \mathfrak{g}(\mathbb{K})$ and $a, b \in A$ we have $Q(f_1 \otimes a, f_2 \otimes b) = K(f_1, f_2) \cdot t(ab)$, where K denotes the Killing form on $\mathfrak{g}(\mathbb{K})$ and $t : A \longrightarrow \mathbb{K}$ is a trace function.

Let us investigate the algebra A. Since A is unital and of dimension 2 over $\mathbb{K}$, one can choose a basis $\{e, 1\}$, where 1 denotes the unit. Moreover, there exist p and q in $\mathbb{K}$ such that $e^2 + pe + q = 0$. Let $\Delta = p^2 - 4q \in \mathbb{K}$. We distinguish the following cases:

(i) Assume $\Delta = 0$. Let $\varepsilon := e + \frac{p}{2}$. Then $\varepsilon^2 = 0$ and $A = \mathbb{K}\varepsilon \oplus \mathbb{K} = \mathbb{K}[\varepsilon]$.

(ii) Assume $\Delta \neq 0$ and has even order as an element of $\mathbb{K}$. This implies that $\Delta = \hbar^{2m}(a_0 + a_1\hbar + a_2\hbar^2 + ...)$, where m is an integer, a_i are complex coefficients and $a_0 \neq 0$.

One can easily check that the equation $x^2 = a_0 + a_1\hbar + a_2\hbar^2 + ...$ has two solutions $\pm x = x_0 + x_1\hbar + x_2\hbar^2 + ...$ in $\mathbb{O}$.

Then $e = -\frac{p}{2} \pm \frac{\hbar^m x}{2}$, which implies that $e \in \mathbb{K}$ and $A = \mathbb{K} \oplus \mathbb{K}$.

(iii) Assume $\Delta \neq 0$ and has odd order as an element of $\mathbb{K}$. We have $\Delta = \hbar^{2m+1}(a_0 + a_1\hbar + a_2\hbar^2 + ...)$, where m is an integer, a_i are complex coefficients and $a_0 \neq 0$.

Again the equation $x^2 = a_0 + a_1\hbar + a_2\hbar^2 + ...$ has two solutions $\pm x = x_0 + x_1\hbar + x_2\hbar^2 + ...$ in $\mathbb{O}$. Since $a_0 \neq 0$, we have $x_0 \neq 0$ and thus x is invertible in $\mathbb{O}$.

Let $j = \hbar^{-m}(2e + p)x^{-1}$. Then $e^2 + pe + q = 0$ is equivalent to $j^2 = \hbar$. Since $A = \mathbb{K}e \oplus \mathbb{K}$ and $2e = \hbar^m xj - p$, $A = \mathbb{K}j \oplus \mathbb{K}$.

We have thus obtained the following result.

Proposition 2.3. *Let $\bar{\delta}$ be an arbitrary Lie bialgebra structure on $\mathfrak{g}(\mathbb{K})$. Then $D(\mathfrak{g}(\mathbb{K}), \bar{\delta}) \cong \mathfrak{g}(\mathbb{K}) \otimes_{\mathbb{K}} A$, where $A = \mathbb{K}[\varepsilon]$ and $\varepsilon^2 = 0$, $A = \mathbb{K} \oplus \mathbb{K}$ or $A = \mathbb{K}[j]$ and $j^2 = \hbar$.*

On the other hand, it is well-known, see for instance [3], that there is a one-to-one correspondence between Lie bialgebra structures on a Lie algebra L and Manin triples $(D(L), L, W)$. For $L = \mathfrak{g}(\mathbb{K})$, this fact implies the following

Proposition 2.4. *There exists a one-to-one correspondence between Lie bialgebra structures on $\mathfrak{g}(\mathbb{K})$ for which the classical double is $\mathfrak{g}(\mathbb{K}) \otimes_{\mathbb{K}} A$ and Lagrangian subalgebras W of $\mathfrak{g}(\mathbb{K}) \otimes_{\mathbb{K}} A$, with respect to the non-degenerate bilinear form Q, and transversal to $\mathfrak{g}(\mathbb{K})$.*

Corollary 2.5. *(i) There exists a one-to-one correspondence between Lie bialgebra structures on $\mathfrak{g}(\mathbb{K})$ for which the classical double is $\mathfrak{g}(\mathbb{K}[\varepsilon])$, $\varepsilon^2 = 0$, and Lagrangian subalgebras W of $\mathfrak{g}(\mathbb{K}[\varepsilon])$, and transversal to $\mathfrak{g}(\mathbb{K})$.*

(ii) There exists a one-to-one correspondence between Lie bialgebra structures on $\mathfrak{g}(\mathbb{K})$ for which the classical double is $\mathfrak{g}(\mathbb{K}) \oplus \mathfrak{g}(\mathbb{K})$ and Lagrangian subalgebras W of $\mathfrak{g}(\mathbb{K}) \oplus \mathfrak{g}(\mathbb{K})$, and transversal to $\mathfrak{g}(\mathbb{K})$, embedded diagonally into $\mathfrak{g}(\mathbb{K}) \oplus \mathfrak{g}(\mathbb{K})$.

(iii) There exists a one-to-one correspondence between Lie bialgebra structures on $\mathfrak{g}(\mathbb{K})$ for which the classical double is $\mathfrak{g}(\mathbb{K}[j])$, where $j^2 = \hbar$, and Lagrangian subalgebras W of $\mathfrak{g}(\mathbb{K}[j])$, and transversal to $\mathfrak{g}(\mathbb{K})$.

3. Lie bialgebra structures in Case I

Here we study the Lie bialgebra structures δ on $\mathfrak{g}(\mathbb{K})$ for which the corresponding Drinfeld double is isomorphic to $\mathfrak{g}(\mathbb{K}[\varepsilon])$, $\varepsilon^2 = 0$. The problem is to find all subalgebras W of $\mathfrak{g}(\mathbb{K}[\varepsilon])$ satisfying the following conditions:

(i) $W \oplus \mathfrak{g}(\mathbb{K}) = \mathfrak{g}(\mathbb{K}[\varepsilon])$.

(ii) $W = W^{\perp}$, with respect to the following non-degenerate symmetric bilinear form:

$$Q(f_1(\hbar) + \varepsilon f_2(\hbar), g_1(\hbar) + \varepsilon g_2(\hbar)) = K(f_1, g_2) + K(f_2, g_1).$$

Proposition 3.1. *Any subalgebra W of $\mathfrak{g}(\mathbb{K}[\varepsilon])$ satisfying conditions (i) and (ii) from above is uniquely defined by a subalgebra L of $\mathfrak{g}(\mathbb{K})$ together with a non-degenerate 2-cocycle B on L.*

Proof. The proof is similar to that of Th. 3.2 and Cor. 3.3 from [10]. □

Remark 3.2. We recall that a Lie algebra is called quasi-Frobenius if there exists a non-degenerate 2-cocycle on it. It is called Frobenius if the corresponding 2-cocycle is a coboundary. Thus we see that the classification problem for the Lagrangian subalgebras we are interested in contains the classification of Frobenius subalgebras of $\mathfrak{g}(\mathbb{K})$. This question is quite complicated, as it is known from studying Frobenius subalgebras of $\mathfrak{g}$. However, for $\mathfrak{g} = sl(2)$ there is only one Frobenius subalgebra, the standard parabolic one.

4. Lie bialgebra structures in Case II and Belavin-Drinfeld cohomologies

Our task is to classify Lie bialgebra structures on $\mathfrak{g}(\mathbb{K})$ for which the associated classical double is isomorphic to $\mathfrak{g}(\mathbb{K}) \oplus \mathfrak{g}(\mathbb{K})$.

Lemma 4.1. *Any Lie bialgebra structure δ on $\mathfrak{g}(\mathbb{K})$ for which the associated classical double is isomorphic to $\mathfrak{g}(\mathbb{K}) \oplus \mathfrak{g}(\mathbb{K})$ is a coboundary $\delta = dr$ given by an r-matrix satisfying $r + r^{21} = f\Omega$, where $f \in \mathbb{K}^*$ and $\mathrm{CYB}(r) = 0$.*

We may suppose that $f = 1$. Naturally, we want to classify all such r up to $\mathrm{Ad}(G(\mathbb{K}))$-equivalence. Here $\mathrm{Ad}(G(\mathbb{K}))$ is a group, which acts naturally on $\mathfrak{g}(\mathbb{K})$.

Let $\overline{\mathbb{K}}$ denote the algebraic closure of $\mathbb{K}$. Any Lie bialgebra structure δ over $\mathbb{K}$ can be extended to a Lie bialgebra structure $\overline{\delta}$ over $\overline{\mathbb{K}}$.

According to [1], Lie bialgebra structures on a simple Lie algebra over an algebraically closed field are coboundaries given by non-skewsymmetric r-matrices. These r-matrices have been classified up to $\mathrm{Ad}(G)$-equivalence and they are given in terms of admissible triples.

Let us fix a Cartan subalgebra $\mathfrak{h}$ of $\mathfrak{g}$ and the associated root system. We choose a system of generators e_α, $e_{-\alpha}$, h_α such that $K(e_\alpha, e_{-\alpha}) = 1$, for any positive root α. Denote by Ω_0 the Cartan part of Ω. Suppose also that $H \subset \mathrm{Ad}(G)$ is a Cartan subgoup with Lie algebra $\mathfrak{h}$.

Let us recall from [1, 3] that any non-skewsymmetric r-matrix depends on certain discrete and continuous parameters. The discrete one is an admissible triple $(\Gamma_1, \Gamma_2, \tau)$, i.e. an isometry $\tau : \Gamma_1 \longrightarrow \Gamma_2$ where $\Gamma_1, \Gamma_2 \subset \Gamma$ such that for any $\alpha \in \Gamma_1$ there exists $k \in \mathbb{N}$ satisfying $\tau^k(\alpha) \notin \Gamma_1$. The continuous parameter is a tensor $r_0 \in \mathfrak{h} \otimes \mathfrak{h}$ satisfying $r_0 + r_0^{21} = \Omega_0$ and $(\tau(\alpha) \otimes 1 + 1 \otimes \alpha)(r_0) = 0$ for any $\alpha \in \Gamma_1$. Then the associated r-matrix is given by the following formula

$$r_{BD} = r_0 + \sum_{\alpha>0} e_\alpha \otimes e_{-\alpha} + \sum_{\alpha \in (Span\Gamma_1)^+} \sum_{k \in \mathbb{N}} e_\alpha \wedge e_{-\tau^k(\alpha)}.$$

Now, let us consider an r-matrix corresponding to a Lie bialgebra on $\mathfrak{g}(\mathbb{K})$. Up to $\mathrm{Ad}(G(\overline{\mathbb{K}}))$-equivalence, we have the Belavin–Drinfeld classification. We may assume that our r-matrix is of the form $r_X = (\mathrm{Ad}_X \otimes \mathrm{Ad}_X)(r_{BD})$, where $X \in G(\overline{\mathbb{K}})$ and r_{BD} satisfies the system $r + r^{21} = \Omega$ and $\mathrm{CYB}(r) = 0$. The corresponding bialgebra structure is $\delta(a) = [r_X, a \otimes 1 + 1 \otimes a]$ for any $a \in \mathfrak{g}(\mathbb{K})$.

Let us take an arbitrary $\sigma \in Gal(\overline{\mathbb{K}}/\mathbb{K})$. Then we have $(\sigma \otimes \sigma)(\delta(a)) = [\sigma(r_X), a \otimes 1 + 1 \otimes a]$ and $(\sigma \otimes \sigma)(\delta(a)) = \delta(a)$, which imply that $\sigma(r_X) = r_X + \lambda\Omega$, for some $\lambda \in \overline{\mathbb{K}}$. Let us show that $\lambda = 0$. Indeed, $\Omega = \sigma(\Omega) = \sigma(r_X) + \sigma(r_X^{21}) = r_X + r_X^{21} + 2\lambda\Omega$. Thus $\lambda = 0$ and $\sigma(r_X) = r_X$. Consequently, we get $(\mathrm{Ad}_{X^{-1}\sigma(X)} \otimes \mathrm{Ad}_{X^{-1}\sigma(X)})(\sigma(r_{BD})) = r_{BD}$. We recall the following

Definition 4.2. Let r be an r-matrix. The *centralizer* $C(r)$ of r is the set of all $X \in G(\overline{\mathbb{K}})$ satisfying $(\mathrm{Ad}_X \otimes \mathrm{Ad}_X)(r) = r$.

Theorem 4.3. *Let r_{BD} be an r-matrix from the Belavin–Drinfeld list for $\mathfrak{g}(\overline{\mathbb{K}})$. Suppose that $(\mathrm{Ad}_{X^{-1}\sigma(X)} \otimes \mathrm{Ad}_{X^{-1}\sigma(X)})(\sigma(r_{BD})) = r_{BD}$. Then $\sigma(r_{BD}) = r_{BD}$ and $X^{-1}\sigma(X) \in C(r_{BD})$.*

Proof. Consider $r = r_{BD}$ which corresponds to an admissible triple $(\Gamma_1, \Gamma_2, \tau)$ and $r_0 \in \mathfrak{h} \otimes \mathfrak{h}$. Denote $Y := X^{-1}\sigma(X)$ and $s := r - r_0$. Then $(\mathrm{Ad}(Y) \otimes \mathrm{Ad}(Y))(s + \sigma(r_0)) = s + r_0$.

Following [7], p. 43–47, let $F(r) : \mathfrak{g} \longrightarrow \mathfrak{g}$ be the operator defined by $F(r)(x) = r'K(r'', x)$, if $r = \sum r' \otimes r''$ and K is the Killing form on $\mathfrak{g}$. Let

$$\mathfrak{g}_r^\lambda = \bigcup_{n>0} \mathrm{Ker}(F(r) - \lambda)^n.$$

Then

$$\mathfrak{g} = \mathfrak{g}_r^0 \oplus \mathfrak{g}_r' \oplus \mathfrak{g}_r^1,$$

where

$$\mathfrak{g}_r' = \bigoplus_{\lambda \neq 0,1} \mathfrak{g}_r^\lambda.$$

In our case, $\mathfrak{n}_- \subseteq \mathfrak{g}^0_{s+r_0} \subseteq \mathfrak{b}_-$, $\mathfrak{n}_+ \subseteq \mathfrak{g}^1_{s+r_0} \subseteq \mathfrak{b}_+$, $\mathfrak{g}'_{s+r_0} \subseteq \mathfrak{h}$, $\mathfrak{g}^0_{s+r_0} + \mathfrak{g}'_{s+r_0} = \mathfrak{b}_-$, $\mathfrak{g}^1_{s+r_0} + \mathfrak{g}'_{s+r_0} = \mathfrak{b}_+$, and similarly for $s + \sigma(r_0)$. It can be easily checked that

$$F(\mathrm{Ad}(Y) \otimes \mathrm{Ad}(Y))(r) = \mathrm{Ad}(Y) \circ F(r) \circ \mathrm{Ad}(Y^{-1}).$$

Hence $\mathrm{Ad}(Y)(\mathfrak{g}^{0,1}_{s+\sigma(r_0)}) = \mathfrak{g}^{0,1}_{s+r_0}$ and $\mathrm{Ad}(Y)(\mathfrak{g}'_{s+\sigma(r_0)}) = \mathfrak{g}'_{s+r_0}$. Therefore $\mathrm{Ad}(Y)(\mathfrak{b}_\pm) = \mathfrak{b}_\pm$ and $\mathrm{Ad}(Y) \in \mathrm{Ad}(H)(\overline{\mathbb{K}})$. Let us analyse the equality

$$\mathrm{Ad}(Y) \otimes \mathrm{Ad}(Y)(s + \sigma(r_0)) = s + r_0.$$

It follows that

$$\mathrm{Ad}(Y) \otimes \mathrm{Ad}(Y)(s) + \sigma(r_0) = s + r_0.$$

Taking into account that $r_0, \sigma(r_0) \in H^{\otimes 2}$ and

$$(\mathrm{Ad}(Y) \otimes \mathrm{Ad}(Y))(s) = \sum_{\alpha>0} e_\alpha \otimes e_{-\alpha} + \sum_{\beta \in (\mathbb{Z}\Gamma_1)^+} \sum_{n>0} k_{\beta,n} e_\beta \wedge e_{-\tau^n(\beta)},$$

for some integers $k_{\beta,n}$, we deduce that $\sigma(r_0) = r_0$. Thus $\sigma(r) = r$ and $\mathrm{Ad}(Y) \in C(r)$. □

In conclusion, $r_X = (\mathrm{Ad}_X \otimes \mathrm{Ad}_X)(r_{BD})$ induces a Lie bialgebra structure on $\mathfrak{g}(\mathbb{K})$ if and only if $X \in G(\overline{\mathbb{K}})$ satisfies the condition $X^{-1}\sigma(X) \in C(r_{BD})$, for any $\sigma \in Gal(\overline{\mathbb{K}}/\mathbb{K})$.

Definition 4.4. Let r_{BD} be a non-skewsymmetric r-matrix from the Belavin–Drinfeld list and $C(r_{BD})$ its centralizer. We say that $X \in G(\overline{\mathbb{K}})$ is a *Belavin–Drinfeld cocycle* associated to r_{BD} if $X^{-1}\sigma(X) \in C(r_{BD})$, for any $\sigma \in Gal(\overline{\mathbb{K}}/\mathbb{K})$.

We denote the set of Belavin–Drinfeld cocycles associated to r_{BD} by $Z(r_{BD})$. This set is non-empty, always contains the identity.

Definition 4.5. Two cocycles X_1 and X_2 in $Z(r_{BD})$ are called *equivalent* if there exists $Q \in G(\mathbb{K})$ and $C \in C(r_{BD})$ such that $X_1 = QX_2C$.

Definition 4.6. Let $H^1_{BD}(r_{BD})$ denote the set of equivalence classes of cocycles from $Z(r_{BD})$. We call this set the *Belavin–Drinfeld cohomology* associated to the r-matrix r_{BD}. The Belavin–Drinfeld cohomology is said to be *trivial* if all cocycles are equivalent to the identity, and *non-trivial* otherwise.

We make the following remarks:

Remark 4.7. Assume that $X \in Z(r_{BD})$. Then for any $\sigma \in Gal(\overline{\mathbb{K}}/\mathbb{K})$, $\sigma(X) = XC$, for some $C \in C(r_{BD})$. We get $(\mathrm{Ad}_{\sigma(X)} \otimes \mathrm{Ad}_{\sigma(X)})(r_{BD}) = (\mathrm{Ad}_X \otimes \mathrm{Ad}_X)(r_{BD})$. Consequently, $(\mathrm{Ad}_X \otimes \mathrm{Ad}_X)(r_{BD})$ induces a Lie bialgebra structure on $\mathfrak{g}(\mathbb{K})$.

Remark 4.8. Assume that X_1 and X_2 in $Z(r_{BD})$ are equivalent. Then $X_1 = QX_2C$, for some $Q \in G(\mathbb{K})$ and $C \in C(r_{BD})$. This implies that $(\mathrm{Ad}_{X_1} \otimes \mathrm{Ad}_{X_1})(r_{BD}) = (\mathrm{Ad}_{QX_2} \otimes \mathrm{Ad}_{QX_2})(r_{BD})$. In other words, the r-matrices $(\mathrm{Ad}_{X_1}\otimes\mathrm{Ad}_{X_1})(r_{BD})$ and $(\mathrm{Ad}_{X_2}\otimes\mathrm{Ad}_{X_2})(r_{BD})$ are gauge equivalent over $\mathbb{K}$ via an element $Q \in G(\mathbb{K})$.

The above remarks imply the following result.

Proposition 4.9. *Let r_{BD} be a non-skewsymmetric r-matrix over $\overline{\mathbb{K}}$. There exists a one-to-one correspondence between $H^1_{BD}(r_{BD})$ and gauge equivalence classes of Lie bialgebra structures on $\mathfrak{g}(\mathbb{K})$ with classical double $\mathfrak{g}(\mathbb{K}) \oplus \mathfrak{g}(\mathbb{K})$ and $\overline{\mathbb{K}}$-isomorphic to $\delta(r_{BD})$.*

5. Belavin-Drinfeld cohomologies for $sl(n)$

Our next goal is to compute $H^1_{BD}(r_{BD})$ for $\mathfrak{g} = sl(n)$. We will first analyse the cohomology associated to the Drinfeld–Jimbo r-matrix r_{DJ}.

Lemma 5.1. *Let $X \in GL(n, \overline{\mathbb{K}})$. Assume that for any $\sigma \in Gal(\overline{\mathbb{K}}/\mathbb{K})$, $X^{-1}\sigma(X) \in \mathrm{diag}(n, \overline{\mathbb{K}})$. Then there exist $Q \in GL(n, \mathbb{K})$ and $D \in \mathrm{diag}(n, \overline{\mathbb{K}})$ such that $X = QD$.*

Proof. Let $\sigma \in Gal(\overline{\mathbb{K}}/\mathbb{K})$ and $\sigma(X) = XD_\sigma$, where $D_\sigma = \mathrm{diag}(d_1, ...d_n)$. Here d_i depend on σ. Then $\sigma(x_{ij}) = x_{ij}d_j$, for any i, j.

On the other hand, in each column of X there exists a nonzero element. Let us denote these elements by $x_{i_1 1}$, ..., $x_{i_n n}$. For $j = 1$, $\sigma(x_{i1}) = x_{i1}d_1$ and $\sigma(x_{i_1 1}) = x_{i_1 1}d_1$. These relations imply that $\sigma(x_{i1}/x_{i_1 1}) = x_{i1}/x_{i_1 1}$ for any $\sigma \in Gal(\overline{\mathbb{K}}/\mathbb{K})$ and thus $x_{i1}/x_{i_1 1} \in \mathbb{K}$, for any i.

Similarly, $x_{i2}/x_{i_2 2} \in \mathbb{K}$,..., $x_{in}/x_{i_n n} \in \mathbb{K}$, for any i. Let $Q = (k_{ij})$ be the matrix whose elements are $k_{ij} = x_{ij}/x_{i_j j}$, for any i and j.

Thus $X = QD$, where $Q \in GL(n, \mathbb{K})$ and $D = \mathrm{diag}(x_{i_1 1}, ..., x_{i_n n})$. □

Proposition 5.2. *For $\mathfrak{g} = sl(n)$, the Belavin-Drinfeld cohomology $H^1_{BD}(r_{DJ})$ associated to r_{DJ} and to the group $GL(n)$ is trivial.*

Proof. It easily follows from the proof of Theorem 4.3 that the center of r_{DJ} is $C(r_{DJ}, GL(n)) = \mathrm{diag}(n, \overline{\mathbb{K}})$. Let us show that any cocycle is equivalent to the identity. Indeed, let $X = (x_{ij})$ be a cocycle from $Z(r_{DJ})$, i.e. $X^{-1}\sigma(X) \in C(r_{DJ})$, for any $\sigma \in Gal(\overline{\mathbb{K}}/\mathbb{K})$.

It follows that $X^{-1}\sigma(X) \in \mathrm{diag}(n, \overline{\mathbb{K}})$. According to Lemma 5.1, there exists $Q \in GL(n, \mathbb{K})$ and $D \in \mathrm{diag}(n, \overline{\mathbb{K}})$ such that $X = QD$. This proves that X is equivalent to the identity. □

It turns out that the above result is true not only for r_{DJ}. Given an arbitrary r-matrix r_{BD} from the Belavin–Drinfeld list, the corresponding cohomology is also trivial. First we will take a closer look to the centralizer $C(r_{BD})$ of an r-matrix r_{BD}. Due to Theorem 4.3, the following result holds.

Lemma 5.3. *Let r_{BD} be an arbitrary r-matrix from the Belavin–Drinfeld list. Then*

$$C(r_{BD}) \subseteq \mathrm{diag}(n, \overline{\mathbb{K}}).$$

Remark 5.4. The above result is not true for $o(2n)$ if we consider $O(2n)$ as the gauge group. However, one can easily show that in this case $C(r_{DJ}, O(2n))$ contains all diagonal matrices of $O(2n)$ (we will describe our presentation of $O(n)$ in Section 6).

For $sl(n)$ we are now able to give the exact description of $C(r_{BD})$.

Lemma 5.5. *$C(r_{BD})$ consists of all diagonal matrices $T = \mathrm{diag}(t_1, ..., t_n)$, such that $t_i = s_i s_{i+1} ... s_n$, where $s_i \in \overline{\mathbb{K}}$ satisfy the condition: $s_i = s_j$ if $\alpha_i \in \Gamma_1$ and $\tau(\alpha_i) = \alpha_j$.*

Proof. Let us assume that r_{BD} is associated to an admissible triple $(\Gamma_1, \Gamma_2, \tau)$, where $\Gamma_1, \Gamma_2 \subset \{\alpha_1, ..., \alpha_{n-1}\}$. Let $T \in C(r_{BD})$. According to Lemma 5.3, $T \in \mathrm{diag}(n, \overline{\mathbb{K}})$, therefore we put $T = \mathrm{diag}(t_1, ..., t_n)$. Now we note that $T \in C(r_{BD})$ if and only if $(\mathrm{Ad}_T \otimes \mathrm{Ad}_T)(e_{\tau^k(\alpha)} \wedge e_{-\alpha}) = e_{\tau^k(\alpha)} \wedge e_{-\alpha}$ for any $\alpha \in \Gamma_1$ and any positive integer k.

For simplicity, let us take an arbitrary $\alpha_i \in \Gamma_1$ and suppose that $\tau(\alpha_i) = \alpha_j$. We then get $t_i t_{i+1}^{-1} = t_j t_{j+1}^{-1}$. Denote $s_j := t_j t_{j+1}^{-1}$ for each $j \le n-1$ and $s_n = t_n$. Then $t_j = s_j s_{j+1} ... s_n$ and moreover $s_i = s_j$.

□

Theorem 5.6. *For $\mathfrak{g} = sl(n)$, the Belavin–Drinfeld cohomology associated to any r_{BD} is trivial. Any Lie bialgebra structure on $\mathfrak{g}(\mathbb{K})$ is of the form $\delta(a) = [r, a \otimes 1 + 1 \otimes a]$, where r is an r-matrix which is $GL(n, \mathbb{K})$-equivalent to a non-skewsymmetric r-matrix from the Belavin–Drinfeld list.*

Proof. Let X be a cocycle associated to r_{BD} which is a fixed r-matrix from the Belavin–Drinfeld list. Thus $X^{-1}\sigma(X)$ belongs to the centralizer of the r_{BD}. On the other hand, according to Lemma 5.3, $C(r_{BD}) \subseteq \mathrm{diag}(n, \overline{\mathbb{K}})$.

We then obtain that for any $\sigma \in Gal(\overline{\mathbb{K}}/\mathbb{K})$, $X^{-1}\sigma(X)$ is diagonal. By Lemma 5.1, we have a decomposition $X = QD$, where $Q \in GL(n, \mathbb{K})$ and $D \in \mathrm{diag}(\overline{\mathbb{K}})$. Since $\sigma(Q) = Q$, we have $X^{-1}\sigma(X) = (QD)^{-1}\sigma(QD) = D^{-1}Q^{-1}Q\sigma(D) = D^{-1}\sigma(D)$. Recall that $X^{-1}\sigma(X) \in C(r_{BD})$. It follows that $D^{-1}\sigma(D) \in C(r_{BD})$.

Let $D = \mathrm{diag}(d_1, ..., d_n)$. Then $\mathrm{diag}(d_1^{-1}\sigma(d_1), ..., d_n^{-1}\sigma(d_n)) \in C(r_{BD})$. Denote $t_i = d_i^{-1}\sigma(d_i)$ and $T = \mathrm{diag}(t_1, ..., t_n)$. According to Lemma 5.5, $T \in C(r_{BD})$ if and only if $t_i t_{i+1}^{-1} = t_j t_{j+1}^{-1}$. Equivalently, $\sigma(d_i^{-1} d_{i+1} d_j d_{j+1}^{-1}) = d_i^{-1} d_{i+1} d_j d_{j+1}^{-1}$. It follows that $d_i^{-1} d_{i+1} d_j d_{j+1}^{-1} \in \mathbb{K}$. Let $s_i := d_i d_{i+1}^{-1}$ for any i and $s_n = d_n$. Then we get $s_j s_i^{-1} \in \mathbb{K}$.

Let us fix a root $\alpha_{i_0} \in \Gamma_1 \setminus \Gamma_2$ and let $\tau^j(\alpha_{i_0}) = \alpha_j$. Then $s_j s_{i_0}^{-1} \in \mathbb{K}$, for any j. Denote $k_j := s_j s_{i_0}^{-1}$. We have $d_j = s_j s_{j+1} ... s_{n-1} s_n = k_j k_{j+1} ... k_n s_{i_0}^{n-j+1}$. Let

$$K := \mathrm{diag}(k_1 k_2 ... k_n, k_2 ... k_n, ..., k_n), \quad C := \mathrm{diag}(s_{i_0}^n, s_{i_0}^{n-1}, ..., s_{i_0}).$$

Note that $D = KC$ and $K \in GL(n, \mathbb{K})$. Moreover, according to Lemma 5.5, $C \in C(r_{BD})$.

Summing up, we have obtained that if X is any cocycle associated to r_{BD}, then $X = QD = QKC$, with $QK \in GL(n, \mathbb{K})$, $C \in C(r_{BD})$. This ends the proof.

□

6. Belavin-Drinfeld cohomologies for orthogonal algebras

The next step in our investigation of Belavin–Drinfeld cohomologies is for orthogonal algebras $o(m)$. We begin with the case of Drinfeld–Jimbo r-matrix. In what follows, we will use the following split form of the orthogonal algebra $o(n, \mathbb{C})$ and $o(n, \mathbb{K})$:

$$o(n) = \{A \in gl(n) : A^T S + SA = 0\}.$$

Here S is the matrix with 1 on the second diagonal and zero elsewhere. The group

$$O(n) = \{X \in GL(n) : X^T SX = S\}$$

naturally acts on $o(n)$. However, the center of the Drinfeld–Jimbo r-matrix might be bigger than the Cartan subalgebra of $O(n)$. On the other hand, it follows from Theorem 4.3 that $C(r_{DJ}, SO(n))$ coincides with the Cartan subgroup of $SO(n)$. Our main result about Belavin-Drinfeld cohomologies for orthogonal algebras is the following:

Theorem 6.1. *Let $\mathfrak{g} = o(m)$ and r_{DJ} be the Drinfeld–Jimbo r-matrix. Then $H^1_{BD}(r_{DJ}, SO(m))$ is trivial. Moreover, if m is odd then both $H^1_{BD}(r_{DJ}, O(m))$ and $H^1_{BD}(r_{DJ}, SO(m))$ are trivial.*

Proof. (i) Assume $m = 2n$. On $\overline{\mathbb{K}}^m$ let us fix the bilinear form

$$B(\mathbf{x}, \mathbf{y}) = \sum_{i=1}^{m} x_i y_{m+1-i}.$$

Let $X \in O(m, \overline{\mathbb{K}})$ be a cocycle associated to r_{DJ}. Thus $X^{-1}\sigma(X) \in C(r_{DJ})$. Recall that $C(r_{DJ}) = \text{diag}(m, \overline{\mathbb{K}}) \cap SO(m, \overline{\mathbb{K}})$. Therefore $X^{-1}\sigma(X) \in \text{diag}(m, \overline{\mathbb{K}})$. By Lemma 5.1, one has the decomposition $X = QD$, where $Q \in GL(m, \mathbb{K})$ and $D \in \text{diag}(m, \overline{\mathbb{K}})$. Let us write $D = \text{diag}(d_1, ..., d_{2n})$ and denote by q_i the columns of Q. Then $X = QD$ is equivalent to $Q^T SQ = D^{-1}SD^{-1}$, which in turn gives that $B(q_i, q_{i'})d_i d_{i'} = \delta_i^{2n+1-i'}$. We get $B(q_i, q_{i'}) = 0$ if $i + i' \neq 2n + 1$ and $B(q_i, q_{2n+1-i})d_i d_{2n+1-i} = 1$. Let $k_i := B(q_i, q_{2n+1-i})$. Since $Q \in GL(2n, \mathbb{K})$, $k_i \in \mathbb{K}$. Because $k_i^{-1} = d_i d_{2n+1-i}$, it follows that $D = Q_1 D_1$, where

$$Q_1 = \text{diag}(k_1^{-1}, ..., k_n^{-1}, 1..., 1) \text{ and}$$

$$D_1 = \text{diag}(d_1 k_1, ..., d_n k_n, d_{n+1}, ..., d_{2n}).$$

We note that $X = (QQ_1)D_1$, $D_1 \in SO(2n)$ and hence, $D_1 \in C(r_{DJ}, SO(2n))$. Then, clearly $QQ_1 \in SO(2n, \mathbb{K})$. which proves that X is equivalent to the identity.

(ii) Assume $m = 2n + 1$. First, we note that if $X \in O(m)$ is such that $det(X) = -1$, then $det(-X) = 1$. Therefore, it follows from Theorem 4.3 that in this case $C(r_{DJ}, O(2n+1))$ consists of diagonal matrices. By Lemma 5.1, we may write again $X = QD$, where $Q \in GL(n, \mathbb{K})$ and $D \in \text{diag}(m, \overline{\mathbb{K}})$.

Let $k_i := B(q_i, q_{2n+2-i}) \in \mathbb{K}$. Repeating the computations as in (i), one gets $k_i^{-1} = d_i d_{2n+2-i}$. If $i = n + 1$, $d_{n+1}^2 = k_{n+1}^{-1} \in \mathbb{K}$. This implies that either $d_{n+1} \in \mathbb{K}$ or $d_{n+1} \in j\mathbb{K}$, where $j^2 = \hbar$.

Actually we can prove that the second case is impossible.

Let us denote $R = Q^{-1}$ and its rows by $r_1, ..., r_{2n+1}$. Then $X^T SX = S$ is equivalent to $RSR^T = DSD$, which in turn gives the following: $B(r_i, r_{i'}) = 0$, for all $i \neq i'$, $B(r_i, r_i) = d_i d_{2n+2-i}$ for all i.

Let us take an arbitrary orthogonal basis $v_1, ..., v_{2n+1}$ in $\mathbb{K}^{2n+1}$ and denote $B(v_i, v_i) = A_i$.

The matrix V with rows v_i satisfies $VSV^T = \text{diag}(A_1, ..., A_{2n+1})$. This relation implies that $A_1 ... A_{2n+1} = (-1)^n det(V)^2 = (i^n det(V))^2$. Therefore $A_1 ... A_{2n+1} = l^2$ is a square of some $l \in \mathbb{K}$.

On the other hand, if M is the change of basis matrix from r_i to v_i, then

$$M^T \mathrm{diag}(A_1, ..., A_{2n+1})M = \mathrm{diag}(d_1 d_{2n+1}, ..., d_{n+1}^2, ..., d_{2n+1}d_1).$$

By taking the determinant on both sides, one obtains

$$det(M)^2 A_1 ... A_{2n+1} = (d_1 d_{2n+1})^2 ... (d_n d_{n+2})^2 d_{n+1}^2$$

which implies that d_{n+1}^2 is a square in $\mathbb{K}$, and consequently, $d_{n+1} \in \mathbb{K}$.

Let us show that X is equivalent to the trivial cocycle. Consider

$$Q_1 = \mathrm{diag}(k_1^{-1}, ..., k_n^{-1}, d_{n+1}, 1, ..., 1)$$

$$D_1 = \mathrm{diag}(d_1 k_1, ..., d_n k_n, 1, d_{n+2} ..., d_{2n+1}).$$

We have $D = Q_1 D_1$ and $D_1 \in O(2n+1, \overline{\mathbb{K}})$. Thus $X = (QQ_1)D_1$, $QQ_1 \in O(2n+1, \mathbb{K})$, $D_1 \in C(r_{DJ})$, i.e. X is equivalent to the trivial cocycle, which completes the proof of triviality of $H_{BD}^1(r_{DJ}, O(m))$.

Finally, the case $H_{BD}^1(r_{DJ}, SO(2n+1))$ can be treated exactly as $H_{BD}^1(r_{DJ}, SO(2n))$. □

We have just seen that the Belavin–Drinfeld cohomology $H_{BD}^1(r_{DJ})$ is trivial. Regarding Belavin–Drinfeld cohomology $H_{BD}^1(r_{BD}, SO(2n))$ for an arbitrary r_{BD}, we can give an example where this set is non-trivial. Let us denote the simple roots of $o(2n)$ by $\alpha_i = \epsilon_i - \epsilon_{i+1}$, for $i < n$, $\alpha_n = \epsilon_{n-1} + \epsilon_n$, where $\{\epsilon_i\}$ is an orthonormal basis of $\mathfrak{h}^*$. Let $\Gamma_1 = \{\alpha_{n-1}\}$, $\Gamma_2 = \{\alpha_n\}$ and $\tau(\alpha_{n-1}) = \alpha_n$. Denote by r_{BD} the r-matrix corresponding to the triple $(\Gamma_1, \Gamma_2, \tau)$ and s, where $s \in \mathfrak{h} \wedge \mathfrak{h}$ satisfies $((\alpha_{n-1} - \alpha_n)) \otimes 1)(2s) = ((\alpha_{n-1} + \alpha_n)) \otimes 1)\Omega_0$.

Lemma 6.2. *The centralizer $C(r_{BD})$ consists of all diagonal matrices of the form*

$$T = \mathrm{diag}(t_1, ..., t_{n-1}, \pm 1, \pm 1, t_{n-1}^{-1}, ..., t_1^{-1}),$$

for arbitrary nonzero $t_1, t_2 \in \overline{\mathbb{K}}$.

Proof. We already know that $C(r_{BD}, SO(2n)) \subseteq \mathrm{diag}(2n, \overline{\mathbb{K}}) \cap O(2n, \overline{\mathbb{K}})$. Let $T \in C(r_{BD})$, where $T = \mathrm{diag}(t_1, ..., t_n, t_n^{-1}, ..., t_1^{-1})$. Since T commutes with r_0 and r_{DJ}, $T \in C(r_{BD})$ if and only if $(\mathrm{Ad}_T \otimes \mathrm{Ad}_T)(e_{\alpha_n} \wedge e_{\alpha_{n-1}}) = e_{\alpha_n} \wedge e_{\alpha_{n-1}}$. One can check that $(\mathrm{Ad}_T \otimes \mathrm{Ad}_T)(e_{\alpha_n} \wedge e_{\alpha_{n-1}}) = t_n^{-2} e_{\alpha_n} \wedge e_{\alpha_{n-1}}$. Therefore we get $t_n^{-2} = 1$ and the conclusion follows. □

Proposition 6.3. *Let $\mathfrak{g} = o(2n)$ and r_{BD} be the r-matrix corresponding to the triple $(\Gamma_1, \Gamma_2, \tau)$ and $s \in \mathfrak{h} \wedge \mathfrak{h}$ as above. Then $H_{BD}^1(r_{BD}, SO(2n))$ is non-trivial.*

Proof. Assume that $X^{-1}\sigma(X) \in C(r_{BD}, SO(2n))$ for all $\sigma \in Gal(\overline{\mathbb{K}}/\mathbb{K})$. By the above lemma, $X^{-1}\sigma(X) = \mathrm{diag}(t_1, ..., t_{n-1}, \pm 1, \pm 1, t_{n-1}^{-1}, ..., t_1^{-1})$.

On the other hand, since $X^{-1}\sigma(X)$ is diagonal, it follows from Theorem 6.1 that there exist $Q \in SO(2n, \mathbb{K})$ and a diagonal matrix $D \in SO(2n, \overline{\mathbb{K}})$ such that $X = QD$. Let $D = \mathrm{diag}(s_1, ..., s_n, s_n^{-1}, ..., s_1^{-1})$. Since $Q \in O(2n, \mathbb{K})$, for any $\sigma \in Gal(\overline{\mathbb{K}}/\mathbb{K})$, $\sigma(Q) = Q$. We obtain $X^{-1}\sigma(X) = D^{-1}Q^{-1}Q\sigma(D) = D^{-1}\sigma(D)$, which is equivalent to the following system: $s_i^{-1}\sigma(s_i) = t_i$, for all $i \leq n-1$ and $s_n^{-1}\sigma(s_n) = \pm 1$.

Assume first that there exists σ such that $\sigma(s_n) = -s_n$. Then $s_n \in j\mathbb{K}$. One can check that X is equivalent to $X_0 = \mathrm{diag}(1, ..., 1, j, j^{-1}, 1, ..., 1)$ which is a non-trivial cocycle.

If $\sigma(s_n) = s_n$ for all $\sigma \in Gal(\overline{\mathbb{K}}/\mathbb{K})$, then $s_n \in \mathbb{K}$. In this case,

$$D = \mathrm{diag}(s_1, ..., s_{n-1}, 1, 1, s_{n-1}^{-1}, ..., s_1^{-1}) \times \mathrm{diag}(1, ..., 1, s_n, s_n^{-1}, 1, ..., 1),$$

where the first matrix is in $C(r_{BD})$ and the second in $SO(2n, \mathbb{K})$. This proves that X is equivalent to the identity cocycle. □

7. Lie bialgebra structures in Case III and twisted Belavin-Drinfeld cohomologies

Here we analyse the Lie bialgebra structures on $\mathfrak{g}(\mathbb{K})$ for which the corresponding Drinfeld double is isomorphic to $\mathfrak{g}(\mathbb{K}[j])$, where $j^2 = \hbar$. The question is to find those subalgebras W of $\mathfrak{g}(\mathbb{K}[j])$ satisfying the following conditions:

(i) $W \oplus \mathfrak{g}(\mathbb{K}) = \mathfrak{g}(\mathbb{K}[j])$.

(ii) $W = W^{\perp}$, with respect to the non-degenerate symmetric bilinear form Q given by

$$Q(f_1(\hbar) + jf_2(\hbar), g_1(\hbar) + jg_2(\hbar)) = K(f_1, g_2) + K(f_2, g_1).$$

We will restrict our discussion to $\mathfrak{g} = sl(n)$. We begin with the following remark. The field $\mathbb{K}[j]$ is endowed with a conjugation. For any element $a = f_1 + jf_2$, its conjugate is $\overline{a} := f_1 - jf_2$. If $A = A_1 + jB_1$ and $B = A_2 + jB_2$ are two matrices in $sl(n, \mathbb{K}[j])$, then $Q(A, B) = Tr(A_1B_2 + B_1A_2)$, i.e. the coefficient of j in $Tr(AB)$.

Lemma 7.1. *Let L be the subalgebra of $sl(n, \mathbb{K}[j])$ which consists of all matrices $Z = (z_{ik})$ satisfying $z_{ik} = \overline{z}_{n+1-i,n+1-k}$. Then L and $sl(n, \mathbb{K})$ are isomorphic via a conjugation of $sl(n, \mathbb{K}[j])$.*

Proof. Assume that $Z = (z_{ik})$ satisfies $z_{ik} = \overline{z}_{n+1-i,n+1-k}$. Then $Z = S\overline{Z}S$, where S is the matrix with 1 on the second diagonal and zero elsewhere.

Choose a matrix $X \in GL(n, \mathbb{K}[j])$ such that $\overline{X} = XS$. Then $\overline{XZX^{-1}} = XS\overline{Z}SX^{-1} = XZX^{-1}$ which implies that $XZX^{-1} \in sl(n, \mathbb{K})$. Conversely, if $A \in sl(n, \mathbb{K})$, then $Z = X^{-1}AX$ satisfies the condition $Z = S\overline{Z}S$. □

From now on we will identify $sl(n, \mathbb{K})$ with L. Let us find a complementary subalgebra to L in $sl(n, \mathbb{K}[j])$. Let us denote by H the Cartan subalgebra of L. If we identify the Cartan subalgebra of $sl(n, \mathbb{K}[j])$ with $\mathbb{K}^{2(n-1)}$, then H is a Lagrangian subspace of $\mathbb{K}^{2(n-1)}$. Choose a Lagrangian subspace H_0 of $\mathbb{K}^{2(n-1)}$ such that H_0 has trivial intersection with H. Let N^+ be the algebra of upper triangular matrices of $sl(n, \mathbb{K}[j])$ with zero diagonal. Consider $W_0 = H_0 \oplus N^+$. We immediately obtain the following

Lemma 7.2. *The subalgebra W_0 as above satisfies conditions (i) and (ii), where $sl(n, \mathbb{K})$ is identified with L as in Lemma 7.1.*

Proposition 7.3. *Any Lie bialgebra structure on $sl(n, \mathbb{K})$ for which the classical double is isomorphic to $sl(n, \mathbb{K}[j])$ is given by an r-matrix which satisfies $CYB(r) = 0$ and $r + r^{21} = j\Omega$.*

Proof. Let W_0 be as in the above lemma. By choosing two dual bases in W_0 and $sl(n, \mathbb{K})$ respectively, one can construct the corresponding r-matrix r_0 over $\overline{\mathbb{K}}$. It is easily seen that r_0 satisfies the system $CYB(r_0) = 0$ and $r_0 + r_0^{21} = j\Omega$.

Let us suppose that W is another subalgebra of $sl(n, \mathbb{K}[j])$, satisfying conditions (i) and (ii). Then the corresponding r-matrix over $\overline{\mathbb{K}}$ is obtained by choosing dual bases in W and $sl(n, \mathbb{K})$ respectively. We have $r + r^{21} = a\Omega$ for some $a \in \mathbb{K}[j]$. On the other hand, the classical double of the Lie bialgebras corresponding to r and r_0 is the same. This implies that r and r_0 are twists of each other and therefore $a = j$. □

Now, we recall that, over $\overline{\mathbb{K}}$, all r-matrices are gauge equivalent to the ones from Belavin–Drinfeld list. It follows that there exists a non-skewsymmetric r-matrix r_{BD} and $X \in GL(n, \overline{\mathbb{K}})$ such that $r = j(\mathrm{Ad}_X \otimes \mathrm{Ad}_X)(r_{BD})$.

Consider an arbitrary $\sigma \in Gal(\overline{\mathbb{K}}/\mathbb{K})$. Since δ is a cobracket on $sl(n, \mathbb{K})$, $(\sigma \otimes \sigma)(\delta(a)) = \delta(a)$ and $(\sigma \otimes \sigma)(\delta(a)) = [\sigma(r), a \otimes 1 + 1 \otimes a]$.

At this point it is worth recalling that $Gal(\overline{\mathbb{K}}/\mathbb{K}) \cong \hat{\mathbb{Z}} = \varprojlim(\mathbb{Z}/n\mathbb{Z})$ (see [9]). Clearly, the subgroup $2\hat{\mathbb{Z}}$ acts trivially on $\mathbb{K}[j]$. Assume that $\sigma \in 2\hat{\mathbb{Z}}$. Exactly as in section 4, it follows that $\sigma(r) = r$ and if $r = (\mathrm{Ad}_X \otimes \mathrm{Ad}_X)(jr_{BD})$ with $X \in GL(n, \overline{\mathbb{K}})$, then $\sigma(X) = XD(\sigma)$.

Since $Gal(\overline{\mathbb{K}}/\mathbb{K}[j]) \cong 2\hat{\mathbb{Z}} \cong \hat{\mathbb{Z}}$, we can use the same arguments as in the proof of Lemma 5.1 to obtain the following result

Lemma 7.4. *Let $X \in GL(n, \overline{\mathbb{K}})$. Assume that for any $\sigma \subset Gal(\overline{\mathbb{K}}/\mathbb{K}[j])$, $X^{-1}\sigma(X) \in \mathrm{diag}(n, \overline{\mathbb{K}})$. Then there exists $P \in GL(n, \mathbb{K}[j])$ and $D \in \mathrm{diag}(n, \overline{\mathbb{K}})$ such that $X = PD$.*

Now let us consider the action of $\sigma_2 \in Gal(\mathbb{K}[j]/\mathbb{K})$, $\sigma_2(a + bj) = a - bj := \overline{a + bj}$. Our identities imply that $\sigma_2(r) = r + \alpha\Omega$, for some $\alpha \in \overline{\mathbb{K}}$. Let us show that $\alpha = -j$. Indeed, since $r + r^{21} = j\Omega$, we also have $\sigma_2(r) + \sigma_2(r^{21}) = -j\Omega$. Combining these relations with $\sigma_2(r) = r + \alpha\Omega$, we get $\alpha = -j$ and therefore $\sigma_2(r) = r - j\Omega = -r_{21}$.

Recall now that $r = j(\mathrm{Ad}_X \otimes \mathrm{Ad}_X)(r_{BD})$. It follows that $X \in GL(n, \overline{\mathbb{K}})$ must satisfy the identity $(\mathrm{Ad}_{X^{-1}\sigma_2(X)} \otimes \mathrm{Ad}_{X^{-1}\sigma_2(X)})(\sigma(r_{BD})) = r^{21}_{BD}$. Using the same arguments as in the proof of Theorem 4.3, we obtain

Proposition 7.5. *Any Lie bialgebra structure on $sl(n, \mathbb{K})$ for which the classical double is $sl(n, \mathbb{K}[j])$ is given by an r-matrix $r = j(\mathrm{Ad}_X \otimes \mathrm{Ad}_X)(r_{BD})$, where r_{BD} is a non-skewsymmetric r-matrix from the Belavin–Drinfeld list and $X \in GL(n, \overline{\mathbb{K}})$ satisfies*

$$(\mathrm{Ad}_{X^{-1}\sigma_2(X)} \otimes \mathrm{Ad}_{X^{-1}\sigma_2(X)})(r_{BD}) = r^{21}_{BD}$$

$$(\mathrm{Ad}_{X^{-1}\sigma(X)} \otimes \mathrm{Ad}_{X^{-1}\sigma(X)})(r_{BD}) = r_{BD},$$

for any $\sigma \in Gal(\overline{\mathbb{K}}/\mathbb{K}[j])$,

Definition 7.6. Let r_{BD} be a non-skewsymmetric r-matrix from the Belavin–Drinfeld list. We call $X \in G(\overline{\mathbb{K}})$ a *Belavin–Drinfeld twisted cocycle* associated to r_{BD} if $(\mathrm{Ad}_{X^{-1}\sigma_2(X)} \otimes \mathrm{Ad}_{X^{-1}\sigma_2(X)})(r_{BD}) = r^{21}_{BD}$ and $(\mathrm{Ad}_{X^{-1}\sigma(X)} \otimes \mathrm{Ad}_{X^{-1}\sigma(X)})(r_{BD}) = r_{BD}$, for any $\sigma \in Gal(\overline{\mathbb{K}}/\mathbb{K}[j])$.

The set of Belavin–Drinfeld twisted cocycle associated to r_{BD} will be denoted by $\overline{Z}(r_{BD})$.

Now, let us restrict ourselves to the case $r_{BD} = r_{DJ}$. In order to continue our investigation, let us prove the following

Lemma 7.7. *Let S be the matrix with 1 on the second diagonal and zero elsewhere. Then*

$$r^{21}_{DJ} = (\mathrm{Ad}_S \otimes \mathrm{Ad}_S) r_{DJ}.$$

Proof. We recall that r_{DJ} is given by the following formula:

$$r_{DJ} = \sum_{\alpha > 0} e_\alpha \otimes e_{-\alpha} + \frac{1}{2}\Omega_0.$$

First note that $(\mathrm{Ad}_S \otimes \mathrm{Ad}_S)(e_{ij} \otimes e_{ji}) = e_{n+1-i,n+1-j} \otimes e_{n+1-j,n+1-i}$, which is a term in r^{21}_{DJ}, if $i > j$. On the other hand, since Ω_0 is the Cartan part of the invariant element Ω, we get $(\mathrm{Ad}_S \otimes \mathrm{Ad}_S)\Omega_0 = \Omega_0$. This could also be proved by using the identity $\Omega_0 = n\sum_{i=1}^{n} e_{ii} \otimes e_{ii} - I \otimes I$, where I denotes the identity matrix of $GL(n, \mathbb{K})$. Then $r^{21}_{DJ} = (\mathrm{Ad}_S \otimes \mathrm{Ad}_S) r_{DJ}$ holds.

□

Remark 7.8. $\overline{Z}(r_{DJ})$ is non-empty. Indeed, choose $X \in GL(n, \mathbb{K}[j])$ such that $\sigma_2(X) = XS$. Then $X \in \overline{Z}(r_{DJ})$.

Corollary 7.9. *Let X be a Belavin–Drinfeld twisted cocycle associated to r_{DJ}. Then $X = PD$, where $P \in GL(n, \mathbb{K}[j])$ and $D \in \mathrm{diag}(n, \overline{\mathbb{K}})$. Moreover, $\sigma_2(P) = PSD_1$, where $D_1 \in \mathrm{diag}(n, \mathbb{K}[j])$.*

Proof. Since X is a twisted cocycle, for any $\sigma \in Gal(\overline{\mathbb{K}}/\mathbb{K}[j])$, $X^{-1}\sigma(X) \in C(r_{DJ})$. Recall that $C(r_{DJ}) = \operatorname{diag}(n, \overline{\mathbb{K}})$. By Lemma 7.4, we have $X = PD$, where $P \in GL(n, \mathbb{K}[j])$ and $D \in \operatorname{diag}(n, \overline{\mathbb{K}})$. Lemma 7.7 implies that $D_2 := S^{-1}X^{-1}\sigma_2(X) \in \operatorname{diag}(n, \overline{\mathbb{K}})$.

Since $S^{-1}D^{-1}P^{-1}\sigma_2(P)\sigma_2(D) = D_2$, we obtain $P^{-1}\sigma_2(P) = DSD_2\sigma_2(D^{-1})$.

Let $D_1 := S^{-1}DSD_2\sigma_2(D^{-1}) \in \operatorname{diag}(n, \overline{\mathbb{K}})$. Then $\sigma_2(P) = PSD_1$ and $D_1 \in \operatorname{diag}(n, \mathbb{K}[j])$. □

Definition 7.10. Let X_1 and X_2 be two Belavin–Drinfeld twisted cocycles associated to r_{DJ}. We say that they are *equivalent* if there exists $Q \in GL(n, \mathbb{K})$ and $D \in \operatorname{diag}(n, \overline{\mathbb{K}})$ such that $X_1 = QX_2D$.

Remark 7.11. Assume that X is a twisted cocycle associated to r_{DJ}. By Corollary 7.9, $X = PD$ and is equivalent to the twisted cocycle $P \in GL(n, \mathbb{K}[j])$.

Definition 7.12. Let $\overline{H}^1_{BD}(r_{DJ})$ denote the set of equivalence classes of twisted cocycles associated to r_{DJ}. We call this set the *Belavin–Drinfeld twisted cohomology* associated to the r-matrix r_{DJ}.

Remark 7.13. If X_1 and X_2 are equivalent twisted cocycles, then the corresponding r-matrices $r_1 = j(\mathrm{Ad}_{X_1} \otimes \mathrm{Ad}_{X_1})(r_{DJ})$ and $r_2 = j(\mathrm{Ad}_{X_2} \otimes \mathrm{Ad}_{X_2})(r_{DJ})$ are gauge equivalent via $Q \in GL(n, \mathbb{K})$.

Proposition 7.14. *There is a one-to-one correspondence between $\overline{H}^1_{BD}(r_{DJ})$ and gauge equivalence classes of Lie bialgebra structures on $sl(n, \mathbb{K})$ with classical double $sl(n, \mathbb{K}[j])$ and $\mathbb{K}$-isomorphic to $\delta(r_{DJ})$.*

Let $m = [\frac{n+1}{2}]$. Denote by J the matrix with elements $a_{kk} = 1$, for $k \leq m$, $a_{kk} = -j$ for $k \geq m+1$, $a_{k,n-k+1} = 1$, for $k \leq m$ and $a_{k,n-k+1} = j$ for $k \geq m+1$.

Theorem 7.15. *For $\mathfrak{g} = sl(n)$, the Belavin–Drinfeld twisted cohomology $\overline{H}^1_{BD}(r_{DJ})$ is non-empty and consists of one element, the class of J.*

Proof. Let X be a twisted cocycle associated to r_{DJ}. By Remark 7.11, X is equivalent to a twisted cocycle $P \in GL(n, \mathbb{K}[j])$, associated to r_{DJ}. We may therefore assume from the beginning that $X \in GL(n, \mathbb{K}[j])$. We will prove that X and J are equivalent, i.e. $X = QJD'$, for some $Q \in GL(n, \mathbb{K})$ and $D' \in \operatorname{diag}(n, \mathbb{K}[j])$. The proof will be done by induction.

For $n = 2$, consider $J = \begin{pmatrix} 1 & 1 \\ j & -j \end{pmatrix}$. Suppose $X = \begin{pmatrix} a & b \\ c & d \end{pmatrix} \in GL(2, \mathbb{K}[j])$ satisfies $\overline{X} = XSD$ with $D = \operatorname{diag}(d_1, d_2) \in GL(2, \mathbb{K}[j])$. The identity is equivalent to the following system: $\overline{a} = bd_1$, $\overline{b} = ad_2$, $\overline{c} = dd_1$, $\overline{d} = cd_2$. Assume that $cd \neq 0$. Let $a/c = a' + b'j$. Then $b/d = a' - b'j$. One can immediately check that $X = QJD'$, where $Q = \begin{pmatrix} a' & b' \\ 1 & 0 \end{pmatrix} \in GL(2, \mathbb{K})$, $D' = \operatorname{diag}(c, d) \in \operatorname{diag}(2, \mathbb{K}[j])$.

For $n = 3$, consider $J = \begin{pmatrix} 1 & 0 & 1 \\ 0 & 1 & 0 \\ j & 0 & -j \end{pmatrix}$ and let $X = (a_{ij}) \in GL(3, \mathbb{K}[j])$ satisfy $\overline{X} = XSD$, with $D = \operatorname{diag}(d_1, d_2, d_3) \in GL(3, \mathbb{K}[j])$. The identity is equivalent to the following system: $\overline{a_{11}} = d_1a_{13}$, $\overline{a_{21}} = d_1a_{23}$, $\overline{a_{31}} = d_1a_{33}$, $\overline{a_{12}} = d_2a_{12}$, $\overline{a_{22}} = d_2a_{22}$, $\overline{a_{32}} = d_2a_{32}$, $\overline{a_{13}} = d_3a_{11}$, $\overline{a_{23}} = d_3a_{21}$, $\overline{a_{33}} = d_3a_{31}$. Assume that $a_{21}a_{22}a_{23} \neq 0$.

Let $a_{11}/a_{21} = b_{11} + b_{13}j$ and $a_{31}/a_{21} = b_{31} + b_{33}j$. Then $a_{13}/a_{23} = b_{11} - b_{13}j$ and $a_{33}/a_{23} = b_{31} - b_{33}j$. On the other hand, let $b_{12} := a_{12}/a_{22}$ and $b_{32} := a_{32}/a_{22}$. Note that $b_{12} \in \mathbb{K}$, $b_{32} \in \mathbb{K}$. One can immediately check that $X = QJD'$, where

$$Q = \begin{pmatrix} b_{11} & b_{12} & b_{13} \\ 1 & 1 & 0 \\ b_{31} & b_{32} & b_{33} \end{pmatrix} \in GL(3, \mathbb{K}),\ D' = \operatorname{diag}(a_{21}, a_{22}, a_{23}) \in \operatorname{diag}(3, \mathbb{K}[j]).$$

For $n > 3$, we proceed by induction. Let us denote our matrix $J \in GL(n, \mathbb{K}[j])$ by J_n. We are going to prove that if $X \in GL(n, \mathbb{K}[j])$ satisfies $\overline{X} = XSD$, then using elementary row operations with entries from $\mathbb{F}$ and multiplying columns by proper elements from $\mathbb{K}[j]$ we can bring X to J_n.

We will need the following operations on a matrix

$$M = \{m_{pq}\} \in \mathrm{Mat}(n):$$

1. $u_n(M) = \{m_{pq},\ p, q = 2, 3, ..., n-1\} \in \mathrm{Mat}(n-2)$;
2. $g_n(M) = \{m_{pq}\} \in \mathrm{Mat}(n+2)$, where m_{pq} are already defined for $p, q = 1, 2, ...n$, $m_{00} = m_{n+1,n+1} = 1$ and the rest $m_{0,a} = m_{a,0} = m_{n+1,a} = m_{a,n+1} = 0$.

It is clear that $u_n(X)$ satisfies the twisted cocycle condition. However, its determinant might vanish. To avoid this complication, we note that columns $2, 3, ..., n-1$ of X are linearly independent. Applying elementary row operations (in fact, they are permutations) we obtain a new cocycle X_1, which is equivalent to X and such that $u_n(X_1)$ is a cocycle in $GL(n-2, \mathbb{K}[j])$. Then, by induction, there exist $Q_{n-2} \in GL(n-2, \mathbb{K})$ and a diagonal matrix D_{n-2} such that

$$Q_{n-2} \cdot u_n(X_1) \cdot D_{n-2} = J_{n-2}.$$

Let us consider $X_n = g_{n-2}(Q_{n-2}) \cdot X_1 \cdot g_{n-2}(D_{n-2})$. Clearly, X_n is a twisted cocycle equivalent to X and $u_n(X_n) = J_{n-2}$.

Applying elementary row operations with entries from $\mathbb{K}$ and multiplying by a proper diagonal matrix, we can obtain a new cocycle $Y_n = (y_{pg})$ equivalent to X with the following properties:

1. $u_n(Y_n) = J_{n-2}$;
2. $y_{12} = y_{13} = ... = y_{1,n-1} = 0$ and $y_{n2} = y_{n3} = ... = y_{n,n-1} = 0$;
3. $y_{11} = y_{1n} = 1$, here we use the fact that if $y_{pq} = 0$, then $y_{p,n+1-q} = 0$.

It follows from the cocycle condition $\overline{Y_n} = Y_n \cdot S \cdot \mathrm{diag}(h_1, ..., h_n)$ that $h_1 = h_n = 1$ and hence, $y_{n1} = \overline{y_{nn}}$.

Now, we can use the first row to achieve $y_{n1} = -y_{nn} = j$ and after that, we use the first and the last rows to annihilate $\{y_{k1},\ k = 2, ..., n-1\}$. Then the set $\{y_{kn},\ k = 2, ..., n-1\}$ will automatically vanish. We have obtained J_n from X and thus, have proved that X is equivalent to J_n.

□

Example 7.16. *For $\mathfrak{g} = sl(2)$, the Belavin-Drinfeld list of non-skewsymmetric constant r-matrices consists of only one class, $r_{DJ} = e \otimes f + \frac{1}{4} h \otimes h$, where $e = e_{12}$, $f = e_{21}$ and $h = e_{11} - e_{22}$. One can easily determine the corresponding class of gauge equivalent Lie bialgebra structures on $sl(2, \mathbb{K})$ with classical double $sl(2, \mathbb{K}[j])$ and $\overline{\mathbb{K}}$-isomorphic to $\delta(r_{DJ})$. Indeed, since any twisted cocycle is equivalent to J, it follows that a class representative is $\delta_0 = dr_0$, where*

$$r_0 = j(\mathrm{Ad}_J \otimes \mathrm{Ad}_J) r_{DJ}.$$

A straightforward computation gives

$$r_0 = \frac{j\Omega}{2} + \frac{1}{4} h \wedge e + \frac{\hbar}{4} f \wedge h.$$

We conclude that any Lie bialgebra structure on $sl(2, \mathbb{K})$ with classical double $sl(2, \mathbb{K}[j])$ is gauge equivalent to that given by dr_0, multiplied by a constant from $\mathbb{K}$.

Remark 7.17. In case $sl(2)$, it follows that r_{DJ} and r_0, multiplied by some constants of $\mathbb{K}$, provide all gauge non-equivalent Lie bialgebra structures on $sl(2, \mathbb{K})$ of types II and III and, consequently, two families of non-isomorphic Hopf algebra structures on $U(sl(2, \mathbb{C}))[[\hbar]]$. Moreover, in some sense, these two structures exhaust all Hopf algebra structures on $U(sl(2, \mathbb{C}))[[\hbar]]$ with a non-trivial Drinfeld associator (see also conjectures below).

Journal of Physics: Conference Series **563** (2014) 012030 doi:10.1088/1742-6596/563/1/012030

Remark 7.18. The next step would be to compute the Belavin–Drinfeld twisted cohomology corresponding to an arbitrary r-matrix r_{BD}. Unlike untwisted cohomology, it might happen that even $\overline{Z}(r_{BD})$ is empty as we will see in next publications.

8. Conjectures

8.1. Belavin–Drinfeld cohomology conjecture

Let $\mathfrak{g}$ be a simple Lie algebra and $G = \mathrm{Ad}(\mathfrak{g})$ be the corresponding adjoint group, which is the connected component of the group unit element modulo its center. Let $C(r_{BD})$ be the subgroup of elements of $G(\overline{\mathbb{K}})$ which act trivially on r_{BD}.

Definition 8.1. We say that $X \in G(\overline{\mathbb{K}})$ is a *Belavin–Drinfeld cocycle* associated to r_{BD} if $X^{-1}\sigma(X) \in C(\overline{\mathbb{K}}, r_{BD})$, for any $\sigma \in Gal(\overline{\mathbb{K}}/\mathbb{K})$.

Definition 8.2. Two Belavin–Drinfeld cocycles X_1 and X_2 are *equivalent* if $X_1 = QX_2C$, where $Q \in G(\mathbb{K})$ and $C \in C(r_{BD})$.

Let us denote the set of equivalence classes by $H^1_{BD}(r_{BD}, G)$.

Conjecture 8.3. *Let $\mathfrak{g}$ be a simple Lie algebra and r_{DJ} the Drinfeld–Jimbo r-matrix. Then $H^1_{BD}(r_{DJ}, G)$ is trivial.*

Remark 8.4. We have already proved the conjecture for $sl(n)$ and $o(n)$.

8.2. Quantization conjecture

Let L be a finite dimensional Lie algebra over $\mathbb{C}$ and δ a Lie bialgebra structure on $L(\mathbb{K})$ such that $\delta = 0(\mathrm{mod}\hbar)$. Let $(U_\hbar(L), \Delta_\hbar)$ be the corresponding quantum group. Let $G(\mathbb{K}) = \mathrm{Ad}(L(\mathbb{K}))$ and $G(\overline{\mathbb{K}}) = \mathrm{Ad}(L(\overline{\mathbb{K}}))$.

Let us define the centralizer $C(\overline{\mathbb{K}}, \delta)$. Consider the classical double $D(L(\mathbb{K}), \delta)$. Clearly, δ can be extended to a Lie bialgebra structure $\overline{\delta}$ on $L(\overline{\mathbb{K}})$ and $D(L(\overline{\mathbb{K}}), \overline{\delta})$ contains $D(L(\mathbb{K}), \delta)$, more precisely $D(L(\overline{\mathbb{K}}), \overline{\delta}) = D(L(\mathbb{K}), \delta) \otimes_{\mathbb{K}} \overline{\mathbb{K}}$. The universal classical r-matrix $r_\delta = \sum e_i \otimes e^i$ is the same for $D(L(\mathbb{K}), \delta)$ and $D(L(\overline{\mathbb{K}}), \overline{\delta})$.

Definition 8.5. The centralizer $C(\overline{\mathbb{K}}, \delta)$ consists of all $X \in G(\overline{\mathbb{K}})$ such that

$$(\mathrm{Ad}_X \otimes \mathrm{Ad}^*_X)(r_\delta) = r_\delta + \alpha\Omega,$$

where Ω is an invariant element of $D(L(\overline{\mathbb{K}}), \overline{\delta})^{\otimes 2}$ and Ad^* is the coadjoint representation on $(L(\overline{\mathbb{K}}))^*$. Equivalently, $(\mathrm{Ad}_X \otimes \mathrm{Ad}_X)\delta(\mathrm{Ad}_X^{-1}(l)) = \delta(l)$, for any $l \in L$.

Definition 8.6. We say that $X \in G(\overline{\mathbb{K}})$ is a *Belavin–Drinfeld cocycle* associated to δ if $\sigma(X) = XC$, where $C \in C(\overline{\mathbb{K}}, \delta)$.

Two cocycles, associated to δ, X_1 and X_2 are *equivalent* if $X_1 = QX_2C$, where $Q \in G(\mathbb{K})$ and $C \in C(\overline{\mathbb{K}}, \delta)$.

The set of equivalence classes will be denoted by $H^1_{BD}(G, \delta)$.

Now let us define quantum Belavin–Drinfeld cohomology. The quantum group $(U_\hbar(L), \Delta_\hbar)$ is defined over $\mathbb{O} = \mathbb{C}[[\hbar]]$. We extend the Hopf structures of $U_\hbar(L)$ to $U_\hbar(L, \mathbb{K}) = U_\hbar(L) \otimes_{\mathbb{O}} \mathbb{K}$ and $U_\hbar(L, \overline{\mathbb{K}}) = U_\hbar(L) \otimes_{\mathbb{K}} \overline{\mathbb{K}}$. By abuse of notation, $\Delta_\hbar$ denotes all three comultiplications.

Definition 8.7. Let P be an invertible element of $U_\hbar(L, \overline{\mathbb{K}})$. We say that it belongs to $C(U_\hbar(L), \Delta_\hbar)$ if, for all $a \in U_\hbar(L)$,

$$(P \otimes P)\Delta_\hbar(P^{-1}aP)(P^{-1} \otimes P^{-1}) = \Delta_\hbar(a).$$

Denote $F_P := (P \otimes P)\Delta_\hbar(P^{-1}) \in U_\hbar(L, \overline{\mathbb{K}})^{\otimes 2}$.

Definition 8.8. P is called a *quantum Belavin–Drinfeld cocycle* if $\sigma(P) = PC$, for any $\sigma \in Gal(\overline{\mathbb{K}}/\mathbb{K})$ and some $C \in C(U_\hbar(L), \Delta_\hbar)$.

Two quantum cocycles P_1 and P_2 are *equivalent* if $P_2 = QP_1C$ where Q is an invertible element of $U_\hbar(L, \mathbb{K})$ and $C \in C(U_\hbar(L), \Delta_\hbar)$.

Remark 8.9. On $U_\hbar(L)$ consider the comultiplications

$$\Delta_{\hbar,P_1}(a) = F_{P_1}\Delta_\hbar(a)F_{P_1}^{-1},\ \Delta_{\hbar,P_2}(a) = F_{P_2}\Delta_\hbar(a)F_{P_2}^{-1}.$$

Clearly,

$$\Delta_{\hbar,P_2}(a) = (Q \otimes Q)\Delta_{\hbar,P_1}(Q^{-1}aQ) \cdot (Q^{-1} \otimes Q^{-1}).$$

Since $Q \in U_\hbar(L(\mathbb{K}))$, it is natural to call $\Delta_{\hbar,P_1}$ and $\Delta_{\hbar,P_2}$ $\mathbb{K}$–equivalent comultiplications.

The set of equivalence classes of quantum Belavin–Drinfeld cocycles will be denoted by $H^1_{q-BD}(\Delta_\hbar)$.

Conjecture 8.10. *There is a natural correspondence between* $H^1_{BD}(G, \delta)$ *and* $H^1_{q-BD}(\Delta_\hbar)$.

Acknowledgment. The authors are grateful to V. Kac, P. Etingof and V. Hinich for valuable suggestions.

References

[1] Belavin A and Drinfeld V 1984 Triangle equations and simple Lie algebras *Math. Phys. Rev.* **4** 93
[2] Benkart G and Zelmanov E 1996 Lie algebras graded by finite root systems and intersection matrix algebras *Invent. math.* **126** 1
[3] Drinfeld V G 1997 Quantum groups *Proceedings ICM (AMS, Berkeley 1996)* **1** 798
[4] Drinfeld V G 1983 Hamiltonian structures on Lie groups, Lie bialgebras and the geometric meaning of classical Yang-Baxter equations. (Russian) *Dokl. Akad. Nauk SSSR* **268** 285
[5] Etingof P and Kazhdan D 1996 Quantization of Lie bialgebras I *Sel. Math. (NS)* **2** 1
[6] Etingof P and Kazhdan D 1998 Quantization of Lie bialgebras II *Sel. Math. (NS)* **4** 213
[7] Etingof P, Schiffmann O 1988 Lectures on Quantum Groups. International Press, Cambridge
[8] Montaner F, Stolin A, Zelmanov E 2010 Classification of Lie bialgebras over current algebras *Sel. Math. (NS)* **16** 935
[9] Serre, J P 1979 Local fields. Springer-Verlag, New York
[10] Stolin A 1999 Some remarks on Lie bialgebra structures on simple complex Lie algebras *Comm. Alg.* **27** (9) 4289

Journal of Physics: Conference Series **563** (2014) 012031 doi:10.1088/1742-6596/563/1/012031

Temporal extension of particles to interpret and intuitively describe quantum symmetries and related quantum and relativistic phenomena

Aron Szucs[1]
Senior R&D Engineer, in ABB Motors & Generators, Helsinki
Associate Professor, University of Pécs, Hungary

E-mail: aron.szucs@fi.abb.com,

Abstract. The paper proposes the theory of temporal extension for fundamental particles as a key to intuitively interpret and understand quantum symmetries and related quantum and relativistic phenomena. While space-time as a principal world view is accepted and utilized in fundamental theories, certain traditional concepts of time have not been challenged before. Such a conception is the view that physical objects can only have a point like feature on the time line. When introducing a more space like quality of time, which allows the temporal extension for physical objects and particles, a fundamentally new world view arises. In this view quantum symmetries and certain relativistic phenomena become more intuitive to describe and deal with and "realism" gets a new definition. As a result the temporal length theory also offers to eliminate the tension between the nonlocality of quantum physics and the locality of relativistic phenomena. The paper provides a detailed description of the theory.

1. Introduction

Intuitivity is not a primary requirement for mainstream theories in Physics, meaning, they do not require an intuitive picture and some even explicitly state to ignore such understanding, like in case of the Copenhagen interpretation of quantum physics. It has lead to the famous objections of great Physicists, but did not stop the trend that intuitivity is regarded secondary to mathematical correctness or even to mathematical "beauty". In light of this fact it is ironic how important intuitivity is considered by some mathematicians. While the major theories do not require an intuitive view, many believe that to develop further from the present status quo in modern physics, it will require fundamentally new understanding and correspondingly new intuitive views of the world.

This paper suggests an alternative for such fundamentally new intuitive world view. It is proposed here that certain observations in QM and Special Relativity point to the intuitive idea of temporal extension of physical objects and fundamental particles. These observations include formulations for quantum symmetries, such as the interpretations of Boost matrices in Lorentz group and similarly in Special relativity the Lorentz factor. In simple words, instead of a point like feature on the time line objects could have a temporal length like a moving train has a length on the train track. The main motivation of this paper when proposing the idea of temporal extension is the hope to provide an inspiration to experts on the field, by throwing light on a new perspective of quantum symmetries and

[1] University of Pécs celebrates its 650th anniversary in 2017: http://www.jubileum.pte.hu/index.php

related quantum and relativistic phenomena. The temporal length idea has recently inspired a numerical technique for coupled numerical simulations by the author [1] who wishes to share his thoughts with the Physics community. The mathematical formulation of this numerical analysis might also serve as a recommendation on how to treat new physical models with temporal extension.

In the followings the paper explains the "Obvious Consequences" of temporal extension, as the "simple view" of the theory. This is followed by the "Emerging consequences" of temporal extension, with its corresponding complexity. The theory and its consequences are shortly discussed for: Quantum Symmetries, Special relativity, the Double slit experiment, the Heisenberg uncertainty principle, Bell's theorem, and potentially provide an intuitive, classical mechanical Spin analogy.

It is also hoped that the paper will inspire a "ping-pong" of thoughts between theoretical physicists and the developers of numerical methods for electromagnetic field computation, when considering suitable mathematical models for the modeling of temporal length either as a physical reality or just as a simulation technique.

The paper is concluded with a summary and a proposal for a roadmap to validate this idea.

2. The intuitive picture of the world and the lack of it.

As it was mentioned in the introduction, mainstream Physics theories do not aim to provide an intuitive picture of the world. Some of these intuitively hard to comprehend but closely connected theories include:

- Wave Particle duality: The Double slit experiment
- Heisenberg Uncertainty principle and the Probability wave function
- Bell's Theorem

and in some level:

- Special relativity

Those "trained in the art" often summarize the greatest challenge in comprehending these ideas as the fight between the views of: "Locality" vs. "Non-locality" and the existence of "Realism" and "Local Realism". Many believe that the need for a new understanding emerges from the present status quo and the corresponding efforts include String theory, E8 theory, and other TOE candidates.

It is inspiring that even some of the inventors of these theories strived to provide an intuitive picture as the quotes from Bob Samples and John Steward Bell demonstrate:

"Albert Einstein called the intuitive or metaphoric mind a sacred gift. He added that the rational mind was a faithful servant. It is paradoxical that in the context of modern life we have begun to worship the servant and defile the divine."
Bob Samples, 1976

"… continue to inspire those who suspect that what is proved by the impossibility proofs is lack of imagination.“
John Stewart Bell

3. Observations in QM and Special Relativity pointing to temporal extension.

The following observations hint, that temporal extension of physical objects and particles might well be a physical reality:

1. Space-Time is a generally accepted concept which suggests that Space and Time cannot be separated.
2. Symmetries in QM are defined for both spatial and temporal coordinates.
3. Special and General relativity shows "strong coupling" between space and time. (e.g.: Lorentz factor)

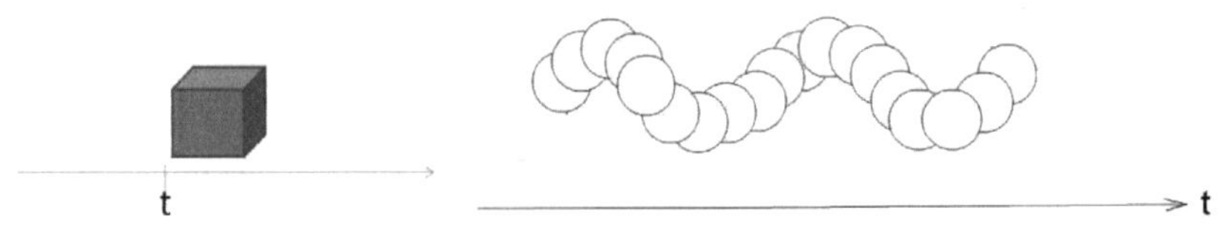

Figure 1. Quantum time and "artistic" representation of temporal extension.

4. Quantum space time - quantized time (and space) theories [2,3,4,5] (Could it mean extension in time?) Figure 1.
5. Particles in QM must occupy several spatial coordinates at the same time. (Probability wave function, Schrödinger, Bell Theorem, Copenhagen interpretation)

Contemplating the above facts, one can wonder, why not to allow an extension in time the same way as we experience in spatial directions?

3.1. Formulations in Quantum Symmetries and Special Relativity pointing to temporal extension.
Contrary to pure rotations which transform spatial coordinates into each other in space time, Boost matrices in Lorentz group describe a transformation of temporal coordinates to spatial and vice versa to describe transformations between frames with relative velocity (1).

Pure rotations | Lorentz Boost –Spacetime rotation

$$\hat{R}_x = \begin{pmatrix} 0 & 0 & 0 & 0 \\ 0 & 1 & 0 & 0 \\ 0 & 0 & \cos(\Delta\theta) & -\sin(\Delta\theta) \\ 0 & 0 & \sin(\Delta\theta) & \cos(\Delta\theta) \end{pmatrix} \qquad \hat{B}_x = \begin{pmatrix} \cosh(\varphi) & \sinh(\varphi) & 0 & 0 \\ \sinh(\varphi) & \cosh(\varphi) & 0 & 0 \\ 0 & 0 & 1 & 0 \\ 0 & 0 & 0 & 1 \end{pmatrix}$$

$$\hat{R}_y = \begin{pmatrix} 0 & 0 & 0 & 0 \\ 0 & \cos(\Delta\theta) & 0 & \sin(\Delta\theta) \\ 0 & 0 & 1 & 0 \\ 0 & -\sin(\Delta\theta) & 0 & \cos(\Delta\theta) \end{pmatrix} \qquad \hat{B}_y = \begin{pmatrix} \cosh(\varphi) & 0 & \sinh(\varphi) & 0 \\ 0 & 1 & 0 & 0 \\ \sinh(\varphi) & 0 & \cosh(\varphi) & 0 \\ 0 & 0 & 0 & 1 \end{pmatrix} \qquad (1)$$

$$\hat{R}_z = \begin{pmatrix} 0 & 0 & 0 & 0 \\ 0 & \cos(\Delta\theta) & -\sin(\Delta\theta) & 0 \\ 0 & \sin(\Delta\theta) & \cos(\Delta\theta) & 0 \\ 0 & 0 & 0 & 1 \end{pmatrix} \qquad \hat{B}_z = \begin{pmatrix} \cosh(\varphi) & 0 & 0 & \sinh(\varphi) \\ 0 & 1 & 0 & 0 \\ 0 & 0 & 1 & 0 \\ \sinh(\varphi) & 0 & 0 & \cosh(\varphi) \end{pmatrix}$$

The Boost transformation would allow – and could also be interpreted as – a transformation of spatial extensions in a certain dimension into temporal length and back between moving frames. Such interpretation also points to the possibility of temporal extension of physical objects.

3.2. Special relativity in Euclidian space time.
Introducing a temporal length of an object and an observer, allows the definition of "perpendicularity" / orthogonality in a space time plane. Even when sticking with a simple Euclidian Space-Time a curious observation can be made. As a consequence of perpendicularity an ordinary wave could be defined to have the same speed in a "space- time aether" relative to observers moving at different speeds as presented in Figure 2.

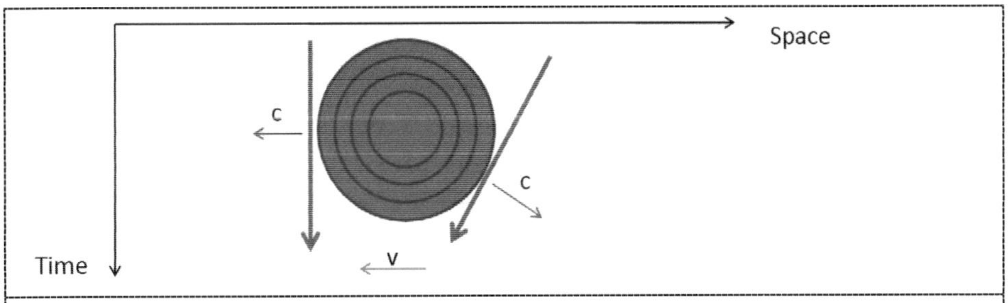

Figure 2. An intuitive picture of special relativity and constant speed of light orthogonally to the observers who are moving at relative speeds to each other.

3.3. An alternative meaning of the Lorentz factor in Special relativity, in Euclidian space time.
If the observer and the observed both have temporal length, an intuitive meaning for the Lorentz factor can be derived with simple geometrical assumptions in Euclidian space-time. Similarly some of the hard to comprehend ideas of special relativity and the Lorentz term can be self explanatory when the temporal length of the observers and observed objects – not only point like features on the time line – is admitted. In this model, two boats are travelling parallel with a speed of "c" on Lake Balaton. Suddenly one of the captains sets his boat on a collision course at a speed of "v" towards the other boat. If the total speed of each boat remains equal and constant c, the parallel distance traveled by the boats associated with the "virtual passing of time", t' can be computed from one boat's perspective of the other. Independently from which boat is used as reference Fig.2 shows that the ratio of the perceived passing of time becomes the Lorenz factor.

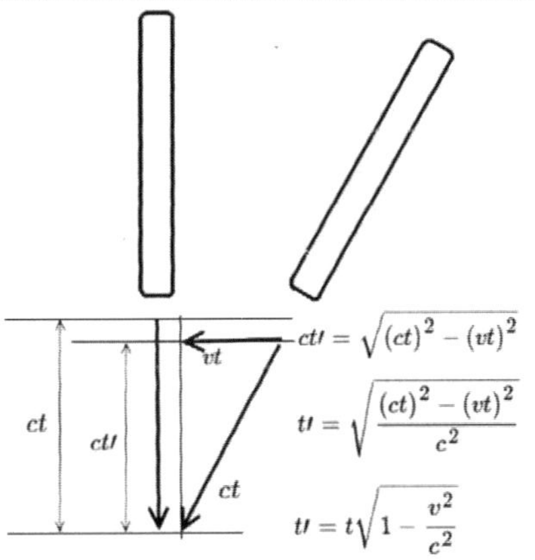

Figure 3. The perceived passing of time when the observer and observed both have temporal length

4. Basic and Emerging Consequences of Temporal length

The consequences of temporal length of particles and physical objects can be grouped as "obvious consequences" and "emerging consequences".

The obvious consequences of temporal length of particles and physical objects can be listed as follows:

1. Particles become 4D physical objects with corresponding 4D symmetries. (SO(4))
2. A particle can actually occupy several spatial locations along its temporal length

Journal of Physics: Conference Series **563** (2014) 012031 doi:10.1088/1742-6596/563/1/012031

3. Particles can interact with each other along their temporal length
4. Quantum symmetries get new intuitive meaning, considering one more "space like" dimension in 4D.

The emerging consequences which can be seen from the symmetry perspective:

1. Local realism gets a whole new meaning: by the spatial distribution along temporal length.
2. The tension between the non-locality of quantum physics and the locality of relativistic phenomena could be reduced / eliminated, together with the re-definition of "Realism".
3. 4D Rotation along planes including the temporal axis keeps temporal symmetry unchanged
4. 4D Rotation along planes including only spatial axes might interchange temporal extension into spatial and vice versa
5. "Coupling" between space and time gets a new intuitive meaning
6. Rotation along temporal axis, or more correctly in mathematical terms: along a spatio-temporal plane is a "new" property. (Or could it actually be related to spin?)
7. "Real 4D cross section" exists intuitively for advanced theories. (E8)

5. The intuitive implications of temporal extension on hard to comprehend theories

We can state that the introduction of temporal length brings new intuitive perspective to interpret not so intuitively explainable phenomena. The following phenomena will be discussed in detail:

1. Quantum symmetries
2. Special relativity
3. Wave particle duality: the Double slit experiment
4. Heisenberg uncertainty principle, Realism and Locality
5. Bell's theorem

5.1. Quantum symmetry and an intuitive perspective: temporal extension.

In light of temporal extension, while "Pure Rotations" transfer length along spatial dimensions "Boost Matrices" in Lorentz group can be seen as transforming spatial extensions to temporal length and vice versa between observation frames. In Lorentz group, Boost matrices could be interpreted as representing temporal extensions mixed into a spatial ones and back.

As demonstrated by the boat example an alternative explanation from Euclidian space can be created by utilizing temporal length to demonstrate certain aspects of the well known "Minkowski topology" consequences:

- Direct correlation to 4D rotations (SO(4)):
- New intuitive meaning of the "time like" dimension
- Spatial and temporal extensions might interchange and intuitively mix up for moving observers.

An intuitive explanation for "Realism" is defined by the spatial multi location along the temporal length. A "quasi - hidden variable" model arises as the spatial distribution along the temporal length characterizes the particle, but when interacting with it, with another 4D object in a measurement, such topology remains "quasi-hidden" from a "3D classical mechanical" point of view.

The symmetry of a particle along rotations about temporal-spatial planes gives an intuitive view for new theories, e.g.: intuitive view of spin. Previously there has been another intuitive view of spin described in [6], where a hyper sphere rolling along the world line of a particle is proposed. However, contrary to that proposal where the use of "world line" for the location of a particle limits its spatio temporal location in one instant, a little more complex picture is proposed in this paper. Consisting of "actual" 4D geometries in all "instants" due to temporal length of particles, does not restrict the particles location in one "instant" of time to a point on the world line, like it is proposed in [6].

The spatial distribution along temporal length offers alternatives for new intuitive ways to view quantum symmetries, e.g.: "hyper" objects such as hyper sphere, 4D spirals, 4D strings, etc.

5.2. Special relativity and temporal extension

As it was explained above in the boat example, calculating the relative passing of time, gave the Lorentz factor as a result. Next to the advantage of getting an intuitive picture introduced by the "temporal length" there is a good chance that with a proper choice and tuning of the following parameters in the Euclidian space time model:

- Speed of "boats"
- Speed of waves in the "space time aether"
- Geometrical surface of the "lake"

we could match special relativistic and maybe even general relativistic predictions. If this would fail there is an alternative to extend the temporal length theory to other, more complex non-Euclidian space time geometries.

5.3. Wave Particle Duality: the Double slit experiment and temporal length.

The new local realism with the possibility of several spatial locations along the temporal extension we can ask the question: How one 4D particle could be perceived as a wave and how it can interact with two slits in 4D? The spatial distribution along the temporal length allows an intuitive perspective to perceive wave like properties of a particle – for example – if it rotates along its temporal extension.

The possibility of temporal length opens a multitude of possibilities for physical interactions between the 4D particle and the 4D pair of slits. The interference picture explained as the interference of probabilistic wave functions can be the result of a complex mechanics resulting from the interactions of two 4D objects.

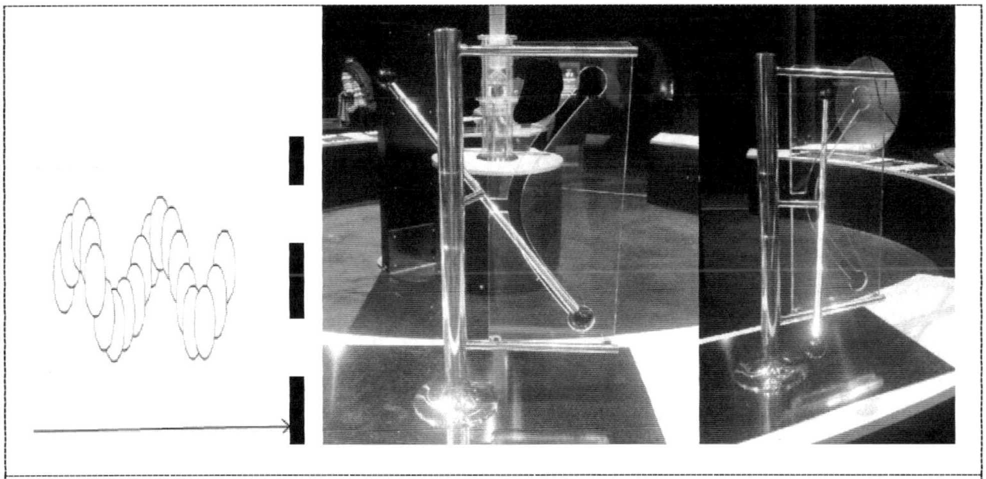

Figure 4. A demonstration device in Heureka Science Center in Helsinki, Finland. A straight object flying through a "curved" hole in a plastic sheet in swiping a 3D surface.

A macro world phenomenon in 3D giving an intuitive hint of such "multi-dimensional" interaction is presented in Heureka Science Center in Helsinki. A straight "1D" object flying through a curved "2D" hole by a rotation swiping a "3D" surface in space.

5.4. Heisenberg uncertainty principle and the probability wave function

The uncertainty when determining a single spatial location and momentum of a particle in a "moment of time" can be translated to an uncertainty of interaction with the spatiotemporal distribution / volume of the given particle with the temporal length theory. Position and momentum are considered as complementary according to the Heisenberg principle. If particles and physical objects would have temporal length, their "location" and "momentum" would be associated with an interaction with another 4D particle in a measurement, which would have consequences on "Realism".

The so called "measurement problem" in quantum mechanics – whether and how the wave function collapses – could also be intuitively represented by the temporal length theory: The measurement – when interacting with the particle at a spatial location along its temporal extension – would directly correlate with the collapse of the wave function to "one" location. This could be very complex, imagining two rotating 4D spirals colliding with one another or the case of two hyper spheres.

The role of the "momentum" associated with the uncertainty principle hints that temporal length of particles might also be related to the mass of particles.

5.5. Bell's inequality and temporal length

Bells theorem has been considered by Henry Stapp as the most profound discovery of science and it unifies the major statements about the nature of quantum physics. However as it was quoted before, John Stewart Bell stated his hope as: "... continue to inspire those who suspect that what is proved by the impossibility proofs is lack of imagination." The theory of "temporal length" of particles could well be a theory providing a new perspective as Bell hoped for. The main question is:

What is the consequence on Bell's theorem and on Realism if the "hidden variable" represented by the spatio-temporal distribution of the particle is actually equivalent with the probability wave function predictions? How the change of definition for Realism affects the measurement?

There have been attempts to re-evaluate the effect of time on the Bell theorem but they were debated or have been fairly complicated. The spatial distribution along the temporal length elegantly creates a "quasi-local" spatial distribution in the time extension and provides a new alternative for the explanation of the outcome of the measurements due to the new "Quasi-Realism".

If the particle's spatial distribution along its temporal length is equivalent to the probability wave function, the Bell inequality could simply be a proof of that shape and corresponding consequences – like spin due to rotation along temporal length as explained above – without further consequences for super luminal speeds of communication between pairs of entangled particles. This interpretation could help to resolve the great divide between the locality and nonlocality perspectives by simply giving an intuitive picture for the hard to comprehend spatial distribution of a single particle along its temporal length and translating Bell inequalities as a consequence of that fact.

There is one more somewhat intuitive picture, which arises from temporal length for the Bell theorem, that the entangled particles could actually be connected along their temporal length. Although such theory might be similarly hard to comprehend as the original Bell statement.

6. A Numerical method inspired by the temporal length idea from QM.

In 2013 at the Compumag 2013 conference in Budapest a numerical technique has been presented [1], based on increased "temporal length" of numerical sub-domain and component models, to handle computational challenges which are typical in engineering for "multiphysics" and system simulation problems. The computational challenges include numerical oscillations when the coupling / interfacing technique between different domains and subsystems is not adequate or not easy to select in advance.

General methods to keep the interaction models in electromagnetic system and multi-physics simulations physically meaningful and correct, within required accuracy, are of great interest.

The paper suggested the introduction of "temporal length" for sub-domain and component models, which reach and "comb" through several time steps at once. These enable new advanced monitoring, error estimation, convergence verification, relaxation and interaction modeling techniques where many consecutive time steps can interact. The temporal length as a simulation technique has been inspired by some thoughts on modern physics.

Figure 5. shows a schematic model of the typical formulation in the time stepping analysis of coupled problems [7, 8, 9, 10]. Traditionally only two consecutive time steps are linked in the formulation to model time derivatives. The figure represents how different domains affect each other in such formulations.

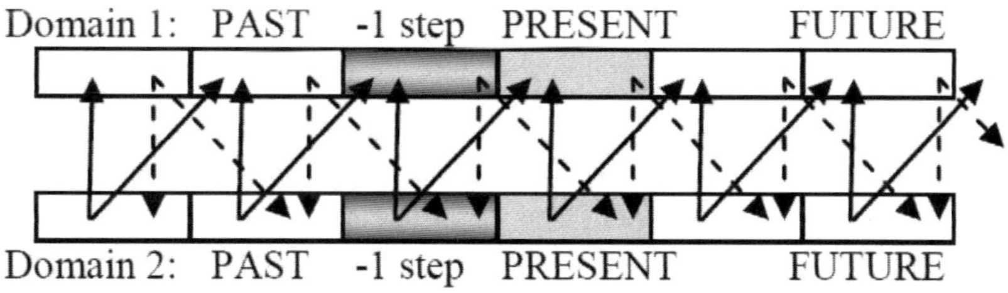

Figure 5. Schematic picture of the traditional view of interaction in coupled problems between domains. Only consecutive time steps affect each other. "Virtual temporal length" of the model: 2 time steps.

It is important to notice that in many cases in electromagnetic field computation – like in eddy current problems and hysteresis modeling – the affecting time step from the past is not only the one before the present, but a "history of the system" affects as well. This does not mean that the formulation must include the values of variables from previous time steps directly, because such history can be carried on in the energy of the computed fields and/or by hysteresis models which also carry the information about such "historical" data.

Introducing longer than 2 time step temporal length can improve monitoring, control and even the formulation of the solution process. As an example numerical oscillations are typically visible when observing at least three or more consecutive time steps. In a 3 time step temporal length model the second derivative of the time function is available momentarily and could warn of numerical oscillations like those which are typical in multi-conductor eddy current problem with Crank-Nicolson time stepping method [11, 12].

Oscillations and divergent solutions can also be caused by unwisely/unluckily calibrated coupling techniques typical in indirect coupling. A schematic picture of a four time step long model is presented in Fig. 6.

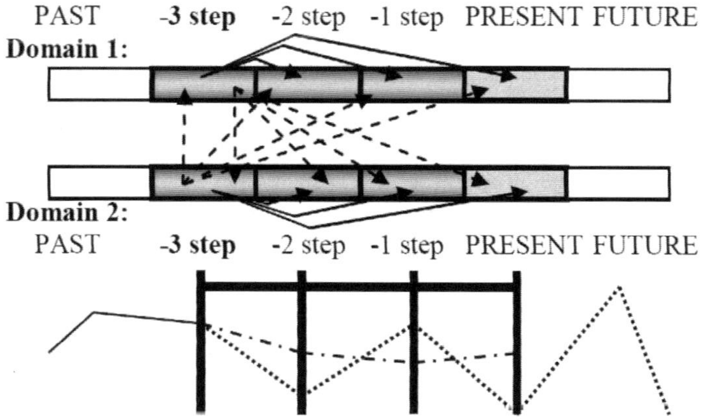

Figure 6. A schematic picture of a 4 time step long model. Only the effects of the time step "-3" are indicated, with arrows, but other time steps can have similar effects. The model with temporal length acts as a "comb" with "4 teeth" smoothing the time functions on the fly, removing oscillations.

Fig.6. hints that the time steps before the present can affect each other and the value of a variable "in the past" can change within the temporal length of the model. When "past time steps" are affected also by the preceding steps, the formulation is like in (2), with a 3 time step temporal length model. Here An and Bn are the traditional system matrices and an and bn are variable vectors of sub domains and/or components in the n-th time step toward the past. The n>1 An and Bn matrices could be unit matrices in some formulations. The C matrices represent the coupling between the domains and the T matrices represent the coupling between different time steps along the temporal length. If the formulation requires that future time steps must affect, and suitably "comb" the past time steps to shape, than the zero matrices in (2) would also be replaced by corresponding T coupling matrices. If the past time steps only affect the present time step, the formulation is like in (1) but TAA23, TBC23, TAC23, TBB23 are 0 matrices.

$$\begin{bmatrix} A1 & C11 & TAA12 & TBC12 & TAA13 & TBC13 \\ C12 & B1 & TAC12 & TBB12 & TAC13 & TBB13 \\ 0 & 0 & A2 & C21 & TAA23 & TBC23 \\ 0 & 0 & C22 & B2 & TAC23 & TBB23 \\ 0 & 0 & 0 & 0 & A3 & C31 \\ 0 & 0 & 0 & 0 & C32 & B3 \end{bmatrix} \begin{bmatrix} a1 \\ b1 \\ a2 \\ b2 \\ a3 \\ b3 \end{bmatrix} = \begin{bmatrix} Ra1 \\ Rb1 \\ Ra2 \\ Rb2 \\ Ra3 \\ Rb3 \end{bmatrix} \quad (2)$$

7. Summary & Proposal for validation

Temporal length of particles and all physical objects is hinted by existing theories including formulations in quantum symmetries and relativity. Wedding QM and Relativity has proven to be a significant challenge and the temporal length of particles seems to promise a way forward. 4D objects and their interaction in space-time might not be that easily intuitive as those in 3D but would offer an alternative to reduce the tension between "locality" and "non-locality" by redefining "Realism".

The temporal length theory can have a couple of strange but verifiable predictions depending on what space time geometry is selected. Supposing Euclidian Space-Time, the temporal length suggests that antiparticles could potentially be actually travelling backward in time – as represented in Feynman diagrams – because the temporal length could provide an "interface" for particles travelling in different directions in space time to interact. The strange prediction is that in a particle accelerator we could be able to observe anti-particles which instead of flying out from the location of a collision should be flying inwards before it happens. This strange prediction however, might not hold if the supposed space time geometry is not Euclidian.

Another prediction is related to neutrino research. As it was stated in the discussion of the Heisenberg uncertainty principle, temporal length might be associated with the mass of the particles also. This means that neutrino oscillations could be equivalent to neutrinos changing their temporal extension to "spatial" in-flight which might be causing them to effect their flight time to a detector.

8. Conclusions

We can conclude that the temporal length hinted by quantum symmetry formulations and special relativity, as presented above, holds promise to provide an intuitive perspective for hard to comprehend theories. While there are a few verifiable predictions, only time will tell whether this idea will inspire new theories. In any case the temporal length idea itself has already motivated the development of numerical techniques for electromagnetic field computation in mutiphysics and system simulation. The author hopes that the temporal length idea will be able to stand the test of time, and it will also inspire other novel ideas and this way it will further benefit the scientific community.

References

[1] Aron Szucs, 2013, Temporal Length for Sub-Domain Models in Multi-Physics and System Analysis, COMPUMAG 2013

[2] Hartland S. Snyder, Quantized Space-Time, Phys. Rev. 71, 38 – Published 1 January 1947

[3] Achim Kempf, Gianpiero Mangano, and Robert B. Mann, 1995, Hilbert space representation of the minimal length uncertainty relation, Phys. Rev. D 52, 1108 – Published 15 July 1995

[4] Sándor Benczik, Lay Nam Chang, Djordje Minic, and Tatsu Takeuchi, 2005, Hydrogen-atom spectrum under a minimal-length hypothesis, Phys. Rev. A 72, 012104 – Published 5 July 2005

[5] M. M. Stetsko and V. M. Tkachuk, 2006, Perturbation hydrogen-atom spectrum in deformed space with minimal length Phys. Rev. A 74, 012101 – Published 5 July 2006

[6] R. Malcor, 1985, Lettere al Nuovo Cimento 16 Aprile 1985, Volume 42, Issue 8, pp 435-437, On the spatio-temporal extension of elementary particles

[7] Aron Szucs, 2000, Macro Elements In the Finite Element Analysis of Multi-Conductor Eddy-Current Problems, IEEE Transactions on Magnetics, Volume 36, Number 4, July 2000

[8] Aron Szucs, 2001, Macro Element Method for Modeling Eddy Currents in the Multi-Conductor Windings of Electrical Machines, Acta Polytechnica Scandinavica, Electrical Engineering Series No. 106

[9] Sami Kanerva, Jukka Kaukonen, Aron Szucs, Terttu Hautamaki, Coupled FEM-Control Simulation in the Analysis of Electrical Machines and Converters, Power Electronics and Motion Control Conference, 2006. EPE-PEMC 2006. 12th International, On page(s): 1925-1930

[10] Aron Szucs, 2012, Macro Element Approach for Electromagnetic Simulations in Multi-physics and System Analysis, at SPEEDAM 2012, "Symposium on Power Electronics, Electrical Drives, Automation & Control", Sorrento, Italy

[11] Yatchev I., Arkkio A., Niemenmaa A., "Eddy-current losses in the stator winding of cage induction motors", Helsinki University of Technology, Laboratory of Electromechanics, Report 47, Espoo 1995, 34 p.

[12] Aron Szucs, Antero Arkkio, 1999 Consideration of Eddy Currents in Multi-Conductor Windings Using the Finite Element Method and the Elimination of Inner Nodes, IEEE Transactions on Magnetics, Volume 35, Number 3, Part I of two parts, May

Aspects of C_3-symmetric calorons from numerical Nahm transform

Atsushi Nakamula[1], **Nobuyuki Sawado**[2] **and Koki Takesue**[3]

[1]Department of Physics, Kitasato University, Sagamihara, Kanagawa 252-0373, Japan
[2,3]Department of Physics, Tokyo University of Science, Noda, Chiba 278-8510, Japan

E-mail: [1]`nakamula@sci.kitasato-u.ac.jp`, [2]`sawado@ph.noda.tus.ac.jp`,
[3]`koki.takesue.phys@gmail.com (Corresponding author)`

Abstract. Calorons are finite action solutions to the anti-selfdual Yang-Mills equations on $\mathbb{R}^3 \times S^1$. They are generally constructed by the so called Nahm construction. We perform the numerical Nahm transform for the Nahm data of 3-calorons with C_3-symmetry, which do not have the monopole limits. Dissimilar to the cases of having monopole limits, we can trace the zero-circumference limit of S^1. It is found that the action density of the calorons tends to fade away as S^1 shrinks.

1. Introduction

Topological solitons in various field theories are vigorously studied in both analytically and numerically [1]. In this paper, we consider the finite action solutions to anti-selfdual Yang-Mills gauge theory in $\mathbb{R}^3 \times S^1$, usually called calorons [2]. As a topological soliton, calorons have connection with many other related objects. In particular, they may have a non-trivial holonomy around the S^1-direction, which links them to Skyrmions [3]. And also, there exists a picture of calorons in which monopoles constitute them [4, 5]. Finally, when the circumference of S^1 becomes large, then calorons will turn into instantons on $\mathbb{R}^4$. Hence the studies on calorons will give insight into the universal understanding of topological solitons. The topological indices of calorons are thus holonomy, monopole charges and instanton charges. Among them, the holonomy and instanton charges do not take integer valued in general [6].

It is well known that calorons can be produced generally through the Nahm construction [7], in which the dual space description of the gauge fields, called the Nahm data, plays central role. In the construction, the transformation from the description on the dual space into that on the ordinary configuration space is crucial, and called the Nahm transform. The Nahm transform for $SU(2)$ caloron of instanton charge $k = 1$ with non-trivial holonomy was studied analytically in [8, 9]. For the higher charges, exact calorons of $k = 2$ were found in [10].

There are many studies for the numerical Nahm transform, *e.g.*, [11, 12]. Recently, the 2-caloron Nahm data of monopole charge $(m_1, m_2) = (2, 2)$ of arbitrary mass with 16 moduli parameters has been proposed [13]. The numerical Nahm transform for the 10 parameter subset of the data has been done in [14]. In this paper we employ the scheme developed in [14], but we make an essential refinement of it to be applicable for general cases.

As mentioned above, calorons are closely related to monopoles and instantons. In fact, the Nahm data of calorons have usually the monopole, or the large scale, limits in addition to the instanton limits. In other words, most of the known Nahm data give examples of the picture in which calorons interpolate between monopoles and instantons [15]. However, it is known that the calorons from Corrigan-Fairlie-t'Hooft ansatz [16], *i.e.*, the Harrington-Shepard type, of higher instanton charges do not have the monopole limits [17]. In addition, it has been shown that there is a one-parameter family of 3-calorons Nahm data with C_3-symmetry which do not have monopole limits [18, 19]. From these facts, it will be interesting to ask that under which conditions the calorons have monopole limits. In this context, we consider the numerical Nahm transform for the C_3-symmetric 3-calorons and show the visualization of their action density. Unlike the cases of having monopole limits, we can trace the zero-circumference limit of S^1, *i.e.*, the limit to $\mathbb{R}^3$. We find that the action density gradually tends to fade away as S^1 becomes smaller.

In this paper, we restrict the gauge group to $SU(2)$ or $U(2)$. In section 2, a brief review on the Nahm construction is given for the case that the holonomy is trivial. In section 3, we give the method of numerical Nahm transform. In section 4, the Nahm data of the C_3-symmetric 3-calorons are introduced. In section 5, the visualization for the action density is performed for the caloron Nahm data given in the previous section. The final section is devoted for summary and concluding remarks.

2. Nahm construction

We give a review of the Nahm construction for the calorons of $U(2)$ gauge theory with trivial holonomy. Let x^μ be the standard coordinates on $\mathbb{R}^3 \times S^1$, where the greek indices run $1, 2, 3$ and 4.

Caloron Nahm data usually consists of two elements. The first is the bulk Nahm data, which are four Hermitian $k \times k$ matrices $T_\mu(s)$ where the $k \in \mathbb{Z}_+$ is the instanton charge of calorons and $s \in [-\mu_0, \mu_0]$ is a dual space coordinate. Here μ_0 determines the fundamental interval of the dual space and the circumference of S^1 is given by π/μ_0. The boundary Nahm data is a k-row vector W of quaternion entries. For the massless case, *i.e.*, the case of the trivial holonomy, the bulk Nahm and the boundary Nahm data satisfy, respectively,

$$\frac{d}{ds}T_i - \frac{i}{2}\varepsilon_{ijm}\left[T_j,\ T_m\right] - i\left[T_4,\ T_i\right] = 0, \tag{1}$$

$$T_i(-\mu_0) - T_i(\mu_0) = \frac{1}{2}\mathrm{tr}_2\left(\sigma_i W^\dagger W\right), \tag{2}$$

where σ_i's are the Pauli matrices, indecies $i, j, m = 1, 2, 3$ and the trace is taken over quaternions.

The caloron gauge fields in the ordinary, or the configuration, space are obtained from the Nahm data through the zero modes of the Weyl equations with impurities at the boundaries. The solutions of the Weyl equations are zero modes with two elements: the bulk zero modes are $2k$-vectors $\mathbf{u}_l$ and the boundary zero modes are 2-vectors $\mathbf{v}_l$, where $l = 1, 2$ indicates the two basis of the solution space of the Weyl equations,

$$\left(\mathbf{1}_{2k}\frac{d}{ds} - i(T_\mu(s) + x_\mu \mathbf{1}_k) \otimes e_\mu\right)\mathbf{u}_l = 0, \tag{3}$$

$$\Delta \mathbf{u}_l := \mathbf{u}_l(-\mu_0) - \mathbf{u}_l(\mu_0) = iW^\dagger \mathbf{v}_l. \tag{4}$$

where $e_\mu = (-i\sigma_i, \mathbf{1}_2)$ are the basis of the quaternion. These are called the bulk Weyl equation and the boundary Weyl equation, respectively. The next step is to find two independent pair of the zero modes $(\mathbf{u}_1, \mathbf{v}_1)$, $(\mathbf{u}_2, \mathbf{v}_2)$, normalized as

$$\int_I \mathbf{u}_a^\dagger \mathbf{u}_b ds + \mathbf{v}_a^\dagger \mathbf{v}_b = \delta_{ab}, \tag{5}$$

where $a, b = 1, 2$ and $I = [-\mu_0, \mu_0]$. From these zero modes, we obtain the gauge connection of calorons as

$$\left(A_\mu(x)\right)_{ab} = \int_I \mathbf{u}_a^\dagger \partial_\mu \mathbf{u}_b ds + \mathbf{v}_a^\dagger \partial_\mu \mathbf{v}_b. \tag{6}$$

In the next section, we shall describe how to numerically solve the Weyl equations.

3. Numerical Nahm transform

In this section, we recast the scheme developed in [14], which is specialized to the case of instanton charge $k = 2$, in order to carry out the numerical Nahm transform for the calorons of higher instanton charge $k \geq 3$. According to [14], we begin with finding the solutions to the bulk Weyl equations (3), followed by solving the boundary Weyl equations (4) with use of a degree of freedom of the linear combinations of the bulk solutions.

3.1. The bulk Weyl equations

The bulk Weyl equations for the charge k calorons are the system of linear ordinary differential equations of rank $2k$, and they can be numerically solved by standard procedure such as Runge-Kutta method, etc..

First, we find out α independent solutions of the Weyl equations by numerically integrating from one endpoint to the direction of another and also from the other to the first one. Thus we obtain 2α numerical solutions $\hat{\mathbf{u}}_i$ totally. Here α, the number of the independent solutions to the bulk Weyl equations from one endpoint, is also equal to the number of shooting parameters on each endpoints ($s = \pm\mu_0$). We then fabricate two independent solutions $\mathbf{u}_l$ for the Weyl equations by taking linear combinations of these 2α solutions,

$$\mathbf{u}_l = (\hat{\mathbf{u}}_1 \ \hat{\mathbf{u}}_2 \ \ldots \ \hat{\mathbf{u}}_{2\alpha}) \cdot \boldsymbol{\omega}_l, \tag{7}$$

where $\boldsymbol{\omega}_l$ are 2α-column vectors, *i.e.* $\boldsymbol{\omega}_l := {}^t({\omega^1}_l \ {\omega^2}_l \ \ldots \ {\omega^{2\alpha}}_l)$, to be fixed as follows. For the monopole construction [20], we impose that these solutions match at the center of the interval. For the caloron, on the other hand, we determine them by the requirement that the bulk solutions are consistent with the boundary Weyl equations (4).

3.2. The boundary Weyl equations

The boundary Weyl equations are given by

$$\Delta\mathbf{u}_l = iW^\dagger \mathbf{v}_l. \iff \begin{pmatrix} \Delta {u^1}_l \\ \Delta {u^2}_l \\ \vdots \\ \Delta {u^{2k}}_l \end{pmatrix} = i \begin{pmatrix} W_1^\dagger \\ W_2^\dagger \\ \vdots \\ W_k^\dagger \end{pmatrix} \begin{pmatrix} {v^1}_l \\ {v^2}_l \end{pmatrix}. \tag{8}$$

where $W = (W_1 \ W_2 \ \ldots W_k)$, $W_i \in \mathbb{H}$ are the boundary Nahm data of k-caloron.

Although the coupled equations (8) look like an *over-determined* system for two unknowns ${v^1}_l$ and ${v^2}_l$, this is not the case because the left-hand-side are still unknown at the present stage. Our goal is to obtain the coefficients of the linear combination $\boldsymbol{\omega}_l$ in (7) and the boundary zero modes $\mathbf{v}_l$ simultaneously from $\hat{\mathbf{u}}_1, \hat{\mathbf{u}}_2, \ldots, \hat{\mathbf{u}}_{2\alpha}$. We can reduce the problem in the form of simple linear algebra as will be shown below. The caloron Nahm data certainly induce $\alpha \geq k$, thus without loss of generality, we choose $\alpha = k$.

If we solve just the upper two rows in (8), we can obtain the boundary zero modes $\mathbf{v}_l$

$$\begin{aligned}\begin{pmatrix} v^1_{\ l} \\ v^2_{\ l} \end{pmatrix} &= -iW_1^{\dagger\ -1}\begin{pmatrix} \Delta u^1_{\ l} \\ \Delta u^2_{\ l} \end{pmatrix} \\ &= \frac{-i}{\mathcal{W}^1_{\ 1}\mathcal{W}^2_{\ 2} - \mathcal{W}^1_{\ 2}\mathcal{W}^2_{\ 1}}\begin{pmatrix} \mathcal{W}^2_{\ 2}\Delta u^1_{\ l} - \mathcal{W}^1_{\ 2}\Delta u^2_{\ l} \\ -\mathcal{W}^2_{\ 1}\Delta u^1_{\ l} + \mathcal{W}^1_{\ 1}\Delta u^2_{\ l} \end{pmatrix},\end{aligned} \tag{9}$$

where $\mathcal{W}^i_{\ j}$ are the components of $W^\dagger$ as a $2k \times 2$ matrix and we used the fact that, in general, $\det W_1 \neq 0$. Substituting (9) into the $2(k-1)$ remaining equations in (8) (*i.e.*, (8) without upper two rows), we obtain $2(k-1)$ conditions. We solve these restraint conditions by using the degree of freedom of the linear combination of the bulk solutions.

From the definition (7), $\Delta\mathbf{u}_l$ is expanded as

$$\Delta\mathbf{u}_l := (\Delta\hat{\mathbf{u}}_1\ \Delta\hat{\mathbf{u}}_2\ \ldots\ \Delta\hat{\mathbf{u}}_{2\alpha}) \cdot \boldsymbol{\omega}_l. \tag{10}$$

where $\Delta\hat{\mathbf{u}}_i := \hat{\mathbf{u}}_i(-\mu_0) - \hat{\mathbf{u}}_i(\mu_0)$. From (9) and (10), the remaining equations of (8) turn out to be,

$$\begin{aligned}&\begin{pmatrix} \Delta u^3_{\ l} \\ \vdots \\ \Delta u^{2k}_{\ l} \end{pmatrix} = i\begin{pmatrix} W_2^\dagger \\ \vdots \\ W_k^\dagger \end{pmatrix}\frac{-i}{\mathcal{W}^1_{\ 1}\mathcal{W}^2_{\ 2} - \mathcal{W}^1_{\ 2}\mathcal{W}^2_{\ 1}}\begin{pmatrix} \mathcal{W}^2_{\ 2}\Delta u^1_{\ l} - \mathcal{W}^1_{\ 2}\Delta u^2_{\ l} \\ -\mathcal{W}^2_{\ 1}\Delta u^1_{\ l} + \mathcal{W}^1_{\ 1}\Delta u^2_{\ l} \end{pmatrix}, \\ \Longleftrightarrow\ &\left[\begin{pmatrix} W_2^\dagger \\ \vdots \\ W_k^\dagger \end{pmatrix}\begin{pmatrix} \mathcal{W}^2_{\ 2}\Delta\hat{u}^1_{\ j} - \mathcal{W}^1_{\ 2}\Delta\hat{u}^2_{\ j} \\ -\mathcal{W}^2_{\ 1}\Delta\hat{u}^1_{\ j} + \mathcal{W}^1_{\ 1}\Delta\hat{u}^2_{\ j} \end{pmatrix} - \det(W_1^\dagger)\begin{pmatrix} \Delta\hat{u}^3_{\ j} \\ \vdots \\ \Delta\hat{u}^2_{\ j} \end{pmatrix}\right]\omega_l^{\ j} = 0, \qquad j = 1, 2, \ldots, 2\alpha, \\ \Longleftrightarrow\ &\underbrace{\left[\mathcal{W}^i_{\ m}\gamma^{mn}_{\ \ op}\mathcal{W}^o_{\ n}\Delta\hat{u}^p_{\ j} - \det(W_1^\dagger)\Delta\hat{u}^i_{\ j}\right]}_{=:A^i_{\ j}}\omega_l^{\ j} = 0, \quad i = 3, 4, \ldots, 2k, \text{ and } m, n, o, p = 1, 2,\end{aligned} \tag{11}$$

where $\omega_l^{\ j}$ are the components of $\boldsymbol{\omega}_l$ and $\gamma^{mn}_{\ \ op}$ are tensor defined as,

$$\gamma^{mn}_{\ \ op} := \begin{cases} 0 & m = n \\ 1 & m \neq n \end{cases} \times \begin{cases} +1 & n = o \\ -1 & n \neq o \end{cases} \times \begin{cases} 0 & o = p \\ 1 & o \neq p \end{cases}.$$

In (11), we define $(2k-2) \times 2\alpha$ matrix A with components $A^i_{\ j}$. We can regard (11) as linear homogeneous equation for $\boldsymbol{\omega}_l$, which determine the basis of the solutions to the bulk Weyl equations. In order to find a basis for the 2-dimensional kernel of A, we perform standard Gauss elimination. Having obtained the independent $\boldsymbol{\omega}_l$, we take the linear combination (7), and find the components of $\mathbf{u}_l$ automatically satisfy the boundary Weyl equations (8). The solutions of the boundary Weyl equations $\mathbf{v}_l$ have already been given by (9).

Finally we notice that the numerical Nahm transform for the calorons in this process is known to cause a singularity lines for the action density [14]. The origin of the singularity lines are the gap of the zero modes, and we can eliminate the singularity lines using the appropriate procedure, which will be given in a forthcoming paper.

4. Nahm data of the C_3-symmetric 3-caloron

The bulk Nahm data of the 3-caloron with C_3-symmetry are given by [18],

$$T_1 = \frac{1}{2}\begin{pmatrix} 0 & f_+ - if_- & f_0 \\ f_+ + if_- & 0 & f_+ - if_- \\ f_0 & f_+ + if_- & 0 \end{pmatrix}, \tag{12a}$$

$$T_2 = \frac{1}{2}\begin{pmatrix} f_0 & -f_+ - if_- & 0 \\ -f_+ + if_- & 0 & f_+ + if_- \\ 0 & f_+ - if_- & -f_0 \end{pmatrix}, \tag{12b}$$

$$T_3 = \frac{1}{4}\begin{pmatrix} -p_2 & 0 & i(p_0 - p_1) \\ 0 & 2p_2 & 0 \\ -i(p_0 - p_1) & 0 & -p_2 \end{pmatrix}, \tag{12c}$$

where $f_\pm := (f_1 \pm f_2)/2$, and $f_j, p_j\,(j = 0,1,2)$ are defined by Jacobi theta functions $\vartheta_\nu(s,q)$ $(0 < q < 1)$ [22],

$$f_0(s) := iC\frac{\sqrt{\vartheta_\nu(s_1)\vartheta_\nu(s_2)}}{\vartheta_\nu(s_0)}, f_1(s) := iC\frac{\sqrt{\vartheta_\nu(s_2)\vartheta_\nu(s_0)}}{\vartheta_\nu(s_1)},$$
$$f_2(s) := iC\frac{\sqrt{\vartheta_\nu(s_0)\vartheta_\nu(s_1)}}{\vartheta_\nu(s_2)}, \tag{13}$$

and

$$p_0(s) := \frac{d}{ds}\log\frac{\vartheta_\nu(s_0)}{\vartheta_\nu(s_2)}, p_1(s) := \frac{d}{ds}\log\frac{\vartheta_\nu(s_1)}{\vartheta_\nu(s_0)},$$
$$p_2(s) := \frac{d}{ds}\log\frac{\vartheta_\nu(s_2)}{\vartheta_\nu(s_1)}. \tag{14}$$

In the expression, we omit the explicit q dependence for simplicity. For the hermiticity of the bulk data, the index of the theta functions have to be $\nu = 0$ or 3, and the constant is chosen such that $C := \vartheta_1'(0)/\vartheta_1(1/3) \in i\mathbb{R}$. In addition, we fix $s_j := \pm s + j/3$ for the periodicity $f_{j+3} = f_j$ and $p_{j+3} = p_j$ due to $\vartheta_\nu(s_{j+3}) = \vartheta_\nu(s_j)$. We also find that the bulk data enjoy the reality conditions $T_i(-s) = {}^tT(s)$. Note that the C_3-symmetry is not manifest in this basis of the bulk data [18].

Next we consider the boundary data W. The boundary data in the charge 3 case are given by 3-row vector of quatenion entries

$$W = (\lambda, \rho, \chi)\,. \tag{15}$$

From the matching conditions (2) and using the bulk data (12), we obtain the C_3-symmetric boundary data,

$$\lambda = i\lambda_1(\sigma_1 + \sigma_2)\,, \rho = -\frac{2}{\lambda_1}g(\mu_0)\mathbf{1}_2,$$
$$\chi = -i\lambda_1(\sigma_1 - \sigma_2)\,, \tag{16}$$

where $g(\mu_0) := -f_-(\mu_0)$ and λ_1 is a real parameter constrained by $2\lambda_1^2 = h(\mu_0) > 0$ with the definition

$$h(\mu_0) := -\frac{1}{2}(p_0(\mu_0) - p_1(\mu_0))\,. \tag{17}$$

Note that the constraint gives a restriction on the bulk data

$$p_0(\mu_0) - p_1(\mu_0) < 0. \tag{18}$$

Taking into account the constraint, we find that the C_3-symmetric 3-caloron has one free parameter q.

For the calorons having monopole limit, *e.g.*, [14, 17], the bulk Nahm data are regular on the fundamental period, say, $s \in (-\ell/2, \ell/2)$, and have simple poles at $s = \pm\ell/2$ where the residues span an irreducible representation of $su(2)$. Hence, μ_0 takes values in the range $(0, \ell/2)$. In other words, it has upper bound which corresponds to the monopole limit. For the C_3-symmetric 3-calorons considering here, however, the Nahm data is regular in $s \in \mathbb{R}$, and consequently, μ_0 will take values from zero to infinity.

In order to satisfy the condition (18), the acceptable theta function depends on the value of μ_0. Introducing the term "sector", which is open interval $M_n := (n/2, (n+1)/2)$ for each $n \in \mathbb{Z}_{\geq 0}$, we find

$$\vartheta_\nu := \begin{cases} \vartheta_3, & \text{for } \mu_0 \in M_{2n}, \\ \vartheta_0, & \text{for } \mu_0 \in M_{2n+1}. \end{cases} \tag{19}$$

Now the sign of the argument of the theta function is arbitrary chosen as $s_j = \pm s + j/3$. From the profile of the bulk Nahm data, we find $h(n/2) = g(n/2) = 0$, $n \in \mathbb{Z}_{\geq 0}$. This leads all of the components of the boundary data W is zero, and the boundary zero mode is not well-defined at the boundary of the sectors, $\mu_0 = n/2$. Consequently, when $\mu_0 = n/2$, the C_3-symmetric 3-caloron Nahm data is not well-defined.

5. The action density

In order to get further insight of the solutions, it is surely worthwhile to visualize the action density, which of course is gauge invariant and positive definite. From the ASD conditions $*F_{\mu\nu} = \pm F_{\mu\nu}$, where $*$ is the Hodge dual on $\mathbb{R}^4$, the action density S of calorons can be written as

$$S(x^i, t) := -\frac{1}{2}\mathrm{tr}_2 F_{\mu\nu}^2 = -2\mathrm{tr}_2\left(F_{12}^2 + F_{23}^2 + F_{31}^2\right). \tag{20}$$

Here the field strength tensor is defined as $F_{\mu\nu} := \partial_\mu A_\nu - \partial_\nu A_\mu + [A_\mu, A_\nu]$. Thanks to the ASD conditions, we do not have to calculate the $t = x^4$ derivative in the field strength so that we can regard t as a parameter in the calculation.

In the following, we fix $t = x^4 = 0$. We employ 151 grid points for the dual space, and $71 \times 71 \times 71$ lattice points for the configuration space, which are sufficient for the numerical convergence. We perform the visualization of the action density mainly by Mathematica [21].

As mentioned in the previous section, the Nahm data of C_3-symmetric 3-caloron has a notable, special feature; the period of the dual space μ_0 takes values in $(0, \infty)$ except for the isolated points $n/2$, $n \in \mathbb{Z}_{\geq 0}$. Thus it will be interesting when we plot the action density for the cases of large period. Since our main concern is to clarify the behavior for the change of μ_0, we plot the action densities for fixed q.

5.1. *The first two sectors: M_0, M_1*

In Fig.1, we present the action isosurface plots for values of $\mu_0 \in M_0 = (0.0,\ 0.5)$, $M_1 = (0.5,\ 1.0)$. For the sector M_0, we use ϑ_3, while for the sector M_1 we use ϑ_0, according to (19). One can manifestly see that all the densities exhibit C_3-symmetry. As was already pointed out, the Nahm data are not well-defined at $\mu_0 = 0.5$, and 1.0. We find that the action isosurface gradually shrinks as μ_0 increases and finally disappears at $\mu_0 = 0.5$. After passing $\mu_0 = 0.5$, it appears again and grows, and repeatedly it reduces and finally vanishes at $\mu_0 = 1.0$. Note that in the figure of $\mu_0 = 0.0$, we use the value $\mu_0 = 0.0000000001$ instead of μ_0 being exactly zero, because we cannot perform the numerical calculation for the case.

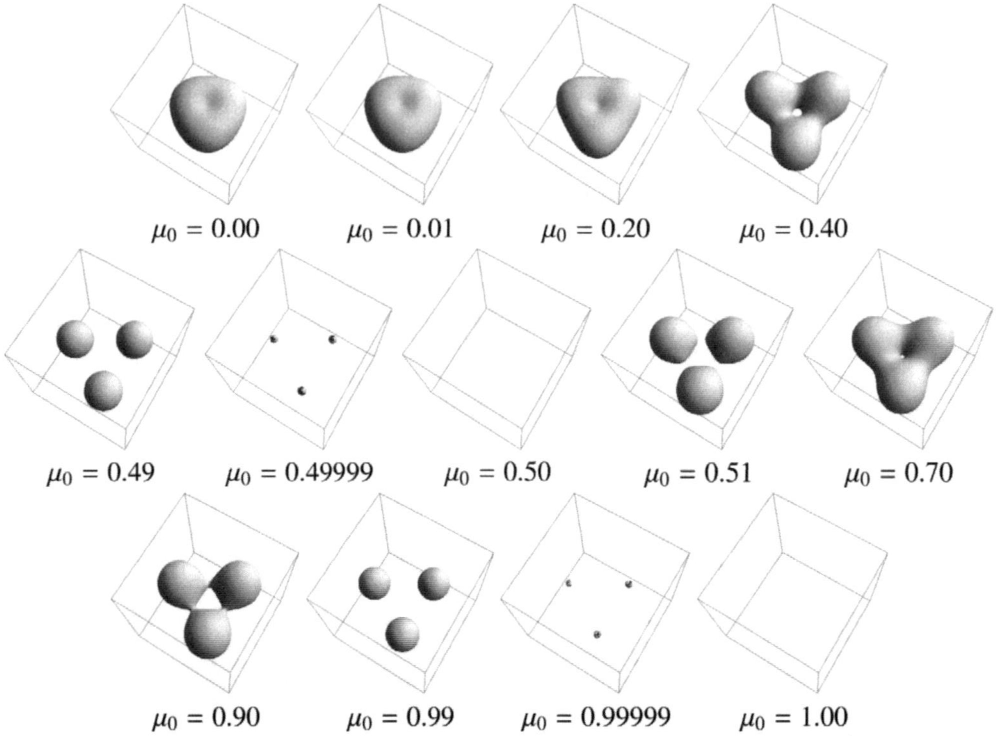

Figure 1. Action isosurface plot of the cyclic 3-caloron with $q = 0.1$, $S(x^i, t) = 7.0$, $t = 0.0$ and $x^i \in (-2.0, 2.0)$ respectively. Note that the plots $\mu_0 = 0.50$ and 1.00 are actually the cases of $\mu_0 = 0.4999999999999999$ and $\mu_0 = 0.999999999999999$, respectively. Because the boundary Nahm data is not well-defined at exactly $\mu_0 = 0.50$ and 1.00. Also, the plot $\mu_0 = 0.00$ is the case $\mu_0 = 0.0000000001$.

5.2. The "exterior" sectors: M_n $(n \geq 2)$

In Figs.2 and 3, we present the isosurface plots of the action density for larger values of μ_0 with fixed q as 0.3 and 0.5. From these plots, we see a periodic behavior of the isosurfaces with fixed $\mu_0(\notin M_0)$ and $\mu_0 + n$, $(n = 1, 2, \cdots)$, namely, the similarity of the plots in the horizontal rows. However, the calculation for $\mu_0 > 1.0$ shows that the action densities are reducing gradually as n increases. We expect the action density tends to vanish for $\mu_0 \to \infty$, which corresponds to the case $\mathbb{R}^3 \times S^1 \to \mathbb{R}^3$ in the configuration space. The detail of this point will be given in elsewhere.

6. conclusion

In this paper, we have constructed a general scheme for numerical Nahm transform of calorons with arbitrary instanton charge. As a first application, we have performed the Nahm transform of the C_3-symmetric 3-calorons and found the quasi-periodic behavior of the action density as μ_0 varies. The bulk Nahm data are given by Jacobi theta functions ϑ_3 and ϑ_0 sector by sector.

Although μ_0, the inverse of the circumference of S^1, can take almost all positive values, the half-integer values $\mu_0 = n/2$, $n \in \mathbb{Z}_{\geq 0}$ are excluded due to the ill-definition of the Nahm data. In particular, the case $\mu_0 = 0.0$, corresponding to the instanton on $\mathbb{R}^4$, is also eliminated from the consideration. At this point, we need complete understanding to the relation with the exact

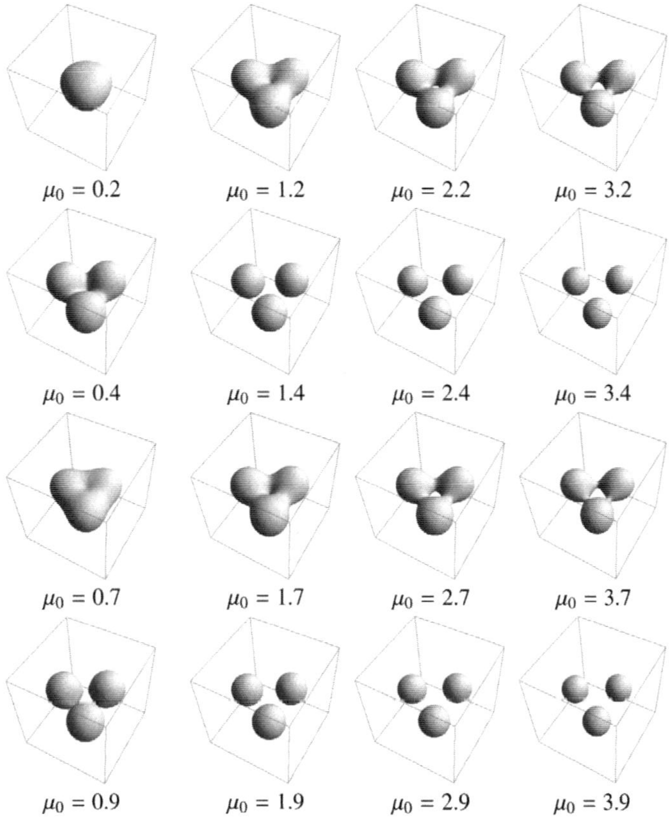

Figure 2. Action isosurface plot with $q = 0.3$, $S(x^i, t) = 7.0$ $t = 0.0$ and $x^i \in (-2.0, 2.0)$, respectively.

ADHM data of C_3-symmetric instanton obtained in [18]. We expect to report the subject in next articles.

Acknowledgement The authors would like to thank D.Muranaka for fruitful discussions and comments. And Koki Takesue is grateful to the conference organizers of ISQS-22 for their kind accommodation and hospitality.

References

[1] N. Manton and P. Sutcliffe, Topological Solitons, Cambridge University Press, Cambridge, 2004.
[2] B.J. Harrington and H. K. Shepard, Phys. Rev. D **17** (1978) 2122-2125; *ibid.* **18** (1978) 2990-2994.
[3] I. Zahed and G. E. Brown, Phys. Rep. **142** (1986) 1-102.
[4] M. Atiyah and N. Hitchin, The Geometry and Dynamics of Magnetic Monopoles, Princeton University Press, Princeton, 1988.
[5] Ya. M. Shnir, Magnetic Monopoles, Springer, Berlin, 2005.
[6] G. Etesi, M. Jardim, Commun.Math.Phys. **280** (2008) 285.
[7] W Nahm, *Self-Dual Monopoles and Calorons*, Springer Lecture Notes in Physics Vol. 201(Springer,New York,1984), pp.189-200.
[8] T. C. Kraan and P. van Baal, Phys. Lett. B **428** (1998) 268-276; *ibid.* **435** (1998) 389-395; Nucl. Phys. B **533** (1998) 627-659.
[9] K. -M. Lee and C. -h. Lu, Phys. Rev. D **58**, 025011 (1998) [hep-th/9802108].
[10] F. Bruckmann, D. Nogradi and P. van Baal, Nucl. Phys. B **666** (2003) 197-229; *ibid.* **698** (2004) 233-254.
[11] A. Gonzalez-Arroyo and C. Pena, JHEP **9809**, 013 (1998) [hep-th/9807172].

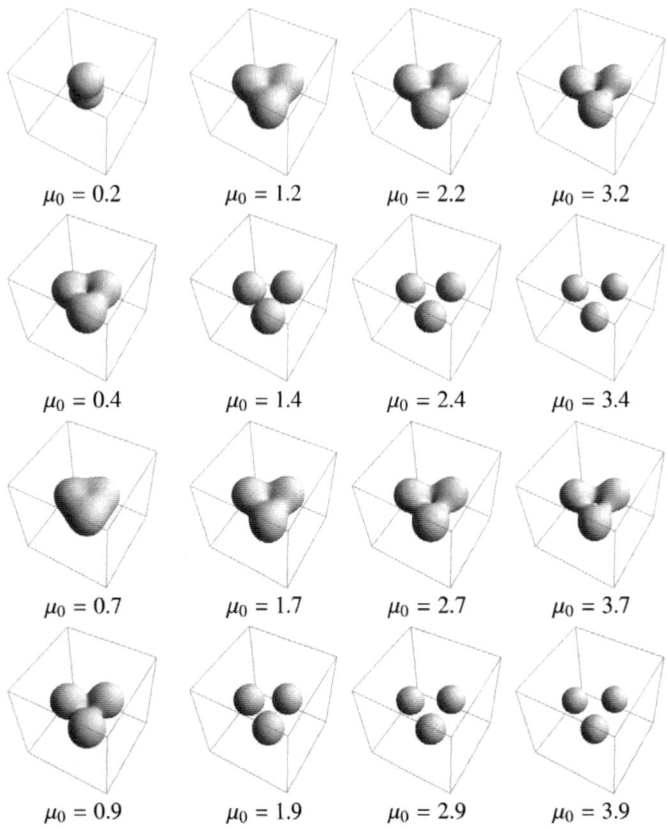

Figure 3. Action isosurface plot with $q = 0.5$, $S(x^i, t) = 7.0$ $t = 0.0$ and $x^i \in (-2.0, 2.0)$, respectively.

[12] M. Garcia Perez, A. Gonzalez-Arroyo, C. Pena and P. van Baal, Nucl. Phys. B **564**, 159 (2000) [hep-th/9905138].
[13] A. Nakamula and J. Sakaguchi, J. Math. Phys. **51** (2010) 043503.
[14] D. Muranaka, A. Nakamula, N. Sawado and K. Toda, Phys.Lett.B **703** (2011) 498.
[15] D. Harland, J. Math. Phys. **48** (2007) 082905.
[16] E. Corrigan and D. B. Fairlie, Phys. Lett. B67 69(1977).
[17] R.S. Ward, Phys.Lett.B **582** (2004) 203.
[18] A. Nakamula and N.Sawado, N.Phys.B **868** (2013) 476-491.
[19] D. Muranaka, A. Nakamula and N.Sawado, J.Phys 411, 012023(2013).
[20] Conor J.Houghton, Paul M.Sutcliffe, Commun. Math. Phys **180** (1996) 343-361.
[21] Wolfram Research, Inc, Mathematica, Version 9.0, 2014.
[22] E. T .Whittaker and G. N. Watson, A course of modern analysis, Cambridge University Press, Cambridge, 1902.

Journal of Physics: Conference Series **563** (2014) 012033 doi:10.1088/1742-6596/563/1/012033

Killing forms and toric Sasaki-Einstein spaces

Vladimir Slesar[1], **Mihai Visinescu**[2] **and Gabriel Eduard Vîlcu**[3,4]

[1] Department of Mathematics, University of Craiova, Str. Al.I. Cuza, Nr. 13, Craiova 200585, Romania

[2] Department of Theoretical Physics, National Institute for Physics and Nuclear Engineering, Magurele, P.O.Box M.G.-6, Romania

[3] Department of Mathematical Modelling, Economic Analysis and Statistics, Petroleum-Gas University of Ploieşti, Bulevardul Bucureşti, Nr. 39, Ploieşti, 100680, Romania

[4] Faculty of Mathematics and Computer Science, Research Center in Geometry, Topology and Algebra, University of Bucharest, Str. Academiei, Nr. 14, Sector 1, Bucharest, 060042, Romania

E-mail: [1]`vlslesar@central.ucv.ro`

E-mail: [2]`mvisin@theory.nipne.ro`

E-mail: [3]`gvilcu@upg-ploiesti.ro`

E-mail: [4]`gvilcu@gta.math.unibuc.ro`

Abstract. The construction of the special Killing forms on toric Sasaki-Einstein manifolds is presented. This goal is achieved using the interplay between complex coordinates of the Calabi-Yau metric cone and the special Killing forms on the toric Sasaki-Einstein space. As a concrete example, we present the complete set of special Killing forms on the five-dimensional Einstein-Sasaki $Y^{p,q}$ spaces. It is pointed out the existence of two additional special Killing forms associated with the complex holomorphic volume form of Calabi-Yau cone manifold.

1. Introduction

In the last time Sasaki-Einstein geometry has become of high interest in connection with some recent developments in mathematics and theoretical physics [1, 2]. Among them, we mention the construction of explicit inhomogeneous Sasaki-Einstein metrics [3, 4]. In particular, in 5-dimensional case an interesting class of toric contact structures on $S^2 \times S^3$ denoted $Y^{p,q}$ (where $q < p$ are positive integers) have been studied by physicists [4]. On the other hand, the Sasaki-Einstein spaces provide supersymmetric backgrounds relevant to the AdS/CFT correspondence [5]. For example Sasaki-Einstein spaces proved to be associated with Euclideanised Kerr-NUT-(A)dS spaces in a certain scaling limit [6]. Moreover, the Kerr-AdS black holes with equal angular momenta in arbitrary odd spacetime dimensions equal to or greater than 5 were considered in [7], the authors showing that Sasaki-Einstein metrics on compact manifolds can be obtained by twisting the Killing vector fields of the black holes. They also discussed the implication of this correspondence to string and M-theory as well as to supersymmetric Yang-Mills theory on Kerr-AdS black hole metrics, focusing on both $AdS_5 \times X_5$ and $AdS_4 \times X_7$ solutions. We recall that $AdS_5 \times X_5$ is a supersymmetric solution of type IIB supergravity that is expected to be dual to a four-dimensional superconformal field theory arising from a stack of D3-branes placed at the tip

Journal of Physics: Conference Series **563** (2014) 012033 doi:10.1088/1742-6596/563/1/012033

of the corresponding Calabi-Yau cone, while $AdS_4 \times X_7$ is a supersymmetric solution of eleven-dimensional supergravity that is expected to be dual to a three-dimensional superconformal field theory arising on a stack of M2-branes sitting at the tip of the corresponding metric cone.

In this paper we want to take a closer look at the special Killing forms on toric Sasaki-Einstein spaces. For this purpose we use foliated coordinates on the metric cone. The interest in the Killing forms is motivated by the significance of symmetries in physics. Killing forms play a fundamental role being related to hidden symmetries, separability of field equations, existence of quantum symmetry operators, supersymmetries, etc.

We shall exemplify the general approach giving an explicit construction of the special Killing forms on 5-dimensional $Y^{p,q}$ spaces.

The paper is organized as follows. In the next Section we describe the Killing forms and emphasize the importance of the special Killing forms in the case of Sasaki spaces. In Section 3 we present the Sasakian geometry and its interrelation with Kähler geometry; also, some particular features of special Killing forms in this framework are emphasized. In Section 4 we exemplify the construction of the special Killing forms on $Y^{p,q}$ spaces. The paper ends with conclusions in Section 5.

2. Special Killing forms

Killing vector fields represent a basic object of differential geometry connected with the infinitesimal isometries. The flow of a Killing vector field preserves a given metric, and there exists a conserved quantity for the geodesic motions. A natural generalization of Killing vector fields is represented by the *conformal Killing vector fields.* In this case the flows preserve a given conformal class of metrics. More generally, one can consider *conformal Killing forms* which are sometimes referred as *twistor forms* or *conformal Killing-Yano tensors.*

Definition. A conformal Killing-Yano tensor of rank p on a n-dimensional Riemannian manifold (M, g) is a p-form Ψ which satisfies

$$\nabla_X \Psi = \frac{1}{p+1} X \lrcorner\, d\Psi - \frac{1}{n-p+1} X^* \wedge d^* \Psi \,, \tag{1}$$

for any vector field X on M.

Here we used the standard conventions: ∇ is the Levi-Civita connection with respect to the metric g, X^* is the 1-form dual to the vector field X, $\lrcorner$ is the operator dual to the wedge product and d^* is the adjoint of the exterior derivative d.

In fact, considering the left hand side of the equation (1) as a natural first order differential operator (called twistor operator), one can also describe the set of conformal Killing-Yano tensors as the kernel of this operator [8]. If Ψ is co-closed in (1), then we obtain the definition of a *Killing-Yano tensor* [9]. Examples of Killing and conformal Killing form can be found on standard sphere, on Sasakian manifolds, on nearly Kähler manifolds [8].

A particular class of Killing forms is represented by the *special Killing forms*:

Definition. A Killing form Ψ is said to be a special Killing form if it satisfies for some constant c the additional equation [10, 8]

$$\nabla_X (d\Psi) = c X^* \wedge \Psi \,, \tag{2}$$

for any vector field X on M.

At the first sight this definition seems to be rather restrictive having in mind that for a space of constant curvature any Killing form is special with $c = R/n(n-1)$ where R is the scalar curvature [10]. Nevertheless, all known examples of Killing forms on Sasaki manifolds are special. Moreover, any Killing form of degree at least 3 on a compact Sasaki-Einstein manifold is actually special and the above result extends on compact 3-Sasakian spaces [11].

In the next Section we see how one can extract these Killing forms in the framework represented by Sasaki-Einstein manifolds

3. Special Killing forms on Sasaki-Einstein manifolds

3.1. Sasaki-Einstein and Kähler structures

A $(2n-1)$-dimensional manifold Y is a *contact manifold* if there exists a 1-form η (called a *contact 1-form*) on M such that

$$\eta \wedge (d\eta)^{n-1} \neq 0\,. \tag{3}$$

For every choice of contact 1-form η there exists a unique vector field K_η, called *Reeb vector field*, that satisfies

$$\eta(K_\eta) = 1 \quad \text{and} \quad K_\eta \lrcorner\, d\eta = 0\,. \tag{4}$$

There are many equivalent definitions of the *Sasakian structures*. A simple and direct definition is the following:

Definition. A compact Riemannian manifold (Y, g) is Sasakian if and only if its metric cone $(X = C(Y) \cong \mathbb{R}_+ \times Y\,,\ \bar{g} = dr^2 + r^2\, g)$ is Kähler.

Here $r \in (0, \infty)$ may be considered as a coordinate on the positive real line $\mathbb{R}_+$. The Sasakian manifold (Y, g) is naturally isometrically embedded into the metric cone via the inclusion $Y = \{r = 1\} = \{1\} \times Y \subset C(Y)$.

Let us denote

$$\tilde{K} \equiv \mathcal{J}\Big(r\frac{\partial}{\partial r}\Big), \tag{5}$$

where $\mathcal{J}$ is the complex structure on the cone manifold. $\tilde{K} - i\mathcal{J}\tilde{K}$ is a holomorphic vector field on $C(Y)$ and the restriction K of $\tilde{K}$ to $Y \subset C(Y)$ is the Reeb vector field on Y.

Now, let Y be a Sasaki-Einstein manifold of dimension $\dim_{\mathbb{R}} Y = 2n-1$ and its Kähler cone $X - C(Y)$ is of dimension $\dim_{\mathbb{R}} X = 2n$, $(\dim_{\mathbb{C}} X = n)$. Sasaki Einstein geometry is naturally "sandwiched" between two Kähler-Einstein geometries as shown in the following proposition [2]:

Proposition. Let (Y, g) be a Sasaki manifold of dimension $2n-1$. Then, the following statements are equivalent

(i) (Y, g) is Sasaki-Einstein with $\mathrm{Ric}_g = 2(n-1)g$;

(ii) The Kähler cone $(C(Y), \bar{g})$ is Ricci-flat ($\mathrm{Ric}_{\bar{g}} = 0$), i. e. Calabi-Yau manifold;

(iii) The transverse Kähler structure to the Reeb foliation $\mathcal{F}_K$ is Kähler-Einstein with $\mathrm{Ric}^T = 2ng^T$.

The Kähler form ω is a 2-form which is exact and homogeneous of degree 2 under the Euler angle $r\frac{\partial}{\partial r}$

$$\omega = -\frac{1}{2}d(r^2\eta) = -r dr \wedge \eta - \frac{1}{2}r^2 d\eta \quad , \quad \mathcal{L}_{r\frac{\partial}{\partial r}}\omega = 2\omega\,. \tag{6}$$

Here η is the Sasakian 1-form of Y. It lifts to $C(Y)$ as

$$\eta = \mathcal{J}\Big(\frac{dr}{r}\Big) = i(\partial - \bar{\partial})\log r\,. \tag{7}$$

We use the same letter η by the abuse of notation. From (5) and (7) it results that $\tilde{K}$ is dual to the 1-form $r^2\eta$. The Kähler form ω can be written as

$$\omega = \frac{1}{2} i\partial\bar{\partial} r^2\,, \tag{8}$$

which means that

$$F = \frac{r^2}{4}\,, \tag{9}$$

is the Kähler potential.

3.2. Special Killing forms and complex coordinates

There is a correspondence between special Killing forms defined on the Sasaki-Einstein manifold Y and the parallel forms defined on the metric cone $C(Y)$. Namely, a p-dimensional differential form Ψ is a special form on Y if and only if the corresponding form

$$\Psi_{cone} := r^p dr \wedge \Psi + \frac{r^{p+1}}{p+1} d\Psi \,, \tag{10}$$

is parallel on $C(M)$.

In particular on a $(2n-1)$-dimensional Sasaki manifold with the contact 1-form η there are the following special Killing forms:

$$\Psi_k = \eta \wedge (d\eta)^k \quad , \quad k = 0, 1, \cdots, n-1 \,. \tag{11}$$

Besides these Killing forms, there are $n-1$ closed conformal Killing forms (also called $*$-Killing forms)

$$\Phi_k = (d\eta)^k \quad , \quad k = 1, \cdots, n-1 \,. \tag{12}$$

Moreover, in the case of the Calabi-Yau cone the holonomy is $SU(n)$ and there are *two additional* parallel forms of degree n. In order to write explicitly the additional Killing forms which correspond to these parallel forms, we shall express the volume form of the metric cone in terms of the Kähler form (6)

$$d\mathcal{V} = \frac{1}{n!} \omega^n \,. \tag{13}$$

Here ω^n is the wedge product of ω with itself n times. The volume of a Kähler manifold can be also written as [8, 12]

$$d\mathcal{V} = \frac{i^n}{2^n} (-1)^{n(n-1)/2} \Omega \wedge \overline{\Omega} \,, \tag{14}$$

where Ω is the complex volume holomorphic $(n,0)$ form of $C(Y)$. The additional (real) parallel forms are given by the real and (respectively) the imaginary part of the complex volume form.

Having in mind the connection between the special Killing forms and the complex volume holomorphic form of the metric cone, it is necessary to consider in the following the complex coordinated defined on a toric Sasaki-Einstein manifold.

To be more specific, let us consider that the n-dimensional metric cone (X, ω) is toric. This means that the real torus $\mathbb{T}^n$ acts effectively on X, preserving the Kähler form ω. We introduce the symplectic coordinates (y^i, ϕ^i) where ϕ^i are the angular coordinates along the orbits of the torus action. The corresponding Kähler metric on $X = C(Y)$ is [13]

$$ds^2 = G_{ij} dy^i dy^j + G^{ij} d\phi^i d\phi^j \,, \tag{15}$$

where the metric coefficients are obtained using the symplectic potential G

$$G_{ij} = \frac{\partial^2 G}{\partial y^i \partial y^j} \,, \tag{16}$$

and $\left(G^{ij}\right) = \left(G_{ij}\right)^{-1}$.

From symplectic coordinates we can pass to complex ones $z^i := x^i + \mathrm{i}\phi^i$, and the metric is written in the form

$$ds^2 = F_{ij} dx^i dx^j + F_{ij} d\phi^i d\phi^j \,. \tag{17}$$

In this setting, the metric coefficients are again obtained using the Hessian of the Kähler potential F, i.e.

$$F_{ij} = \frac{\partial^2 F}{\partial x^i \partial x^j} \,. \tag{18}$$

The symplectic potential G and the Kähler potential F are related by the Legendre transform

$$F(x) = \left(y^i \frac{\partial G}{\partial y^i} - G \right) \ (y = \partial F / \partial x) \ . \tag{19}$$

From the above considerations it follows that $F_{ij} = G^{ij}$ $(y = \partial F / \partial x)$.

Our interest for complex coordinates on Sasakian toric manifolds is motivated by the fact that using this particular type of coordinates it is possible to describe a complex volume form in a very convenient way. More exactly, on these manifolds the complex volume form Ω can be written as [13]

$$\Omega = e^{z^1} dz^1 \wedge .. \wedge dz^n \, .$$

The above relation makes possible the calculation of the special Killing forms on manifolds of Sasaki-Einstein type.

3.3. *Complex volume form and foliated coordinates on geometric cone*

Within this subsection, we show that working with foliated coordinates we can locally extract the special Killing forms on a Sasaki-Einstein manifold.

In order to study the foliated structure on the metric cone $C(Y)$, we consider from now on the existence of (local) foliated coordinates $\left(r; f^2, .., f^n, \phi^1, .., \phi^n\right)$. r will stands for the *transverse coordinate*, while $f^2, ..., f^n, \phi^1, .., \phi^n$ will stand for *leafwise coordinates*; for general topics concerning foliated structures we refer to [14]. As the coordinates (x^i, ϕ^i) are assumed independent, the initial local coordinates $\left(x^i, \phi^i\right)$ and the foliated coordinates $\left(r; f^i, \phi^i\right)$ are related in the following manner

$$\begin{cases} x^i = x^i \left(r, f^2, .., f^n\right) \, , \\ \phi^i = \phi^i \, . \end{cases} \tag{20}$$

Consequently, for the coframes $\left(dx^i, d\phi^i\right)$ and $\left(dr; df^i, d\phi^i\right)$ we get

$$\begin{cases} dx^i = \frac{\partial x^i}{\partial r} dr + \frac{\partial x^i}{\partial f^j} df^j \, , \\ d\phi^i = d\phi^i \, . \end{cases} \tag{21}$$

Using foliated coordinates, the complex volume form Ω is:

$$\begin{aligned} \Omega =& e^{z^1} \left(\frac{\partial x^1}{\partial r} dr + \frac{\partial x^1}{\partial f^j} df^j + \mathrm{i} d\phi^1 \right) \wedge \ldots \\ & \cdots \wedge \left(\frac{\partial x^n}{\partial r} dr + \frac{\partial x^n}{\partial f^j} df^j + \mathrm{i} d\phi^n \right) \, . \end{aligned} \tag{22}$$

With this preparative, and using (10), we are ready to extract the complex differential form ω^Y which corresponds to the complex volume form Ω:

$$\Omega = r^{n-1} dr \wedge \omega^Y + \frac{r^n}{n} d\omega^Y \, . \tag{23}$$

The explicit expression of ω^Y in the foliated coordinates approach is given in [15]. In the next Section we shall exemplify this construction in the case of the 5-dimensional Sasaki-Einstein space $Y^{p,q}$.

Journal of Physics: Conference Series **563** (2014) 012033 doi:10.1088/1742-6596/563/1/012033

4. An application: Special Killing 2-forms on $Y^{p,q}$ space

The special Killing 2-forms on $Y^{p,q}$ space were previously obtained in [16] using a different approach. In what follow we shall construct these special Killing forms working with foliated coordinates.

For the beginning, we consider the explicit local metric of the 5-dimensional $Y^{p,q}$ manifold given by the line element [17]

$$\begin{aligned} ds^2 = & \frac{1-cy}{6}(d\theta^2+\sin^2\theta\, d\phi^2)+\frac{1}{w(y)q(y)}dy^2+\frac{q(y)}{9}(d\psi-\cos\theta\, d\phi)^2 \\ & + w(y)\left[d\alpha+\frac{ac-2y+cy^2}{6(a-y^2)}[d\psi-\cos\theta\, d\phi]\right]^2 , \end{aligned} \tag{24}$$

where

$$\begin{aligned} w(y) &= \frac{2(a-y^2)}{1-cy} , \\ q(y) &= \frac{a-3y^2+2cy^3}{a-y^2} . \end{aligned} \tag{25}$$

The constant c can be rescaled by a diffeomorphism, so in what follows we take $c=1$. For $0<\alpha<1$ we can take the range of the angular coordinates (θ,ϕ,ψ) to be $0\leq\theta\leq 2\pi\,, 0\leq \phi\leq 2\pi\,, 0\leq\psi\leq 2\pi$. Choosing $0<a<1$, the range of the coordinate y is taken between the negative and the smallest positive roots of the cubic equation

$$a-3y^2+2y^3=0 . \tag{26}$$

Following [17, 13], for this particular space we take the complex coordinates

$$\begin{aligned} z^1 :=& \log\left(r^3\sin\theta\sqrt{\frac{p(y)(1-y)}{2}}e^{i\psi}\right) \\ =& 3\ln r+\ln\sin\theta+\frac{1}{2}\ln\frac{p(y)(1-y)}{2}+i\psi' , \\ z^2 :=& \frac{1}{3\sqrt{3}}\log\left(\tan\frac{\theta}{2}e^{i\phi}\right) \\ =& \frac{1}{3\sqrt{3}}\ln\tan\frac{\theta}{2}+i\phi' , \\ z^3 :=& \frac{1}{6}\log\left(\frac{1}{\sin\theta}\sqrt{(y-y_1)^{-\frac{1}{y^1}}(y_2-y)^{-\frac{1}{y^2}}(y_3-y)^{-\frac{1}{y^3}}}e^{-6i\alpha-i\psi}\right) \\ =& -\frac{1}{6}\ln\sin\theta-\frac{1}{2}\ln\left((y-y_1)^{-\frac{1}{y^1}}(y_2-y)^{-\frac{1}{y^2}}(y_3-y)^{-\frac{1}{y^3}}\right) \\ & +i\beta' , \end{aligned} \tag{27}$$

where we denote (see also [16])

$$\begin{aligned} p(y) &:= w(y)\cdot q(y) = \frac{2\left(a-3y^2+2y^3\right)}{a-y^2} , \\ \beta' &:= -\alpha-\frac{1}{6}\psi , \\ \phi' &:= \frac{1}{3\sqrt{3}}\phi , \\ \psi' &:= \psi \end{aligned} \tag{28}$$

From (27) we can extract the real parts of the complex coordinates:

$$\begin{aligned} x^1 &= 3\ln r + \ln\sin\theta + \frac{1}{2}\ln\frac{p(y)(1-y)}{2}\,,\\ x^2 &= \frac{1}{3\sqrt{3}}\ln\tan\frac{\theta}{2}\,,\\ x^3 &= -\frac{1}{6}\ln\sin\theta - \frac{1}{2}\ln\left((y-y_1)^{-\frac{1}{y^1}}(y_2-y)^{-\frac{1}{y^2}}(y_3-y)^{-\frac{1}{y^3}}\right)\,, \end{aligned} \tag{29}$$

while for the toric coordinates we have

$$\begin{aligned} \phi^1 &= \psi'\,,\\ \phi^2 &= \phi'\,,\\ \phi^3 &= \beta'\,. \end{aligned} \tag{30}$$

From the relation (29) and (30) it follows that we can consider the foliated coordinates $(r;\theta,y,\psi',\phi',\beta')$. We regard r as transverse coordinate and $\theta,y,\psi',\phi',\beta'$ as leafwise coordinates, describing the immersed submanifolds.

It is easy to verify that

$$\Omega = e^{z^1}dz^1\wedge dz^2\wedge dz^3\,, \tag{31}$$

is the correct complex volume 3-form.

Using the procedure described in the previous Section, after some calculations we finally get the 2- form ω^Y on the Sasaki-Einstein space $Y^{p,q}$:

$$\begin{aligned} \omega^Y = e^{\mathrm{i}\psi'}\sqrt{\frac{1-y}{6p(y)}}\Big(&d\theta\wedge dy + \mathrm{i}\frac{\sqrt{3}}{2}\cos\theta p(y)d\theta\wedge d\phi'\\ &-\mathrm{i}3\sqrt{3}\sin\theta dy\wedge d\phi' + \mathrm{i}p(y)d\theta\wedge d\beta'\\ &-3\sqrt{3}p(y)\sin\theta d\phi'\wedge d\beta'\Big)\,. \end{aligned} \tag{32}$$

Calculating the real and imaginary parts, we end up with the desired special Killing 2-forms on the Sasaki-Einstein $Y^{p,q}$ space:

$$\begin{aligned} \Xi = \sqrt{\frac{1-y}{6p(y)}}\Big(&\cos\psi'\left[-dy\wedge d\theta + 3\sqrt{3}p(y)\sin\theta d\beta'\wedge d\phi'\right]\\ &-\sin\psi'\Big[-3\sqrt{3}\sin\theta dy\wedge d\phi' - p(y)d\beta'\wedge d\theta\\ &+\frac{\sqrt{3}}{2}\cos\theta p(y)d\theta\wedge d\phi'\Big]\Big)\,, \end{aligned} \tag{33}$$

and respectively

$$\begin{aligned} \Upsilon = \sqrt{\frac{1-y}{6p(y)}}\Big(&\cos\psi'\Big[-3\sqrt{3}\sin\theta dy\wedge d\phi' - p(y)d\beta'\wedge d\theta\\ &+\frac{\sqrt{3}}{2}\cos\theta p(y)d\theta\wedge d\phi'\Big] + \sin\psi'\Big[-dy\wedge d\theta\\ &+3\sqrt{3}p(y)\sin\theta d\beta'\wedge d\phi'\Big]\Big)\,. \end{aligned} \tag{34}$$

Finally, we calculate the Killing forms described in (11-12) using the contact 1-form η. We obtain

$$\begin{aligned}\eta &= \frac{1}{3}\left(d\psi' - (1-y)\cos\theta d\phi' + yd\beta'\right) ,\\ d\eta &= \frac{1}{3}\left((1-y)\sin\theta d\theta\wedge d\phi' + dy\wedge d\beta' + \cos\theta dy\wedge d\phi'\right) .\end{aligned} \tag{35}$$

Then, using (11), (12) and (35), for $k = 1$ and 2 we get (see also [16])

$$\begin{aligned}\Psi_1 =& \frac{1}{9}\left((1-y)\sin\theta d\theta\wedge d\phi'\wedge d\psi' + dy\wedge d\beta'\wedge d\psi\right.\\ &+ \cos\theta dy\wedge d\phi'\wedge d\psi' - \cos\theta dy\wedge d\beta'\wedge d\phi'\\ &\left.+ (1-y)\,y\sin\theta d\beta'\wedge d\theta\wedge d\phi'\right) ,\\ \Psi_2 =& \frac{2}{27}\left((1-y)\sin\theta dy\wedge d\beta'\wedge d\theta\wedge d\phi'\wedge d\psi'\right) ,\end{aligned} \tag{36}$$

and respectively

$$\begin{aligned}\Phi_1 =& \frac{1}{3}\left((1-y)\sin\theta d\theta\wedge d\phi' + dy\wedge d\beta' + \cos\theta dy\wedge d\phi'\right) ,\\ \Phi_2 =& \frac{2}{9}\,(1-y)\sin\theta dy\wedge d\beta'\wedge d\theta\wedge d\phi' .\end{aligned} \tag{37}$$

5. Conclusions

In this expository paper we present the complete set of Killing forms on 5-dimensional Sasaki-Einstein spaces $Y^{p,q}$. Furthermore, a general scheme to construct the special Killing forms on toric Sasaki-Einstein manifolds is described. In fact, using toric geometry many examples of Sasaki-Einstein manifolds can be constructed, and these spaces are highly relevant for the predictions of the AdS/CFT correspondence [13].

Concerning Killing forms, let us stress the fact that a family of Stäckel-Killing tensors can be associated with these differential forms, implying the superintegrability of the geodesic motions. Also, using the third rank Killing-Yano tensors on $Y^{p,q}$ a geometrical interpretation of the Lax representation can be provided [18, 19, 20].

On the other hand, concerning the potential of the general scheme presented in the paper, we emphasize the fact that many non-trivial examples of toric Sasaki-Einstein manifolds occur in the literature; we only refer here to [21], where authors study two infinite families $Y^{p,q}$ of Sasaki-Einstein 7-manifolds; these are lens-space bundles $S^3/\mathbb{Z}_p$ over CP^2 and $CP^1 \times CP^1$, respectively, and it turns out that the metric cones of these Sasaki-Einstein manifolds are toric.

Acknowledgments

The work of M. V. has been partly supported by the joint Romanian-LIT, JINR, Dubna Research Project, theme no. 05-6-1119-2014/2016 and partly by the program NUCLEU PN-09370102, Romania. G. E. V. was supported by National Research Council - Executive Agency for Higher Education Research and Innovation Funding (CNCS-UEFISCDI), project number PN-II-ID-PCE-2011-3-0118.

References

[1] Boyer C and Galicki K 2010 Sasakian geometry, holonomy, and supersymmetry *Handbook of pseudo-Riemannian geometry and supersymmetry* (IRMA Lect. Math. Theor. Phys.) vol 16 (Zürich: Eur. Math. Soc.) p 39 (*Preprint* math/0703231)
[2] Sparks J 2011 Sasaki Einstein Manifolds *Surv. Diff. Geom.* **16** 265 (*Preprint* 1004.2461)

[3] Gauntlett J P, Martelli D, Sparks J and Waldram D 2004 Supersymmetric AdS_5 solutions of M-theory *Class. Quant. Grav.* **21** 4335 (*Preprint* hep-th/0402153)
[4] Gauntlett J P, Martelli D, Sparks J and Waldram D 2004 Sasaki-Einstein metrics on $S^2 \times S^3$ *Adv. Theor. Math. Phys.* **8** 711 (*Preprint* hep-th/0403002)
[5] Maldacena J M 1998 The large N limit of superconformal field theories and supergravity *Adv. Theor. Math. Phys.* **2** 231
[6] Kubizňák D 2009 On the supersymmetric limit of Kerr-NUT-AdS metrics *Phys. Lett. B* **675** 110 (*Preprint* 0902.1999)
[7] Hashimoto Y, Sakaguchi M and Yasui Y 2004 Sasaki-Einstein twist of Kerr-AdS black holes *Phys. Lett. B* **600** 270 (*Preprint* 1103.5573)
[8] Semmelmann U 2003 Conformal Killing forms on Riemannian manifolds *Math. Z.* **245** 503 (*Preprint* math/0206117)
[9] Yano K 1952 Some remarks on tensor fields and curvature *Ann. of Math.* **55** 328
[10] Tachibana S and Yu W N 1970 On a Riemannian space admitting more than one Sasakian structures *Tôhoku. Math. J.* **22** 536
[11] Stromenger C 2011 *Sasakian manifolds: Differential Forms, Curvature and Conformal Killing Forms* (Ph.D. Thesis, University of Köln)
[12] Visinescu M and Vîlcu G E 2012 Hidden symmetries of Euclideanised Kerr-NUT-(A)dS metrics in certain scaling limit *SIGMA* **8** 058 (*Preprint* 1205.6036)
[13] Martelli D, Sparks J and Yau S T 2005 The geometric dual of a-maximisation for toric Sasaki-Einstein manifolds *Commun. Math. Phys.* **268** 39 (*Preprint* hep-th/0503183)
[14] Tondeur Ph 1997 *Geometry of Foliations* (Basel: Birkhäuser)
[15] Slesar V, Visinescu M and Vîlcu G E 2014 Special Killing forms on Sasaki-Einstein manifolds *Physica Scripta* (in press) (*Preprint* 1403.1015)
[16] Visinescu M 2012 Killing forms on the five-dimensional Einstein-Sasaki $Y(p,q)$ spaces *Mod. Phys. Lett. A* **27** 1250217 (*Preprint* 1207.2581)
[17] Martelli D and Sparks J 2006 Toric geometry, Sasaki-Einstein manifolds and a new infinite class of AdS/CFT duals *Commun. Math. Phys.* **262** 51 (*Preprint* hep-th/0411238)
[18] Rosquist K and Goliath M 1998 Lax pair tensors and integrable spacetimes *Gen. Rel. Grav.* **30** 1521 (*Preprint* gr-qc/9707003)
[19] Goliath M, Karlovini M and Rosquist K 1999 Lax pair tensors in arbitrary dimensions *J. Phys. A: Math. Gen.* **32** 3377 (*Preprint* solv-int/9810011)
[20] Visinescu A and Visinescu M 2012 Killing-Yano tensors of rank three and Lax pair tensors *Rom. J. Phys.* **57** 1002
[21] Martelli D and Sparks J 2008 Notes on toric Sasaki-Einstein seven-manifolds and AdS4/CFT3 *J. High Energy Phys.* **11** 016 (*Preprint* 0808.0904)

Journal of Physics: Conference Series **563** (2014) 012034 doi:10.1088/1742-6596/563/1/012034

Function algebras on a 2-dimensional quantum complex plane

I Cohen[1] and E Wagner[2]

[1] Instituto de Física y Matemáticas, Universidad Michoacana, Morelia, Mexico, and Centro de Ciencias Matemáticas, UNAM, Morelia, Mexico
[2] Instituto de Física y Matemáticas, Universidad Michoacana, Morelia, Mexico

E-mail: ismaelcohen10@gmail.com elmar@ifm.umich.mx

Abstract. The well-behaved representations of the coordinate algebra of a 2-dimensional quantum complex plane are classified and a C*-algebra is defined which can be viewed as the algebra of continuous functions on the 2-dimensional quantum complex plane vanishing at infinity.

1. Introduction

The general purpose of this paper is to study non-compact quantum spaces in the C*-algebraic framework. Usually quantum spaces arising in Quantum Group Theory are given by generators and relations. The (*-)algebra obtained in this way can then be viewed as the coordinate ring of polynomial functions on the quantum space. For compact quantum spaces, there is a general procedure to assign a unital C*-algebra to the quantum space: one considers the universal C*-norm defined as the supremum of the operator norms of all bounded *-representation of the coordinate ring and takes the closure with respect to this norm. Here, having a compact quantum space is essentially synonymous to the existence of the universal C*-norm.

The non-compact situation is characterized by the fact that the *-algebra admits unbounded *-representations and that the universal C*-norm might not exist. For this setting, S. L. Woronowicz developed a theory of C*-algebras generated by unbounded elements [10, 13]. However, this method is not constructive, the unbounded operators and the C*-algebra have to be given at the beginning, one only proves that the unbounded operators actually generate the C*-algebra. Since we are more interested in having an explicit C*-algebra at hand than proving technical details, we prefer to construct a non-commutative C*-algebra by analogy to the classical C*-algebra of continuous functions vanishing at infinity on the corresponding locally compact space. The analogy to the classical case involves concrete Hilbert space representations of the coordinate ring and can therefore be done only by a case to case study. In the present paper, we will do it for a 2-dimensional quantum complex plane, the 1-dimensional version has already been treated in [3,9]. Similar construction of function algebras on non-compact quantum spaces can be found, for instance, in [1, 5, 6, 11, 12] but none of these papers touches the C*-algebra framework.

Let us briefly outline our construction. First we classify all well-behaved Hilbert space representations of the coordinate ring $\mathcal{O}(\mathbb{C}_q^2)$. It is important to have knowledge about all possible representations because it turns out that different representations correspond to different domains of the quantum complex plane. The next step is to realize these representations on a function space ($\mathcal{L}_2$-space) such that modulus of each generator (the non-negative self adjoint part in its polar decomposition) acts as a multiplication operator. Furthermore, the measures are chosen in such a way that the partial isometries from the polar decompositions are given on the same footing: they act as multiplicative q-shifts on functions. In this manner we obtain very simple commutation relations between the multiplication operators and the partial isometries. Then we consider an auxiliary *-algebra of bounded operators generated by continuous functions of the moduli of the generators (represented by multiplication operators) and powers of the partial isometries and their adjoints. For the interpretation as continuous functions on the 2-dimensional quantum complex plane vanishing at infinity, we require that the continuous functions belong to $C_0([0,\infty)\times[0,\infty))$ and that these functions, when evaluated at 0, do not depend on the phases (the partial isometries from the polar decompositions). Moreover, in order not to "miss any points", we consider some sort of universal representation, where the involved measures have the largest possible support. Finally, the C*-algebra of continuous functions vanishing at infinity is defined by taking the C*-closure of the auxiliary algebra in the operator norm.

An advantage of our approach is that it allows a geometric interpretation of the different representations. As usual, a nontrivial 1-dimensional representation corresponds to a classical point, in our case to the origin of $\mathbb{C}_q^2$. Setting one generator to zero, we get a copy of $\mathbb{C}_q$ inserted into the quantum space $\mathbb{C}_q^2$. Last but not least, there is a family of faithful representations that describe a 2-dimensional quantum complex plane, where the copy $\mathbb{C}_q$ from the previous representation is shrunk to a point. Therefore, restricting oneself (as quite customary) to the family of faithful representations will not yield the whole 2-dimensional quantum complex plane.

2. Preliminaries

Throughout this paper, q stands for a real number in the interval $(0,1)$. The coordinate ring $\mathcal{O}(\mathbb{C}_q^2)$ of polynomial functions on 2-dimensional quantum complex plane is the *-algebra over $\mathbb{C}$ generated by z_1 and z_2 satisfying the (overcomplete) relations [4]

$$z_2z_1 = qz_1z_2\,, \qquad z_1^*z_2^* = qz_2^*z_1^*\,, \tag{1}$$
$$z_2z_1^* = qz_1^*z_2\,, \qquad z_1z_2^* = qz_2^*z_1\,, \tag{2}$$
$$z_2z_2^* = q^2z_2^*z_2\,, \qquad z_1z_1^* = q^2z_1^*z_1 - (1-q^2)z_2^*z_2\,. \tag{3}$$

By a slight abuse of notation, we will use in the following sections the same letter to denote a generator of the coordinate ring $\mathcal{O}(\mathbb{C}_q^2)$ and its representation as a Hilbert space operator.

We adopt the convention that $\mathbb{N}=\{1,2,\ldots\}$ and $\mathbb{N}_0=\{0,1,2,\ldots\}$. Given an at most countable index set I and a Hilbert space $\mathcal{H}_0$, consider the orthogonal sum $\mathcal{H}=\oplus_{i\in I}\mathcal{H}_0$. We write η_i for the vector in $\mathcal{H}$ which has the element $\eta\in\mathcal{H}_0$ as its i-th component and zero otherwise. It is understood that $\eta_i=0$ whenever $i\notin I$.

For a subset $A\subset[0,\infty)$, the indicator function $\chi_A:[0,\infty)\to\mathbb{C}$ is defined by

$$\chi_A(t) := \begin{cases} 1, & t\in A\,, \\ 0, & t\notin A\,. \end{cases} \tag{4}$$

3. Hilbert space representations of the 2-dimensional quantum complex plane

In this section, we give a complete description of "good" *-representations of $\mathcal{O}(\mathbb{C}_q^2)$. Here "good" means that, in order to avoid pathological cases, we impose in Definition 1 some natural regularity conditions on the unbounded operators. These representations will be called *well-behaved*, see [8]. To motivate the regularity conditions, we start with formal algebraic manipulations. These algebraic relations, together with the regularity conditions of Definition 1, will allow us to classify in Theorem 1 all well-behaved representations of $\mathcal{O}(\mathbb{C}_q^2)$.

Let z_1 and z_2 be densely defined closed operators on a Hilbert space $\mathcal{H}$ satisfying the relations (1)–(3) on a common dense domain. Set $Q := z_2^* z_2$. From the relations (1)–(3) within the algebra $\mathcal{O}(\mathbb{C}_q^2)$, we get

$$z_1 Q = Q z_1, \quad z_1^* Q = Q z_1^*, \quad z_2 Q = q^2 Q z_2, \quad z_2^* Q = q^{-2} Q z_2^*, \tag{5}$$

and for any polynomial p in one variable, (5) yields

$$z_1 p(Q) = p(Q) z_1, \quad z_1^* p(Q) = p(Q) z_1^*, \quad z_2 p(Q) = p(q^2 Q) z_2, \quad z_2^* p(Q) = p(q^{-2} Q) z_2^*. \tag{6}$$

Let us assume that (6) holds for all bounded Borel measurable functions on spec(Q), where $p(Q) = \int p(\lambda)\, \mathrm{d}E(\lambda)$ is defined by the spectral theorem with the unique projection-valued measure E of Q. Then $\ker(Q) = E(\{0\})\mathcal{H}$ and $\ker(Q)^\perp = E((0,\infty))\mathcal{H}$ are invariant under the actions of the generators of $\mathcal{O}(\mathbb{C}_q^2)$. On $\ker(Q)$, we have $Q = z_2^* z_2 = 0$, thus $z_2 = z_2^* = 0$, and (3) becomes

$$z_1 z_1^* = q^2 z_1^* z_1. \tag{3'}$$

On $\ker(Q)^\perp$, the operator $\sqrt{Q}^{-1} = \int \frac{1}{\sqrt{\lambda}} \mathrm{d}E(\lambda)$ is well-defined. Consider at the moment the abstract element

$$w := \sqrt{Q}^{-1} z_1 = z_1 \sqrt{Q}^{-1}. \tag{7}$$

Inserting (7) into the second relation of (3) yields formally

$$ww^* - q^2 w^* w = -(1-q^2) z_2^* z_2 Q^{-1} = -(1-q^2). \tag{8}$$

Note that, by (5), we have $z_1^* z_1\, z_2^* z_2 = z_2^* z_2\, z_1^* z_1$. This relation together with (8), (6), (3'), and the first equation in (3) motivate the following definition of well-behaved *-representations of $\mathcal{O}(\mathbb{C}_q^2)$.

Definition 1. A well-behaved *-representations of $\mathcal{O}(\mathbb{C}_q^2)$ is given by densely defined closed operators z_1 and z_2 satisfying (1)–(3) on a common dense domain such that

(i) The self-adjoint operators $z_1^* z_1$ and $z_2^* z_2$ strongly commute.

(ii) z_2 is a q-normal operator, i.e., it satisfies the operator equation

$$z_2 z_2^* = q^2 z_2^* z_2.$$

(iii) For all bounded Borel measurable functions f on spec(Q), the operator relations

$$f(Q) z_1 \subset z_1 f(Q), \quad f(Q) z_1^* \subset z_1^* f(Q), \quad f(Q) z_2 \subset z_2 f(q^2 Q), \quad f(Q) z_2^* \subset z_2^* f(q^2 Q)$$

hold.

(iv) On $\ker(Q)$, z_1 is a q-normal operator, i.e.,

$$z_1 z_1^* = q^2 z_1^* z_1.$$

(v) On $\ker(Q)^\perp$, z_1 commutes with $\sqrt{Q}^{-1}$ and setting $w := \sqrt{Q}^{-1} z_1 = z_1 \sqrt{Q}^{-1}$ defines a densely defined closed operator fulfilling the operator equation

$$ww^* = q^2 w^* w - (1-q^2). \tag{9}$$

Here, the equality of operators on both sides of the equations includes the equality of their domains

The well-behaved representations of q-normal operators have been studied in [2] and [3]. By [3, Corollary 2.2], any q-normal operator ζ on a Hilbert space $\mathcal{G}$ admits the following representation:

$$\mathcal{G} = \ker(\zeta) \oplus (\oplus_{n\in\mathbb{Z}}\,\mathcal{G}_0), \qquad \zeta = 0 \ \text{ on } \ \ker(\zeta), \qquad \zeta\, g_n = q^n Z g_{n-1} \ \text{ on } \ \oplus_{n\in\mathbb{Z}}\,\mathcal{G}_0, \tag{10}$$

where Z denotes a self-adjoint operator on $\mathcal{G}_0$ sucht that $\mathrm{spec}(Z) \subset [q,1]$ and q is not an eigenvalue of Z.

Furthermore, the representations of operators w satisfying (9) on a Hilbert space $\mathcal{G}$ have been classified in [5, Lemma 2.3]. It follows from this lemma that $\mathcal{G}$ can be written as a direct sum $\mathcal{G} = \oplus_{m\in\mathbb{N}}\,\mathcal{G}_0$, and the actions of w and w^* are determined by

$$w\, g_m = \sqrt{q^{-2m}-1}\, g_{m+1}, \quad w^* g_m = \sqrt{q^{-2(m-1)}-1}\, g_{m-1}, \quad g\in\mathcal{G}_0, \ \ m\in\mathbb{N}. \tag{11}$$

Equations (10) and (11) are all we need for the classification of the well-behaved representations of $\mathcal{O}(\mathbb{C}_q^2)$.

Theorem 1. *Any well-behaved Hilbert space representation of $\mathcal{O}(\mathbb{C}_q^2)$ is unitarily equivalent to a representation given by the following formulas: Let $\mathcal{H}_0$, $\mathcal{H}_{00}$ and $\mathcal{N}$ be Hilbert spaces, and let A and B be self-adjoint operators on $\mathcal{H}_0$ and $\mathcal{H}_{00}$, respectively, such that their spectrum belongs to $[q,1]$ and q is not an eigenvalue. Then the Hilbert space $\mathcal{H}$ of the representation decomposes into the direct sum*

$$\mathcal{H} = \mathcal{N} \oplus (\oplus_{k\in\mathbb{Z}}\mathcal{H}_0) \oplus (\oplus_{n\in\mathbb{Z}}\oplus_{m\in\mathbb{N}}\,\mathcal{H}_{00}),$$

and the actions of z_1 and z_2 are determined by

$$z_1 = z_2 = 0 \ \ \text{on} \ \ \mathcal{N}, \tag{12}$$

$$z_1\, h_k = q^k A\, h_{k-1}, \quad z_2 = 0 \ \ \text{on} \ \ \oplus_{k\in\mathbb{Z}}\,\mathcal{H}_0, \tag{13}$$

$$z_1\, h_{n,m} = \sqrt{q^{-2m}-1}\, q^n B\, h_{n,m+1}, \quad z_2\, h_{n,m} = q^n B\, h_{n-1,m} \ \ \text{on} \ \ \oplus_{n\in\mathbb{Z}}\oplus_{m\in\mathbb{N}}\mathcal{H}_{00}. \tag{14}$$

A common dense domain is obtained by considering the subspace of those elements of $\mathcal{H}$ which have at most a finite number of non-zero components in the direct sum. Only the representation (14) is faithful. A representation is irreducible if and only if one of the Hilbert spaces $\mathcal{H}_0$, $\mathcal{H}_{00}$ and $\mathcal{N}$ is isomorphic to $\mathbb{C}$ and the others are zero.

Proof. As the sets $\{0\}$ and $(0,\infty)$ are invariant under multiplication with powers of q, it follows from Definition 1(iii) that $\mathcal{K} := E(\{0\})\mathcal{H}$ and $\mathcal{G} := E((0,\infty))\mathcal{H}$ are invariant under the actions of z_1 and z_2. Clearly, $\mathcal{H} = \mathcal{K}\oplus\mathcal{G}$. Since $\mathcal{K} = \ker(Q) = \ker(z_2^* z_2)$, we have $z_2 = 0$ on $\mathcal{K}$. By Definition 1(iv), the restriction of z_1 to $\mathcal{K}$ is a q-normal operator, therefore its representation is given by (10). Setting $\mathcal{N} := \ker(z_1)$, $\mathcal{H}_0 := \mathcal{G}_0$ and $A := Z$, we obtain (12) y (13) from (10).

By Definition 1(ii) and the definition of $\mathcal{G}$, z_2 is a q-normal operator on $\mathcal{G}$ with $\ker(z_2) = \{0\}$. Therefore z_2 acts on $\mathcal{G} = \oplus_{n\in\mathbb{Z}}\mathcal{G}_0$ by the formulas on right hand side of (10). Note that

$$Q\, g_n = q^{2n} Z^2 g_n \quad \text{on} \quad \mathcal{G}_n := \{g_n : g\in\mathcal{G}_0\}, \tag{15}$$

with $\mathrm{spec}(q^{2n}Z^2) \subset [q^{2n+2}, q^{2n}]$ and q^{2n+2} is not an eigenvalue. Considering the disjoint union $(0,\infty) = \cup_{n\in\mathbb{Z}}\,(q^{2n+2}, q^{2n}]$, one readily sees that $\mathcal{G}_n = E((q^{2n+2}, q^{2n}])\mathcal{G}$. From Definition 1(iii), it follows that

$$E((q^{2n+2},q^{2n}])z_1 \subset z_1 E((q^{2n+2},q^{2n}]), \quad E((0,\infty)\setminus(q^{2n+2},q^{2n}])z_1 \subset z_1 E((0,\infty)\setminus(q^{2n+2},q^{2n}]),$$

and the same holds for z_1 replaced by z_1^*. Since $\sqrt{Q}^{-1}$ trivially commutes with $E((q^{2(n+1)}, q^{2n}])$, we conclude that $w := \sqrt{Q}^{-1} z_1$ and w^* leave $\mathcal{G}_n$ invariant. On $\mathcal{G}_n$, w still satisfies (9), thus its representation is given by (11). Therefore we can write $\mathcal{G}_n = \oplus_{m \in \mathbb{N}} \mathcal{H}_{n0}$ and

$$w\, h_{n,m} = \sqrt{q^{-2m} - 1}\, h_{n,m+1}, \tag{16}$$

where $h_{n,m}$ belongs to the m-th position in the direct sum $\oplus_{m\in\mathbb{N}}\mathcal{H}_{n0}$. But $\mathcal{G}_n$ is just a copy of $\mathcal{G}_0$, so $\mathcal{H}_{n0} = \mathcal{H}_{00}$ for all $n \in \mathbb{Z}$. Equation (16) yields

$$w^* w\, h_{n,m} = (q^{-2m} - 1)\, h_{n,m}, \tag{17}$$

hence $\mathcal{H}_{nm} := \{h_{n,m} : h \in \mathcal{H}_{00}\}$ is the eigenspace for the eigenvalue $q^{-2m} - 1$ of the restriction of w^*w to $\mathcal{G}_n$. Definition 1(i) implies that w^*w and Q strongly commute. Therefore the restrictions of w^*w and Q to $\mathcal{G}_n = E((q^{2n+2}, q^{2n}])\mathcal{G}$ also strongly commute. As a consequence, the self-adjoint operator Z from (15) leaves the eigenspaces $\mathcal{H}_{nm}$ invariant. Denote the restriction of Z to $\mathcal{H}_{00}$ by B. Since $\mathcal{H}_{nm}$ is an identical copy of $\mathcal{H}_{00}$ in the m-th position of the direct sum $\oplus_{m\in\mathbb{N}}\mathcal{H}_{n0}$, we get

$$Z\, h_{n,m} = B\, h_{n,m} \quad \text{for all} \quad h_{n,m} \in \mathcal{H}_{nm}. \tag{18}$$

Moreover, B inherits the spectral properties from Z as required in Theorem 1. Finally, (10) and (18) give

$$z_2\, h_{n,m} = q^n Z\, h_{n-1,m} = q^n B\, h_{n-1,m}, \qquad h_{n,m} \in \mathcal{H}_{nm}, \tag{19}$$

and from (15) and (16), we get

$$z_1\, h_{n,m} = \sqrt{Q}\, w\, h_{n,m} = \sqrt{q^{-2m} - 1}\sqrt{q^{2n} Z^2}\, h_{n,m+1} = \sqrt{q^{-2m} - 1}\, q^n B h_{n,m+1}$$

for all $h_{n,m} \in \mathcal{H}_{nm}$. This proves (14).

That the representation (14) is faithful follows from the fact that the representations of z_2 and ω are faithful, see [2] and [5], respectively. The statement about irreducible representations is obvious since writing any of the Hilbert spaces $\mathcal{H}_0$, $\mathcal{H}_{00}$ and $\mathcal{N}$ as an orthogonal sum of two non-zero subspaces will result in an orthogonal sum of non-trivial representations. □

4. Hilbert space representations on function spaces

Note that the decomposition of $\mathcal{H}$ in Theorem 1 is determined by the spectral properties of the self-adjoint operators Q and w^*w. As well-known [7, Theorem VII.3], each self-adjoint operator T on a separable Hilbert space is unitarily equivalent to a direct sum of multiplication operators on $\mathcal{L}_2(\mathrm{spec}(T), \mu)$. We will use this fact to realize the representations of Theorem 1 on $\mathcal{L}_2$-spaces which will be the basis for studying function algebras on the 2-dimensional quantum complex plane in the next section.

The direct sum of Hilbert spaces $\oplus_{n\in\mathbb{Z}} \oplus_{m\in\mathbb{N}} \mathcal{H}_{00}$ in Theorem 1 is isomorphic to the tensor product $\ell_2(\mathbb{N}) \otimes (\oplus_{n\in\mathbb{Z}}\mathcal{H}_{00})$. Let ζ be a q-normal operator acting on $\oplus_{n\in\mathbb{Z}}\mathcal{H}_{00}$ by the formulas on the right hand side of (10), and let ω act on $\ell_2(\mathbb{N})$ by the formulas in (11) with $\mathcal{G}_0 = \mathbb{C}$. Then z_2 from (19) and w from (16) can be written $z_2 = \mathrm{id} \otimes \zeta$ and $w = \omega \otimes \mathrm{id}$, respectively. It has been shown in [2, Theorem 1] that any q-normal operator ζ is unitarily equivalent to a direct sum of operators of the following form: There exists a q-invariant Borel measure μ on $[0, \infty)$ such that ζ and ζ^* act on $\mathcal{H} = \mathcal{L}_2([0, \infty), \mu)$ by

$$\zeta f(t) = qtf(qt), \;\; \zeta^* f(t) = tf(q^{-1}t), \;\; f \in \mathrm{dom}(\zeta) := \{h \in \mathcal{L}_2([0,\infty), \mu) : \textstyle\int t^2 |h(t)|^2 \mathrm{d}\mu(t) < \infty\}. \tag{20}$$

Here, the q-invariance of the measure means that $\mu(qS) = \mu(S)$ for all Borel subsets S of $[0, \infty)$. Note that $\ker(\zeta) = \{0\}$ if and only if $\mu(\{0\}) = 0$. Therefore, in order to obtain a representation of the form (14), we have to assume that $\mu(\{0\}) = 0$.

To turn $\ell_2(\mathbb{N})$ into an $\mathcal{L}_2$-space, we consider the operator $y := \sqrt{\omega^*\omega + 1}$ on $\ell_2(\mathbb{N})$. Denoting by $\{e_n : n \in \mathbb{N}\}$ the standard basis of $\ell_2(\mathbb{N})$, we have

$$y e_n = q^{-n} e_n, \quad n \in \mathbb{N}. \tag{21}$$

Since the set of eigenvalues of y is discrete, y can be realized as a multiplication operator on $\mathcal{L}_2(\mathrm{spec}(y), \sigma) \cong \ell_2(\mathbb{N})$ by choosing the counting measure $\sigma(\{q^{-n}\}) = 1$ on $\mathrm{spec}(y)$. Extending σ to a Borel measure on $[0, \infty)$ by setting $\sigma([0, \infty) \setminus \mathrm{spec}(y)) := 0$, we get $yg(s) = sg(s)$. The set

$$\{e_n := \chi_{\{q^{-n}\}}(s) : n \in \mathbb{N}\}$$

is an orthonormal basis of $\mathcal{L}_2(\mathrm{spec}(y), \sigma)$, where $\chi_{\{q^{-n}\}}$ denotes the indicator function (4). Note that

$$\chi_{\{q^{-n}\}}(qs) = \chi_{\{q^{-(n+1)}\}}(s) = e_{n+1} \ \text{ and } \ \sqrt{(qs)^2 - 1}\,\chi_{\{q^{-(n+1)}\}}(s) = \sqrt{q^{-2n} - 1}\,\chi_{\{q^{-(n+1)}\}}(s),$$

where we used $f(t)\chi_{\{t_0\}}(t) = f(t_0)\chi_{\{t_0\}}(t)$ in the second equation. Hence

$$\omega g(s) = \sqrt{(qs)^2 - 1}\, g(qs) \quad \text{and} \quad \omega^* g(s) = \sqrt{s^2 - 1}\, g(q^{-1}s) \tag{22}$$

for $g \in \mathrm{dom}(\omega) = \mathrm{dom}(\omega^*) := \{h \in \mathcal{L}_2([0, \infty), \sigma) : \int s^2 |h(s)|^2 \mathrm{d}\sigma(s) < \infty\}$). En particular,

$$\omega^* e_1 = \sqrt{s^2 - 1}\, \chi_{\{q^{-1}\}}(q^{-1}s) = \sqrt{1^2 - 1}\, \chi_{\{q^{-1}\}}(q^{-1}s) = 0, \tag{23}$$

as required. Also, although $||\chi_{\{1\}}(s)|| = 0$ and $\chi_{\{1\}}(qs) = \chi_{\{q^{-1}\}}(s) = e_1$, we have

$$\sqrt{(qs)^2 - 1}\, \chi_{\{1\}}(qs) = \sqrt{(qq^{-1})^2 - 1}\, \chi_{\{1\}}(qs) = 0,$$

so that (22) remains consistent.

Now, under the isomorphism $\mathcal{L}_2([0, \infty), \sigma) \otimes \mathcal{L}_2([0, \infty), \mu) \cong \mathcal{L}_2([0, \infty) \times [0, \infty), \sigma \otimes \mu)$, we obtain from (20) and (22) the following representation of $z_1 = \sqrt{Q}w = \omega \otimes \sqrt{\zeta^*\zeta}$ and $z_2 = \mathrm{id} \otimes \zeta$,

$$z_1 h(s, t) = \sqrt{(qs)^2 - 1}\, t h(qs, t), \qquad z_2 g(s, t) = qt g(s, qt), \tag{24}$$

where $h \in \mathrm{dom}(\omega) \otimes_{\mathrm{alg}} \mathrm{dom}(\zeta)$ and $g \in \mathcal{L}_2([0, \infty), \sigma) \otimes_{\mathrm{alg}} \mathrm{dom}(\zeta)$. To sum up, we have shown that the representations from (14) are unitarily equivalent to a direct sum of representations of the type described in (24).

Recall that $\mathcal{N} \oplus (\oplus_{k \in \mathbb{Z}} \mathcal{K}_0)$ in Theorem 1 corresponds to the kernel of the q-normal operator z_2, and that a q-normal operator in the representation (20) has a trivial kernel if and only if $\mu(\{0\}) = 0$. Since μ from the last paragraph was assumed to satisfy $\mu(\{0\}) = 0$, we will now add a point measure δ_0 centred at 0 to it. By unitary equivalence, we may assume that $\delta_0(\{0\}) = 1$. Then the representation of z_1 on $\oplus_{k \in \mathbb{Z}} \mathcal{K}_0$ is again unitarily equivalent to a direct sum of representations of the type (20). To realize these representations on our $\mathcal{L}_2$-space, we choose a q-invariant measure on $[0, \infty)$, say ν, assume again $\nu(\{0\}) = 0$, take the product measure $\nu \otimes \delta_0$, and add it to $\sigma \otimes \mu$. Then

$$\mathcal{L}_2([0, \infty) \times [0, \infty), \sigma \otimes \mu + \nu \otimes \delta_0) \cong \mathcal{L}_2([0, \infty) \times [0, \infty), \sigma \otimes \mu) \oplus \mathcal{L}_2([0, \infty) \times [0, \infty), \nu \otimes \delta_0),$$

and on $\mathcal{L}_2([0, \infty) \times [0, \infty), \nu \otimes \delta_0)$, we have the representation

$$z_1 h(s, t) = qs h(qs, t) = q\chi_{\{0\}}(t) s h(qs, t), \quad z_2 g(s, t) = qt g(s, qt) = 0 \tag{25}$$

for all h such that $\int s^2|h(0,s)|^2 \mathrm{d}\nu(s) < \infty$ and for all g. Here, for functions depending on the second variable t, we used the fact that $\mathrm{supp}(\delta_0) = \{0\}$. Again, by [2, Theorem 1] and the same argumentation as above, the representations from (13) are unitarily equivalent to a direct sum of representations of the type described in (25).

Finally, to obtain a non-trivial component $\mathcal{N} = \ker z_1 \cap \ker z_2$, we add to $\sigma\otimes\mu+\nu\otimes\delta_0$ the point measure $\epsilon\delta_0 \otimes \delta_0$, where $\epsilon = 0$ or $\epsilon = 1$ depending on whether $\mathcal{N} = \{0\}$ or $\mathcal{N} \neq \{0\}$. Summarizing, we have proven the following theorem:

Theorem 2. *The Hilbert space representations of $\mathcal{O}(\mathbb{C}_q^2)$ from Theorem 1 are unitarily equivalent to a direct sum of representations of the following type: Let μ and ν be q-invariant Borel measures on $[0,\infty)$ such that $\mu(\{0\}) = \nu(\{0\}) = 0$. Denote by δ_0 the Dirac measure centred at 0, and define a Borel measure σ on $[0,\infty)$ by setting $\sigma(\{q^{-n}\}) := 1$ for all $n \in \mathbb{N}$ and $\sigma([0,\infty)\setminus\{q^{-n} : n\in\mathbb{N}\}) := 0$. For $\epsilon \in \{0,1\}$, consider the Hilbert space*

$$\mathcal{H} := \mathcal{L}_2\big([0,\infty)\times[0,\infty),\, \sigma\otimes\mu+\nu\otimes\delta_0 + \epsilon\delta_0\otimes\delta_0\big), \tag{26}$$

and set

$$\mathrm{dom}(z_1) := \{h \in \mathcal{H} \,:\, sth \in \mathcal{H} \text{ and } s\chi_{\{0\}}(t)h \in \mathcal{H}\}, \quad \mathrm{dom}(z_2) := \{g \in \mathcal{H} \,:\, tg \in \mathcal{H}\},$$

where $\chi_{\{0\}}$ denotes the indicator function from (4). For $h \in \mathrm{dom}(z_1)$ and $g \in \mathrm{dom}(z_2)$, the actions of the generators of $\mathcal{O}(\mathbb{C}_q^2)$ are given by

$$z_1 h(s,t) = \sqrt{(qs)^2-1}\,t h(qs,t) + q\chi_{\{0\}}(t)sh(qs,t), \qquad z_2 g(s,t) = qtg(s,qt), \tag{27}$$

$$z_1^* h(s,t) = \sqrt{s^2-1}\,t h(q^{-1}s,t) + \chi_{\{0\}}(t)sh(q^{-1}s,t), \qquad z_2^* g(s,t) = tg(s,q^{-1}t). \tag{28}$$

Note that

$$\begin{aligned}\mathcal{H} &= \mathcal{L}_2\big([0,\infty)\times[0,\infty),\, \epsilon\delta_0\otimes\delta_0\big) \oplus \mathcal{L}_2\big([0,\infty)\times[0,\infty),\, \nu\otimes\delta_0\big) \oplus \mathcal{L}_2\big([0,\infty)\times[0,\infty),\, \sigma\otimes\mu\big)\\ &= \mathcal{L}_2\big(\{0\}\times\{0\},\, \epsilon\delta_0\otimes\delta_0\big) \oplus \mathcal{L}_2\big([0,\infty)\times\{0\},\, \nu\otimes\delta_0\big) \oplus \mathcal{L}_2\big(\{q^{-n} : n\in\mathbb{N}\}\times[0,\infty),\, \sigma\otimes\mu\big)\end{aligned} \tag{29}$$

and that the restriction of the representation (27) to one of the orthogonal components corresponds to one of the representations from (12)–(14). Of course, we could have formulated Theorem 2 for each of the orthogonal subspaces separately. The reason why we prefer to work with a single Hilbert space on the domain $[0,\infty)\times[0,\infty)$ will become clear in the next section.

5. C*-algebra of continuous functions vanishing at infinity

The aim of this section is to define a C*-algebra which can be viewed as the algebra of continuous functions on the 2-dimensional quantum complex plane vanishing at infinity. The definition will be motivated by a similar construction for the 1-dimensional quantum complex plane [3]. As a point of departure, we first look for an auxiliary *-algebra, where the commutation relations are considerable simple.

For the convenience of the reader, we recall the construction of the C*-algebra $C_0(\mathbb{C}_q)$ of continuous functions vanishing at infinity on the 1-dimensional quantum complex plane [3]. Given a representation of the type (20), consider the following *-subalgebra of $\mathrm{B}(\mathcal{L}_2([0,\infty),\mu))$:

$$\text{*-alg}\{C_0(\mathrm{spec}(|\zeta|), U\} := \Big\{ \sum_{\text{finite}} f_k(|\zeta|)U^k \,:\, k\in\mathbb{Z},\ f_k\in C_0(\mathrm{spec}(|\zeta|),\ f_k(0)=0 \text{ if } k\neq 0\Big\}, \tag{30}$$

where $\mu(\{0\}) = 0$ and U denotes the unitary operator from the polar decomposition $\zeta = U|\zeta|$. For all bounded continuous functions f on spec$(|\zeta|)$, the operators $f(|\zeta|)$ and U satisfy the commutation relation

$$U f(|\zeta|) = f(q|\zeta|)U.$$

In [3], a representation of the type (20) of a q-normal operator $Z = U|Z|$ was said to be *universal* if spec$(|Z|) = [0, \infty)$, or equivalently if supp$(\mu) = [0, \infty)$. It has the universal property that

$$\text{*-alg}\{C_0(\text{spec}(|Z|), U\} \ni \sum_{\text{finite}} f_k(|Z|)U^k \longmapsto \sum_{\text{finite}} f_k(|\zeta|)U^k \in \text{*-alg}\{C_0(\text{spec}(|\zeta|), U\} \quad (31)$$

yields always a well-defined surjective *-homomorphism. Although the exact definition in [3] is slightly abstract, [3, Theorem 3.3] states that $C_0(\mathbb{C}_q)$ is isomorphic to the norm closure of *-alg$\{C_0(\text{spec}(|Z|), U\}$ in $\mathrm{B}(\mathcal{L}_2([0, \infty), \mu))$.

Motivated by the previous description, we call a representation from Theorem 2 *universal* if $\epsilon = 1$ and supp$(\mu) =$ supp$(\nu) = [0, \infty)$. Such q-invariant measures can be obtained, for instance, by taking the Lebesgue measure λ on $(q, 1]$ and setting

$$\mu(M) = \sum_{k\in\mathbb{Z}} \lambda(q^{-k}(M \cap (q^{k+1}, q^k])).$$

Given a universal representation, consider the polar decompositions $z_1 = U|z_1|$ and $z_2 = V|z_2|$. For all $h \in \text{dom}(|z_1|) = \text{dom}(z_1)$ and $g \in \text{dom}(|z_2|) = \text{dom}(z_2)$, (27) and (28) imply

$$|z_1|h(t,s) = \big(\chi_{[q^{-1},\infty)}(s)\sqrt{s^2-1}\,t + s\,\chi_{\{0\}}(t)\big)h(t,s), \qquad |z_2|g(t,s) = tg(t,s). \quad (32)$$

Since

$$\begin{aligned}\text{ran}(|z_1|) &= \ker(|z_1|)^\perp = \text{ran}\big(\chi_{(0,\infty)}(t)\,\chi_{[q^{-1},\infty)}(s) + \chi_{\{0\}}(t)\,\chi_{(0,\infty)}(s)\big),\\ \text{ran}(|z_2|) &= \ker(|z_2|)^\perp = \text{ran}\big(\chi_{(0,\infty)}(t)\big),\end{aligned}$$

it follows from (27) that

$$Uh(s,t) = \big(\chi_{(0,\infty)}(t)\,\chi_{[q^{-1},\infty)}(qs) + \chi_{\{0\}}(t)\,\chi_{(0,\infty)}(s)\big)h(qs,t), \quad (33)$$
$$Vh(s,t) = \chi_{(0,\infty)}(t)h(s,qt), \quad (34)$$

for all $h \in \mathcal{H}$, where we used $\chi_{(0,\infty)}(qr) = \chi_{(0,\infty)}(r)$. Their adjoints act on $\mathcal{H}$ by

$$U^*h(s,t) = \big(\chi_{(0,\infty)}(t)\,\chi_{[q^{-1},\infty)}(s) + \chi_{\{0\}}(t)\,\chi_{(0,\infty)}(s)\big)h(q^{-1}s,t), \quad (35)$$
$$V^*h(s,t) = \chi_{(0,\infty)}(t)h(s,q^{-1}t). \quad (36)$$

From (33)–(36), we get

$$U^*U = \chi_{(0,\infty)}(t)\,\chi_{[q^{-1},\infty)}(s) + \chi_{\{0\}}(t)\,\chi_{(0,\infty)}(s), \quad (37)$$
$$UU^* = \chi_{(0,\infty)}(t)\,\chi_{[q^{-1},\infty)}(qs) + \chi_{\{0\}}(t)\,\chi_{(0,\infty)}(s), \quad (38)$$
$$V^*V = \chi_{(0,\infty)}(t) = VV^*. \quad (39)$$

Using again $\chi_{(0,\infty)}(q^{\pm1}t) = \chi_{(0,\infty)}(t)$ and $\chi_{\{0\}}(q^{\pm1}t) = \chi_{\{0\}}(t)$, one easily sees that

$$UV = VU, \quad UV^* = V^*U, \quad U^*V = VU^*, \quad U^*V^* = V^*U^*. \quad (40)$$

Considering Borel measurable functions f on $[0,\infty)\times[0,\infty)$ as multiplication operators by

$$f\,h(s,t) := f(s,t)\,h(s,t),$$

we obtain from (33)–(36) the following simple commutation relations:

$$U\,f(s,t) = f(qs,t)U, \qquad U^*\,f(s,t) = f(q^{-1}s,t)U^*, \tag{41}$$

$$V\,f(s,t) = f(s,qt)V, \qquad V^*\,f(s,t) = f(s,q^{-1}t)V. \tag{42}$$

In fact, the reason for choosing $|\zeta|$ from (10) and y from (21) as multiplication operators was to obtain such simple commutation relations between functions and the phases from the polar decompositions of the generators of $\mathcal{O}(\mathbb{C}_q^2)$. As a consequence,

$$\mathrm{Fun}(\mathbb{C}_q^2) := \Big\{ \sum_{\text{finite}} f_{nm}(s,t)U^{\#n}V^{\#m} \;:\; f \in \mathcal{L}_\infty([0,\infty)\times[0,\infty)) \Big\} \tag{43}$$

is a *-subalgebra of $\mathrm{B}(\mathcal{H})$, where

$$U^{\#n} := \begin{cases} U^n, & n \geq 0, \\ U^{*n}, & n < 0, \end{cases} \qquad V^{\#n} := \begin{cases} V^n, & n \geq 0, \\ V^{*n}, & n < 0, \end{cases} \qquad n \in \mathbb{Z}. \tag{44}$$

Moreover, by (32) and the previous commutation relations, there exists for all $k,l,m,n \in \mathbb{N}_0$ a Borel measurable function p_{klmn} on $[0,\infty)\times[0,\infty)$ such that

$$z_1^k z_1^{*l} z_2^m z_2^{*n} = p_{klmn}(s,t)\,U^{\#k-l}V^{\#m-n}. \tag{45}$$

Equations (43) and (45) are the motivation for the construction of the C*-algebra of continuous functions on $\mathbb{C}_q^2$ vanishing at infinity. Before treating the quantum case, let us briefly review the classical C*-algebra $C_0(\mathbb{C}^2)$. In analogy to the polar decomposition of the generators, write $z_1 = \mathrm{e}^{\mathrm{i}\varphi}|z_1|$ and $z_2 = \mathrm{e}^{\mathrm{i}\theta}|z_2|$. Let $n,m \in \mathbb{Z}$. Given a function $f_{nm} \in C_0([0,\infty)\times[0,\infty))$, the assignment

$$\mathbb{C}^2 \ni (\mathrm{e}^{\mathrm{i}\varphi}|z_1|, \mathrm{e}^{\mathrm{i}\theta}|z_2|) \longmapsto f_{nm}(|z_1|,|z_2|)\,\mathrm{e}^{\mathrm{i}\varphi n}\mathrm{e}^{\mathrm{i}\theta m} \in \mathbb{C}$$

defines a function in $C_0(\mathbb{C}^2)$ if and only if

(a) $f_{nm}(0,|z_2|)\mathrm{e}^{\mathrm{i}\varphi n}\mathrm{e}^{\mathrm{i}\theta m}$ does not depend on $\varphi \implies f_{nm}(0,|z_2|) = 0$ for $n \neq 0$,

(b) $f_{nm}(|z_1|,0)\mathrm{e}^{\mathrm{i}\varphi n}\mathrm{e}^{\mathrm{i}\theta m}$ does not depend on $\theta \implies f_{nm}(|z_1|,0) = 0$ for $m \neq 0$.

Moreover, the following *-subalgebra of $C_0(\mathbb{C}^2)$,

$$\mathcal{C}_0(\mathbb{C}^2) := \Big\{ \sum_{\text{finite}} f_{nm}(|z_1|,|z_2|)\mathrm{e}^{\mathrm{i}\varphi n}\mathrm{e}^{\mathrm{i}\theta m} \;:\; f_{nm} \in C_0([0,\infty)\times[0,\infty)),\; n,m \in \mathbb{Z}, \\ f_{nm}(0,|z_2|) = 0 \;\text{ if }\; n \neq 0,\;\; f_{nm}(|z_1|,0) = 0 \;\text{ if }\; m \neq 0 \Big\}, \tag{46}$$

separates the points of $\mathbb{C}^2$. By the Stone–Weierstraß theorem, its norm closure yields $C_0(\mathbb{C}^2)$.

To pass from the classical to the quantum case, we start with a universal representation from Theorem 2. For all bounded continuous functions g on $[0,\infty)$, the operators $g(|z_1|), g(|z_2|) \in \mathrm{B}(\mathcal{H})$ are well-defined by the spectral theorem, and (32) shows that

$$g(|z_1|)\,h(s,t) = g\big(\chi_{[q^{-1},\infty)}(s)\sqrt{s^2-1}\,t + s\chi_{\{0\}}(t)\big)\,h(s,t), \qquad g(|z_2|)\,h(s,t) = g(t)\,h(s,t).$$

By the universality of the representation, we have $\|g(|z_i|)\| = \|g\|_\infty$, $i=1,2$. In particular, the norm does not depend on the chosen measures of a universal representation. Similarly, for $f \in C_0([0,\infty)\times[0,\infty))$, the formula

$$f(|z_1|,|z_2|)\,h(s,t) := f\big(\chi_{[q^{-1},\infty)}(s)\sqrt{s^2-1}\,t + s\chi_{\{0\}}(t), t\big)\,h(s,t), \qquad h \in \mathcal{H}, \tag{47}$$

yields a well-defined operator in $\mathrm{B}(\mathcal{H})$ with $\|f(|z_1|,|z_2|)\| = \|f\|_\infty$. These observations lead to the following definition of $C_0(\mathbb{C}_q^2)$.

Definition 2. Given a universal representation of $\mathcal{O}(\mathbb{C}_q^2)$ from Theorem 2, let $z_1 = U|z_1|$ and $z_2 = V|z_2|$ be the polar decompositions of z_1 and z_2, respectively. The C*-algebra $C_0(\mathbb{C}_q^2)$ of continuous functions on the 2-dimensional quantum complex plane vanishing at infinity is defined as the norm closure of

$$\mathcal{C}_0(\mathbb{C}_q^2) := \Big\{ \sum_{\text{finite}} f_{nm}\big(\chi_{[q^{-1},\infty)}(s)\sqrt{s^2-1}\,t + s\chi_{\{0\}}(t)\,,\,t\big)\,U^{\#n}V^{\#m} \,:\, n,m\in\mathbb{Z}, \quad f_{nm}\in C_0([0,\infty)\times[0,\infty)),\ f_{nm}(0,t)=0 \text{ if } n\neq 0,\ f_{nm}(s,0)=0 \text{ if } m\neq 0\Big\} \tag{48}$$

in $\mathrm{B}(\mathcal{H})$.

Apart from the non-commutativity in (41) and (42), the main difference to the classical case is the unusual expression in the first argument of the function f_{nm}. However, if we look at the representation on the orthogonal components of (29) separately, our formulas have a natural geometric interpretation. First note that the function $h_{00}(s,t) := \chi_{\{0\}}(s)\chi_{\{0\}}(t) \in \mathcal{H}$ generates the 1-dimensional invariant subspace $\mathcal{L}_2([0,\infty)\times[0,\infty), \epsilon\delta_0\otimes\delta_0)$, and the representation of $\mathcal{C}_0(\mathbb{C}_q^2)$ on it reads

$$\sum_{\text{finite}} f_{nm}(\chi_{[q^{-1},\infty)}(s)\sqrt{s^2-1}\,t + s\chi_{\{0\}}(t)\,,\,t)\,U^{\#n}V^{\#m}\,h_{00}(s,t) = f_{00}(0,0)\,h_{00}(s,t) \tag{49}$$

since $f_{nm}(0,0) = 0$ if $n \neq 0$ or $m \neq 0$. Obviously, (49) corresponds to evaluating functions on 2-dimensional complex plane at $(0,0)$. This 1-dimensional representation describes the only classical point $(0,0)$ of $\mathbb{C}_q^2$.

Next, on $\mathcal{L}_2([0,\infty)\times[0,\infty), \nu\otimes\delta_0)$, we have $z_2 = 0$ and can thus write

$$\sum_{n,m} f_{nm}(\chi_{[q^{-1},\infty)}(s)\sqrt{s^2-1}\,t + s\chi_{\{0\}}(t)\,,\,t)\,U^{\#n}V^{\#m} = \sum_n f_{n0}(s,0)\,U^{\#n}. \tag{50}$$

Recalling that z_1 acts on $\mathcal{L}_2([0,\infty)\times[0,\infty), \nu\otimes\delta_0)$ as a q-normal operator and comparing (50) with (30) shows that the restriction of $\mathcal{C}_0(\mathbb{C}_q^2)$ to $\mathcal{L}_2([0,\infty)\times[0,\infty), \nu\otimes\delta_0)$ generates $C_0(\mathbb{C}_q)$. This representation corresponds to an inclusion $\mathbb{C}_q\times\{0\}\subset\mathbb{C}_q^2$.

Finally, on $\mathcal{L}_2\big([0,\infty)\times[0,\infty), \sigma\otimes\mu\big)$, we have $t = |z_2| > 0$ and $\chi_{[q^{-1},\infty)}(s)\sqrt{s^2-1} = |\omega|$, see (22). Thus the representation of the functions from (48) can be written

$$\sum_{\text{finite}} f_{nm}(|\omega|\,t\,,\,t)\,U^{\#n}V^{\#m}.$$

Classically we get, for all $|\omega| \geq 0$,

$$\sum_{\text{finite}} f_{nm}(|\omega|\,t\,,\,t)\mathrm{e}^{\mathrm{i}\varphi n}\mathrm{e}^{\mathrm{i}\theta m}\Big|_{t=0} = \sum_{\text{finite}} f_{nm}(0,0)\mathrm{e}^{\mathrm{i}\varphi n}\mathrm{e}^{\mathrm{i}\theta m} = f_{00}(0,0).$$

Therefore these functions separate only the points of $\mathbb{C}^2\backslash\mathbb{C}\times\{0\}$ and the whole subspace $\mathbb{C}\times\{0\}$ gets identified with the single point $(0,0)$. Geometrically, this corresponds to a 2-dimensional complex plane, where $\mathbb{C}\times\{0\}$ is shrunk to one point.

Arguing backwards, we can say that the representation from (24) corresponds to a 2-dimensional quantum complex plane, where $\mathbb{C}_q\times\{0\}$ is shrunk to a point, and that $\mathbb{C}_q\times\{0\}$ gets glued into this space by the representation (25). Moreover, the origin of the 2-dimensional quantum complex plane is the only classical point described by the 1-dimensional representation (49).

References

[1] Bershtein O and Sinel'shchikov S 2012 J. Geom. Phys. **62** 1323
[2] Cimprič J, Savchuk Y and Schmüdgen K 2014 Trans. Amer. Math. Soc. **366** 135
[3] Cohen I and Wagner E 2012 Banach Center Publ. **98** 55
[4] Klimyk A U and Schmüdgen K 1998 *Quantum Groups and their Representations* (New York: Springer)
[5] Kürsten K-D and Wagner E 2007 Publ. Res. Inst. Math. Sci. **43** 1
[6] Osuna Castro O and Wagner E 2012 Int. J. Geom. Methods Mod. Phys. **9** 1
[7] Reed M and Simon B 1980 *Methods of modern mathematical physics* (New York: Academic Press)
[8] Schmüdgen K 2002 J. Operator Theory **48** 487
[9] Soltan P M 2010 J. Math. Anal. Appl. **372** 224
[10] Woronowicz S L 1991 Comm. Math. Phys. **136** 399
[11] Shklyarov D L, Sinel'shchikov S D and Vaksman L L 1997 Mat. Fiz. Anal. Geom. **4** 286
[12] Shklyarov D L, Sinel'shchikov S D and Vaksman L L Czech. J. Phys. **50** 175
[13] Woronowicz S L 1995 Rev. Math. Phys. **7** 481

Journal of Physics: Conference Series **563** (2014) 012035 doi:10.1088/1742-6596/563/1/012035

Exposition on classification of finite dimensional irreducible representations of the Lie superalgebra $B(m,n)$

Hiroyuki Yamane

Department of Mathematics, Faculty of Science, University of Toyama, Gofuku 3190, Toyama-shi, Toyama, 930-8555, Japan

E-mail: hiroyuki@sci.u-toyama.ac.jp

Abstract. We demonstrate a new proof (see also [1]) for the Kac's classification of finite dimensional irreducible representations of the simple Lie superalgebra $B(m,n)$.

1. Lie superalgebras

In [3], V. Kac classified finite dimensional irreducible representations of simple Lie superalgebras. In [1], we classified those of quantum superalgebras and recovered his result by using some reduced expressions of Weyl groupoids; some fundamental results of Weyl groupoids were given by [2]. In this note, we give an elemental proof of classification of finite dimensional irreducible representations of the Lie superalgebra $B(m,n)$, which would demonstrate an essence of argument in [1].

For $x, y \in \mathbb{R}$, let $J_{x,y} := \{r \in \mathbb{Z} | x \leq r \leq y\}$.

Let $\mathfrak{g} = \mathfrak{g}(0) \oplus \mathfrak{g}(1)$ be a Lie superalgebra over $\mathbb{C}$, that is, $\mathfrak{g}$ is a linear space over $\mathbb{C}$ and there exists a homogeneous bilinear map $[\,,\,] : \mathfrak{g} \times \mathfrak{g} \to \mathfrak{g}$ for which $[\mathfrak{g}(i), \mathfrak{g}(j)] \subset \mathfrak{g}(i+j)$, where $\mathfrak{g}(i+2) := \mathfrak{g}(i)$, and

$$[Y,X] = -(-1)^{ij}[X,Y], \quad [X,[Y,Z]] = [[X,Y],Z] + (-1)^{ij}[Y,[X,Z]],$$

where $X \in \mathfrak{g}_i$, $Y \in \mathfrak{g}_j$, $Z \in \mathfrak{g}$.

For $n \in \mathbb{N}$, let $\mathrm{M}_n(\mathbb{C})$ be the $\mathbb{C}$-algebra of square matrices of size n. Let $E_{ij} \in \mathrm{M}_n(\mathbb{C})$ be the matrix unit whose (i,j)-component is 1. For m. $n \in \mathbb{Z}_{\geq 0}$, let $\mathfrak{gl}(m|n) := \mathrm{M}_{m+n}(\mathbb{C})$ as a $\mathbb{C}$-linear space, and regard it as the Lie superalgebra by

$$\begin{aligned}
&\mathfrak{gl}(m|n)(0) := (\oplus_{(i,j)\in(J_{1,m}\times J_{1,m})\cup(J_{m+1,m+n}\times J_{m+1,m+n})}\mathbb{C}E_{ij}),\\
&\mathfrak{gl}(m|n)(1) := (\oplus_{(i,j)\in(J_{1,m}\times J_{m+1,m+n})\cup(J_{m+1,m+n}\times J_{1,m})}\mathbb{C}E_{ij}),\\
&[X,Y] := XY - (-1)^{ij}YX \quad (X \in \mathfrak{gl}(m|n)(i), Y \in \mathfrak{gl}(m|n)(j)).
\end{aligned}$$

Let $m, n \in \mathbb{Z}_{\geq 0}$. Let $\mathcal{D}_{m|n}$ denote the set of maps p from $J_{1,m+n}$ to $J_{0,1}$ with $\sum_{i=1}^{m+n} p(i) = n$. Let $\mathfrak{gl}(p)$ be $\mathrm{M}_{m+n}(\mathbb{C})$ as a $\mathbb{C}$-linear space. For $r \in J_{0,m+n}$, let $\mathfrak{gl}(p)(r)$ be the $\mathbb{C}$-subspace of $\mathfrak{gl}(p)$ formed by E_{ij} with $(-1)^{p(i)+p(j)} = (-1)^r$. Then we can regard $\mathfrak{gl}(p)$ as the Lie superalgebra by $[X,Y] := XY - (-1)^{ij}YX$ ($X \in \mathfrak{gl}(p)(i)$, $Y \in \mathfrak{gl}(p)(j)$). Then $\mathfrak{gl}(p)$ is isomorphic to $\mathfrak{gl}(m|n)$.

Let $m, n \in \mathbb{Z}_{\geq 0}$ be such that $m+n \geq 1$. Define the map $\theta : J_{1,m+2n} \to J_{1,m+2n}$ by $\theta(i) := m+2n+1-i$. Let $\mathcal{D}^\theta_{m|2n}$ be the subset of $\mathcal{D}_{m|2n}$ formed by the elements p for which $p(\theta(i)) = p(i)$ for all i. Let $p \in \mathcal{D}^\theta_{m|2n}$. Then if m is an odd integer, $p(n+\frac{m+1}{2}) = 0$. Let $\mathcal{K}(p)$ be the set of maps $g : J_{1,m+2n} \to \{-1,1\}$ with $g(i) = 1$ for all $i \in J_{1,m+2n}$ with $p(i) = 0$, and $g(j)g(\theta(j)) = -1$ for all $j \in J_{1,m+2n}$ with $p(j) = 1$. Let $g \in \mathcal{K}(p)$. Let $\Omega : \mathfrak{gl}(p) \to \mathfrak{gl}(p)$ be the isomorphism defined by $\Omega(E_{ij}) := -(-1)^{p(i)p(j)+p(j)}g(i)g(j)E_{\theta(j)\theta(i)}$. Let $\mathfrak{osp}(g) := \{X \in \mathfrak{gl}(p) | \Omega(X) = X\}$. If $g(i) = -1$ for $i \in J_{1,n}$ and $g(j) = 1$ for $j \in J_{n+1,m+2n}$, we also denote $\mathfrak{osp}(g)$ by $\mathfrak{osp}(m|2n)$. If $m = 2m'-1$ for some $m' \in \mathbb{N}$, then we also denote $\mathfrak{osp}(m|2n)$ by $B(m'-1,n)$. It is clear that for $p' \in \mathcal{D}^\theta_{m|2n}$, and $g' \in \mathcal{K}(p')$, $\mathfrak{osp}(g)$ and $\mathfrak{osp}(g')$ are isomorphic, as Lie superalgebras. Let $\mathcal{G}_{m|2n} := \cup_{p\in\mathcal{D}^\theta_{m|2n}} \mathcal{K}(p)$.

2. Finite dimensional irreducible representations of $B(m,n)$

From now on until end of this paper, let m, $n \in \mathbb{Z}_{\geq 0}$, $g \in \mathcal{G}_{2m+1|2n}$, and $I := J_{1,m+n}$, and let $p \in \mathcal{D}^\theta_{m|2n}$ be such that $g \in \mathcal{K}(p)$.

Let $i \in I$. Define the map $\tau_i : \mathcal{G}_{2m+1|2n} \to \mathcal{G}_{2m+1|2n}$ as follows. If $i \in I \setminus \{m+n\}$, define $(\tau_i(g))(i) := g(i+1)$, $(\tau_i(g))(i+1) := g(i)$, $(\tau_i(g))(m+n+1-i) := g(2m+2n+2-i)$, $(\tau_i(g))(2m+2n+2-i) := g(2m+2n+1-i)$, and $(\tau_i(g))(j) := g(j)$ for $j \in J_{1,2m+2n+1} \setminus \{i, i+1, 2m+2n+1-i, 2m+2n+2-i\}$. Define $(\tau_{m+n}(g))(m+n) := g(m+n+2)$, $(\tau_{m+n}(g))(m+n+2) := g(m+n)$, and $(\tau_{m+n}(g))(j) := g(j)$ for $j \in J_{1,2m+2n+1} \setminus \{m+n, m+n+2\}$. Define the map $\tau_i : \mathcal{D}^\theta_{2m+1|2n} \to \mathcal{D}^\theta_{2m+1|2n}$ as follow. If $i \in I \setminus \{m+n\}$, define $(\tau_i(p))(i) := i+1$, $(\tau_i(p))(i+1) := i$, and $(\tau_i(p))(j) := j$ for $j \in I \setminus \{i, i+1\}$. Let $\tau_{m+n}(p) := p$. Then $\tau_i(g) \in \mathcal{K}(\tau_i(p))$.

Let $i \in I$. Define the elements H^g_i, E^g_i, F^g_i of $\mathfrak{osp}(g)$ as follows. Let p be a unique element of $\mathcal{D}^\theta_{2m+1|2n}$ with $g \in \mathcal{K}(p)$. Let

$$H^g_i := \begin{cases} E_{i,i} - (-1)^{p(i)+p(i+1)}E_{i+1,i+1} \\ \quad +(-1)^{p(i)+p(i+1)}E_{2m+2n+1-i,2m+2n+1-i} - E_{2m+2n+2-i,2m+2n+2-i} \\ \qquad \text{if } i \in J_{1,m+n-1}, \\ 2E_{m+n,m+n} - 2E_{m+n+2,m+n+2} \quad \text{if } i = m+n, \end{cases}$$

$$E^g_i := \begin{cases} E_{i,i+1} - (-1)^{p(i)p(i+1)+p(i+1)}g(i)g(i+1)E_{2m+2n+1-i,2m+2n+2-i} \\ \qquad \text{if } i \in I \setminus \{m+n\}, \\ 2E_{m+n,m+n+1} \\ \quad -2(-1)^{p(m+n)p(m+n+1)+p(m+n+1)}g(m+n)g(m+n+1)E_{m+n+1,m+n+2} \\ \qquad \text{if } i = m+n, \end{cases}$$

and

$$F^g_i := E_{i+1,i} - (-1)^{p(i+1)p(i)+p(i)}g(i+1)g(i)E_{2m+2n+2-i,2m+2n+1-i} \text{ for } i \in I.$$

For i, $j \in I$, define $a^g_{ij} \in \mathbb{Z}$ as follows. For $i \in I \setminus \{m+n\}$, let $a^g_{ii} := 1 + (-1)^{p(i)+p(i+1)}$. Let $a^g_{m+n,m+n} := 2$. For $i \in I \setminus \{m+n\}$, let $a^g_{i,i+1} := -(-1)^{p(i)+p(i+1)}$. For $i \in I \setminus \{m+n-1, m+n\}$, let $a^g_{i+1,i} := -1$. Let $a^g_{m+n,m+n-1} := -2$. For i, $j \in I$ with $|i-j| \geq 2$, let $a^g_{i,j} := 0$. Then we have

$$[H^g_i, H^g_j] = 0, \ [H^g_i, E^g_j] = a^g_{ij}E^g_j, \ [H^g_i, F^g_j] = -a^g_{ij}F^g_j, \\ [E^g_i, F^g_j] = \delta_{ij}H^g_i.$$

For $r \in J_{1,2m+2n}$, let $S'_r := E_{r,r+1} + E_{r+1,r} + \sum_{k\in J_{1,2m+2n}-\{r,r+1\}} E_{kk} \in \mathrm{M}_{2m+2n+1}(\mathbb{C})$. For $i \in I$, define $S_i \in \mathrm{M}_{2m+2n+1}(\mathbb{C})$ as follows. For $i \in I \setminus \{m+n\}$, let $S_i := S'_iS'_{2m+2n+1-i}$. Let $S_{m+n} := S'_{m+n}S'_{m+n+1}S'_{m+n}$. Then $S^2_i = \mathrm{id}$, $S_jS_{j+1}S_j = S_{j+1}S_jS_{j+1}$ $(j \in I \setminus \{m+n\})$, $S_{m+n}S_{m+n+1}S_{m+n}S_{m+n+1} = S_{m+n+1}S_{m+n}S_{m+n+1}S_{m+n}$, $S_kS_r = S_rS_k$ $(|k-r| \geq 2)$.

Let $i \in I$. Define the Lie superalgebra isomorphism $T^g_i : \mathfrak{osp}(g) \to \mathfrak{osp}(\tau_i(g))$ by $T^g_i(X) := S_i X S_i$. Then $T^{\tau_i(g)}_i \circ T^g_i = \mathrm{id}_{\mathfrak{osp}(g)}$. Let $p \in \mathcal{D}^\theta_{m|2n}$ be such that $g \in \mathcal{K}(p)$. Then we have the following equations. If $i \in I \setminus \{m+n\}$, $T^g_i(H^g_i) = -(-1)^{\tau_i(p)(i)+\tau_i(p)(i+1)} H^{\tau_i(g)}_i$. If $i = m+n$, $T^g_i(H^g_i) = -H^{\tau_i(g)}_i$. If $i \in I\setminus\{m+n-1, m+n\}$, $T^g_i(H^g_{i+1}) = H^{\tau_i(g)}_i + (-1)^{\tau_i(p)(i)+\tau_i(p)(i+1)} H^{\tau_i(g)}_{i+1}$. If $i = m+n-1$, $T^g_i(H^g_{i+1}) = 2H^{\tau_i(g)}_i + (-1)^{\tau_i(p)(i)+\tau_i(p)(i+1)} H^{\tau_i(g)}_{i+1}$. If $i \in I\setminus\{1\}$, $T^g_i(H^g_{i-1}) = H^{\tau_i(g)}_{i-1} + (-1)^{\tau_i(p)(i-1)+\tau_i(p)(i)} H^{\tau_i(g)}_i$. If $i \in I \setminus \{m+n\}$, $T^g_i(E^g_i) = F^{\tau_i(g)}_i$. If $i = m+n$, $T^g_i(E^g_i) = 2F^{\tau_i(g)}_i$. If $i \in I \setminus \{m+n\}$, $T^g_i(E^g_{i+1}) = [E^{\tau_i(g)}_i, E^{\tau_i(g)}_{i+1}]$. If $i \in I \setminus \{1, m+n\}$, $T^g_i(E^g_{i-1}) = [E^{\tau_i(g)}_{i-1}, E^{\tau_i(g)}_i]$. If $i = m+n$, $T^g_i(E^g_{i-1}) = -\frac{1}{4}(-1)^{\tau_i(p)(i)\tau_i(p)(i+1)+\tau_i(p)(i+1)}\tau_i(g)(i)\tau_i(g)(i+1)[[E^{\tau_i(g)}_{i-1}, E^{\tau_i(g)}_i], E^{\tau_i(g)}_i]$. For $j \in I$ with $|i-j| \geq 2$, we have $T^g_i(H^g_j) = H^{\tau_i(g)}_j$ and $T^g_i(E^g_j) = E^{\tau_i(g)}_j$.

Let Λ be the set of maps from I to $\mathbb{C}$. For $\lambda \in \Lambda$, let V^g_λ be a non-zero irreducible left $\mathfrak{osp}(g)$-module such that there exists $v^g_\lambda \in V^g_\lambda \setminus \{0\}$ with $H^g_i v^g_\lambda = \lambda(i) v^g_\lambda$ and $E^g_i v^g_\lambda = 0$ for all $i \in I$. For $k \in \mathbb{N}$, we have

$$E^g_i (F^g_i)^k v^g_\lambda = \begin{cases} (-\frac{k(k-1)}{2} a^g_{ii} + k\lambda(i))(F^g_i)^{k-1} v^g_\lambda & \text{if } (-1)^{p(i)+p(i+1)} = 1, \\ (-\frac{k}{2} a^g_{ii})(F^g_i)^{k-1} v^g_\lambda & \text{if } (-1)^{p(i)+p(i+1)} = -1 \text{ and } k \in 2\mathbb{N}, \\ (-\frac{k-1}{2} a^g_{ii} + \lambda(i))(F^g_i)^{k-1} v^g_\lambda & \text{if } (-1)^{p(i)+p(i+1)} = -1 \text{ and } k \in 2\mathbb{N}-1. \end{cases} \tag{1}$$

For $\lambda \in \Lambda$ and $i \in I$, define $\lambda^{g;i} \in \Lambda$ as follows. If $(-1)^{p(i)+p(i+1)} = 1$, let $\lambda^{g;i} := \lambda$, so $\lambda^{g;m+n} = \lambda$. If $(-1)^{p(i)+p(i+1)} = -1$ (so $i \in I \setminus \{m+n\}$), let $\lambda^{g;i}(i) := \lambda(i)$, $\lambda^{g;i}(i-1) := (-1)^{\tau_i(p)(i-1)+\tau_i(p)(i)}(1 - \delta_{\lambda(i),0} + \lambda(i)) + \lambda(i-1)$ (if $i \in I \setminus \{1, m+n\}$), and $\lambda^{g;i}(i+1) := 2^{\delta_{i,m+n-1}}\lambda(i) + (-1)^{\tau_i(p)(i)+\tau_i(p)(i+1)}(1 - \delta_{\lambda(i),0} + \lambda(i))$. Then $(\lambda^{g;i})^{\tau_i(g);i} = \lambda$. Let

$$h^g_{\lambda.i} := \begin{cases} \infty & \text{if } (F^g_i)^r v^g_\lambda \neq 0 \text{ for all } r \in \mathbb{Z}_{\geq 0}, \\ \max\{r \in \mathbb{Z}_{\geq 0} | (F^g_i)^r v^g_\lambda \neq 0\} & \text{otherwise.} \end{cases}$$

By (1) we can see

Lemma. *It follows that $h^g_{\lambda.i} < \infty$ if and only if one of the following* (x), (y) *and* (z) *holds;* (x) $(-1)^{p(i)+p(i+1)} = 1$ *and* $\lambda(i) \in \mathbb{Z}_{\geq 0}$; (y) $(-1)^{p(i)+p(i+1)} = -1$ *and* $a^g_{ii} = 0$; (z) $(-1)^{p(i)+p(i+1)} = -1$ *and* $\lambda(i) \in 2\mathbb{Z}_{\geq 0} + 1$. *If this is the case, there exists a linear isomorphism* $\hat{T}^g_i : V^g_\lambda \to V^{\tau_i(g)}_{\lambda^{g;i}}$ *such that* $\hat{T}^g_i(X v^g_\lambda) = T^g_i(X)(F^{\tau_i(g)}_i)^{h^g_{\lambda.i}} v^{\tau_i(g)}_{\lambda^{g;i}}$ *for all* $X \in U(\mathfrak{osp}(g))$, *and we have* $\hat{T}^{\tau_i(g)}_i \circ \hat{T}^g_i = x \cdot \mathrm{id}_{V^g_\lambda}$ *for some* $x \in \mathbb{C}^\times$, *where* $\mathbb{C}^\times := \mathbb{C} \setminus \{0\}$.

Let $N := (m+n)^2$. Define the map $f : J_{1,N} \to I$ as follows. For $t \in J_{1,m}$, let $f(t) := n+t$. For $t \in J_{m+1,m^2}$, let $f(t) := f(t-m)$. For $t \in J_{m^2+1,m^2+m+n}$, let $f(t) := t - m^2$. For $t \in J_{m^2+m+n+1,m^2+2m+n}$, let $f(t) := m^2+2m+2n-t$. For $t \in J_{m^2+2m+n+1,k}$, let $f(t) := f(t-2m-n)$. Then $S_{f(1)}S_{f(2)} \cdots S_{f(N)} = \sum_{i=1}^{2m+2n+1} E_{i,2m+2n+2-i}$. Let $g_0 \in \mathcal{G}_{2m+1|2n}$ be such that $g_0(i) = -1$ ($i \in J_{1,n}$) and $g_0(j) = 1$ ($j \in J_{n+1,2m+n+1}$), so $\mathfrak{osp}(g_0) = \mathfrak{osp}(2m+1|2n)$. For $t \in J_{1,k}$, let $g_t := \tau_{f(t)}(g_{t-1})$. Let $\widetilde{T} := T^{g_1}_{f(1)} \circ \cdots \circ T^{g_N}_{f(N)}$. Then $\widetilde{T}(H^{g_N}_i) = -H^{g_0}_i$, $\widetilde{T}(E^{g_N}_i) \in \mathbb{C}^\times F^{g_0}_i$ and $\widetilde{T}(F^{g_N}_i) \in \mathbb{C}^\times E^{g_0}_i$. For $t \in J_{0,N}$, let $p_t \in \mathcal{D}^\theta_{m|2n}$ be such that $g_t \in \mathcal{K}(p_t)$. For $k \in J_{0,m}$, let $\bar{p}_k := p_{m^2+n-1+k}$. In particular, $p_0(i) = \delta_{i,n}$. We have $p_t = p_0 = \bar{p}_0$ for $t \in J_{1,m^2+n-1}$. For $k \in J_{1,m-1}$, we have $\bar{p}_k(k) = \delta_{i,n-1} + \delta_{i,n+k-1} + \delta_{i,n+k}$. We have $p_t = p_{2m^2+2m+2n-1-t}$ for $t \in J_{m^2+m+n,m^2+2m+n}$. We have $p_t = p_{t-2m-n}$ for $t \in J_{m^2+2m+n+1,N}$. Define the map $\pi : \Lambda \to \Lambda$ by $\pi(\lambda)(i) := \lambda(i)$ for $i \in I \setminus \{n\}$, and $\pi(\lambda)(n) := \lambda(n) - \frac{\lambda(m+n)}{2} - \sum_{i=m+1}^{m+n-1} \lambda(j)$. If the linear isomorphism $\check{T} := \hat{T}^{g_1}_{f(1)} \circ \cdots \circ \hat{T}^{g_N}_{f(N)}$ whose end space is V^g_λ is defined, then $\dim V^{g_0}_\lambda < \infty$. Let

$M := m^2 + m + n$. If $\check{T}' := \hat{T}^{g_1}_{f(1)} \circ \cdots \circ \hat{T}^{g_M}_{f(M)}$ is defined, then $\check{T}$ is also defined. Conversely if $\dim V^{g_0}_\lambda < \infty$, then $\check{T}'$ has to be defined. Thus using $\check{T}$, we recover a result by Kac.

Theorem. (See also [3]) *It follows that* $\dim V^{g_0}_\lambda < \infty$ *if and only if both of the following* (x) *and* (y) *hold;* (x) $\pi(\lambda)(i) \in \mathbb{Z}_{\geq 0}$ *for all* $i \in I$*;* (y) *if* $r := \pi(\lambda)(n) \in J_{0,m-1}$, $\pi(\lambda)(j) = 0$ *for all* $j \in J_{n+r+1,m+n}$.

References

[1] S Azam, H Yamane and M Yousofzadeh (*Preprint* arXiv:1105.0160 [math-ph])
[2] I. Heckenberger and H. Yamane 2008 *Math. Z.* **259** 255
[3] V.G. Kac 1977 *Adv. in Math.* **26** 8

Journal of Physics: Conference Series **563** (2014) 012036 doi:10.1088/1742-6596/563/1/012036

Homogeneous grading affine Toda quantum solitons

Alexander Zuevsky

Institute of Mathematics of the Academy of Sciences of the Czech Republic, Prague, Czech Republic

E-mail: zuevsky@yahoo.com

Abstract. We construct quantum soliton solutions to the homogeneous grading case of the affine Toda models, in particular, the sine–Gordon equation.

1. Introduction

The Lie-algebraic way to construct non-linear exactly solvable models in classical regions is very well known and elaborated [10]. Applying the zero-curvature conditions on elements of connection containing Lie algebra generators in appropriate grading subspaces, we obtain systems of equations of motion associated to a specific Lie algebra. In [4] the higher grading generalization to the conformal affine Toda models was considered. Elements of the higher (then number one) grading subspaces are taking into account while connection elements are constructed. The main example of [4] is the principal grading case. In this paper we consider an alternative, the homogeneous grading case. We derive the systems of equations generalizing the case of the sine–Gordon equation and provide quantum group solutions.

2. Homogeneous higher grading generalization of the affine Toda model

2.1. Classical case

We start with the equations (14–17) of [4] (see subsection 5.2 in Appendix). Consider the case $l = 1$. In the principal grading we obtain from (14) the sin–Gordon equation. Recall that In the homogeneous grading of $\widetilde{\mathcal{G}}$ the grading subspaces are $\widetilde{\mathcal{G}}_n = \left\{H^n, E^n_\pm\right\}$. We take

$$E_1 = E^1_+ + E^1_-, \quad E_{-1} = E^{-1}_+ + E^1_-. \tag{1}$$

Consider a particular case when we parameterize the group element b as

$$b = e^{\phi H^0}. \tag{2}$$

Then, substituting (1) and (2) into (14–16) we get the following system of equations

$$\partial_\pm \phi = e^\eta \left(e^{-2\phi} - e^{2\phi}\right), \quad \partial_\pm \nu = -e^\eta \left(e^{2\phi} + e^{-2\phi}\right), \quad \partial_\pm \eta = 0,$$

i.e., in the first equation is again the sine–Gordon equation. The solution to the field ϕ is then the standard classical solution (19), [12] (see subsection 5.2 of Appendix).

Now consider the case $l = 2$. The equations corresponding to the principal grading can be found in [4]. Here again we take $b = e^{\phi H^0}$ though it this is not the most general choice of the group element parameterization since it does not contain dependence on the $E^0_\pm$ elements from $\widehat{\mathcal{G}}$, i.e., we have send corresponding fields near those generators to zero. Let us also put

$$F_1^+ = \kappa^+ H^1 + f_+^+ E_+^1 + f_-^+ E_-^1, \quad F_1^- = \kappa^- H^1 + f_+^- E_+^{-1} + f_-^- E_-^{-1}.$$

Then the system (14–16) gives the following system of equations:

$$\begin{array}{l} \partial_\pm \phi = e^\eta \left(e^{-2\phi} - e^{2\phi}\right) + e^\eta \left(f_-^+ f_+^- e^{-2\phi} - f_-^- f_+^+ e^{2\phi}\right), \\ \partial_\pm \nu = 2e^\eta \left(e^{2\phi} - e^{-2\phi}\right) + e^\eta \left(-2\kappa^+\kappa^- + f_-^- f_+^+ e^{2\phi} - f_-^+ f_+^- e^{-2\phi}\right), \; \partial_\pm \eta = 0 \end{array} \tag{3}$$

$$\begin{array}{l} \partial_+ \kappa^- = -e^\eta \left(f_-^+ e^{-2\phi} - f_+^+ e^{2\phi}\right), \quad \partial_- \kappa^+ = e^\eta \left(f_-^- e^{-2\phi} - f_+^- e^{2\phi}\right), \\ \partial_+ f_+^- = 2e^\eta \kappa^+, \; \partial_- f_+^+ = -2e^\eta \kappa^-, \; \partial_+ f_-^- = -2e^\eta \kappa^+, \; \partial_- f_-^+ = 2e^\eta \kappa^-, \\ e^{2\phi} f_+^+ \kappa^- = \kappa^+ f_+^-, \quad e^{-2\phi} f_-^+ \kappa^- = \kappa^+ f_-^-. \end{array} \tag{4}$$

The formal general solution to (3–4) age given in [4]:

$$e^{-\phi} = e^{\phi_0^+ - \phi_0} \frac{\langle \Lambda_1 | \mu_+^{-1} \mu_- | \Lambda_1 \rangle}{\langle \Lambda_0 | \mu_+^{-1} \mu_- | \Lambda_0 \rangle^{m_1}}, \tag{5}$$

for the ϕ field and for the $F_1^\pm$ elements

$$\langle i | F_i^+ | i; i \rangle = e^{k_{i1}(\phi_{0i}^- - \phi_{0i})} \partial_+ \left(\langle i | \gamma_0^- b \left(\gamma_0^+\right)^{-1} \mu_+^{-1} \mu_- | i; i \rangle \right).$$

In the homogeneous grading case, taking into account the parameterization of b element, we get

$$\langle 1 | F_i^+ | 1; 1 \rangle = e^{2(\phi_0^- - \phi_0)} \partial_+ \left(\langle 1 | e^{\phi_0^- H^0} e^{\phi H^0} e^{-\phi_0^+ H^0} \mu_+^{-1} \mu_- | 1; 1 \rangle \right).$$

Here we have made use of the properties of the i-th fundamental representation (corresponding to the homogeneous grading) of the Lie algebra $\widehat{sl}_2$. Thus, using (5) we get

$$\langle 1 | F_i^+ | 1; 1 \rangle = e^{2(\phi_0^- - \phi_0)} \partial_+ \left(\langle 1 | \mu_+^{-1} \mu_- | 1; 1 \rangle \cdot \frac{\langle \Lambda_0 | \mu_+^{-1} \mu_- | \Lambda_0^{m_1} \rangle}{\langle \Lambda_1 | \mu_+^{-1} \mu_- | \Lambda_1 \rangle} \right).$$

3. Solitonic solutions from general solutions

In [12] it was shown how to extract solitonic solutions from the formal general solutions of the affine Toda field equations. Let's take $\gamma_0^\pm = 1$ in (19) to be a constant function. Then the mappings $\mu_\pm$ are $\mu_\pm = \mu_\pm^0 e^{z_\pm \mathcal{E}_\pm}$ with $\mu_\pm^0$ being some fixed mappings independent of $z_\pm$. Next take $\tilde{\mathcal{E}}_\pm$ in (21) as $\mathcal{E}_\pm \equiv E_{\pm l} + \sum_{N=1}^{l-1} c_N^\pm E_{\pm N}$ where $E_\pm$ are elements of a Heisenberg subalgebra of $\widehat{\mathcal{G}}$, namely $[\mathcal{E}_+, \mathcal{E}_-] = \Omega C$. One can consider principal of homogeneous Heisenberg subalgebras for that purpose. In this paper we only deal with the principal case while the homogeneous case will be discussed elsewhere. Thus, we arrive at a special solution to (18)

$$e^{-\beta\lambda_i \cdot \phi} = e^{-\beta\lambda_i \cdot \phi_0} \frac{{}^{(1)}\langle \lambda_i | e^{x_\pm \mathcal{E}_\pm} \mu^0 e^{x_\pm \mathcal{E}_\pm} | \lambda_i \rangle^{(1)}}{\left({}^{(1)}\langle \lambda_0 | e^{x_\pm \mathcal{E}_\pm} \mu^0 e^{x_\pm \mathcal{E}_\pm} | \lambda_0 \rangle^{(1)}\right)^{m_j}}. \tag{6}$$

In order to compute these solutions explicitly we have to remove $\mathcal{E}_\pm$-dependence from (6) moving $\mathcal{E}_+$ to the right and $\mathcal{E}_-$ to the left . Then we should find such a $\mu_0 = \prod_{i=1}^N e^{\mathcal{V}_i}$ so that $\mathcal{V}_i$ would be eigenvectors with respect to the adjoint action of $\mathcal{E}_\pm$, i.e., $[\mathcal{E}_\pm, \mathcal{V}_i] = \omega_\pm^{(i)} \mathcal{V}_i$. Then it turns out [12] that resulting expressions provide us with solitonic solutions to the equations under considerations while parameters $\omega_\pm^{(i)}$ characterize solitons.

4. Quantum group soliton solution for sine–Gordon in homogeneous grading

As in [11] one can show that the affine Toda models are co-invariant with respect to the light-cone quantization. Namely, the equation of motion are preserved in form though a standard normal ordering has to be introduced as well as some infinite constant coming from quantum versions of Lax pair to generate equations using Lie algebra elements in quantum case. At the same time infinite constants do not appear in final formal solutions to the light-cone quantized versions of equations. In order to find quantum solutions, one has to replace [3], [8], [9] group elements as well as state vectors formal general solutions by their quantum group counterparts.

In this subsection we write examples of quantum group solutions to the quantized affine Toda model in the specific case of the higher grading sine–Gordon equation (the cases $l = 1, 2, 3$). Recall [13], that the homogeneous grading subspaces of $U_q(\widehat{sl}_2)$ are ${}_q\widehat{\mathcal{G}}_0 = \{K, \gamma, x_0^+, x_0^-\}$, ${}_q\widehat{\mathcal{G}}_n = \{x_n^+, x_n^-, a_n, n \in \{\mathbf{Z} - 0\}\}$.

4.1. The case $l = 1$

From the commutation relations for $x_m^\pm$ and a_m (see subsection 6.1 of Appendix) it follows that in this realization of the quantum group $U_q(\widehat{sl}_2)$, the generators $x_m^\pm, a_m \in \mathcal{G}_m$, $x_0^\pm \in \mathcal{G}_0$. The solution

$$e^{-\beta\lambda_j\cdot\phi} = e^{-\beta\lambda_j\cdot\phi_0}\frac{\langle\Lambda_j|e^{-a_1 z_+}e^{Q\tilde{\Phi}_-(\zeta)}e^{a_{-1}z_-}|\Lambda_j\rangle}{\langle\Lambda_0|e^{-a_1 z_+}e^{Q\tilde{\Phi}_-(\zeta)}e^{a_1 z_-}|\Lambda_0\rangle^{m_j}},$$

where $|\Lambda_0\rangle = |1\otimes 1\rangle$, $|\Lambda_1\rangle = |1\otimes e^{\frac{\alpha}{2}}\rangle$ and the homogeneous grading quantum vertex operator is

$$\tilde{\Phi}_-(\zeta) = exp\left(\sum_{k=1}^{\infty}\frac{a_{-k}}{[2k]}q^{\frac{7k}{2}}\zeta^k\right)exp\left(-\sum_{k=1}^{\infty}\frac{a_k}{[2k]}q^{-\frac{5k}{2}}\zeta^{-k}\right)\otimes e^{\frac{\alpha}{2}}(-q^3\zeta)^{\frac{(\partial\alpha+i)}{2}}.$$

Using the fact that [5] $[a_k, \tilde{\Phi}_-(\zeta)] = q^{\frac{7k}{2}}\frac{[k]}{k}\zeta^k$, $\tilde{\Phi}_-(\zeta), k > 0$, $[a_k, \tilde{\Phi}_-(\zeta)] = q^{-\frac{5k}{2}}\frac{[k]}{k}\zeta^{-k}$, $\tilde{\Phi}_-(\zeta), k > 0$, we commute $\exp(-a_1 z_+)$ with $\exp(Q\tilde{\Phi}_-(\zeta))$ to the right and $\exp(Q\tilde{\Phi}_-(\zeta))$ with $\exp(a_{-1}z_-)$ to the left. The commutation of $\exp(-a_1 z_+)$ withexp$(a_{-1}z_-)$ gives $\exp(-z_+z_-[2]))$. Thus it follows that

$$\langle\Lambda_j|e^{-a_1 z_+}e^{Q\tilde{\Phi}_-(\zeta)}e^{a_1 z_-}|\Lambda_j\rangle = \langle\Lambda_j|exp\left(Qe^{-q^{\frac{7}{2}}z_+\zeta - q^{-\frac{5}{2}}z_-\zeta^{-1}}\tilde{\Phi}_-(\zeta)\right)exp\left(-z_+z_-[2]\right)|\Lambda_j\rangle,$$

(recall that $\exp(-a_1 z_+)$ and $\exp(a_{-1}z_-)$ act on $|\Lambda_j\rangle$ and $\langle\Lambda_j|$ as identities). Then we expand $exp\left(Qe^{-q^{\frac{7}{2}}z_+\zeta - q^{-\frac{5}{2}}z_-\zeta^{-1}}\tilde{\Phi}_-(\zeta)\right)$ as a series and apply the action of powers of operators $exp\left(\sum_{k=1}^{\infty}\frac{a_{-k}}{[2k]}q^{\frac{7k}{2}}\zeta^k\right)$ and $exp\left(-\sum_{k=1}^{\infty}\frac{a_k}{[2k]}q^{-\frac{5k}{2}}\zeta^{-k}\right)$, to the left and to the right. Powers of operators $\tilde{\Phi}_-(\zeta)$ act on the second part of tensor product as follows:

$$\left(\tilde{\Phi}_-(\zeta)\right)^n|1\otimes e^{\frac{\alpha}{2}}\rangle = (-q^3\zeta)^{\frac{i+n}{2}}|1\otimes e^{\frac{\alpha(n+1)}{2}}\rangle,\ \left(\tilde{\Phi}_-(\zeta)\right)^n|1\otimes 1\rangle = (-q^3\zeta)^{i+\frac{n-1}{2}}|1\otimes e^{\frac{\alpha n}{2}}\rangle.$$

Thus we have

$$\begin{aligned}
e^{-\beta\lambda_j\cdot\phi} &= e^{-\beta\lambda_j\cdot\phi_0}\frac{\langle\Lambda_j|e^{-z_+z_-[2]}\sum_{n=1}^{\infty}\frac{1}{n!}\left(Qe^{-q^{\frac{7}{2}}z_+\zeta - q^{-\frac{5}{2}}z_-\zeta^{-1}}\right)^n(-q^3\zeta)^{\frac{i+n}{2}}|1\otimes e^{\frac{\alpha(n+1)}{2}}\rangle}{\langle\Lambda_0|e^{-z_+z_-[2]}\sum_{n=1}^{\infty}\frac{1}{n!}\left(Qe^{-q^{\frac{7}{2}}z_+\zeta - q^{-\frac{5}{2}}z_-\zeta^{-1}}\right)^n(-q^3\zeta)^{i+\frac{n-1}{2}}|1\otimes e^{\frac{\alpha n}{2}}\rangle^{m_j}} \\
&= e^{-\beta\lambda_j\cdot\phi_0}\frac{\langle\Lambda_j|e^{-z_+z_-[2]}exp\left(Qe^{-q^{\frac{7}{2}}z_+\zeta - q^{-\frac{5}{2}}z_-\zeta^{-1}}(-q^3\zeta)^{\frac{i}{2}}\right)(-q^3\zeta)^{\frac{i}{2}}\otimes exp\left(e^{\frac{\alpha}{2}}\right)|\Lambda_j\rangle}{\langle\Lambda_0|e^{-z_+z_-[2]}exp\left(Qe^{-q^{\frac{7}{2}}z_+\zeta - q^{-\frac{5}{2}}z_-\zeta^{-1}}(-q^3\zeta)^{\frac{1}{2}}\right)(-q^3\zeta)^{i-\frac{1}{2}}\otimes exp\left(e^{\frac{\alpha}{2}}\right)|\Lambda_0\rangle^{m_j}}.
\end{aligned} \tag{7}$$

In the limit $q \to 1$ we obtain ordinary soliton solutions.

4.2. The case $l = 2$

As in [13], if we put $\widehat{\phi}^{\pm}_{x,1} = 0$, then $E_{\pm} = a_{\pm 2} + a_{\pm 1}$, and one can integrate the equations for ${}_q\mu_{\pm}$ to obtain ${}_q\mu_{\pm}(z^{\pm}) = {}_q\mu_{\pm}(0)e^{(a_{\pm 2}+a_{\pm 1})z^{\pm}}$. Then the quantum soliton solution to the quantized (3) is

$$: e^{-\beta\widehat{\phi}(z^+,z^-)} :=: e^{-\beta\widehat{\phi}_0(z^+,z^-)} : \frac{{}_q\langle\Lambda_1|e^{(a_{+1}+a_{+2})z^+}\,{}_q\mu(0)e^{(a_{-1}+a_{-2})z^-}|\Lambda_1\rangle_q}{{}_q\langle\Lambda_0|e^{(a_{+1}+a_{+2})z^+}\,{}_q\mu(0)e^{(a_{-1}+a_{-2})z^-}|\Lambda_0\rangle_q},$$

where ${}_q\mu(0)$ should be chosen the same as in [13]. Then we have

$$: e^{-\beta\widehat{\phi}(z^+,z^-)} :=: e^{-\beta\widehat{\phi}_0(z^+,z^-)} : \frac{{}_q\langle\Lambda_1|e^{-\frac{\alpha}{2}}exp\big(i(-1)^{\partial_\alpha+1}Q\, W_2\cdot{}_q\Phi(\zeta)\big)e^{\frac{\alpha}{2}}\zeta^{\frac{1}{2}\partial_\alpha}|\Lambda_1\rangle_q}{{}_q\langle\Lambda_0|e^{-\frac{\alpha}{2}}exp\big(i(-1)^{\partial_\alpha+1}Q\, W_2\cdot{}_q\Phi(\zeta)\big)e^{\frac{\alpha}{2}}\zeta^{\frac{1}{2}\partial_\alpha}|\Lambda_0\rangle_q}$$

$$=: e^{-\beta\widehat{\phi}_0(z^+,z^-)} : \frac{1+iW_2Q}{1-iW_2Q}\zeta^{\frac{1}{2}},$$

where $W_2 = \exp\left(\sum_{k=1}^{2}\frac{7k}{2}\frac{[k]}{k}\zeta^k z^+ - \sum_{k=1}^{2}\frac{5k}{2}\frac{[k]}{k}\zeta^{-k}z^-\right)$. Similarly,

$${}_q\langle 1|F_1^+|1;1\rangle_q = e^{2(\widehat{\phi}_0^- - \widehat{\phi})}$$

$$\times\partial_+\left({}_q\langle 1|e^{(a_{+1}+a_{+2})z^+}\,{}_q\mu(0)e^{(a_{-1}+a_{-2})z^-}|1;1\rangle_q \times \frac{{}_q\langle\Lambda_1|e^{(a_{+1}+a_{+2})z^+}\,{}_q\mu(0)e^{(a_{-1}+a_{-2})z^-}|\Lambda_1\rangle_q}{{}_q\langle\Lambda_0|e^{(a_{+1}+a_{+2})z^+}\,{}_q\mu(0)e^{(a_{-1}+a_{-2})z^-}|\Lambda_0\rangle_q}\right).$$

Thus,

$${}_q\langle 1|F_1^+|1;1\rangle_q = e^{2(\widehat{\phi}_0^- - \widehat{\phi})}$$

$$\times\partial_+\Big({}_q\langle 1|e^{(a_{+1}+a_{+2})z^+}e^{-\frac{\alpha}{2}}exp\left(i(-1)^{\partial_\alpha+1}Q\, W_2\cdot{}_q\Phi(\zeta)\right)e^{\frac{\alpha}{2}}\zeta^{\frac{1}{2}\partial_\alpha}e^{(a_{-1}+a_{-2})z^-}|1;1\rangle_q$$
$$\times\frac{{}_q\langle\Lambda_1|e^{-\frac{\alpha}{2}}exp\big(i(-1)^{\partial_\alpha+1}Q\, W_2\cdot{}_q\Phi(\zeta)\big)e^{\frac{\alpha}{2}}\zeta^{\frac{1}{2}\partial_\alpha}|\Lambda_1\rangle_q}{{}_q\langle\Lambda_0|e^{-\frac{\alpha}{2}}exp\big(i(-1)^{\partial_\alpha+1}QW_2\cdot{}_q\Phi(\zeta)\big)e^{\frac{\alpha}{2}}\zeta^{\frac{1}{2}\partial_\alpha}|\Lambda_0\rangle_q}\Big).$$

Finally,

$${}_q\langle 1|F_1^+|1;1\rangle_q = e^{2(\widehat{\phi}_0^- - \widehat{\phi})}\partial_+\left((1+iW_2Q[2])\cdot\frac{1-iW_2Q}{1+iW_2Q}\zeta^{-\frac{1}{2}}\right).$$

4.3. Case l=3

The states

$$|\Lambda_0\rangle^{(m)} = |\prod_{k=1}^{m+1}a_{-(m-k)}\otimes 1\rangle,\ |\Lambda_1\rangle^{(m)} = |\prod_{k=1}^{m+1}a_{-(m-k)}\otimes e^{\frac{\alpha}{2}}\rangle,\ |\Lambda_0\rangle^{(1)} = |\Lambda_0\rangle,\ |\Lambda_1\rangle^{(1)} = |\Lambda_1\rangle,$$

are annihilated by the action of $\mathcal{G}_n, n \geq m$. Therefore for F_m^+ we have

$${}^{(1)}\langle\Lambda_1|e^{-\sum_{k=1}^{3}a_kz_+}e^{Q\tilde{\Phi}_-(\zeta)}e^{\sum_{k=1}^{3}a_{-k}z_-}|\Lambda_1\rangle^{(m)}$$

$$= {}^{(1)}\langle\Lambda_1|exp\left(Qe^{-q^{\frac{7}{2}}z_+\zeta-q^{\frac{-5}{2}}z_-\zeta^{-1}}\tilde{\Phi}_-(\zeta)\right)exp\left(-z_+z_-\sum_{k=1}^{3}\frac{[2k]}{k}[k]\right)e^{-\sum_{k=1}^{3}a_kz_+}|\Lambda_1\rangle^{(m)}.$$

Action by the operators $e^{-\sum_{k=1}^{3}a_kz_+}$ on $|\Lambda_1\rangle^{(m)}$, $m = 1, 2, 3$ we get for instance,

$$e^{-\sum_{k=1}^{3}a_kz_+}|\Lambda_1\rangle^{(3)} = |\Lambda_1\rangle^{(3)} - z_+(C_2 + a_{-2})|\Lambda_1\rangle^{(1)} + z_+^2C_1C_2|\Lambda_1\rangle^{(1)},$$

where $C_k = \frac{[2k]}{k}[k]$. Then we expand $e^{\tilde{\Phi}_-(\zeta)}$ again and act on the states. Therefore we get an infinite series over $|\Lambda_1\rangle^{(3)}, |\Lambda_1\rangle^{(2)}, |\Lambda_1\rangle^{(1)}$ which contain $C_k, (k = 1,2,3)$, z_+ and tensor $\otimes$-part due to powers of $e^{\frac{\alpha}{2}}(-q^3\zeta)^{\frac{(\partial_\alpha+i)}{2}}$.

5. Appendix

5.1. *Affine Kac–Moody algebras*

Here we recall facts about affine Kac–Moody algebras [7], [4]. Consider an untwisted affine Kac-Moody algebra $\widehat{\mathcal{G}}$ endowed with an integral grading $\widehat{\mathcal{G}} = \bigoplus_{n\in\mathbb{Z}} \widehat{\mathcal{G}}_n$, and denote $\widehat{\mathcal{G}}_\pm = \bigoplus_{n>0} \widehat{\mathcal{G}}_{\pm n}$. By an affine Lie algebra we mean a loop algebra corresponding to a finite dimensional simple Lie algebra $\mathcal{G}$ of rank r, extended by the center C and the derivation D. According to Tkac, integral gradings of $\widehat{\mathcal{G}}$ are labelled by a set of co-prime integers $\mathbf{s} = (s_0, s_1, \dots s_r)$, and the grading operators are given by

$$Q_{\mathbf{s}} \equiv H_{\mathbf{s}} + N_{\mathbf{s}}\, D - \frac{1}{2N_{\mathbf{s}}} \operatorname{Tr}\,(H_{\mathbf{s}})^2\, C\,. \tag{8}$$

Here $H_{\mathbf{s}} \equiv \sum_{a=1}^r s_a \lambda_a^v \cdot H^0$, $N_{\mathbf{s}} \equiv \sum_{i=0}^r s_i m_i^\psi$, $\psi = \sum_{a=1}^r m_a^\psi \alpha_a\,,\quad m_0^\psi = 1$. H^0 is an element of Cartan subalgebra of $\mathcal{G}$; α_a, $a = 1, 2, \dots r$, are its simple roots; ψ is its maximal root; m_a^ψ the integers in expansion $\psi = \sum_{a=1}^r m_a^\psi \alpha_a$; and λ_a^v are the fundamental co–weights satisfying the relation $\alpha_a \cdot \lambda_b^v = \delta_{ab}$.

The principal grading operator Q_{ppal} is given by (8) where $N_{\mathbf{s}} = h$ is Coxeter number. Therefore $\widehat{\mathcal{G}}_0 = \{H_a^0, a = 1, 2, \dots r\,; C; Q_{\text{ppal}}\}$, $\widehat{\mathcal{G}}_m = \{E^0_{\alpha^{(m)}}, E^1_{-\alpha^{(h-m)}}\}$, $\widehat{\mathcal{G}}_{-m} = \{E^0_{-\alpha^{(m)}}, E^{-1}_{\alpha^{(h-m)}}\}$ where $0 < m < h$, and $\alpha^{(m)}$ are positive roots of height m. The element B is parameterized as $B = e^{\varphi\cdot \tilde{H}^0}\, e^{\nu\, C}\, e^{\eta Q_{\text{ppal}}} = e^{\varphi\cdot H^0}\, e^{\tilde{\nu}\, C}\, e^{\eta Q_{\text{ppal}}}$, where $\tilde{H}^0$ was defined in [4] as $\widetilde{H}_a^0 = H_a^0 - \frac{1}{N_{\mathbf{s}}} \operatorname{Tr}\left(H_{\mathbf{s}}\, H_a^0\right) C = H_a^0 - \frac{2}{\alpha_a^2} \frac{s_a}{N_{\mathbf{s}}}\, C$, and $\tilde{\nu} = \nu - \frac{2}{h}\widehat{\delta}\cdot\varphi$, with $\widehat{\delta} = \sum_{a=1}^r \frac{\lambda_a}{\alpha_a^2}$, and λ_a being the fundamental weights of $\mathcal{G}$. Let us denote by H^n, $E^n_\pm$, D, C the Chevalley basis generators of $\widehat{sl}_2$. The commutation relations are

$$\begin{aligned} [H^m\,,\, H^n] &= 2\, m\, C\, \delta_{m+n,0}, \quad [H^m\,,\, E^n_\pm] = \pm 2\, E^{m+n}_\pm, \\ [E^m_+\,,\, E^n_-] &= H^{m+n} + m\, C\, \delta_{m+n,0}, \quad [D\,,\, T^m] = m\, T^m\,, \quad T^m \equiv H^m, E^m_\pm. \end{aligned}$$

The grading operator for the principal grading ($\mathbf{s} = (1,1)$) is $Q \equiv \frac{1}{2}H^0 + 2D$. Then the eigensubspaces are $\widehat{\mathcal{G}}_0 = \{H^0, C, Q\}$, $\widehat{\mathcal{G}}_{2n+1} = \{E^n_+, E^{n+1}_-\}$, $n \in \mathbb{Z}$, $\widehat{\mathcal{G}}_{2n} = \{H^n\}, n \in \{\mathbb{Z} - 0\}$.

5.2. *Higher grading affine Toda system*

In this and the next sections we recall [4] the affine Toda system consrtuction. Consider a two dimensional manifold $\mathcal{M}$ with local coordinates $z_\pm$. Up to a gauge transformation, $(1,0)$-component lying in (see subsection 5.1 of Appendix) $\bigoplus_{n=0}^l \widehat{\mathcal{G}}_{+n}$ and $(0,1)$-component in $\bigoplus_{n=0}^l \widehat{\mathcal{G}}_{-n}$ of a flat connection $\mathcal{A}$ in the trivial holomorphic principal fibre bundle $\mathcal{M}\times\widehat{G} \longrightarrow \mathcal{M}$ ($l > 0$ is fixed integer) satisfy the zero curvature condition

$$\partial_+ A_- - \partial_- A_+ + [A_+\,,\, A_-] = 0. \tag{9}$$

The components $A_\pm$ are the following (we keep notations of [4])

$$A_+ = -B\, F^+\, B^{-1}\,, \qquad A_- = -\partial_- B\, B^{-1} + F^-. \tag{10}$$

Here B is a mapping $\mathcal{M} \longrightarrow \widehat{G}_0$ ($\widehat{G}_0$ is a group with the Lie algebra $\widehat{\mathcal{G}}_0$) and F^+ $(1 \le m \le l-1)$ are mappings to $\bigoplus_{n=1}^l \widehat{\mathcal{G}}_{\pm n}$

$$F^\pm = E_{\pm l} + \sum_{m=1}^{l-1} F^\pm_m,$$

where $E_{\pm l}$ are some fixed elements of $\widehat{\mathcal{G}}_{\pm l}$ and $F_m^{\pm} \in \widehat{\mathcal{G}}_{\pm m}$, $(1 \le m \le l-1)$. Substituting (10) into (9) one arrives at the equations of motion

$$\begin{aligned}
\partial_+ \left(\partial_- B\, B^{-1}\right) &= [E_{-l}\,,\, B\, E_l\, B^{-1}] + \sum_{n=1}^{l-1} [F_n^-\,,\, B\, F_n^+\, B^{-1}]\,, && (11)\\
\partial_- F_m^+ &= [E_l\,,\, B^{-1}\, F_{l-m}^-\, B] + \sum_{n=1}^{l-m-1} [F_{n+m}^+\,,\, B^{-1}\, F_n^-\, B]\,, && (12)\\
\partial_+ F_m^- &= -[E_{-l}\,,\, B\, F_{l-m}^+\, B^{-1}] - \sum_{n=1}^{l-m-1} [F_{n+m}^-\,,\, B\, F_n^+\, B^{-1}]\,. && (13)
\end{aligned}$$

Since $Q_{\mathbf{s}}$, $C \in \widehat{\mathcal{G}}_0$ then B can be parameterized as $B = b\, e^{\eta\, Q_{\mathbf{s}}}\, e^{\nu\, C}$ where b is a mapping to G_0, the subgroup of $\widehat{G}_0$ generated by all elements of $\widehat{\mathcal{G}}_0$ other than $Q_{\mathbf{s}}$ and C. Substituting B into the equations of motion (11–13) one has

$$\begin{aligned}
\partial_+ \left(\partial_- b b^{-1}\right) &+ \partial_+\partial_-\, \nu\, C = e^{l\eta}[E_{-l}\,,\, b\, E_l\, b^{-1}] + \sum_{n=1}^{l-1} e^{n\eta}\, [F_n^-\,,\, b\, F_n^+ b^{-1}]\,, && (14)\\
\partial_- F_m^+ &= e^{(l-m)\eta}\, [E_l\,,\, b^{-1}\, F_{l-m}^-\, b] + \sum_{n=1}^{l-m-1} e^{n\eta} [F_{m+n}^+\,,\, b^{-1}\, F_n^-\, b]\,, && (15)\\
\partial_+ F_m^- &= -e^{(l-m)\eta}\, [E_{-l}\,,\, b\, F_{l-m}^+\, b^{-1}] - \sum_{n=1}^{l-m-1} e^{n\eta} [F_{m+n}^-\,,\, b\, F_n^+\, b^{-1}]\,, && (16)\\
\partial_+\partial_-\, \eta\, Q_{\mathbf{s}} &= 0\,. && (17)
\end{aligned}$$

Now consider the case $l = 1$. Let us parameterize the element B in the homogeneous grading of $\widehat{\mathcal{G}}$, [4]. From the equations (14–17) for an infinite dimensional Lie algebra $\widehat{\mathcal{G}}$ in the principal grading we obtain the affine Toda field theory systems of equations

$$\partial_+\partial_-\phi + \tfrac{4\mu}{\beta} \sum_{i=1}^{r} \left(m_i \tfrac{\alpha_i}{\alpha_i^2} \exp(\beta\alpha_i \cdot \phi) - \tfrac{\psi}{2} \exp(-\beta \cdot \phi) \right) = 0. \qquad (18)$$

The formal general solution to the above equation was introduced in [12]:

$$e^{-\beta\lambda_i\cdot\phi} = e^{-\beta\lambda_i\cdot\phi_0} \frac{{}^{(1)}\langle\lambda_i|(\gamma_0^+)^{-1}\mu_+^{-1}(z_+)\mu_-(z_-)(\gamma_0^-)|\lambda_i\rangle^{(1)}}{\left({}^{(1)}\langle\lambda_0|(\gamma_0^+)^{-1}\mu_+^{-1}(z_+)\mu_-(z_-)(\gamma_0^-)|\lambda_0\rangle^{(1)}\right)^{m_j}} = e^{-\beta\lambda_i\cdot\phi_0} \frac{(1)\langle\lambda_i|B^{-1}|\lambda_i\rangle^{(1)}}{\left({}^{(1)}\langle\lambda_0|B^{-1}|\lambda_0\rangle^{(1)}\right)^{m_j}}, \qquad (19)$$

The general solutions to the matter fields $F_i^{\pm}$ may be written in the following form. For $m = 1$ in (14–17) one has [4]

$$\langle i|F_1^+|i;i\rangle = f_i^+ = e^{\sum_{l=0}^{r} k_{il}(\phi^- - \phi)_l}\, e^{\nu_0}\partial_+ \left(\langle i|\mu_+^{-1}\mu_-|i;i\rangle \frac{\langle 0|\mu_+^{-1}\mu_-|0\rangle^{m_i}}{\langle i|\mu_+^{-1}\mu_-|i\rangle} \right).$$

Here $|i;i\rangle$ denotes an element of the Verma module which is result of the action of the lowering generator on the highest state vector. The fact that (19) is indeed a solution to (18) may be checked by using the representation theory of $\widehat{\mathcal{G}}$. A map $g : \mathcal{M} \longrightarrow G$ appearing in the gradient form of the flat connection $A_{\pm} = g^{-1}\partial_{\pm} g$, may be factorized (according to the Lie algebra decomposition $\mathcal{G} = \mathcal{G}_- \oplus \mathcal{G}_0 \oplus \mathcal{G}_+$) by the modified Gauss decomposition $g = \mu_-\nu_+\gamma_{0-}$

or $g = \mu_+\nu_-\gamma_{0+}$ with maps $\gamma_{0\pm} : \mathcal{M} \longrightarrow G_0$, $\mu_\pm, \nu_\pm : \mathcal{M} \longrightarrow G_\pm$. The grading condition provides the holomorphic property of $\mu_\pm$, i.e., they satisfy the initial value problem

$$\partial_\pm \mu_\pm(z_\pm) = \mu_\pm(z_\pm)\tilde{\mathcal{E}}_\pm(z_\pm), \tag{20}$$

$$\tilde{\mathcal{E}}_\pm(z_\pm) = \sum_{m=1}^{M} \tilde{\mathcal{E}}_m^\pm(\Phi^\pm), \qquad \tilde{\mathcal{E}}_m^\pm(\Phi^\pm) = \sum_{\alpha\in\Delta_m^+} \Phi_\alpha^{\pm m}(z_\pm)X_{\pm\alpha}, \tag{21}$$

with arbitrary functions $\Phi_\alpha^{\pm m}(z_\pm)$ determining the general solution to the system. Note that the summations in (21) are performed over the set of positive roots Δ_m^+ of $\mathcal{G} = \sum_{m\in\mathbb{Z}} \mathcal{G}_m$ in the subspace $\mathcal{G}_m$.

6. Soliton solution for the sine–Gordon in homogeneous grading

Another way to construct soliton solutions [13] to the sine–Gordon equation is to consider the formal general solution (18) in the homogeneous grading and to use vertex operators [7] which are related to the homogeneous Heisenberg subalgebra of $\widehat{sl_2}$. Take the general solution (19) to the affine Toda system (18). In the homogeneous grading the mappings $\gamma_\pm$ can be parameterized as $\gamma_\pm = e^{d\phi_d}e^{c\phi_c}e^{\phi_0^\pm x_0^\pm}$, where d is the grading operator, c is the center of $\widehat{sl_2}$ and $x_k^\pm$ are generators of the subspaces $\widehat{\mathcal{G}}_k$ corresponding to the homogeneous grading. The mappings $\mu_\pm$ satisfy (20) where $\kappa_\pm(z^\pm) = a_{\pm1} + \phi^\pm x_1^\pm$. In order to obtain a soliton solution we put $\phi^\pm = 0$, $\phi_0^\pm = 0$. Then the general solution reduces to

$$e^{-\beta\phi(z^+,z^-)} = \langle\Lambda_1|e^{a_{+1}z^+}\mu(0)e^{a_{-1}z^-}|\Lambda_1\rangle\left(\langle\Lambda_0|e^{a_{+1}z^+}\mu(0)e^{a_{-1}z^-}|\Lambda_0\rangle\right)^{-1}. \tag{22}$$

The following group element $\mu(0)$ in (22)

$$\mu(0) = e^{-\frac{\alpha}{2}N}\prod_{n=1}^{N}\left[exp\left((-1)^{\partial_\alpha+1}iQ_n\Phi(\zeta_n)\right)e^{\frac{\alpha}{2}}\zeta_n^{\frac{1}{2}\partial_\alpha}\right],$$

generates an N–soliton solution. Here the action of the operators $\frac{1}{2}\partial_\alpha$ and $e^{\frac{\alpha}{2}}$ on the highest vectors $|\Lambda_n\rangle = |1\otimes e^{\frac{\alpha}{2}n}\rangle$, $n = 0, 1$ is the same as in the case of $U_q'(\widehat{sl_2})$ [7] when $q = 1$. The operator $\Phi(\zeta)$ is given by

$$\Phi(\zeta) = exp\left(\sum_{k=1}^{\infty}\frac{a_{-n}}{n}\zeta^n\right)exp\left(-\sum_{k=1}^{\infty}\frac{a_{+n}}{n}\zeta^{-n}\right),$$

and diagonalises the action of $a_{\pm k}$, $k \in \mathbb{N}$, i.e., $[a_{\pm k}, \Phi(\zeta)] = \zeta^{\pm k}\Phi(\zeta)$. The product of two vertex operators can be normal ordered as

$$\Phi(\zeta_1)\Phi(\zeta_2) = X(x) : \Phi(\zeta_1)\Phi(\zeta_2) :,$$

where $X(x) = exp(-\sum_{n=1}^{\infty} x^{2n}/n) = \exp\left(\log(1-x^2)\right)$. When $x = 1$, $X(x)$ vanishes which results in $\Phi(\zeta)\cdot\Phi(\zeta) = 0$. Therefore the exponential of $\Phi(\zeta)$ operator terminates after the first order.

In the limit $q \longrightarrow 1$ soliton–soliton, antisoliton–antisoliton and soliton–-antisoliton scattering reduce to the classical case, i.e.,

$$F^a(\zeta_1)F^b(\zeta_2) = \frac{1}{x}\frac{X(x)}{X(x^{-1})}F^b(\zeta_2)F^a(\zeta_1),$$

where $x^2 = \zeta_2/\zeta_1$, a, b denote soliton (antisoliton), and the factor $1/x$ comes from the commutation of $e^{\frac{\alpha}{2}}\zeta_1^{\frac{1}{2}\partial_\alpha}$ and $e^{\frac{\alpha}{2}}\zeta_2^{\frac{1}{2}\partial_\alpha}$ operators. Therefore the vertex operator generating a classical soliton solution is

$$F(\zeta) = Q\,\Phi(\zeta)\,e^{\frac{\alpha}{2}}\,\zeta_2^{\frac{1}{2}\partial_\alpha}.$$

Taking into account the properties of the operator $F(\zeta)$ we rewrite the solution (22) as

$$e^{-\beta\phi(z^+,z^-)} = \frac{\langle\Lambda_1|\big(1+(-1)^{\partial_\alpha+1}iQ\Phi(\zeta)\big)e^{\frac{\alpha}{2}}\zeta^{\frac{1}{2}\partial_\alpha}|\Lambda_1\rangle}{\langle\Lambda_0|\big(1+(-1)^{\partial_\alpha+1}iQ\Phi(\zeta)\big)e^{\frac{\alpha}{2}}\zeta^{\frac{1}{2}\partial_\alpha}|\Lambda_0\rangle}$$
$$= \left(1 + iQe^{\zeta z^+ - \zeta^{-1}z^-}\right)\left(1 - iQe^{\zeta z^+ - \zeta^{-1}z^-}\right)\zeta.$$

The antisoliton solution can be associated with the vertex operator

$$\bar{F}(\zeta) = -Q\,\Phi(\zeta)\,e^{\frac{\alpha}{2}}\,\zeta^{\frac{1}{2}\partial_\alpha}.$$

6.1. *Quantized universal enveloping algebra $U'q(\widehat{sl_2})$*

In the spirit of [2], [5], the quantised enveloping algebra $U_q(sl_2)$ is an associative algebra generated by X^+, X^-, H with q-deformed commutation relations

$$X^+X^- - X^-X^+ = \left(q^H - q^{-H}\right)\left(q - q^{-1}\right)^{-1}, \qquad HX^\pm - X^\pm H = \pm 2X^\pm.$$

It possesses a Hopf algebra structure with the deformed adjoint action

$$(ad_{X^\pm})_q a = X^\pm a q^{H/2} - q^{\mp 1} q^{H/2} a X^\pm, \qquad (ad_H)_q a = Ha - aH,$$

for all $a \in U_q(sl_2)$. Let us recall the second Drinfeld realization of the quantized universal enveloping algebra $U_q'(\widehat{sl_2})$, (i.e., $U_q(\widehat{sl_2})$ without grading operator) [2], [6], which is a natural quantum analogue of the algebra $\widehat{sl_2}$ in the loop realizations. $U_q'(\widehat{sl_2})$ is an associative algebra generated by $\{x_k^\pm, k \in \mathbb{Z}; a_n, n \in \{\mathbb{Z} - 0\}; \gamma^{\pm\frac{1}{2}}, K\}$, where $\gamma^{\pm\frac{1}{2}}$ belong to the center of the algebra, satisfying the commutation relations

$$[K, a_k] = 0, \quad [a_k, a_l] = \delta_{k,-l}\frac{[2k]}{k}\frac{\gamma^k - \gamma^{-k}}{q - q^{-1}}, \quad Kx_k^\pm K^{-1} = q^{\pm 2}x_k^\pm,$$
$$\left[a_n, x_k^\pm\right] = \pm\frac{[2n]}{n}\gamma^{\mp\frac{|n|}{2}}x_{n+k}^\pm, \quad \left[x_k^+, x_n^-\right] = \frac{\gamma^{(k-n)/2}\psi_{k+n} - \gamma^{(n-k)/2}\phi_{k+n}}{q - q^{-1}},$$
$$x_{k+l}^\pm x_l^\pm - q^{\pm 2}x_l^\pm x_k^\pm = q^{\pm 2}x_k^\pm x_{l+1}^\pm - x_{l+1}^\pm x_k^\pm.$$

The generators ϕ_k and ψ_{-k}, $k \in \mathbb{Z}_+$ are related to a_k and a_{-k} by means of the expressions $\sum_{k=0}^{\infty}\psi_m z^{-m} = Kexp\left((q - q^{-1})\sum_{k=1}^{\infty}a_k z^{-k}\right)$, $\sum_{k=0}^{\infty}\phi_{-m}z^m = K^{-1}exp\left(-(q - q^{-1})\sum_{k=1}^{\infty}a_{-k}z^k\right)$, i.e., $\psi_m = 0$, $m < 0$; $\phi_m = 0$, $m > 0$. Here $[k] = \frac{q^k - q^{-k}}{q - q^{-1}}$.

It is easy to define the grading operators corresponding to the principal and homogeneous grading of $U_q'(\widehat{sl_2})$ by analogy with the grading of $U_q'(\mathcal{G})$ where $\mathcal{G}$ is a simple Lie algebra. The principal grading can be realized with the help of the operator $D_p x = \frac{1}{2}qK^{-1}\left(\frac{d}{dq}(KxK^{-1})\right)K + 2\lambda\frac{d}{d\lambda}x$, where $x \in U_q(\widehat{sl_2})$ and λ is an affinization parameter. The power of λ is denoted by the subscript of $U_q'(\widehat{sl_2})$ generators. Then the grading subspaces are ${}_q\widehat{\mathcal{G}}_0 = \{K, \gamma\}$, ${}_q\widehat{\mathcal{G}}_{2n+1} = \{x_n^+, x_{n+1}^-, n \in \mathbb{Z}\}$, ${}_q\widehat{\mathcal{G}}_{2n} = \{a_n, n \in \{\mathbb{Z} - 0\}\}$. The grading operator for the homogeneous grading is $D_h x = 2\lambda \frac{d}{d\lambda} x$, so that the grading subspaces are ${}_q\widehat{\mathcal{G}}_0 = \{K, \gamma, x_0^+, x_0^-\}$, ${}_q\widehat{\mathcal{G}}_n = \{x_n^+, x_n^-, a_n, n \in \{\mathbb{Z} - 0\}\}$.

The level one irreducible integrable highest weight representation of $U_q'(\widehat{sl_2})$ can be constructed in the following way [6]. Let $P = \mathbb{Z}\frac{\alpha}{2}$, $Q = \mathbb{Z}\alpha$ be the weight/root lattice of sl_2. Consider the group algebras $F[P]$, $F[Q]$ of P and Q. The multiplicative basis of $F[P]$ is formed by $e^{\frac{\alpha}{2}n}$, $n \in \mathbb{Z}$. The $F[Q]$-module is split into $F[P] = F[P]_0 \oplus F[P]_1$ where $F[P]_n = F[Q]e^{\frac{\alpha}{2}n}$. The sl_2-module structure on the space $W = F[a_{-1}, a_{-2}, ...] \otimes F[P]$ is given by the action of the $a_k, k \in \{\mathbb{Z} - 0\}$ and e^α, $\partial_\alpha = a_0$ generators in accordance with the rules

$$a_k(f \otimes e^\beta) = (a_k f \otimes e^\beta), \quad k < 0, \; a_k(f \otimes e^\beta) = ([a_k, f] \otimes e^\beta), \quad k > 0,$$
$$e^\alpha(f \otimes e^\beta) = (f \otimes e^{\alpha+\beta}), \quad \partial_\alpha(f \otimes e^\beta) = (\alpha, \beta)(f \otimes e^\beta),$$
$$K = 1 \otimes q^{\partial_\alpha}, \quad \gamma = q \otimes id.$$

Then W is a $U_q'(\widehat{sl_2})$-module. Its submodules are isomorphic to irreducible highest weight modules $V(\Lambda_n)$ with the highest vectors $|\Lambda_n\rangle = |1 \otimes e^{\frac{\alpha n}{2}}\rangle$, $n = 0, 1$.

Acknowledgements

We would like to thank the organizers of the International Conference on Integrable Systems and Quantum symmetries (ISQS) Praha, Czech Republic, June 23–29, 2014, and especially Prof. C. Burdik.

References

[1] Delius G W and Hueffmann A 1996 J. Phys.A29
[2] Drinfeld V G 1985 Sov. Math. Dokl. 32
[3] Fedoseev I A and Leznov A N 1984 Phys. Lett. 141B, 1-2 , 100–104
[4] Ferreira L A, Gervais J-L, Guillen, J S and Saveliev M V 1996 Nuclear Phys. B 470 1-2
[5] Jimbo M 1985 Lett. Math. Phys. 10, 63
[6] Jimbo M, Miki K, Miwa T and Nakayashiki A 1992 Phys. Let. A. 168 256–263
[7] Kac V G 1990 *Infinite dimensional Lie algebras.* Third edition (Cambridge university Press, Cambridge)
[8] Leznov A N and Fedoseev I A 1982 Theor. Math. Phys. 53 3, 1175
[9] Leznov A N and Mukhtarov M A 1987 Theor. Math. Phys. 71 1
[10] Leznov A N and Saveliev M V 1992 *Group-Theoretical Methods for Integration of Non-Linear Dynamical Systems.* Progress in Physics Series, v. 15, (Birkhauser–Verlag, Basel)
[11] Mansfield P 1983 Nucl. Phys. B 222
[12] Olive D I Turok N and Underwood J W R 1993 Nucl. Phys. B401
[13] Saveliev M V and Zuevsky A B 2000 Internat. J. Modern Phys. A 15 24

Institute of Physics
Dirac House, Temple Back
Bristol BS1 6BE UK

CURRAN ASSOCIATES INC
proceedings.com